Horticulture for Nutrition and Income Security

NIPA® GENX ELECTRONIC RESOURCES & SOLUTIONS P. LTD.
New Delhi-110 034

About the Editors

Dr. Ravindra Singh, joined his ARS services in ICAR in 1998 as a scientist in the discipline of Agronomy. He has published more than hundred research papers in referred national, international journals, published many books and having a vast experience of protected cultivation, precision water and nutrient management, weed management and sensor application in different horticultural and field crops. Dr. Singh went to USA for training in the area of "Sensor based applications of Nitrogen" at the Oklahoma State University, Department of Plant and Soil Sciences, 368 Agricultural Hall, Stillwater, OK, USA for three months. He had earned Young Scientist award of ISA, Bharat Shiksha Ratan, Manad Upadhi and many more awards in different national international events.

Dr. Sharda Choudhary, working as a Senior Scientist (Plant Biotechnology) at ICAR-NRCSS, Ajmer, Rajasthan. She has published more than 40 research papers in reputed national international journals; she has published four books with national and international publishers. She has published more than 50 technical bulletins/technical articles. She has been recognized with Young Women Scientist Award-2019, Scientist of the Year Award- 2017, Fellow ISSS, Young Scientist Award- 2016. She has handled more than eight research projects and completed DUS and molecular characterization of seed spices crops, released one variety of fennel. She has done training in the area of "Marker Assisted Selection (Horticulture)" at the University of Arkansas, Fayetteville, Arkansas, USA for three months. She has published first transcriptome data on Indian coriander which will help in understanding the molecular pathways and mechanism in the crop for different diseases.

Dr. Arvind Kumar Verma, secured all India 3rd rank in ICAR-JRF exam in Horticulture group and completed M.Sc. and Ph.D., degree from ICAR-IARI, New Delhi. He qualified ARS-2012 in Horticulture (SPMA) and joined ICAR in 2014 and posted at ICAR- NRCSS, Ajmer, Rajasthan. He is currently working on genetic improvement of seed spices. He has published more than 30 research papers, 40 popular articles, 10 book chapters, 15 bulletins/ training manuals and two books. He is the recipient Fellow of CHAI and Fellow of SHRD.

Mr. Ravi Y., received his Bachelor's degree in Horticulture and his Master's degree in Spices, Plantation, Medicinal and Aromatic Crops (Horticulture) from Kittur Rani Chennamma College of Horticulture, Arabhavi, Karnataka. During 2017, he has been selected as an ARS scientist in the SPMAP subject and began working as a scientist at ICAR-NRCSS, Ajmer, Rajasthan. He is currently working on Metabolomic approaches in seed spices improvement

Horticulture for Nutrition and Income Security

Ravindra Singh
Principal Scientist (Agronomy)
ICAR- National Research Centre on Seed Spices
Ajmer, Rajasthan-305206, India

Sharda Choudhary
Senior Scientist (Biotechnology)
ICAR- National Research Centre on Seed Spices
Ajmer, Rajasthan-305206, India

Arvind Kumar Verma
Scientist (Horticulture)
ICAR- National Research Centre on Seed Spices
Ajmer, Rajasthan-305206, India

Ravi Y.
Scientist (Horticulture)
ICAR- National Research Centre on Seed Spices
Ajmer, Rajasthan-305206, India

NIPA® GENX ELECTRONIC RESOURCES & SOLUTIONS P. LTD.
New Delhi-110 034

NIPA® GENX ELECTRONIC RESOURCES & SOLUTIONS P. LTD.

101,103, Vikas Surya Plaza, CU Block
L.S.C. Market, Pitam Pura, New Delhi-110 034
Ph : +91 11 27341616, 27341717, 27341718
E-mail: newindiapublishingagency@gmail.com
Website: www.nipabooks.com

For customer assistance, please contact
Phone: + 91-11-27 34 17 17
Fax: + 91-11-27 34 16 16
E-Mail: feedbacks@nipabooks.com

ISBN: 978-93-95319-60-7

Composed and Designed by NIPA®.

Foreword

Horticulture contributes to more than one third of the agricultural GDP of the country occupying less than one fifth of the cultivated area. During the year 2019-20, the country recorded its highest ever horticultural production of 320.77 million tons from an area of 25.66 million hectares. As per the 1st Advance Estimates for 2021-22 the total horticultural production in the country was 333.33 lakh MT from an area of 27.50 lakh ha. Keeping in view the huge potential and role of the horticulture sector in increasing farmer's income, the Government of India allocated Rs. 7183 Crore in 2022-23 for development of horticulture sector in the country.

This stems from the fact that there is very high demand and a continued growth in consumption of fresh produce and processed products of fruits and vegetables. Floriculture is another growing subsector of Horticulture which is gaining popularity because of its ornamental and aesthetic value apart from commercial significance. Medicinal and aromatic plants (MAPs) are the richest bioresource of drugs for traditional systems of medicine, modern medicines, folk medicines, pharmaceutical intermediates food supplements, nutraceuticals and chemical entities for synthetic drugs, fragrances, flavours, cosmeceuticals, health beverages and chemical terpenes. Therefore, improving the income of the rural populace engaged in horticultural cultivation is a necessity in order to improve the economy of India in general and farmer's income in particular. The per-unit earning capacity of farmers growing horticultural crops is much higher than those engaged in production of food grains. Moreover, it also addresses environmental concerns.

Hence, situation specific modern technologies need to be identified/developed, refined/validated and popularized. High cost of input in horticulture crops, prevalence of old and senile orchards, unorganized supply chain are the major bottle neck to meet the growing demands, therefore, adoption of organic agriculture practices and farming system approach should be promoted for decreasing input cost and ultimately increasing the income. Moreover, rejuvenation of old and unproductive orchards is required to enhance the production and productivity these crops.

This book is a compilation of chapters contributed by the scientists and professors from ICAR and SAU's working on different aspects of horticultural crops. The book covers production technology, protected cultivation, precision farming, organic production, disease management, INM, IPM, agronomic interventions, value addition, biotechnology, etc. in different horticultural crops. I am confident that students, professionals, farmers and different stakeholders in horticulture sector will find this book really valuable because it contains a wealth of information in one location. The information contained in this book would be useful to the horticultural farmers and help them to enhance their income through application of modern technologies explained therein.

Phundan Singh
Former Director
ICAR-Central Institute for Cotton Research
Nagpur, Maharashtra, India

Preface

Horticultural crops are an important component of agriculture and account for a significant portion of a country's total agricultural production. The significance of horticulture sector in India is demonstrated by its benefits such as high export value, high yield and return per unit area yield, best utilisation of wasteland, provision of raw materials for industries, full-fledged engagement of labours or human resources, production of more food per unit area than field crops, and many more. Doubling the farmers' income is one of the avowed objectives of the Indian government as well as research and development organizations considering the mounting agrarian distress in the country. Though technological interventions have helped farmers secure significantly higher productivity, the realization of desirable income gain remains a major challenge. The horticulture sector has emerged as the most economically beneficial and viable choice for agricultural diversification. Further, the shift in food habits favouring horticulture crops and value added products has resulted in increased demand for these commodities. These crops are export oriented crops and scalable technologies for horticulture crops production and management will lead to more income generation for the farmers. Therefore, this book "Horticulture for Nutrition and Income Security" will go a long way in addressing these important aspects of doubling farmers' income through scalable technologies in horticultural crops.

The book has covered important horticultural technologies that have the potential to enhance farmer's income. All of the 36 chapters are written by renowned professors and scientists from agricultural universities and the ICAR institutes.

We hope that this book will be very valuable to scientists, students, farmers, and other stakeholders. We editors are thankful to the book chapter contributors and publisher.

Editors

Contents

List of Colour Plates

1

Taro (*Colocasia esculenta* (L.) Schoff.) for Nutritional Security and Health Benefits

***Vijay Bahadur Singh Chauhan*[1]*, Samarendra Narayan Mallick*[1]*, Pinki Mohapatra*[1]*, Kalidas Pati*[1]*, R. Arutselvan*[1] *M.Nedunchezhiyan*[1]*and Arvind Kumar Verma*[2]**

[1]*Regional Centre, ICAR-Central Tuber Crops Research Institute Bhubaneswar, Odisha, India*
[2]*ICAR-National Research Centre on Seed Spices, Ajmer, Rajasthan, India*

1. Introduction

Tuber crops are important group of crop species which produce tubers that are used for human food and animal feed. The popular tuber crops include cassava, sweet potato, greater yam, white yam, lesser yam, taro, tannia, elephant foot yam, yam bean, coleus etc. They contribute substantial amount of cheap energy coupled with high quality nutrition for more than two million people living in developing countries. Taro is one of the oldest cultivated tuber crop grown for its edible corms and leaves. It is an important tropical tuber crop, used as a staple food or subsistence food by millions of people in the developing countries in Asia, Africa and Central America. World-wide it is the fifth most consumed root vegetable with over 25% produced in Oceania and South-East Asia. It is widely used as a tuber vegetable in India, whereas it is the staple food and also very closely associated with culture in many of the South Pacific islands. Taro is ordinarily grown in the homestead garden and its cormels, petiole and leaves serve the important purpose as an instant vegetable (Wheatley et al., 1995). Starchy root and tuber crops are the second important sources of carbohydrates used globally after the cereal crops.They give a significant area of the planet food supply and are likewise a significant wellspring of creature feed and handled items for human utilization and modern use. Bland roots and tubers are plants that store palatable starch material in underground stems, roots, rhizomes, corms, and tubers and are begun from broadened natural sources (Chandrasekhara and Kumar, 2016).Taro (*Colocasia esculenta* (L.)

Schoff.), one of the tuberous plant which stores edible starch in root tubers. Sweet potatoes and yams are tubers, whereas taro is derived from corms, underground stems, and swollen hypocotyls. Cassava and sweet potatoes are storage roots and canna and arrowroots are edible rhizomes. The taro plant can be propagated by vegetative parts like side shoots, stolons, or corm heads. Nutritionally, roots and tubers have a great potential to provide economical sources of dietary energy, in the form of carbohydrates. Taro is a good source of potassium. Roots and tubers are deficient in most other vitamins and minerals but contain significant amounts of dietary fibre (FAO, 1990).Conventional medication frameworks play had a huge influence in worldwide medical care since the period of time. Plants have assumed an extremely fundamental part in the reasonable improvement of human existence as the plants served people well for prescriptions, preparing, refreshments, beauty care products, and colors. Numerous customary mending spices and their parts have been displayed to have restorative worth and can be utilized to forestall, reduce or fix a few human illnesses. It is assessed that 70-80% of individuals overall depend predominantly on conventional, generally natural medication to meet their essential medical care needs (Farnsworth and Soejarto, 1991; Shengji, 2001) Indian Systems of Medicine being one of the biggest and most established frameworks comprises every one of the meds which have their starting point in India. Because of this reality, the emphasis on plant research has expanded from one side of the planet to the other. Restorative plants or natural medications and their concentrates containing detached mixtures have displayed a wide range of organic exercises. One such plant with wide applications is *C. esculenta.*

2. Taxonomic Classification

Kingdom-Plantae

Subkingdom-Tracheobionta (Vascular plants)

Super division-Spermatophytes (Seeded plants)

Division-Magnoliophyta (Flowering plants)

Class-Liliopsida (Monocotyledons)

Subclass-Arecidae

Order-Allismatales

Family-Araceae

Genus-*Colocasia* Schott

Species-*Colocasia esculenta* (L.) Schott

Vernacular names

English: Taro, Cocoyam

Sanskrit: Alupam, Alukam

Hindi: Arvi, Kachalu

Odia- Saru

Bengali: AltiKachu, Kachu

Marathi: Alluu

Malayalam: Chempu, Madantha, Chempakizhnna

Gujarati: Aalavi, Patarveliya

Tamil: Sempu

Kannada: Kesavedantu

Fig. 1: Taro plants in field and corms

3. Plant Botany

Taro experimentally known as (*Colocasia esculenta* L. Schott) is broadly developed plant for utilization of the two leaves and tubers. It is a suffering, maritime and semi-maritime herbaceous types of the family Araceae which started in North-Eastern India and afterward Asia and a short time later dynamically spread from one side of the planet to the other (Macharia et al., 2014).The species is developed for the most part for its eatable beefy tubers called corms or cormels, while the leaves are consumed as green vegetables (Aregheore and Perera, 2003).

The taro (*Colocasia esculenta* L. Schott) plant is a tall herb having tuberous or a bold short caudex, with lots of long heart or bolt focuses shaped leaves that point earthwards with flexible petioles and blooming. The stem is available over the ground and enlarged marginally at the foundation of the leaf-sheaths, emerging from a hard tightening rhizome. The stolons create at side during development of cormels and a tuberous rhizome with presence of suckers. Leaves are basic with a heavy petiole, praise cordate or sagittate-cordate, lamina peltate. The tremendous green leaves of taro are known as 'Elephant ear' and which is 1-2 m high. The leaf peduncle is more limited than the petiole. Spadix, the inflorescence of taro is more limited than the petiole and much it is more limited than the spathe rather than slim. Index a lot more limited than the inflorescence, style exceptionally short;

disgrace discoid. It is erect, stretch conelike or fusiform, subulate or shorten. The spathe lamina is barely lanceolate, tangle, taper and bended somewhat in reverse in blossom. Female inflorescence of taro is short while male inflorescence is long, barrel shaped and typically mediated fixes between the two. Seeds are elongated, sulcate with bountiful egg whites, undeveloped organism in axile position. The tuberous root is the truly edible piece of the yield with the presence of erect stems that may be green, red dim or variegated only a few meters high. The plant contains central corm which is the underground part underneath the soil surface that foster the leaves upwards and roots downwards (Ahmed et al., 2013).

4. Nutritional Potential and Health Benefits

Taro is wealthy in supplements that can give significant medical advantages. A one-cup serving of taro contains a suggested measure of manganese, which adds to great digestion, bone wellbeing, and blood coagulating. It has been expressed by numerous others that the presence of protein content of *C. esculenta* is higher than the other root crops in leaves and tuber individually. Many workers (Hedges and Lister, 2006; Onwume, 1999; Englis et al., 1996) suggested that corm and leaves of taro contain good amount of protein, fibers, starch, fats in good amount which are nutritionally gives health benefit to human being (Table 1) Presence of undeniable degrees of nutrients like L-ascorbic acid, vitamin B6, and vitamin E assists with advancing sound vision, skin, dissemination, and invulnerable framework work. Taro root has more dietary fibre than potato which works on stomach related capacity and assists with getting help in issues like stoppage, runs, stomach ulcers, and heart burn. The carb content in taro root is known as a safe starch which can possibly balance out glucose, helps in weight the executives and to diminish the gamble of diabetes. These safe starches are additionally reasonable for low-carb and keto slims down. This starch is with low calories that advance a solid stomach and diminish the opportunity of coronary illness. Taro pull is advantageous for glucose the executives, fiber and safe starch which can't be consumed by blood which result no effect on glucose levels. Presence of undeniable degrees of potassium in taro root, assists with controlling hypertension by separating over abundance salt. This potassium additionally lessens the weight on your cardiovascular framework and furthermore help to forestall improvement of ongoing heart issues. Taro root and its palatable leaves are loaded with cell reinforcements. Quercetin, present more in onion, apples, which come from the purple colour of taro, is a strong cell reinforcement that safeguards your body from free extremists and lessens the opportunity of malignant growth. Its undeniable degrees of likewise help to keep a sound invulnerable framework and may kill free revolutionaries. The Zinc inadequacy is far and wide and influences the wellbeing and prosperity of populaces around the world. The taro is one of the plant-based and non-creature wellsprings of zinc, which use should accordingly be sought after to help in the lightening of zinc inadequacy which is related to hindering (Temesgen and Ratta, 2015).

Table 1: Proximate composition of taro corm on fresh weight basis

Composition	Hedges & Lister, (2006)	Onwume (1999)	Englis et al., (1996)
Moisture (%)	68.1	63-85	65.7
Carbohydrate (%)	26.8	13-29	26.3
Protein (%)	0.34	1.4-3	1.1
Fat (%)	0.11	0.16-0.36	0.4
Crud Fiber (%)	2.5	0.6-1.18	2.1
Ash (%)	1.91	0.6-1.3	2.1
Vitamin C (mg/100g)	14.3	7-9	<1
Thiamin (mg/100g)	0.028	0.18	0.11
Riboflavin (mg/100g)	0.029	0.04	0.02
Niacin (mg/100g)	0.78	0.9	1.3

5. Ethnomedicinal Uses

Entire plant parts like leaves, stem, and tubers of *C. esculenta* have the capability of various restorative properties. The entire plant is utilized as an antimicrobial, antihepatotoxic, against diabetic, hostile to lipid peroxidative activity, antimetastatic, antifungal, calming. The new leaf juice is used to keep up with solid layers, skin, visual perception. It likewise goes about as a decent energizer, expectorant, astringent, tidbit, and otalgia (Ferdous et al., 2015). The leaf stalks with salt are utilized as a retentive to control aroused organs and buboes while the corm juice is utilized in instances of alopecia. It is additionally utilized as a counteractant against wasps stings of wasps and different bugs (Prajapati et al., 2011). Taro is likewise used to get the shield from oral hole tumours and lung illnesses. A hydro alcoholic concentrate of the leaf juice is utilized as a stimulant, anxiolytic, narcotic, and smooth muscle relaxant. The corm juice is utilized to outer alopecia and to lessen body torment. Inside, admission of taro goes about as a diuretic, demulcent, anaesthetic, galactagogue, and is additionally utilized in instances of heaps and blockage of the entry framework (Keshav et al., 2019). The juice of corm is useful for the treatment of sparseness (Wiersema et al., 2015). The petiole juice is styptic which is utilized to control blood vessel drain. The decoction of the strip is utilized as antidiarrheal, expands body weight, and forestalls inordinate discharge of sputum in asthmatic people (Lim, 2012). In Asia and Africa, this species is additionally utilized in conventionalmedication to treat blood vessel hypertension, liver issues,ulcers, snakebites, and stiffness. Leaves are utilized infever, for torment, to forestall stone development in the urinary lot, to capture regular pee and on bubbles andulcer. Munda clans use corm is utilized by individuals of the Munda clan as a solution for body throb (Prajapati et al., 2011).

5.1. Anti-Cancer Activity

Malignant growth is a main source of death around the world, and it is for the most part connected with unfortunate food propensities and way of life. It is critical to track down ways of diminishing and forestall the gamble of malignant growth through dietary parts, which are available in plant food varieties. Malignant growth is a multistage illness condition and tapping at any underlying stage could assist with weakening the infection condition. Root and tuber phytochemicals have exhibited anticancer impacts in a few sorts of carcinoma cell lines and creature models (Huang et al., 2004).Brown et al. (2005) hatched the dissolvable concentrates of Taro at 100mg/ml in-vitro to concentrate on enemy of proliferative action against the rodent YYT colon malignant growth cell line which showed the counter disease impact of the taro plant on colonic adenocarcinoma cells. During the 3H thymidine consolidation study by utilizing taro separate, the plant showed restraint result to multiply these disease cells. At the point when plant remove was brooded following 2 days end with the YYT cells accordingly these YYT cells showed apoptotic changes. The plant improved the multiplication of ordinary mouse splenocyte control cells, which recommended that the plant isn't poisonous to all cells yet the plant has a positive safe stimulatory job. Consequently, the plant showed the anticancer activity by two unmistakable systems: One is inciting apoptosis inside colon malignant growth cells and second one is vaguely actuating lymphocytes which thusly can lyse dangerous cells. In this examination results show that that poi might have novel growth explicit enemy of disease exercises and future exploration is recommended with creature studies and human clinical preliminaries (Brown et al., 2005).A water-dissolvable concentrate of taro (TE) strongly restrained lung-colonizing capacity and unconstrained metastasis from mammary organ embedded cancers (Kundu et al., 2012).

5.2. Antimicrobial and Antifungal Activity

The *in-vitro* antimicrobial movement in watery concentrate of *Colocasia esculenta* (AECE) leaves was considered against gram-positive bacterial strains like. *Streptococcus mutans* (MTCC-890), *Bacillus subtillis* (MTCC-121), gram-negative bacterial strains for example *Klebsiella pneumoniae, Pseudomonas fragi, Escherichia coli* and parasitic strains as,*Candida albicans, Aspergillus niger. C. esculenta* showed most extreme action at low fixation against microorganisms *Streptococcus mutans*. Watery taro leaf extricate showed antibacterial movement *in-vitro* against *Vibrio alginolyticus, V. cholerae , V. harveyi , V. parahaemolyticus* and *V. vulnificus* yet not the stem and corm extricates showed the outcome (Lee et al., 2004).From the review it has been accounted for that the antimicrobial action of the concentrate of *C. esculenta* was more compelling in microscopic organisms than growth. (Singh et al., 2011).

The *in vitro* antifungal movement of Colocasia esculenta was tested by the food contamination strategy technique against two contagious species *Alternaria solani* and *Alternaria ricini*. The alcoholic leaf remove showed great antifungal movement than the watery concentrate of Colocasia esculenta. Alcoholic concentrate of *Colocasia esculenta* showed 100 percent antifungal activity against *Alternaria solani* and keeping in mind that showed 25% antifungal activity against *Alternaria ricini* (Mengane, 2015).

5.3. Antidiabetic Activity

An *in vitro* cancer prevention agent test of watery concentrate of *C. esculenta* corm extricate displayed solid cancer prevention agent potential and free extremist rummaging limit. *C. esculenta* corm extricate showed its cell reinforcement likely when the concentrate was examined by seven distinct measures, viz. complete phenolic content, all out flavonoid content, absolute flavonol content, diminishing power assessment as well as 1,1-diphenyl-2-picrylhydrazyl (DPPH), nitric oxide (NO) and 2,2'- azino-bis(3-ethylbenzothiazoline-6-sulphonic corrosive) (ABTS) revolutionary searching exercises. The cell reinforcement capability of *C. esculenta* corm is basically because of the presence of its phytoconstituents tannins, saponins, flavonoids, steroids, sugars, proteins, and glycosides (Yadav et al., 2017). Leong et al. (2010) reported that the presence of catechol moiety in the B-ring of isoorientin, orientin and luteolin 7-O -sophoroside, disconnected from taro shoot framework showed solid cell reinforcement action with various instruments of activity in DPPH revolutionary rummaging, β-carotene dying and superoxide extremist hindrance measures (Leong et al., 2010). *C. esculenta* entire leaf squeeze additionally forestalled the inspire of lipid peroxidative responses caused because of the presence of free extremists created by the hepatotoxins CCl4 and acetaminophen in vitro rodent liver cut model (Patil and Ageely, 2011a) .

5.4. Antdiabetic Activity

The antidiabetic movement of the ethanol concentrate of *C. esculenta* (EECE) leaves was accounted for in an alloxan-incited diabetes model. In this concentrate on EECE at the portion level of 100, 200, and 400 mg/kg were controlled orally metformin 450 mg/kg was utilized as a norm against alloxan (120 mg/kg, i.p.) initiated diabetic rodents. The beginning of decrease of blood glucose of EECE was seen at 4 h (96 mg/dl), top at 6 h (120 mg/dl) yet antihyperglycaemic impact melted away at 24 h. In the subacute review, the outcomes exhibit the greatest decrease in blood glucose was seen at a portion of 400 mg/kg on the fourteenth day showed huge enemy of hyperglycemic movement in alloxan-initiated diabetic rodents (Kumawat et al., 2010). Trial studies showed that oral organization of ethanolic concentrates of taro leaves, petrol ether and chloroform essentially displayed antidiabetic movement in Streptozotocin (STZ)- initiated diabetic rodents by lessening and normalizing the raised fasting blood glucose levels when contrasted with those of STZ control bunch (Tripathi and Koli, 2013).

5.5. Anti-Inflammatory Activity

The calming action property of *C. esculenta* leaf extricate was shown on the carrageenan-actuated intense paw edema model and the cotton pellet granuloma technique. Shah et al. (2007) explored mitigating action of the ethanolic concentrate of the leaves of *Colocasia esculenta* in Wistar rodents hindered carrageenan-prompted rodent paw edema. The plant removes restrained leukocyte movement, decreased the pleural exudates, and diminished the granuloma weight in the cotton pellet granuloma technique when contrasted and standard medication (Shah et al., 2007)

5.6. Anthelmintic Activity

The anthelmintic action examination of *Colocasia esculenta* fluid concentrate and ethanolic leaf separate was displayed portion subordinate anthelmintic movement at the focus scope of 10-50 mg/ml against night crawler. The unrefined concentrates of *C. esculenta* incapacitate and causes the demise of worms when worms are in higher fixations (Kubde et al., 2010). The ethanolic taro leaf separates caused loss of motion of 16 minutes and season of death of 42 minutes, while water removes the uncovered loss of motion of 31.25 minutes and passing of 60.34 minutes against the worm. Phytochemical screening of the unrefined concentrates uncovered the presence of riboflavin, niacin, oxalic corrosive, and sapotoxin; anthocyanins, perlargonidin 3-glucoside, cyanidin 3-rhamnoside and cyanidin 3-glucoside; and saponins.

5.7. Neuropharmacological Activity

C. esculenta is customarily utilized for the treatment of uneasiness, despondency, and other CNS problems. Kalariya et al. (2010) depicted the neuropharmacological impact of hydroalcoholic leaves concentrate of *Colocasia esculenta* at uneasiness, misery, thiopental-instigated dozing time, and rotarod execution assessment. The leaves concentrate of taro showed critical portion subordinate anxiolytic, antidepression, gentle neurosedative, and skeletal muscle relaxant impact. The presence of flavonoids, β-sitosterol, and steroids in C.esculenta are liable for neuropharmacological exercises ((Kalariya et al., 2010)).

5.8. Wound Healing Activity

C. esculenta has the injury mending potential against oxidative/nitrosative harm and anticipation of hyaluronic corrosive corruption through injuries (Gonçalves et al., 2013). The review showed that *C. esculenta* assortments had the option to rummage a few oxidant animal categories and to repress hyaluronidase because of the presence of different metabolites other than phenolics compound.

5.9. Hepatoprotective Activity

An *in vitro* leaf juice investigation of *C. esculenta* showed hepatoprotective and antihepatotoxic movement against the event of lipid peroxidative responses brought about by paracetamol, acetaminophen and hepatotoxins $CCl4$ initiated hepatic harm (Patil and Ageely, 2011b). At the same time-stamped heights and counteraction of consumption of complete tissue, glutathione were seen within the sight of taro leaf juice. Taro leaf squeeze likewise showed antihepatotoxic and hepatoprotective viability against hepatotoxicity incited by CCl 4 and paracetamol making oxidative pressure and harming different cell organelles thus bringing about injury to the hepatocytes (Patil and Ageely, 2011a). Taro leaf squeeze fundamentally decreased the spillage of marker compounds AST, ALT and ALP in the medium demonstrating hepatocyte respectability.

6. Conclusion

Roots and tubers are significant eating regimen parts for people and changeup it. Notwithstanding the fundamental job as an energy giver, they give various alluring wholesome and medical advantages, for example, antioxidative, hypoglycemic, hypocholesterolemic, antimicrobial, and immune modulatory exercises *C. esculenta* is one of the staple root and tuber crops developed for different purposes, give quantities of attractive nourishing and medical advantages like anticancer action, phenolic corrosive, phytochemicals, antidiabetic, cancer prevention agent action. This matter gives significant data about the dietary and medical advantages of taro corms and leaves. Taro is utilized as a staple food or means food by a huge number of individuals in a significant number of the emerging nations where the corms of taro are utilized as vegetables and considered as a rich wellspring of sugars, proteins, minerals, and nutrients. Taro tubers contain little granules of starch which are profoundly edible. Taro can be developed as a root crop, like a verdant vegetable, as a fancy and as a therapeutic plant. Taro leaves, like higher plants, is wealthy in protein supplements the high sugar content of the tubers. Underutilized yields, for example, taro have a lot of significance in guaranteeing food security, in making unfamiliar cash just like a money crop and furthermore as a method for the rustic turn of events. Most rustic people groups experience the ill effects of unhealthiness for the financial status as well as because of failure to use the accessible nutritious underutilized yields, for example, taro to meet their day by day prerequisites. To fulfil the healthful need there should be a need to see the financial and dietary effect of native new underutilized yields like taro for its creation and utilization in provincial networks. Accordingly, it tends to be better for unhealthiness and food instability for rustic people groups as provincial food sources.

References

Ahmed, I., Matthews, P.J., Biggs, P.J., Naeem,M, McLenachan, P.A., Lockhart, P.J. 2013. Identification of chloroplast genome loci suitable for high-resolution phylogeographic studies of *Colocasia esculenta* (L.) Schott (Araceae) and closely related taxa. Mol. Ecol. Res.13(5): 929-937.

Aregheore, E.Perera, D. 2003. Dry matter, nutrient composition and palatability/acridity of eight exotic cultivars of cocoyams-taro (*Colocasia esculenta*) in Samoa. Plant Foods Hum. Nutr. 58: 1–8.

Brown, A.C.,Reitzenstein, J.E., Liu, J.,Jadus, M.R.2005. The anticancer effects of poi (*Colocasia esculenta*) on colonic adenocarcinoma cells In-vitro. Phytotherapy Research 19(9): 767-771.

Chandrasekhara, A. Kumar, T.J. 2016. Roots and Tuber Crops as Functional Foods: A Review on Phytochemical Constituents and Their Potential Health Benefits, Int. J. Food Sci. 3: 1-15.

Ferdous, R.U., Arman,S.I., Tanvir, M.I., Sumi, S., Siddique, M.R., Billah M. 2015. Biologically Potential for Pharmacologicals and Phytochemicals of Medicinal Plants of *Colocasia esculenta*: A Comprehensive Review. American J. of Clinic. Expt. Med.3(3):7-11.

Food and Agriculture Organization (FAO).1990.*Roots, Tubers, Plantains and Bananas in Human Nutrition*, Food and Nutrition Series, Food and Agriculture Organization, Rome, Italy: 24.

Gonçalves, R.F., Silva, A.M., Silva,A.M., Valentão, P., Ferreres, F., Gil-Izquierdo, A., Silva, J.B., Santos, D. Andrade, P.B.2013. Influence of taro (*Colocasia esculenta* (L.) Shott) growth conditions on the phenolic composition and biological properties. Food Chem. 141(4):3480–3485.

Hedges, L.J.,Lister, C.E. 2006. Health attributes of roots and tubers. Crop and Food Research Confidential Report No. 1569. New Zealand Institute for Crop and Food Research Limited Private Bag 4704, Christchurch, New Zealand.

Huang, D.J., Lin, C.D., Chen, H.J, Lin, Y.H.2004. Antioxidant and anti-proliferative activities of sweet potato (*Ipomoea batatas* (L.) Lam. "Tainong 57" constituents, Bot. Bull.of Academia Sinica. 45(3):179-186.

Kalariya, K., Parmar, S., Sheth, N. 2010. Neuropharmacological activity of hydro alcoholic extract of leaves of *Colocasia esculenta*. Pharma. Biol. 48(11):1207– 1212.

Keshav, A., Sharma, A., Mazumdar, B. 2019. Phytochemical Analysis and Antioxidant Activity of *Colocasia esculenta* (L) leaves. Int. J. Chem. and Mol. Engg. 13(1):20-23.

Kumawat, N.S., Chaudhari, S.P., Wani,N.S., Deshmukh, T.A., Patil, V.R.2010.Antidiabetic activity of ethanol extract of *Colocasia esculenta* leaves in alloxan induced diabetic rats. Int. J. of Pharm Tech. 2(2):1246-1249.

Kundu, N., Campbell, P., Hampton, B., Lin, C.Y., Ma, X., Ambulos, N., Zhao, X.F., Goloubeva, O., Holt, D, Fulton, A.M. 2012. Antimetastatic activity isolated from *Colocasia esculenta* (taro). Anticancer Drugs23(2):200–211

Lee, S.W., Najiah, M., Wee, W. 2004. *In vitro* antimicrobial activities of *Colocasia esculenta* extract against Vibrio spp. – short communication. Agric. Sci J. 1:5–11

Lim, T.K. 2012. Edible Medicinal and Non Medicinal plants. 1, Springer, Netherlands, 1023.

Macharia,W.M,Nuro, M.S., Muchugi, A.N., Palapala, V.2014. Genetic Structure and diversityof East African Taro *Colocasia esculenta* L.Afr. J. Biotechnol. 139:2950-2955.

Mengane, S. 2015. Antifungal activity of the crude extracts of *Colacasia esculenta* leaves in vitro on plant pathogenic fungi. Int. Res.J. Pharm. 6(10):713-714.

Onwueme, I., 1999. Taro cultivation in Asia and the Pacific. Published in RAP Publication 1999/16. Published by Food and Agriculture Organization of the United Nations, Regional Office for Asia and the Pacific, Bangkok.

Patil, B.R.,Ageely, H.M. 2011a. Anti-Lipid Peroxidative Activity of *Colocasia esculenta* leaf Juice against CCl4 and Acetaminophen Mediated Cell Damage. Int. J.Pharma Appl. 2(3):141- 149

Patil, B.R., Ageely, H.M. 2011b. Anti-lipid peroxidative activity of *Colocasia esculenta* leaf juice against CCL 4 and acetaminophen mediated cell damage. Int J Pharm Appl. 2(3):141–149.

Shah, B.N., Nayak,B.S., Bhatt, S.P., Jalalpure, S.S., Sheth, A.K. 2007.The anti-inflammatory activity of the leaves of *Colocasia esculenta*. Sau. Pharm J. 15:3-4.

Singh, B., Namrata, K.L., Dwivedi, S.C. 2011. Antibacterial and antifungal activity of *Colocasia esculenta* aqueous extract: An edible plant. J. of Pharm. Res. 4(5):1459-1460.

Temesgen, M.,Retta, N.2015. Nutritional potential, health and food security benefits of taro-*Colocasia esculenta* (L.): A Review, Food Sci. and Quality Mgmt. 36:23-30

Tripathi, A.K., Kohli, S. 2013. Phytochemical screening and evaluation of antidiabetic activity of *Colocasia esculenta* (L) leaves on STZ induced diabetic rats. Adv. Pharmacol. Toxicol. 14(2):1–12.

Wheatley, C., Scott, G.J., Best, B.,Wiersema, S. 1995. Adding value to root and tuber crops. A manual on product development, International Centre for Tropical Agriculture, 247, Columbia.

Wiersema, J.H., Leon, Méditerranée, B. 1999. Regards Croisés sur la Domestication, Dialogue Entre la Biologie.Ethnobiologie, Biologie des Populations, Université Montpellier 2, Mo World Economic Plants, A standard reference. New York: CRC Press; 143.

Yadav, M., Kushawaha, D.K., Chatterji, S., Watal, G. 2017. Assessment of antioxidant activity and phytochemical screening of *Colocasia esculenta* corm. Int. J of Pharma. Sci. Res. 8(4):1758-1764.

2

Cultivation of Export Oriented Cumin Crop

Arvind Kumar Verma[1], Ravi, Y.[1], Ravindra Singh[1], Sharda Choudhary[1], Pranava Pandey[2] and Vijay Bahadur Singh Chauhan[3]

[1]ICAR- National Research Centre on Seed Spices, Ajmer, Rajasthan India
[2]V.K.S. College of Agriculture, Buxer, Bihar Agricultural University, Sabour Bhagalpur, Bihar, India
[3]Regional Centre, Bhubaneswar, ICAR-Central Tuber Crops Research Institute, Odisha, India

1. Introduction

Cumin (*Cuminum cyminum* L.) crop belongs to the family Apiaceae and having diploid chromosome number 2n =14. It is also known as *Jeera* in Hindi. This crop is originated from Mediterranean region. It is an erect or sub erect, glabrous annual herb which grows to a height of 30-50cm. The leaves are alternate, compound and green, the blade consists of three slender filiform leaflets. The inflorescence is a compound umbel with white or pinkish flowers (Verma et al., 2018a). The typical pleasant aroma of the seeds is due to their essential oil and the main constituent in essential oil is cuminaldehyde. It is mainly cultivated in arid and semi arid parts of Gujarat and Rajasthan. The total area is about 12 lakh hectare with the production of about 8.5 lakh tons and total export of cumin is worth of more than Rs. 4200 crore during 2020-21 (Fig. 1 & 2). Cumin seeds are generally used to flavour meals in cooking and the cumin oil is used to add aroma in food and cosmetics. Cumin is a key ingredient in curry and chilli powders, and it's also used to flavour a wide range of commercial foods. The dried seed used for steam distillation and it yields 3-6% of volatile oil depending upon variety and location. The processed products include essential oil, oleoresins and fixed oil, encapsulated oil, cumin power, cumin based drinking products, cookies. It is also used to flavor alcoholic beverages, desserts, and condiments. Beside this, it is used as a fragrant component of creams, lotions, and perfumes. In Ayurvedic system of medicine, dried cumin seeds are used for therapeutic purposes (Choudhary et al., 2017). It is known for its activities like enhancing appetite, taste perception, digestion, vision, strength and lactation. It is used to treat diseases like fever, loss of appetite, diarrhea, vomiting, abdominal distension, edema and puerperal disorders. The typical pleasant aroma of the seeds is due to their volatile oil content, the principal constituent of which is cuminaldehyde. Cumin is believed to be native of Mediterranean region, but cultivated in India, Egypt, Libya, Iran, Pakistan, Mexico and Japan. Cumin seed has a significant demand as a spice all around the globe especially in the places where spicy food is preferred. Cumin seed mainly exported to Vietanam, Bangladesh, USA, Egypt, Pakistan, Nepal, UAE, UK etc.

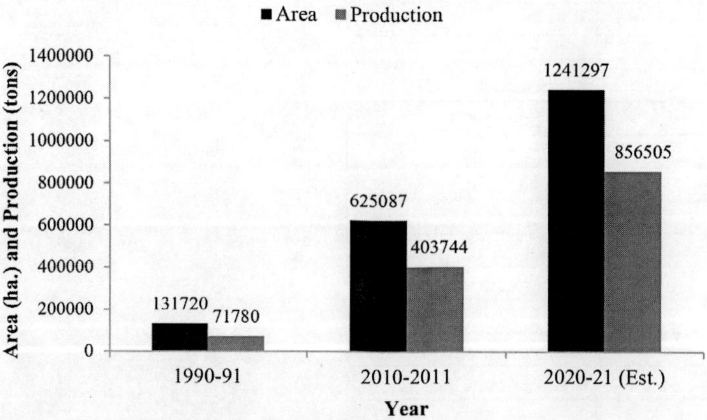

Fig. 1: Area and production of cumin from 1990-91 to 2020-2021 (Source: Spice Board India, Kochi and DASD, Calicut)

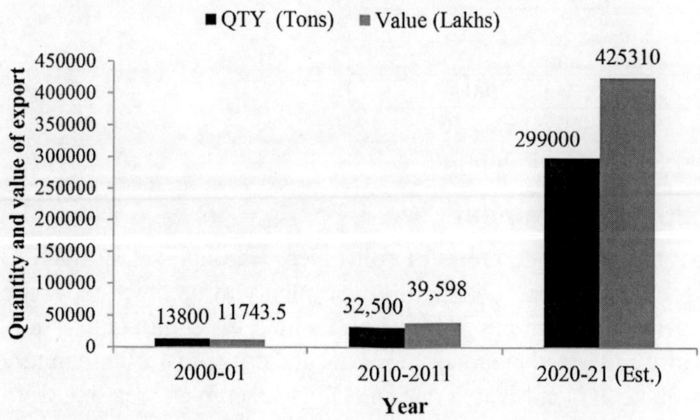

Fig. 1: Quantity and value of cumin export from 2010-11 to 2020-2021 (*Source*: Spice Board India, Kochi and DASD, Calicut)

2. Chemical Composition of Cumin Seed

Cumin seed has many uses but little information is available on the physical and chemical properties of cumin seed. Cumin seeds have sevreal chemical cmompounds and these compounds might be measured as interesting bioactive natural substances that may be used in several fields, such as nutraceuticals, cosmetics and agro-food industry. Due to its very selective dry and cool climate requirement, cumin is mainly being grown in western regions of India. The physical and chemical properties of cumin seed are give in Table 1.

Table 1: The physico-chemical composition of cumin seeds

Length (mm)	5.2
Breath (mm)	1.58
Moisture	6.2 per cent
Protein	17.7 per cent
Fat	23.8 per cent
Carbohydrate	35.5 per cent
Crude fibre	9.1 per cent
Mineral matters	7.7 per cent
Calcium	0.9 per cent
Phosphorus	0.45 per cent
Iron	0.048 per cent
Sodium	0.16 per cent
Potassium	2.1 per cent
Vitamins	mg per 100 g
Vit B-1	0.73 mg
Vit B-2	0.38 mg
Niacin	2.5 mg
Vit. C 17	.2mg
Vit A	175 I.U. /100 g
Food energy	460 calories/ 100 g

(*Source*: Varshney et al., 2005; Lal et al., 2014)

3. Soil and Climatic Requirement

Cumin can be cultivated in a wide range of soil types, although sandy loam to medium heavy soils which having good fertility and plenty of organic matter are the best. Generally pH range from 6.8 to 8.3 is best suited for cumin cultivation. The crop is successfully cultivated in moderately cool and dry climate, temperature ranging between 25-30°C. It is cultivated during Rabi season in an area free from severe frost during flowering and seed formation and there should be no rain during harvesting time (Sastry and Anandaraj, 2013). It does not prefer humidity in the atmosphere during flowering and seed setting stage. The recurrent rain during flowering and fruit setting stage and at the time of maturity period is highly undesirable. These factors are highly favourable for the serious diseases like blight and powdery mildew (Lal et al., 2014). Cloudy weather during flowering and fruiting stages increases incidences of insect-pests and diseases resulting into poor quality produce.

4. Improved Varieties

RZ – 19: Erect in growth behavior, bears pink coloured flowers and bold pubescent grains matures in 120–140 days and gives an average yield of 5–6 q/ha.

RZ – 209: Variety has shown tolerance to wilt, matures in 140–150 days and gives average seed yield of 6.5 q/ha.

RZ – 223: Variety possesses resistance to wilt, essential oil content is 3.23% and gives seed yield of 6.0 q/ha.

RZ-345: This variety suitable for Rajasthan and Gujarat. Its seeds bold, attractive, higher volatile oil content. Medium maturity group 120-130 days and give average yield of 607 kg/ha.

GC- 1: Plants are erect with pink flowers and seeds are bold, linear, oblong and ash brown coloured can withstand shattering and lodging. It matures in 105 -110 days and average yield is 7.0 q / ha..

GC – 2: Plants are bushy with good branching habit and attractive grains. Its matures in 100 days and gives average seed yield of 7.0 q/ha.

GC – 3: Plants are bushy with good branching habit and attractive grains. Its matures in 100 days and gives average seed yield of 7.0 q/ha. It also wilt resistant variety have essential oil content is 3.3 %.

GC – 4: Very popular variety of cumin because of its capacity to resistant to *Fusarium* wilt and higher yield performance (average yield is 8.75 q / ha).

5. Time of Sowing and Seed Rate

Sowing time varies based on the local environmental condition. Generally cumin sowing done from the last week of October to last week of November. Although best sowing time for cumin cultivation is from 1-15 November. In Gujarat, farmers sown cumin crop little early as compared to Rajasthan. The seed rate of cumin crop is about 10-12 kg/ha. The line sowing recommended for the cumin cultivation and 25-30 cm row to row and 5-10 cm plant to plant depending on varieties is ideal for the crop. The seed should not sow deeper than 1.5 cm otherwise the germination percentage reduced. The seed should be treated with bavistin @ 2.5 g/kg seed to control the seed borne diseases. Application of *Trichoderma harzianum* strains @ 8-10 gm per kg seed and @ 40 kg/ha is beneficial to control wilt and blight diseases (Deepak et al., 2008). Cumin seed having the slow germination problem so soaking the seed in water for 6-8 h is recommended for fast germination (Saxena et al., 2015). Farmers advised to follow the crop rotation to prevent wilt disease in cumin.

6. Manure and Fertilizers

The requirement of fertilizers depends upon the fertility status of the soil. Hence, fertilizer should be applied based on soil testing report. Judicious application of manures and fertilizers in appropriate dose are important with respect to plant growth, seed development, yield and quality of produce (Lal et al., 2014). 10-15 MT FYM, 30 kg nitrogen, 20 kg Phosphorus and 20 kg potassium per ha is recommended for cumin crop. Whole quantity of FYM should be mixed into the soil at the time of land preparation and 15 kg nitrogen and whole quantity of phosphorus and potassium should be applied as basal dose. Another, 15 kg nitrogen should be applied as topdressing one month after germination of seeds. Besides this, application of micronutrients is utilized by crop in higher quantity and lead to improve the quality of cumin. Average uptake of micronutrients by cumin is Fe-439.8 g ha^{-1}, Mn-201.8g ha^{-1}, Cu-107.6g ha^{-1}, Zn-22.2g ha^{-1} and B-37.4g ha^{-1} (Harisha et al., 2017).

7. Irrigation

Cumin crop is low input loving crop and required very less water as compared to other seed spices and other crops. The frequency of irrigation should be given based on oils type and weather condition. This crops grown in area where the scarcity of water is there. So, use of drip irrigation of sprinkler irrigation is advisable. For successful germination, mild irrigation should be applied immediately after the seed sowing and again after 8-10 days. After seed germination, normally the crop should be irrigated at 20-30 days intervals. The last irrigation should be slightly heavy at the time of grain formation to supplement moisture requirements during the crop ripening stage. Irrigation during the maturity stage of the crop should be avoided since it will degrade seed quality and increase the risk of powdery mildew and blight disease. Five irrigations i.e. first at sowing followed by 10, 30, 55 and 80 days after sowing should be given (Lal et al., 2014). Patel et al. (1991) observed that irrigation at 0.60 IW: CPE ratio (four irrigations each of 50 mm depth) significantly increased seed yield of cumin (529 kg/ha). This levels of scheduling irrigation also showed significantly higher N removal by seed (13.25 kg/ha) and stover (7.06 kg/ha) of cumin compared with lower ratio of 0.30 (2 irrigation) .

8. Weed Management

The cumin crop has very slow germination process so before the germination of cumin weeds come first. So, the cumin crop should be kept free from weeds for proper growth and development. The main weed of cumin includes bathua (*Chenopodium album*), kharbathua (*Chenopodium murale*), Piaji (*Asphodellus tenifolius*), zeeri (*Plantago pumila*; wild), motha (*Cyperus rotundus*), krishnaneel (*Anagallis arvensis*), pilisanjee (*Melilotus indica*) etc. Deep summer ploughing and soil solarization are also advisable to the cumin growers to reduce the weed problem. The first weeding/hoeing should be done 30-40 days after sowing and generally 2-3 hand weeding re required to keep the cumin field free from weeds. In case of zeeri escapes than the same should be removed manually at the stage of its flowering. Pre-emergence weedicides like oxadiargyl should be applied @ 75 g/ha with water in the field to keep cumin field free from weeds. Weedicides should be applied after sowing but before the germination of the weeds (Lal et al., 2014).

9. Insects Management

The indiscriminate use of chemical pesticides in curative mode has resulted in rejection of cumin export consignments. Therefore, technologies targeting integrated protection by preventing resistance to pesticides and to avoid varietal resistance to reduce biotic pressure needs to be addressed (Lamichhane et al., 2017). The rationale of the present study was to develop bio-intensive integrated pest management (IPM) module encompassing host plant resistance, cultural, biological and the minimum prophylactic use of chemicals to promote sustainable production with reduced environmental pollution.

9.1. Aphids

The aphis species *Myzus persicae* and *Aphis gossypii* is the major aphids species reported in cumin. The massive spread of aphids on cumin seeds occurs during December to first week of March which caused yield losses of over 50% in unprotected plants. Adults and nymphs suck the sap from plants and produce honeydew secretions, on which shooty moulds develop and lead to the failing seed production. If aphid incidence occurs during flowering and fruiting, the fruits will not form, and if they do, they will wrinkle and deteriorate in quality. Higher losses in yield could be caused by a small number of aphids infesting the crop at the beginning of flowering than by a large number of aphids at the grain filling stage (Verma et al., 2018b). To control the aphid population in cumin fields follow the crop rotation. Installed the yellow sticky card @ 50-60 numbers per hectare. Apply neem based commercial formulation like Neemarin at 1% and seed extract of neem (*Azadirachata indica*) to reduce the aphids population by 50% within 7 days of application. Coccienellied is consisting of large predators that feed on various sucking pests of seed spices. Major coccienellid found predating on seed spices crops are, *Coccinella septumpunctata* L. *Bromoides suturalis* F. *Menoochilus sesxmacalatus* and *Adonia* sp. In cumin crop spraying of 0.03% phosphomidon, 0.03% monocrotophos at 2 week intervals after first appearance of aphids give effective protection. It has also been found two spray of dimethoat 0.03%, thiomethoxam 0.025%, imidachlorprid 0.005% and acephate 0.037% decrease more than 90 % of aphid (*Myzus persicae*) population in cumin (Jadon et al., 2020.

9.2. Thrips

Thrips are other important sucking pests of many seed spice crops particularly cumin. The nymphs are slender, yellowish-brown and look similar to adults but are wingless and slightly smaller in size. The adults are yellowish-brown and measure about 1 mm in length. The males are wingless while the females have long, narrow fringed wings. Female live for 12-30 days and lay 50-60 kidney shaped eggs singly inside leaf tissue with its sharp ovipositor. The incubation period is 4-10 days and the nymphs feed on the plant sap by lacerating the leaf tissues. Usually they congregate at the leaf sheath or in the flowers. Nymphs mature in 4-5 days after passing through four instars and pupate in the soil. There is a pre-papal stage, which lasts for 1-2 days, and pupal period is 2-4 days. Several generations are completed in a year. Control measure includes, follow the crop rotation followed by use of botanicals to control the thrips at early stage of crop growth. Two sprays of 1250 ml/ha of Carbosulfan 25EC proved to be the optimal dose for effective control of sucking pest populations such as aphid and whitefly. Use of carbosulfan did not adversely affect the natural enemy population and left no phytotoxic effects on the cumin plants. (Verma et al., 2018b).

10. Disease Management

10.1. Wilt Disease

Cumin wilt is responsible to cause a loss up to 60% with a 25.7% incidence. In severe infestation crop may damage up to 90%. This disease caused by fungus known as *Fusarium oxysporum* f. sp. *cumini*. The main symptoms of this disease are leaves and tip fall-down and later, the whole plant may die. Typical symptoms of the infected plants are leaf-tip dropping and overall canopy, which certainly lead to the death of the plant. The disease can be occurred at any stage of growth stage, however, from the germination to younger plants are more prone to wilt severity. The roots exhibit browning of the vascular region when it split-open. Continuing to grow cumin crop in the same field for longer time will increase the inoculation of pathogens in the field, so a three-year crop rotation should be followed to prevent the spread of wilt disease. Few cumin cultivars such as RZ-223 and GC-4 are tolerant to *Fusarium* wilt can be grown. Summer ploughing, soil solarization with 25 micron polythene, use of disease free seeds, seed treatment with suitable fungicides or bioagents are recommended for management of this disease. Organic amendments of soil with mustard cake (1%), incorporation of mustard residues in soil reduce *Fusarium* propogules. Use of *richoderma viride* @8-10g/kg seed for also control the disease. A common method for reducing the incidence of wilt disease is seed treatment with 2.5-3.0 g / kg chillum or captan or 2 g / kg carbendazim.

10.2 Alternaria Blight

This disease is widespread and causes serious losses in cumin growing areas of India (Khare et al., 2014; Deepak et al., 2008). This is very severe disease of cumin crop and the losses may be up to 100%. The cumin blight disease is caused by a fungus *Aleternria burnsii*. Cumin plants are usually attacked by fungi after sowing for 60 days. It is spread by seeds, air and soil and the main symptoms of this disease are that the plant has small brownish necrotic spots that later turn black. The diseased plants could not produce seeds, or if seeds are produced the quality of seeds is not good. Currently, it is a common disease in all cumin-growing areas, which having humid and cloudy weather during cropping season. It has since been observes to be widespread and causing serious losses in Rajasthan and Gujarat states. The pathogen survives for about 10-12 months in both seeds and debris. A temperature of around 23-28°C is optimal for the onset of illness. Crops sown in December had the lowest incidence of disease. The highest incidence of corruption was in October (Verma et al., 2018b; Verma et al., 2022). Moreover, raining during the off-season, a cloudy span of weather, dew, and temperature 23–28 °C helps the pathogen to incite conidia germination and encourage the disease (Gemawat and Prasad, 1972). To get he higher yield and good qulity seed, there is need to cntrol blight disease. So far no variety available in the market which is resistance to cumin blight disease. Strictly follow the crop rotation and deep summer fallowing is effective to reduce the disease. Stop irrigation under cloudy and high dew condition to avoid blight disease. This disease can be controlled by the application of 0.2% solution of Dithane-M-4 and Carbeddazim 0.1%. Its spraying can be done 4 times at 10-15 days interval starting 50 days after sowing. Recently propiconazole (0.1%), carbndazim + iprodione (0.2%), chleorothalonil (0.2%) are giving least disease incidence and higher yield. A schedule comprising foliar application of mancozeb followed by propiconazole proves effective to combat the blight disease of cumin (Sharma, 2016).

10.3. Powdery Mildew (PM)

PM is caused by fungus a *Erysiphe polygoni*. The disease appears in February and March at flowering time. The disease spreads fast under warm (27-35°C) and moist conditions. In the early stages of the disease occurrence, whitish spots appear on the surface of the leaves, petioles, stems and seeds of the diseased plant. Gradually the seeds become whiter, wrinkled and lighter. Avoid late sowing of cumin crop, as late sowing crop under irrigated conditions are greatly affected. In a serious infestation, complete crop failure was observed. Prevention of this disease can be achieved by spraying the plant with 300 mesh of sulfur powder @ 25 kg per hectare, as soon as the symptoms are noticed. Single dusting of 300 meshes sulphur 20-25 kg/ha at the time of flowering in January is essential. Dinocap (0.1%), carbendazim (0.1%), tridemorph (0.05%) and wettable sulphur (0.2%) are also effective (Khare et al., 2014).

10.4. Emerging Disease of Cumin

Yellowing of leaves is a recent problem in cumin growing areas which causing significant loses to cumin growers. Although exact cause of problem not reported but as per NRCSS, Ajmer it is due to potyvirus and the disease spread by the aphid.

11. Harvesting Time and Methods

Harvesting is one of the major important factors that determine the quality of the produce. It is hardly possible to produce a good cumin crop from low quality harvested material. Cumin crop harvested they when fully mature. The immature and over mature/ ripened are not good as far as quality is concerned. The colour turning stage from green to brown/light brown is a good indicator for proper time of harvest. Improved method of harvesting with the help of improved/ new implements and harvesting machines at proper stage of maturity and proper drying of harvested material offers quality produce as well as overcomes the problem of contamination.

12. Yield

Harvesting of cumin starts from the end of February depending upon sowing date. Adoption of mechanized threshing preferably by seed spice threshers can further improve the quality and reduce the post harvest losses. Cumin crops take 110-120 days to reach at maturity. Under scientific management condition 8-12 q/ ha cumin seed of improved varieties can be obtained. The volatile oil content of cumin seeds is lost due to delayed harvesting & drying of crop in sun. Further, crop should be harvested in early morning to avoid shattering of seeds and thereby reduce the losses of seed yield. Seed can be separated by beaten and trampled on clean threshing floor of by cumin threshing machine (Lal et al., 2014; Verma et al., 2018b).

13. Conclusion

Cumin is an important spice crops grown in India and having huge potential for its cultivation in arid and semi arid regions. This crop requires low input as compared to other crops and having ability to thrive in poor soil. Hence, the cumin cultivation is more profitable than any other crop cultivation as it has huge export potential and in recent post the export demand rise exponentially. Farmers can earn more net income by adopting the advanced production technology of cumin. Hence, the less information available on scientific cultivation and problem associated with cumin cultivation. In this chapter all the aspect of cumin cultivation has been covered like importance, export potential, climatic requirement, soil, seeds rate fertilizers, irrigation, weed management and insects and disease management harvesting time and methods.

References

Choudhary, S., Pereira, A., Basu, S., Verma, A.K. 2017. Differential antioxidant composition and potential of some commonly used Indian spices. Journal of Agrisearch 4(3):160-166.

Deepak, Arora, D.K., Saran P.L., Lal, G. 2008. Evaluation of cumin varieties against blight and wilt diseases with time of sowing. Annals of Plant Protection Sciences 16(2): 441-43.

Harisha, C.B., Diwakar, Y., Aishwath, O.P. Singh, R., Asangi H. 2017. Response of *Cuminum cyminum* for various modes of micronutrients application including uptake and their availability in soil. International J. Seed Spices 7(1): 48-51.

Jadon, K.S., Singh, S.K., Patel, N., Sharma, A.K. 2020. Integrated pest management module for cumin (*Cuminum cyminum*). production under arid environment. Indian Journal of Agricultural Sciences 90(6): 1120-1124.

Khare, M. N., Tiwari, S. P. and Sharma, Y. K. 2014. Disease problems in the cultivation of I. cumin (*Cuminum cyminum* L.) II. caraway (*Carum carvi* L.) and their management leading to the production of high quality pathogen free seed. International Journal Seed Spices 4(1): 1–8.

Lal, G., Saran, P.L., Devi, G., Deepak, Raj, R. 2014. Production technology of cumin (*Cuminum cyminum* L.). In book: Advances in vegetable agronomy. Publisher: Post Graduate School, IARI, New Delhi and Indian Council of Agricultural Research Department of Agricultural Research and Education Ministry of Agriculture, Government of India, New Delhi. Pp. 223-231.

Lamichhane, J.R., Bischoff, S.M., Blueme, S., Dachbrodt, S.S., Dreux, L., Jansen, J.P., Kiss, J., Kohl, J., Kudsk, P., Malausa, T., Messean, A, Nicot, P.C., Ricci, P., Thibierge, J. Villeneuve, F. 2017. Identifying obstacles and ranking common biological control research priorities for Europe to manage most economically important pests in arable, vegetable and perennial crops. Pest Management Science 73: 14–21

Patel, K.S, Patel, J.C., Patel B.S., Sadaria, S.G. 1991. Indian Journal of Agronomy, 36:4: 627-629.

Sastry, E.V.D., Anandaraj, M. 2013. Cumin, fennel and fenugreek. Soils, plant growth and crop production. Encyclopedia of Life Support Systems Pp. 1-10.

Saxena, S.N., Kakani, R.K., Sharma, L.K., Agrawal, D., Rathore, S.S. 2015. Usefulness of hydro-matrix seed priming in cumin (*Cuminum cyminum* L.) for hastening germination. International Journal of Seed Spices 5, 24-28.

Sharma, Y.K. 2016. Disease management in cumin and coriander including from damage. In: Training Manual on Sustainable Production Technology for Seed Spices. Ravindra, S., Gopal, L., Sharda, C. and Arvind K. V. Eds. ICAR NRCSS: Ajmer.

Varshney, A.K., Vyas, D.M., Akbari, S.H. 2005. Some physical and engineering properties of cumin (*Cuminum cyminum* L.) seed. Agricultural Engineering Today 28(3-4): 7-10.

Verma, A.K., Choudhary, S., Meena, R.S., Lal, G., Sharma, M. 2018a. Improvement of seed spices. In: Phytochemistry of Fruits and Vegetables (Peter, K.V. Editor). Brillion Publishing, New Delhi– 110005, India. Pp. 287-314.

Verma, A.K., Ghatak, A., Meena, R.D, Sharma, M., Kumar, R., Vinod and Balodi U.R. 2022. Epidemiology and management of important diseases of seed spices. In book: *Diseases of Horticultural Crops Diagnosis and Management*. Vol. 3: Ornamental Plants and Spice Crops (Editors: J.N. Srivastava and A.K. Singh, P). Apple Academic Press Inc., USA. Pp. 279-317.

Verma, A.K., Singh, R., Choudhary, S. and Lal, G. 2018b. Cultivation of dollar earning cumin crop for higher income. Acta Scientific Agriculture 2(3): 46-48.

3

Chrysanthemum: A Versatile Crop of Commercial Significance

Madhu Bala

Department of Floriculture and Landscaping
College of Horticulture and Forestry, Punjab Agricultural University
Ludhiana, Punjab, India

1. Introduction

Chrysanthemum is commonly known as 'Guldaoudi', "Queen of the East", "Glory of East" or "Autumn Queen". It is one of the most beautiful and widely cultivated flowers that can be grown for its aesthetic value as well as for its commercial potential. The word chrysanthemum is derived from two Greek words, viz. "*Chrysos*" means gold and "*Anthemon*" means 'flower' i.e. golden flower. Chrysanthemum is the only crop that offers a great genetic diversity of flower colour, shape, forms, size, easy to propagate, growing habit, longer flowering duration and excellent vase life. Based on the shape of flower heads and floret arrangement of petals it can be used as pot plants for indoors and outdoors decoration as well as exhibition purpose; as cut flowers for flower arrangements, bouquets making and vase decoration; loose flower for making garlands, worshipping purpose and for bedding and garden decoration. Tall cultivars are suitable for background planting in borders and the dwarf and compact can be planted in front rows. Chrysanthemum is gaining popularity as the most cultivated plants which is highly appreciated and priced by flower lovers throughout the world. It ranks next to rose most of the times in the international flower market as cut flower and holds good position as pot plant also due to wide diversity. The prolific fall blooms casually referred to as "mums", known for their brightly colored heads of tightly packed petals. *Chrysanthemum* is a versatile genus, consisting of more than 300 species of herbs and shrubs, among them a few are having insecticidal properties known as Pyrethrum. Several species are having ornamental potential and are grown for commercial production as well as decoration purpose. Chrysanthemum is gaining popularity among the flower lovers, farmers and amateurs as it is easy to propagate and has long flowering duration with excellent vase life. Chrysanthemum can be classified into various types such as incurved, reflex, decorative, button/pompon, single / semi/double, Korean, spoon, anemone, spider, quill, thistle, etc.

1.1. Origin and History

Chrysanthemum is native to northern hemisphere, chiefly Europe and Asia. The first cultivation of chrysanthemum as a flowering herb has been reported in China and is described in writings in 15th Century B.C. In Japan several cultural events or symbols are famous, including 'Chrysanthemum Day' i.e. festival of happiness to honor the chrysanthemum.

The Chinese city of Chu-Hsien (means Chrysanthemum City) was named to honor the flower. Chrysanthemum as "Yellow Glory" has been found in the writings of Confucius, a great Chinese philosopher. The first cultivation is in China, it is generally associated with Japan and believed to have reached there through Korea, as early as 386 AD. Chrysanthemum is proclaimed as national flower of Japan. It's a symbol of Royalty in Japan and "The Imperial Order of Chrysanthemum" is the highest order of chivalry in Japan which has a National Chrysanthemum Day known as the "Festival of Happiness". In India, chrysanthemum flower was first reported in Marathi literature by Sant Geneyswar in the 13th century. The standard type, large flowered chrysanthemums were introduced by Britishers in India during 19th century AD. Now-a-days many new cultivars have been developed with diverse range of colours, shape and sizes with novel characters.

2. Importance and Uses

Chrysanthemum is one of the most popular traditional flowers among the top ten ranking flowers in the world due to its high ornamental value. Chrysanthemum is a multipurpose crop and flowers can be utilized in different ways.

2.1. Trade Value

Chrysanthemum rooted cuttings can be sold and good profit can be made by selling rooted cuttings at Rs. 8- 10 / cutting. Chrysanthemum pot can also be sold at the rate of Rs. 100-400 per pot and by selling 1000 pots, the income of 1- 2 lakh rupees can be obtained. Alternatively, other loose flower varieties can also be planted in field and flowers harvested at fully open stage can be sent to the market for sale and about 1.5-2.0 lakh rupees/ acre income can be obtained.

2.2. Ornamental Value

Varieties with compact and medium sized are mainly used for loose flower purpose, for garlands making, floral decoration, worship purpose, in religious and social ceremonies and for making ornaments like jaimala, veni, bracelets, necklace etc. Vigorous growing varieties are used for bedding and garden decoration which gives beautiful mass effect, if maintained properly. Standard varieties with extra large bloom can be used for exhibition purpose and interior decoration. Varieties with strong and sturdy stems are used as cut flowers for bouquet making and flower arrangements in combination with other flowers and foliage plants. Dwarf, compact and pot mum varieties are used for pot culture for interior and exterior decoration. Compact varieties form a very good pot plant which can be used for decoration of windows, corners and the central table.

Uses of Chrysanthemums

Floral Rangoli

Pots for Exhibition

Cut Stems for Bouquet

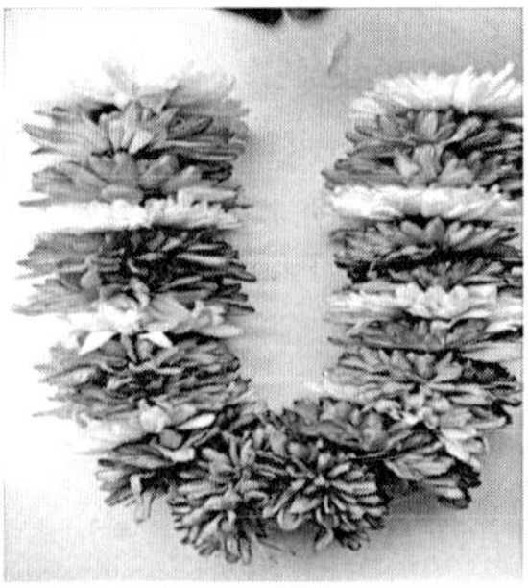

Garland Making

Flower Arrangement

Multiuser chrysanthemums

2.3. Edible / Nutritional Value

Chrysanthemum flowers are good source of nutrients, minerals and vitamins. Some minerals elements such as Calcium, Magnesium, Manganese, Iron, Copper and potassium are also present in chrysanthemum that are necessary for a healthy immune system. It also contains vitamin A, C and B9. Yellow or white chrysanthemum flowers are used to make sweet "chrysanthemum tea" in some Asian countries due to its medicinal properties. The flower heads are dried and packaged as leaf tea or sometimes blended with some other dried flowers and tea leaves. In Korea, chrysanthemum is used to flavor rice wine. Steamed or boiled leaves are used as greens in Chinese cuisine. The flowers can also be added to dishes and soups to add aroma. Small flowers are used for garnishing of food items in Japan. The flowers can also be eaten raw, but may develop a bitter flavor when mature.

2.4. Insecticidal Value

Pyrethrin is extracted from the flowers of Pyrethrum *(Chrysanthemum cinerariifolium)* which has insecticidal properties and is also sold in the market under the same name as an oleoresin. The insecticide, used against all types of insects including beneficial ones as well as mosquitoes in suspension in oil or water, or in powdered form. It is much less toxic to birds and humans as compared to synthetic insecticides but is more harmful to fishes. It attacks the nervous system of all insects which cause their death and also inhibits female mosquitoes from biting. Being biodegradable, it gets decomposed easily on exposure to light rays. Thus, it is considered as the safest insecticides and considered under botanical insecticides.

2.5. Phytoremediation Value

Chrysanthemum plants help to reduce air pollution when placed in an appropriate location due to gaseous exchange. Chrysanthemum is among the top ten indoor plants that improves the air quality. It also helps in the removal of ammonia, benzene, formaldehyde, xylene and trichloroethylene from indoor air.

2.6. Medicinal Value

Chrysanthemum tea can be useful to cure sore throat and to reduce fever as per traditional Chinese medicine. It is also used to treat blurring, spots in front of the eyes, diminished vision, and dizziness.

Chrysanthemum morifolium flower is widely used in as a food, beverage, and medicine for in China and Japan (Yang et al., 2019). Its boiled roots are also useful as a remedy for headache, young leaves and petals can be eaten as salad and leaves are fermented for a festive drink (Bose et al., 2003).

3. Botany

Chrysanthemum belongs to family *'Asteraceae'* and more than 200 species are reported within the genus Chrysanthemum that are very complex (Anderson, 1987). The complete flowering stem or spray is of cyme type with multiple inflorescences (Cockshull, 1985). Chrysanthemum inflorescence comprises of hermaphrodite florets in central disc (both pistillate and staminate) and the outer ray florets having female flowers (pistillate) with inferior ovaries (Cockshull, 1985). The cultivars are classified on the basis of flower form, type and size. The common flower types are irregular incurve, reflex, regular incurve, decorative, intermediate incurve, pompon/button, anemone, singles and double Koreans, spiders, quills, spoons, brush/thistle type etc. Chrysanthemum is an herbaceous perennial, shallow rooted plant with fibrous root system. Plants grow 50-150 cm tall, with deeply lobed leaves and large flower heads of different colours. *Chrysanthemum* species has a complex base number of $x = 9$ (Shimotomai and Takemoto, 1940), and the cultivated or garden species is *Dendranthema x grandiflora* Tzvelv. Synon. *Chrysanthemum x morifolium* Ramat.

4. Types of Chrysanthemums

Worldwide more than 20,000 chrysanthemum cultivars have been reported and about 7,000 are in China (Wang et al., 2014). Chrysanthemum is broadly classified into two categories on the basis of size and form of blooms i.e., Standard and Spray type.

4.1 Standard Type

Also known as Japanese type having large flowers (usually grown as standard with 1-3 stems, bearing a single flower each). Large flowered or standard chrysanthemums have genetic potential to produce large size bloom on a single stalk.

4.2 Spray Type

These are also known as Korean type and small flowered (grown as bushes with multiple numbers of branches bearing more number of flowers per plant).

National Chrysanthemum Society (NCS) of America has classified the chrysanthemum on the basis of number, size, form and arrangement of disc and ray florets in to 13 classes and this classification system in followed in all over the world. The details are given as below:

Class-I: Irregular Incurve: These are the standard flower group. The petals are loosely incurving with closed centers and the basal florets are asymmetrical in appearance. Flowers are 6-8 inches in size.Examples: Crimson Tide, Shirley Perfection.	
Class- II: Reflex: The petals are curving downward and overlap with each other. The center of these type is full, but somewhat flattened. Flower size is 4-6 inches. Examples: Star of India, Playmate, Dareen Hall.	
Class- III: Regular Incurve: Flowers are incurved smoothly, globular in shape and equal in breadth and depth forming a ball like structure. Flowers are 4-6 inches in size. Examples: Snow Ball, Sonar Bangla, Chandarama	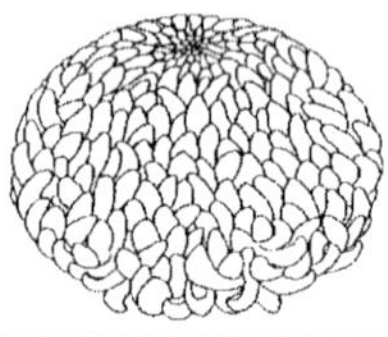
Class- IV: Decorative: Flowers are flattened with short petals. The upper florets curved in wards incurve, but the lower ones are of reflex type. Flower are of 5 inches or more in size. Examples: Punjab Shingar, Ratlam Selection, Purnima, Flirt	

Class- V: Intermediate Incurve: Flower is smaller than class-I, florets are short and partially incurving with complete centre. Flower size is 6 inches or greater. Examples: Alfred Simpson and Kokka Senkin.	
Class- VI: Pompon/Button: Flowers are globular, flat in early stage but fully round or globular when developed properly. Ray florets are broad, smooth, firm, incurved and fully cover the centre. Examples: Punjab Shyamli, White Bouquet, Kelvin Mandarin, Kelvin Victory, Kelvin Tattoo. Buttons are compact, small and hemi spherical. Example: Lemon Baby Drop, Liliput, King Fisher	
Class- VII: Single and Semi Double: Single or more than one row of ray florets at right angle to the stems, disc may be slightly rounded. Examples: Yellow Charm, Punjab Mohini, Pusa Sona, Reagan Emperor, Reagan White, Punjab Gold.	
Class- VIII: Anemone: Ray florets are flattened, twisted or quilled, pointed at tips having unequal length. Disc florets are well developed and prominent forming a raised cushion like structure. Examples: Mother Teresa, Mars, Royal Purple, Anmol, Autumn Eyes, Dark Eyes.	
Class- IX: Spoon: Ray florets are regular, tubular, straight, disc portion open, flattened and spoon like at the tips. The central disc is open and visible. Examples: Garden Beauty, Winter Queen, Anokha.	
Class- X: Quill: Ray florets are tubular, straight and not coiled, tips may be closed and pointed, open and speculate. The blooms are fully double without disc. Examples: Golden Crystal, Snow Crystal, Ravi Kiran, Kasturba Gandhi.	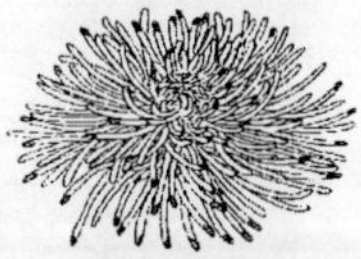
Class- XI: Spider: Ray florets are long, tubular which may coils or hooks at the tip portion. Examples: Swan Dance, Rosy Whisker, Coronation.	
Class- XII: Brush or Thistle: Fine tubular ray florets that grow parallel to the stem in brush or thistle manner. The ray florets are flattened, twisted and drooping. Examples: Whisp of Pink, Saga, White Ice.	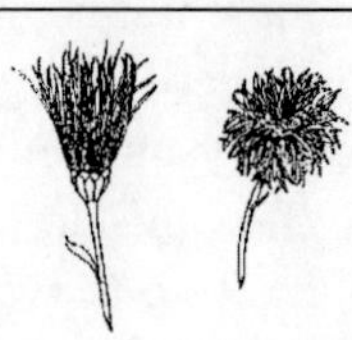

Class- XIII: Unclassified: The blooms or class that do not fall in none of other classes. Ray florets are twisted and also exhibit characteristics of more than one bloom class. Examples: Ronaldo, Bronze Robert, Jhalar.	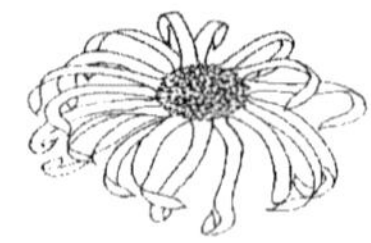

Classes of Chrysanthemum

Irregular IncurveReflex

Regular

Incurve

Decorative

Intermediate Incurve

Pompon/Button

Single and Semi Double

Anemone

Spoon

Quill

Spider

Brush or Thistle

Table1: Chrysanthemum varieties recommended by various institutes

Sr. No.	Institute	Recommended varieties
1.	PAU, Ludhiana	Punjab Gold, Baggi, Ratlam Selection, Punjab Shingar, Punjab Mohini, Punjab Shyamli, Anmol, Royal Purple, Autumn Joy, Garden Beauty, Winter Queen
2.	IARI, New Delhi	Pusa Sona, Pusa Guldasta, Pusa Sundri, Pusa Shwet
3.	IIHR, Bangalore	Arka Kirti, Arka Chandrika, Arka Yellow Gold
4.	NBRI, Lucknow	Nanako, Gauri, Maghi, Himanshu, NBRI Kusum

5. Soil and Climate

Chrysanthemums can grow well in well drained soil and fully sunny days. Plants grown under low light intensity will become be very weak with a few flower buds. Chrysanthemums are shallow rooted and frequent watering is essential during hot and dry months of growth and development. Chrysanthemum is very sensitive to water logged soils so good drainage is necessary. Light to heavy textured well drained, sandy loam soil rich in organic matter mixed with well rotten FYM having pH of 6.2 -6.7 is ideal soil for chrysanthemum. Addition of good amount of organic matter is essential for chrysanthemum cultivation since it has a shallow and fibrous root system and very sensitive to moist soil conditions.

Photoperiodically chrysanthemum is a short day plant and it requires long days for growth and short days for development of flower buds. The critical day length is 14-15 hours for floral bud initiation and 13-14 hours for flower development. The humidity level ranges from 70 to 90 per cent during the complete phase of plants.The optimum day light requirement for good quality flower production is 13.5 hours. The optimum temperature requirement for this crop is 20- 28 ^{o}C during the day and 16- 20 ^{o}C during night. The effect of night temperature is more pronounces than day temperate. High temperature may cause distortion of flowers and very low temperature may cause discoloration of petals.

Chrysanthemum varieties are grouped in to three groups according to temperature requirement for flowering (Cathey, 1955):

i. Thermopositive cultivars: Varieties that require at least 16 ^{o}C temperature for flowering and 27 ^{o}C for rapid but delayed flowering.
ii. Thermonegative cultivars: When bud initiation occurs at temperature of 10-27 ^{o}C, but high temperature delays flowering.
iii. Thermozero cultivars: Varieties which bloom between 10-27 ^{o}C temperature but most constant at 16 ^{o}C night temperature.

6. Propagation

Chrysanthemum is generally propagated vegetatively either through terminal cuttings and root suckers. Other methods of propagation are through seeds for breeding of new varieties and tissue culture for mass propagation of plants.

6.1. Terminal Cuttings

This is commercial method and generally adopted by the flower growers. Plants raised from the cuttings are preferred as they are healthy and sturdy. Terminal cuttings (5-7 cm) from the healthy mother plants free from insect pest and diseases are generally taken in the mid June to end July and basal leaves are removed. These cuttings are planted either on sand beds or propagation trays for large scale multiplication. The common river sand is the best and cheapest rooting medium for most of the varieties. Fresh river sand may be filled in any shallow pan, earthen pot or wooden box for the rooting of cuttings, but in all cases the drainage of excess water must be assured. The rooting of terminal cuttings can also be done in burnt rice husk as rooting medium due to its good water holding capacity and growth of new roots. Burnt rice husk is partially sterilized and free from most of the disease causing organisms. The cuttings are planted after dipping in NAA or IBA (500 mg/l, quick dip i.e. less than 5 sec) in the trays, nursery beds or pots, which are placed under partial shade to provide protection from direct sunlight and high temperature conditions for successful rooting. The terminal cuttings can also be dipped in rooting hormone Seradix or Rootex powder readily available in the market. The cuttings should be watered 2-3 times daily to maintain high humidity and prevent desiccation of cuttings for formation of roots. Rooting takes place in 2-3 weeks and cuttings can be planted in the main field. To avoid rotting of cuttings, apply Captan (0.3%) or Dithane M-45 (0.2%) along with sprinkling of water for drenching.

6.2. Suckers

Plants are cut above the ground after flowering for the production of suckers. This induces the formation of just side suckers which are separated from the mother plants when they are 15-20 cm high and then directly planted in the field or pot. The advantage of this method is easy establishment of the plant without mortality. Sometimes, parental diseases are transmitted that results in lack of uniformity and poor quality of flower production. The optimum time of separation of side suckers is February-March. Before planting, basal leaves and long roots should be trimmed.

6.3. Micro Propagation

Micro propagation is used for large scale multiplication of plants by using shoot apex, peduncle, leaves and other floral parts as explants. MS medium (Murashige and Skoog, 1962) along with various amendments like BA, IBA, NAA etc. are used as substrate. The advantages of micro propagation are year round production of disease free propagules on large scale.

6.4. Seeds

The seeds are collected after flower has completely matured and stored in dry places. Seeds can be sown in February March or June –July in beds and fine compost is used to cover the seeds. Seedlings are ready for transplanting after 30-40 days of sowing. Seeds are not true to type as chrysanthemum is cross pollinated crop and normally used by the breeders to develop new varieties and hybrids.

7. Planting and Nutrition

The best time of preparation of cutting is from mid June to end July. These cuttings form roots in 2-3 weeks time and ready for transplanting in pots or beds. Rooted cuttings should be transplanted from mid of July to mid of August in the evening hours of the day. The cuttings should be planted 30 x 30 cm apart in the field. Standard varieties are mostly planted at 20 x 20 cm spacing. The rooted cutting can also be transplanted in the pots of 8 inch size having 3-4 rooted cuttings. In case of standard varieties, one rooted cutting should be transplanted in one pot.

Chrysanthemum is a very heavy feeder crop and requirement of nitrogen, phosphorous and potassium is more to get good quality flowers. The field should be ploughed 3-4 times, at the time of final ploughing, 20 tons of well rotten farm yard manure should be added per acre of the land before transplanting of rooted cuttings. Basal dose of fertilizer @ 40 kg N (160 Kg Calcium Ammonium Nitrate or 87 kg Urea), 80 kg of P_2O_5 (500 Kg Single Super Phosphate) and 80 kg K_2O (133 Kg Muriate of Potash) per acre should also be incorporated at the time of final ploughing. Supplemental dose of 20 kg N (80 kg CAN or 44 kg Urea) should be supplied after one and two months of transplanting of rooted cuttings. Application of liquid fertilization is beneficial, especially for pot grown varieties. A fertilizer mixture containing 10 g potassium nitrate, 30 g ammonium sulphate and 30 g of super phosphate per 10 liters of water should be applied at an interval of 10-15 days till the bus are about to open. Liquid manure can also be used by decomposing fresh cow dung and oil cake about 1-2 kg of each in 10 liters of water to tea colour and applied @500 ml/plant. During early growth phase the crop should be kept weed free as the crop weed competition is known to be more intense in the early stages of crop growth i.e. vegetative stage that adversely affect plant growth and quality flower production. About 3 to 4 hand weeding to keep the crop free from weeds may be necessary during the growing season.

8. Growing Chrysanthemum in Pots

Chrysanthemum can also be planted as pot plant to decorate the corners of houses, offices, verandas, windows and in the lawn areas. Different type of earthen and plastic pots can be used for pot culture. The potting mixture consists of garden soil, sand, well rotten farm yard manure, leaf mould (ratio of 2:1:1:1). To maintain the proper drainage, some brick pieces or small stones are kept at the bottom of the pot and then the potting mixture is filled in the pots. The rooted cuttings are transplanted in these pots in the end of July to first week of August and immediately watering is done. Watering is done depending on the requirements by the plant. Required small doses of fertilizers and nutrients are also given to the plants for quality growth and flowering.

9. Irrigation and Weeding

Plants require enough moisture in the soil during dry and cool months of September and October. During this phase plants grow very fast and vigorously therefore, frequent irrigations should be given at regular intervals. The best method of irrigation is flooding of field. Overwatering should be avoided as this crop is very sensitive to water logging. In pot plants, extra water should drain out through drainage whole and overwatering should be avoided. After transplanting of rooted cuttings, irrigation should be done to firm the soil for establishment of roots in the soil. Watering of pot plants through sprinkling cane should always be preferred. Chrysanthemum needs 3-4 hand weeding to keep the crop free from weeds and to get quality flowers. Weeds can be control through mulching on the beds in order to retain the soil moisture. Heavy or deep mulching may cause stem rot. Though this crop flowers during cold months of the year, covering of plants through mulch material is considered to be the best option.

10. Pinching, Disbudding and Staking

Pinching and disbudding are two major operations in chrysanthemum which should be done very carefully at appropriate time. Pinching is the removal of terminal shoots to encourage the development of side shoots and it should be repeated to facilitate more number of side shoots. The main purpose of pinching is to overcome apical dominance. Normally a cutting is ready for pinching when 5-7 cm of new growth has occurred. All the cuttings in a pot should be pinched at the same to promote uniform leveled growth of the whole plant. Pinching is done twice in spray varieties after 4 and 7 weeks of transplanting of rooted cuttings. Pinching leads to emergence of side shoots and increase the number of flowers/ plant.

On the other side, disbudding is the removal of flower buds which develop from the side shoots and auxiliary branches so that number of blooms per plants are limited and blooms of good quality can be obtained. Disbudding is done for better bloom size mainly in standard varieties in August and September to allow the development of single bloom with large size.

Staking of plants can be done in field and pots with the help of sticks or plastic nets to provide support and to avoid lodging. The net levels can be raised up to different heights likewise, 15 cm, 30 cm and 45 cm from the ground level in field area. The standard varieties with big-sized blooms may require single staking with stick. In spray varieties, more than 2-3 stakes are required that are given in the periphery of the pot to give support to the outer growing branches.

Pinching

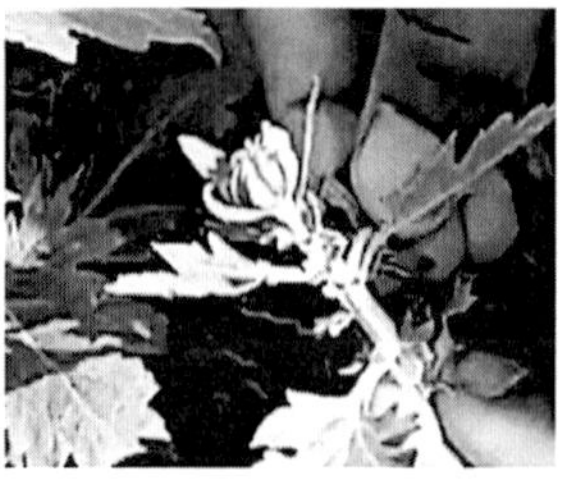

Disbudding

Staking with nets in field

11. Pests

11.1. Bihar Hairy Caterpillar (*Spilosoma oblique = Spilarctia oblique*)

This insect laid eggs in clusters on leaves during March-April and September-November. The larvae make big holes on the flowers and leaves which result in deterioration of flower quality. The hairy caterpillar at early stage should be destroyed by picking and application of quinalophos 25 % EC @ 2ml/l of water will be effective.

11.2. Aphids (*Macrosiphoniella sanbroni*)

Aphids commonly suck the sap from the shoots, underside of the leaves, flower buds and the flowers. The infested plants remain small and flower buds fail to open. Spray of dimethoate 30 % EC @ 2 ml or Imidacloprid 17.8 % SL @ 0.4 ml/l of water will be effective.

12. Diseases

12.1. Leaf Spot

The leaf spot disease caused by species of *Septoria* (*S. chrysanthemella* and *S. obesa*) is the major disease infecting chrysanthemum and symptoms appear as small circular bronze areas on the upper surface of leaves that enlarge and form bigger more irregular spots. The colour of the spots changes to dark brown to black and on lower surface of the leaves, the colour is grayish-brown. This disease can be controlled by spraying mancozeb 75%WP @ 2g/l of water at 7-10 day intervals.

12.2. Stem Rot

The causal organism of this disease is *Rhizoctonia solani* and very common in propagation beds. Rotting of stem may occur due to high moist and warm weather conditions. Soil sterilization and application of pencycuron 22.9% SC @ 2 ml/l of water may be done as a control measure.

12.3. Powdery Mildew

Infection is more severe in under humid conditions. The appearance of white powdery fungus on the leaves and the leaves turn yellow and dry out. The growth of plants may remains stunted and fails to flower. This disease can be controlled with Sulphur fungicides or Captan (0.2%) or any fungicide. Maintenance of proper hygienic and good ventilation is recommended.

13. Harvesting and Yield

Right time, method and stage are of utmost importance for longer after harvest keeping quality of cut stems. Flowers should always be harvested when temperature is mild because high temperatures lead to faster the rate of respiration and excessive transpiration losses. Flowers should therefore, be harvested either in the morning or in the evening. In the morning, flowers are fully turgid due to low transpiration loss of water during night. Evening harvest is also advised because of more sugar in the cut stem due to increased activity of photosynthesis during day time.

Proper method should be adopted to harvest the flowers. The stems should always be harvest with sharp knives or secateurs and care should be taken not to crush the stem. Cut stems should not be harvested near to the ground level as the stem close to the soil is lignified and hard which allows lesser water uptake. Chrysanthemum inflorescence occurs in various shapes and forms. The stages of harvest also vary in these different forms of inflorescences, distance from market and time of planting of varieties.

14. Standard Cultivars

Flowers fully open but before the central disc florets in the diameter of 1.5-2.5 cm begin to elongate.

14.1. Spray Cultivars

Flowers fully open but before the shedding of pollens or when 2-3 flowers are fully open and others are showing colour.

14.2. Anemones

Flowers open but before the central disc florets of the topmost flower begin to elongate.

14.3. Pompons and Decorative

Centre of the flower fully open, flowers should always be harvested when temperatures are mild because high temperatures increase respiration and also promote water loss. Chrysanthemums are highly prone to water stress and should preferably be harvested in the morning or evening.

15. Yield

The yield varies from 80,000-90,000 cut flowers /spray stems per acre, whereas for loose varieties, 15-65 q/acre flowers can be produced.

16. Post Harvest Handling and Storage

Grading of chrysanthemum stems is based on the flower diameter and length of the stem. The Society of American Florists has suggested the following grades for standard Chrysanthemum in U.S.A.

Grade	Minimum flower diameter (mm)	Minimum length overall (cm)
Blue	140	76
Red	121	76
Green	102	61

The standards are bunched in groups of 10 or 12. The flowers can also be packed as single units in plastic sleeves or paper bags to protect the florets from physical injury. Spray chrysanthemums are made into bunches of 200-300 g or having 10, 15 or 20 stems are placed in sleeves. The bunches of flowers are wrapped in cone shaped plastic sleeves and packed in fiberboard boxes.

The flowers should be transported in pre-cooled moisture retentive boxes. The bunches wrapped in plastic sleeves can also be transported under ambient conditions (20-25°C) for shorter distances without major loss of vase life. Box sizes are typical as in chrysanthemum 80cm x 50cm x15 cm with 15 kg in weight. Loose flowers are packed in gunny bags or bamboo baskets for marketing. Chrysanthemum flowers can be stored dry at 1°C for 3 weeks, whereas, flowers are sealed in moisture-retentive plastic bags to prevent the loss of moisture. In wet conditions, however stems are stored at 4°C for 2 weeks with their bases dipping in water or solution of floral preservative, usually biocidal compound to check the growth of microbes on the stem surface as well as in water in the vase.

After harvest, the cut stems should be immediately placed in water or preservative solutions to prevent water loss from the stem. The most common practice for keeping the flowers at farmer's level is to keep them in water. However, the vase solution containing sucrose (1.5%) plus 8-HQC (200 ppm) or silver nitrate (25 ppm) + citric acid (75 ppm) improves vase life of cut chrysanthemum stems. The buds of standard chrysanthemum cut at 5 cm size can also be made to open in chemical solution containing sucrose (2%) + 8-HQC (200 ppm) or silver nitrate (25 ppm) + citric acid (75 ppm). Another solution containing silver nitrate (25 ppm) and Citric Acid (75 ppm) along with sucrose (3%) can also be used.

17. Maintenance of Mother Stock Plants

Healthy suckers sprouts from base of the plants after flowering in the month of February-March. The dry flowering shoots are removed and healthy suckers uprooted from the main field are replanted on raised beds at 30cm x 30cm spacing. These mother stock plants grow very vigorously in March to maintain height and for production of more number of lateral branches for terminal cuttings. The pinching of new shoots is necessary during April and May to get sufficient number of terminal cutting in the month of June-July. Mother stock plants are maintained under partial shady conditions and plants are fertilized and irrigation should be given depending upon the growth, soil and weather conditions.

18. Value Added Products

The value added products from chrysanthemum can be in the form of cut flowers, flower bouquets, flower arrangement, tinting of flowers, garlands, wreaths, floral ornament like, venis, bracelets, necklace, jaimala, greeting cards using petal-embedded craft papers, dry flowers, etc. Edible chrysanthemums as salad, in soups and dishes to enhance flavor, garnishing of dishes etc.

19. Conclusion

Chrysanthemum is one of the most beautiful flower crops grown for its aesthetic value as well as commercial significance. This flower is gaining popularity and is highly appreciated by the flower lovers all over the world. Dwarf and compact varieties can be used for pot culture for interior and exterior decoration as well as for exhibition purpose; varieties with strong and sturdy stem are used as cut flowers for bouquets making, arrangements and indoor vase decoration; fully open flowers for making garlands, worshipping, in religious and social ceremonies and floral ornaments like jaimala, veni, bracelets, necklace, greeting card etc. Chrysanthemum cultivation is an important venture and profitable business and demand is increasing day by day due to more urbanization. However, its commercial production needs proper management of inputs, quality planting material, irrigation, weeding, insect pest disease management and good market facilities to earn profits. Various research institutes have developed chrysanthemum varieties for different purposes that can be exploited for commercial cultivation to gain much more profit in Indian Floriculture industry.

References

Anderson, N.O. 1987. Reclassifications of the genus *Chrysanthemum* L. Hort Sci. 22(2):313.

Bose, T.K., Yadav, L.P., Pal, V.A., Parthasarathy, Das, P. 2003. Commercial Flowers. Vol. II. – Naya Udyog, Kolkata, India.

Cathey, H.M. 1955. Temperature guide to chrysanthemum varieties, Bull. N. Y. St. Flower grower, No. 119.

Cockshull, K.E. 1985. *Chrysanthemum morifolium*. In: A Halevy (ed.) CRC handbook of flowering, 2: 236-257.

Murashige, T., Skoog, F. 1962. A Revised Medium for Rapid Growth and Bio Assays with Tobacco Tissue Cultures. Plant Physiology 15, 473-497.

Shimotomai, N., Takemoto, T. 1940. Hybrids between *Chrysanthemum wakasaense* and some species of chrysanthemum and the increase in chromosome number of hybrids. Bot. & Zool. 8(4): 61-66.

Wang, F., Zhang, F.J., Chen, F.D., Fang, W.M., Teng, N.J. 2014. Identification of chrysanthemum (*Chrysanthemum morifolium*) self-incompatibility. – Hindawi Article ID 625658.

Yang, P.F., Yang, Y.N., He, C.Y., Chen, Z.F., Yuan, Q.S., Zhao, S.C., Fu, Y.F., Zhang, P.C., Mao. D.B. 2019. New caffeoylquinic acid derivatives and flavanone glycoside from the flowers of *Chrysanthemum morifolium* and their bioactivities.Molecules 24: 850.

4

Karonda Ethnobotanical

Jyoti Kanwar, R.N. Kanpure, Praveen Kumar Ausari

Rajmata Vijayaraje Scindia Krishi Vishwa Vidyalaya, Gwalior Madhya Pradesh, India

1. Introduction

Karonda is widespread throughout India, primarily in arid and semi-arid regions. As a hedge plant, the Karonda tree is frequently grown in residential gardens, farmer's fields, and orchards. Karonda is widely used as a condiment or ingredient to Indian pickle and spices. It is a commonly growing indigenous shrub in India that may thrive in marginal and wasteland areas where other commercially important crops are unsuited. Karonda fruits are used in processed goods like as jam, jelly, squash, syrup, and chutney, and they are in high demand on the international market. Karonda is a rich source of nutrition and have highest Iron. The karonda fruits also contain vitamin C and it has antiscorbutic properties and helps to cure anaemia (Table 1). The fruits of Karonda are also used in many ayurvedic formulations. Root extract is used to reduce Chest pain. The extract of leaves is very useful in case of fever. An experiment conducted at Indian Horticultural Research Institute, Bangalore revealed that the fruit of karonda are rich in vitamins like Thiamine, Riboflavin, Pantothenic acid, Pyridoxine, Biotinand Folic acid.

Table 1: Nutritional value of karonda fruits

Nutrient	Nutritional Value (100 gram)	
	Fresh	Dried
Moisture (%)	91	18.2
Energy (Calorie)	42	364
Fat (%)	2.9	9.6
Carbohydrates (%)	2.9	67.1
Protein (%)	1.1	2.3
Phosphorus (mg)	2.8	60
Vitamin-C (mg)	200-500	1
Iron (mg)	-	39
Mineral (%)	-	2.8
Calcium (mg)	2.1	160

Source: Tripathi et. al. (2014)

1.1. Karonda

Karondais known as christ's thorn or bengal currant in South India (Imran et al., 2012). Locally, it is known as *karaunda, karumcha* (Hindi), *karamardaka* (Sanskrit), *koromcha* (Bengali), *vakkay* (Telugu), *kauliHannu* (Kannada), *kilaakkaai* (Tamil), and *karjatenga* (Asam). In Konkan region, it is called as "Dongarchi Kali Maina" (Divate et al., 2019). Karonda scientifically known as *Carissa carandas* (L.) and having chromosome number 2n=22. Karonda belongs to family Apocynaceae/Elaeocarpaceae which is also known as dogbane family.

1.2. Distribution

This species originated from Indian subcontinent, and was introduced to Java. It grows widely in South Africa, Nepal, Afghanistan, Myanmar, Thailand, Sri Lanka and Malaysia. In India it grows naturally in the Himalayas, West Karbi Anglong (Assam), Jowai(Meghalaya) and Western Ghats of 300 to 1800 meters of elevation (Sarma et al., 2020). It found in deciduous forest from plains to 900m. In India, it is naturally occurred in Konkan area of Maharashtra, Bihar, West Bengal and parts of Southern India (Malik et al., 2010). Commercially it is planted in Varanasi district of Uttar Pradesh, Madhya Pradesh and Rajasthan (Banik et al., 2012). In Rajasthan, it is widely distributed in Aravali Hills mainly the Mount Abu, Udaipur, Chittorgarh, Sirohi, and some other parts of Jodhpur, Jaipur, Pali, Ajmer, Bhilwara etc. (Mehra and Arora, 1982; Meghwal and Singh, 2015). Massive collection of *Carissa* species have been made for fruit attributes in eight districts of Maharashtra mainly from Kolhapur, Ratnagiri and Pune districts and one of Goa (Sawant et al., 2013 and Ghate et al., 1997). Its germplasm is being maintained at various centres such as Kolhapur, MPKV, Rahuri; CISH, Lucknow; CAZRI, Jodhpur; ANDUAT, Faizabad, CCSHAU, Regional Research Station, Bawal; NBPGR, Regional Station, Jodhpur; CIAH, Bikaner.

2. Soil and Climatic Requirements

The shrub of Karonda is hardy, easily survive in drought and hot semi-arid eco region. Mostly prevalent in dryland area. It grows well in sub tropical and tropical climatic zones of India (Panda et al., 2014). Karondais also suitable to wasteland and also flourishing well as a rainfed crop, cultivation of karonda proved a goodsupportto tribal communities lies in Indian states like Bihar, Uttar Pradesh West Bengal, and South India (Banik et al., 2012). It flourishes well in regions with high temperatures ($48^0C\underline{+}2$), nevertheless the plants of karonda is very sensitive for frostiness (Peter, 2008). Plant intensification is exaggerated in heavy rain fall and drenched areas (Tripathi et al., 2014). In India, it can be cultivated on wide range of soils and has potential for commercial planting. Being hardy in nature and varied utility, it can be cultivated in wasteland where soils are saline and sodic in nature with pH value up to 5.0 to 8.0 (Bankar et al., 1994; Chundawat, 1995; Bose et al., 1999; Shafeeq et al., 2014). In India, it grows wildly on the rockiest soils. It is fruitful in deep, rich, well-drained soil, but excessive vegetative growth and poor fruit output result if the soil is too wet. (Kumar et al., 2013).

3. Botanical Description

Karonda is a medium sized evergreen deciduous, sprawling semi vine shrub which grows up to 2-4 m height. Meghwal et al., (2012) recorded variation in plant height, canopy area, growth habit, precocity, morpho-physico-chemical characteristics of leaves, fruits and fruit yield. The stem produces white latex. It has numerous branches with forked spines upto 1-3 cm long. The bark is smooth gray in colour. The bark is thick, soft and of red color from inside. Its leaves are opposite, oblong and conical, 10-15 cm long and 5-8 cm wide. Appearances of leaves are leathery green with polished above whiledreary green colour on beneath surface. Being evergreen deciduous in nature old leaves keep shedding throughout the year.

It is self pollinated fruit crop. Greyish brown twigs at the end produces axillary corymbose cymes. Flowers are perfect and appear terminally. They are solitary and appear as cymose (Randle, 1925). Flowers are white in colour with 3-5 cm diameter. Karale et al. (1990) found variability in floral morphology and fruit set in population which revealed that variation are found in number of flower buds (14.6-22.6) per umbel, petal number (4-6) and number of umbel segments (1-6).

Fruit is a globose, ovoid berry produced in clusters of 3-10 fruits, and immature fruits are green in colour and become reddish purple after attaining maturity. These are sweet, mild acidic, juicy when fully ripe. Ripe fruit of karonda varies in colourlike white, green and pinkish red depends on the genotype. Fruit contains 3-5 Seed, blackish brown in colour, flat in shape and elliptical light in weight (Shafeeq et al., 2014; Sontakke, 2014; Reshu et al., 2017; Singh et al., 2014). According to Singh & Singh (1998), fruits from the Red-green type were much larger (weight, length, and width) than those from the White yellow variety, particularly at the mature stage. Seed size variations followed the same trend.

When compared to Red-green fruits, the width and length of White-yellow fruits differed more between the two maturation stages.

4. Phenology

Leafy buds keep sprouting throughout the year, although more for the duration of spring (March-April) which reaches to maturity in August-September. Karonda flowers from February to June in Northern India. Fruit appear from March to August and starts ripening between May to December.

5. Uses

Carissa carandas L. grow for its eye-catchingcoloredsuitable for eating fruits and it is also used as live fencing due to the presence of spines which can provide protection against wander away and wild animals.

5.1. Traditional Uses

The fruit is generally used as an additive to Indian spices and cold beverages. The fruits with sweet tastemay eaten raw and fruits having high acidic content are best stewed with plenty of sugar (Arif and Fareed, 2011).Karonda Shrub commonly used for making beautiful hedge for orchards as well as in the home gardens because it possess axiliary spines so it can be a very good bio-fence and are (Reshu et al., 2017; Sharma and Banyal, 2010). In Rajasthan, tribes used to sell it's leaves for rolling tobacco to beedi manufacturers. In some other states,roots are used to kill the worms in cattle wounds, and to treat rheumatism.

5.2. Role in Nutritional Security

Fruits possess significant nutritional properties viz., vitamin C (1.6-17.9 mg/100g, protein (1.1-2.25%)) and minerals especially calcium (21mg/100g), iron (39.1mg/100g), vitamin-A (1619IU/100g), phosphorus (38mg/100g) and carbohydrates (67.10 mg/100g) (Anon., 1950; Anon., 1979; Julia et al., 1987; Kumar and Singh, 1993; Dhaka and Fageria, 2004; Malik et al., 2010). Fruits of karonda were used to cure anaemia due to its antiscorbutic properties.

5.3. Role in Livelihood Security

Karonda is having hard wood, straight-grained and can use for firewood purpose. Karonda pickles are useful for rural communities to generate additional income and employment for sustainable livelihoods.

Unripe karonda fruits are used as an astringent, appetiser, antipyretic, and antidiabetic in Ayurveda under the name 'Karmarda' and a preparation called 'Marmagutica' (Iyer and Dubhash 2006, Itankar et al. 2011). Maheshwari et al. define Karonda as having phyto-therapeutic qualities (2012). The fruits have traditionally been used to cure malaria, epilepsy, nerve disorders, pain and headache relief, fever, blood purifier, myopatic spams, dog bites, cough, colds, itches, and leprosy (Das et al. 2013; Rahmatullah et al., 2009). Its leaves are used to treat fever, diarrhoea, and earaches, while the roots are used as a stomachic, vermifuge, itching cure, and insect repellant (Malik et al., 2010). The roots are used to treat intestinal worms, scabies, and itch, as well as as an anthelmintic, stomachic, and antiscorbutic substance (Warrier et al., 1993). C. carandas plant extract has been shown to have cardioprotective, antipyretic, and antiviral properties (Rajasekaran et al., 1999; Taylor et al., 1996).

6. Potential Status

Karonda has a high potential for value addition, and numerous value-added products can be made from both unripe and ripe fruits (Srivastava et al., 2017). It is potentially very useful drug with wide range of medicinal utility in various health ailments.

7. Processing Possibilities

Ripe fruits with a distinct aroma can be served as a dessert. Mature fruits contain a substantial quantity of jelly grade pectin and can be used to make a variety of goods such as jelly, jam, squash, sauce, syrup, and so on (Burkill, 1935), which have a high export potential in the worldwide market. Mature karonda fruits are used to make pickles, chutney, preserves, and candies, while ripe maroon karonda fruits can be processed into high-quality syrup and jelly (Singh, 2001). Unripe fruits are sour and astringent, making them ideal for pickles and chutney (Bose et al., 1999).

Various work has been done on product preparation of karonda by researchers. Divate et al., (2020) standardized karonda crush recipes i.e. 35 and 40% juice with 55^0 Brix T.S.S. and 0.8% acidity. Shaheel et al., (2015) recorded highest TSS (10.35%), reducing sugars (6.96%) and organoleptic score (7.42) from blend of 25% karonda juice + 75% pineapple juice. The ripe fruit of karonda, exudates gummy latex, upon cooking, it converted into red colour juice and becomes clear when it allowed to cool, it can be used as a refreshing drink in summer (Singh and Bhatnagar, 2019). The ripe karonda fruits were used to create a natural 'food colourant and nutraceuticals supplement.' 'Lalima' was the name given to the formulation. 1 ml of this pigment suspension formulation is enough to colour one serving of any colourless beverage (100 ml), such as lemonade, a gorgeous crimson. One serving of such enhanced beverage may also contain 469.2 g anthocyanin, 12.7 mg flavonoids, and 14.1 mg phenol, for a total antioxidant activity of 390 μMTrolox Equivalent (Mohammed et al., 2016).

The dried fruits can be a substitute of raisins (Cheema and Cheema, 1971; Nalawadi and Jayasheela, 1975).Karonda fruit powder can produce after drying (Patil et al., 2014). Karonda fruit powder can prepared by two methods of drying, i.e., Sun and cabinet drying.The candy can prepared with the combination of fruit pieces embedded with cane sugar (TSS 60 °Brix)+ 0.5% citric acid + solar drying and it is very economical as compared to recipe prepared in cabinet drying(Saxena et al., 2016). The maximum yield (21.7%) of powder was obtained in the sun-dried samples of Pant Sudarshan. Highest ascorbic acid content (30.45 mg/100 g) was found in the cabinet dried samples of 70-day-old fruits of cv. Pant Sudarshan. Phosphorus (0.447%), potassium (18.73%), iron (0.365 mg/100 g), copper (0.012 mg/100 g), and manganese (0.193 mg/100 g) contents were higher in the powder prepared from the fruits of cv. Pant Suvarna. Among the drying methods, cabinet drying resulted in better retention of nutrients and less non-enzymatic browning.

According to Chaudhary et al. (2007), the biochemical variations in karonda jelly after storage were determined to be organoleptic properties acceptable for 4-5 months. After then, the jelly began to turn brownish at room temperature.

Karonda comes in two varieties, pink and green, and undergoes qualitative changes over a four-month storage period. During storage, pH, T.S.S./acid ratio, ascorbic acid, non-reducing sugars, and organoleptic parameters declined, but acidity, reducing sugar, and browning increased (Manivasagan et al. 2007). TSS and total sugars decreased in pickles created from pink and green karonda, and it was discovered that pickles prepared from pink karonda were of higher quality than sweet pickles. Wine made from ripe fruits contains approximately 14.5 to 15% alcohol and is popular among wine enthusiasts (Nalawadi and Jayasheela, 1975).

8. Genetic Diversity

Carissa L. genus contains about thirty species being native of Africa, Australia, Asia and China (Li et al., 1995). *Carissa carandas* L., *C. inermis, C. paucinerv* and *Carissa spinarum* L. are native to India while *Carissa grandiflora* is native to South Africa. *C. edulis* introduced from USA (Singh et al., 1982). *C. bispinosa, C. ovata*are other cultivated species (Singh and Arora, 1978). Hooker (1882) described five *Carissa* L. species from the Indian subcontinent, noting that the individuals in the area are difficult to classify and may be condensed to one or two varied types. Cooke (1957) named four species from Bombay Presidency, naming *Carissa carandas* L. as the most abundant species found throughout the region, while *Carissa inermis* Vahl, *Carissa spinarum* L., and *Carissa suavissima* Bedd. are exclusively found in the south. Haine (1919) revealed variety within and between Indian *Carissa* L. species, where five unique species were recognized and varietal variability was described. Fruit based diversity within and between the species has also been reported by Ghate et al., (1997). Karthikeyan et al. (2009) listed about 10 species (including 1 cultivated) and four varieties from India. Some other species are *C. arduina, C. suavissima, C. bispinosa, C. acuminate, C. paucinerria, C. congesta, C. ibispinosa, C. edulis, C. ovate, C. grandiflora* (Syn. *Arduinagrandiflora*), *C. macrophylla,* etc. it opens the doors of scope for improvement in karonda by selections (Joshi et al., 1986).

The cultivated types are classified according to fruit colour *i.e.*, green fruited, pink fruited and white fruited. Singh (1967) reported that pink and white fruited are ornamental in nature. in earlier times two distinct varieties: *C. Carandas* var. amara which has fruits of oval, dark-purple, red-fleshed and acid flavour; While,*C. carandas*var. dulcis has round, reddish purple, with pink flesh and sweet-sub-acid flavour were identified (Azeez et al., 2016).At Rahuri in Maharashtra, some promising types No. 13, 16, 12 and 3 have been identified (Karaleet al., 1990).In the recent past, karonda cultivars, viz. Pant Sudarshan, Pant Manohar, Pant Suvarna, White Pink Blush and Maroon colouredwere also developed for commercial cultivation (Mishra, 2007; Singh et al., 2010; Kashyap et al., 2015). Karonda Accession-3 was released by Regional Fruit Research Station, Vengurle for commercial cultivation, as it gives good yield, has large fruit size, higher fruit weight (16.23 g/fruit), better pulp content (92%) and low acidity (0.21%) (Shukla and Kumar, 2009). Promising karonda accessions from different parts of the country were identified and collected in field gene bank at ICAR-CISH, Lucknow and ICAR-CIAH, CHES, Godhra. Thar Kamal from CHES, Godhra and Maru Gaurav from ICAR-CAZRI, Jodhpur has been identified (Singh et al., 2013; Krishna, et al., 2015).CISH Kr-11 developed through selection at CISH, Lucknow (Singh et al., 2010).Azeez et al. (2016) suggested that varieties Konkan Bold and CHES-K-II-7 having health promoting polyphenols and antioxidant activity.

A wide range of genetic diversity is found throughout the country which could be exploiting for selection of better genotypes.

9. Objectives for Crop Improvement

i. To develop a cultivar with high yield potential and producing big sized fruits with attractive skin colour.

ii. To develop a cultivar producing fruits of good quality having higher pulp content, less number of seeds, high TSS with low in acidity and high in ascorbic acid.

iii. To develop a cultivar of dwarf stature with less spine intensity.

iv. To develop a cultivar having drought resistance/free from major diseases and pests.

10. Propagation

Karonda can be propagate either through sexual or asexual method of propagation. Mostly sexual method is practiced to raise plants. *Carrisacarandas* L. is usually propagated by seeds and viability of seeds is short that's why sown immediately after extraction. The longevity of seeds can considerably be increased by storing them in certain well defined conditions (Sharma and Singh, 1997; Naidu and Amritphali, 1994). Freshly extracted seed can be sown in July-August. One year old seedlings are transplanted and bearing started in plants raised from seeds is three years after planting (Peter, 2008). The seeds of karonda are quite hard and having low germination percent, Considering the constraints of low seed germination percentage, Bhavya et al. (2017) conducted experiment and observed that the treatments with water soaking for eight hours showed maximum percent of germination (83.30).

Vegetative propagation can be done by budding, inarching, grafting, air layering and cuttings have given variable results which are not very satisfactory (Mehta and Singh, 1960; Misra and Jaiswal, 1993; Misra and Singh, 1990; Tyagi et al., 1999; Peter, 2008).

Cuttings from a fruiting tree do not root well, although those from young nursery plants root successfully. *Karamarda*has proved to be a superior rootstock for *Carissa* L. (Kumar et al., 2013). It is performed in rainy season. Under mist chamber, treatment of cutting using IBA (3000 ppm) produced better rooting. After 6-8 months, cuttings are finally shifted to fields. Tripathi et al. (2014) stated that semi hard wood cuttings are appropriate for reproduction of plants. Approximately, 25-30 cm long and 1' diameter cuttings may be used for multiplication of plants. The suitable time for planting of cutting is Monsoon Season.

Air- layering is very successful in Karamarda. It is done in the month of junejuly. After rooting, layers can be detached and it takes normally 3 months after layering but it can also be multiplied by hard wood cuttings with the use of growth regulators (Kumar et al., 2013; Peter, 2008). Mishra and Jaisawal, 1993 reported that IBA and NAA concentrations significantly enhanced rooting in air layers. With 5000 ppm, the maximum proportion of rooting was obtained. IBA was more effective than NAA at encouraging roots in air layers. Raut et al. (2015) found that varying concentrations of IBA had a substantial effect on shoot and root characteristics of karonda air layers. IBA 5000 ppm was successful in terms of rooting time (10.73 days), number of roots per layer (102.55), height of rooted layers (44.53 cm), and number of leaves (145.77). Furthermore, in 5000 ppm IBA, the root to shoot ratio (1.17) and maximum surviving percentage (90.82%) were reported. However, the control therapy was related with the lowest values for these characters.

Propagation by Shield budding is also practiced in karonda, One year old seedlings may useful for budding (Bankar et al., 1994). Singh and Ravishankar (2010) attempted softwood grafting with success rate of 40-50%.

These methods, on the other hand, are season specific and require a long period to multiply. Micro propagation, on the other hand, is the technique of growing hundreds of plantlets in a relatively short amount of time using the tissue culture method. Rai (2001) standardised the methodology for C. carandas micropropagation. Furthermore, Rai and Mishra (2005) demonstrated the ability of "Pant Sudarshan," an outstanding cultivar of C. carandas, to regenerate through shoot tip culture on Murashige and Skoog's (MS) basal medium enriched with BA and IBA during different seasons. In addition, protocol developed for in-vitro multiple shoot induction through node and shoot apex (Imran et al., 2012). Furthermore, Hasmah et al. (2013), developed protocol for C. *carandas* using nodal segment as the explants.

11. Cultivation Practices

11.1. Planting

The first monsoon shower is proposed time of planting. Planting is done in the month of June to July. Planting should be done in the evening hours followed by irrigation. Tripathi et al. (2014) suggested 3X3 m^2 spacing with square system of planting for table purpose whereas for hedge purpose 2X2 m^2 distance is ideal.

11.2. Training and Pruning

Staking is needed for bearing crop load. Thekaronda plant has tendency to produce quite a lot of branch near to ground surface. Unwanted laterals or cris-cross should be removed to get good crop. The plant must be trained in the early stages to conform to the provided framework. Three to four shoots from 30-45 cm above the ground are permitted to produce the suitable frame. The centre should be trained to increase output. Karonda plants survived slowly in desert places and required less pruning, however in wet and tropical regions, plants develop rapidly and require significant pruning every year. Pruning is typically done in October under Coorg conditions. It is not necessary to train hedges in this manner, as 3-4 stems are allowed to grow near the earth's surface to form a dense hedge. After gathering the fruits, the hedges are pruned. When a plant becomes senile, it should be beheaded.

11.3. Fertilization & Irrigation

The plant has hardly needs care even gives yield within least management. Once the plants established, become very hardy and survive without irrigation and much care, but require manuring regularly. Ornamental karonda does not require manuring and fertilization. For fruit purpose, one year old plant required 5 kg of FYM and 100 gm mixture of N, P and K, which lasts up to 3 years. Three year onwards, plants should be fertilized regularly after harvesting of fruits with four times higher than given dosage. Water requirement of Karonda is quite low. Irrigation required after planting and fertilizer application is indispensable. Plantation once established does not need much water. Juvenile plants required irrigation at 10-15 days interval in the winter and 6-7 days in summer season (Chadha, 2003; Tripathi et al., 2014;Kumar et al., 2013).

11.4. Harvesting and Postharvest Handling

Karonda plant comes into bearing 3 year after transplanting. The fruits of karondamature 100-110 days after fruit set and then develop their natural color of fruit. For this, taking about 120 days (after fruit set) when fruits become soft and attain dark purple/maroon/ red colour. The change in colour is major maturity indices. The fruits start ripening from May to July in north India. All the fruit does not ripe at one time therefore complete harvesting of fruits requires 3-4 picking.

It is generally done by manually. Fruits harvested with stash helps to minimize the emission of latex by fruits and retain value and storage of fruits. Fruits can be stored for a weak at room temperature. After packing of fruits, they are kept in shade. Fruits can be preserved / stored for 6 months in SO_2 solution (2000 ppm). Karonda yields 4-5 kg fruits/plant. The orchard planted with capable lines may yield 10-15 kg/tree. Karonda fruits can be last up to 3-4 days under room temperature. From January-February flowering, become ready for harvesting by March-May.

12. Availability in Local Markets

The bright reddish purple fruits of karonda, found available in market during the summer months from May to July in the north and year-round in the south states. In Rajasthan, there are two fruiting seasons- January-February and August-September (Singh et al., 1982).

13. Products

Fruits in markets available may be used as dessert when ripe and can used in the preparation of fruit products such as pickle, candies, jelly, squash, chutney and fruit powder.

14. Future Prospects for Marketing

Karonda with their attributes of hardiness, nutritional richness, therapeutic and medicinal values, cost effective cultivation, and great scope of value addition, can be popularized under different agro ecological region of the country. Good agricultural practices for karonda can be develop for site specific cultivation throughout the tropical and subtropical region of the country. It 's commercial cultivation can be a alternative income generating source. The karonda fruits are not suitable as a table fruit when freshed due to presence of strong acidic and astringent properties and they do not have market values (Srivastava et al., 2017). Hence it has great potential in value addition. Emphasis on post harvest management could be given to develop proper value addition chain. There is need to establish small scale processing units to exploit fruit potential on commercial level by securing nutritional values. Uneducated and unemployed youth of tribal and rural region should appoint themselves in the preparation of quality food and other products from karonda fruits as a source of income (Mohammad et al., 2016).

15. Constraints in Popularization

i. Lack of knowledge about various species.

ii. Non availability of detailed information on good agricultural practices, processing, packaging, storage technology and transport.

iii. Absence of scientific resources for cultivation and post harvest management of karonda.

iv. Lack of awareness on food value, nutritive and medicinal values.
v. Lack of market network to popularize crop.
vi. Lack of varietal development for specific utility.
vii. Unavailability of standardized propagation methods
viii. Low yield
ix. Highly perishable nature
x. Available at local markets only

16. Research and Developmental

There is a great need to strengthen karonda varieties for both table and processing purposes. The study on genetic diversity using many features such as morphological, biochemical, and molecular markers was successful in classifying karonda accessions into groups depending on sweetness, fruit colour, and geographic region (Kanpuriya et al., 2019). The conjunction of all three characterisation approaches demonstrated that there is significant variability among the accessions. According to Bose et al. (1999), crop improvement can be performed through two methods: introduction and selection. Mehra and Arora (1982) also noted that diversity in Carissa occurs in North Western India, specifically in Mount Abu, the Western Ghats, and Khandala in Maharashtra, where seedling selection might improve karonda. As a result of its economic and health care importance, research on varietal development for selecting among wild species and manufacture of commonly accepted goods is required for this modest fruit crop.

17. Conclusion

The country is undergoing a horticultural revolution. Karonda is an important horticultural crop in India due to its occupied, production, value addition, and employment generation in both rural and urban areas. Karonda is becoming a significant fruit crop in the Malwa area of central India. Karonda agriculture is primarily concentrated in western India. Its cultivation is now limited to the tropical and peninsular belts, but it has begun to expand into other areas with several promising cultivars. Our primary goal would be to use Karonda culture and enology scientific knowledge to ensure comprehensive and sustainable Karonda manufacturing through crop improvement, assessment, refinement, and adoption of the most appropriate and precise technologies in Karonda manufacturing and value addition, thereby increasing financial profits to Karonda growers and all stakeholders involved in the industry.

References

Anonymous. 1950. The Wealth of India. A Dictionary of Indian Raw Materials and Industrial Production. Vol.3, CSIR, New Delhi, p.22.

Anonymous. 1979. Extension Bulletin, IIHR, Bangalore, No. II, pp.34-35.

Arif, M., Fareed, S. 2011. Pharmacognostical and preliminary phytochemical analysis of *Carissa carandas* L. fruits. JMAPS, 33:53–58.

Azeez, S., Karunakaran, G., Tripathi, Shivashankara, P.C., Roy, T.K. 2016. Evaluation of antioxidant activity, total phenolics and phytochemical content of selected varieties of karonda fruits (*Carissa carandas* L.). Indian Journal of Agricultural Sciences 86(6): 815–822.

Banik, B.C., Ghosh, S.N., Singh, S.R. 2012. Research and development in karonda (*Carissa carandas* L.), a semi wild fruit in India. Acta Horticulturae 948: 61-69.

Banker, G.J., Verma, S.K., Prasad, R.N. 1994. Indian Horticulture, April-June, pp.46-47.

Barua, U., Das, R.P., Gogoi, B., Baruah, S.R. 2019. Underutilized Fruits of Assam for Livelihood and Nutritional Security. Agricultural Reviews 40(3): 175-184.

Bose, T.K., Mitra, S.K., Farooki, A.A., Sadhu, M.K. 1999. Karonda. In: Sub-Tropical Horticulture, Vol 2. Naya Prokash, Kolkata, p.721-723.

Burkill, I.H. 1935. A dictionary of economic product of Malay peninsula. Ministry of Agriculture, Malaysia, pp.464–465.

Chadha, K.L. 2003. Karonda. In: Handbook of Horticulture. Directorate ofInformation and Publications of Agriculture, ICAR, New Delhi, p.200.

Chaudhary, R., Yadav, M., Singh, D.B. 2007. Changes in physico-chemicalcharacteristics of karonda jelly during storage period. Plant Archives 7(2):885-887.

Cheema , G.S., Cheema, B.K. 1971. Farm Journal, Calcutta, 12:24.

Cooke,T. 1904. Flora of the Presidency of Bombay, vol. 2, Taylor and Francis, London, pp. 123-126.

Das, S.C., Prakash, J., Deb, A.K., Biswas, T. 2013. Medicinal value of underutilized fruits in hilly Tripura. Acta Horticulturae 972: 135–41.

Dhaka, R.S., Saroj, P.L., Chouhan, N., Verma, P. 2004. Karonda. In: production technology of underutilized fruit crops (Eds. Dashora, L.K., Sen, N.L. and Maurya, I.B.), Yash Publishing House, Bikaner, pp.142-153.

Divate, S.M., Savale, B.G., Patil, N.B., Relekar P.P. 2020. Study of preparation and standardization of Karonda (*Carissa congesta* L.) crush. The Pharma Innovation Journal,9(1): 176-179.

Divekar, P.B. 1984. M.Sc. (Ag) Thesis submitted to Konkan Krishi Vidhyapeeth, Dapoli.

Ghate, V.S., Kulkarni, D.K., Upadhye, A.S. 1997. Karvanda (*Carissa* L.): an underutilized minor fruit of India. Plant Genetic Resources Newsletter, 109:20-21.

Haines, H.H. 1919. Indian Species of *Carissa*. The Indian Forester, 45 (7):375-388.

Hasmah, Nor, S., Bhatt, A.,Lai Keng C. 2013. Micropropagation of Asam Karanda (*Carissa carandas* L.). Pertanika Journal of Tropical Agricultural Science36(1):89-98.

Hooker, J.D. 1982. Flora of British India, vol. 3, L Reeve & Co, London, pp.630-632.

http://www.ciah.ernet.in/publication/natural_colorant.pdf.

Imran, M.A., Begum G., Sujatha K., Mallaiah B. 2012 Effect of adenine sulphate (ads) with cytokinins on multiple shoot production in *carissacarandas,* International Journal of Pharma and Bio Sciences 3(1):473- 480.

Itankar, P.R., Lokhande, P.J., Verma, P.R., Arora, S.K., Sahu, R.A., Patil, A.T. 2011. Antidiabetic potential of unripe *Carrisacarandas* Linn. fruit extract. Journal of Ethnopharmacology 135: 430–3.

Iyer, C.M.,Dubhash, P.J. 2006. Anthocyanin of Karwand (*CarrisacarandusL.*) and studies on its stability in model systems. Journal of Food Science and Technology. 30: 246–8.

Joshi, G.D., Prabhudesai, V.G.,Salri, M.J. 1986. Physico-chemical characteristics of karonda (*Carissa carandas* L.). Maharashtra Journal of Horticulture 3: 39-44.

Julia, F., Morton, F.L., Morton, J. 1987. Karonda, in: fruits of warm climates, pp. 422–424

Kanupriya, K C., Tripathi, P.C., Singh, P., Venugopalan, R., Radhika, V. 2019. Analysis of morphological, biochemical and molecular diversity in karonda (*Carissa carandas* L.) germplasm. Fruits 74(3): 130–140.

Karale, A.R., Keskar, B.G., Dhawale, B.C., Kale, P.N.,Choudhari, K.G. 1990. Variability studies in floral morphology and fruit set in seedling population of karonda. Journal of Maharashtra Agriculture University 15(1): 109-110.

Karthikeyan, S., Sanjappa, M., Moorthy, S. 2009. Flowering plants of India, vol.1, Dicotyledons, Botanical Survey of India, Kolkata, p. 365.

Kashyap, P., Kansal, A., Prusty, A.K., Singh, S.P. 2015. Karonda is multipurpose plant for more return. *Indian Horticulture* 60(1): 28-29.

Kumar, S., Gupta, P., Virupaksha Gupta, K.L. 2013. A Critical Review on Karamarda (*Carissa carandas* L.). International Journal of Pharmaceutical & Biological Archives 4(4): 637–642.

Kumar, S., Singh, I.S. 1993. Variation in quality traits of karonda (*Carissa carandas* L.) germplasm. South Indian Horticulture, 41(2): 108-109.

Li, P., Leeuwenberg, A.J.M., Middleton, D.J. 1995. Apocynaceae. In: Flora of China, (Eds. Z.Y. Wu, P.H. Raven), Science Press and St. Louis: Missouri Botanical Garden Press, Beijing, vol. 16, pp.143-188.

Maheshwari, R., Sharma, A., Verma, D. 2012. Phytotherapeutic significance of Karunda. Journal of Environment and Pharmacology Life Science1 (12):34–36.

Malik, S.K., Chaudhury, R., Dhariwaln, O.P., Bhandari, D.C. 2010. Genetic Resources of Tropical Underutilized Fruits in India, NBPGR, New Delhi, p-178.

Manivasagan, S., Rana, G.S., Joon, M.S.,Godara, A.K. 2007. Study on qualitativechanges in pickles prepared from karonda (*Carissa carandas*L.) during storage.Haryana Journal of Horticulture Sciences 36(1/2):44-45.

Maurya, I.B., Singh, J. 2008. Tree-Borne Vegetables. In: Underutilized and Underexploited Horticultural Crops (Eds. K.V. Peter), New India Publishing Agency, New Delhi, pp-217-218.

Meghwal, P.R., Singh, A. 2015. Underutilized Fruits Research in Arid Regions: A Review. Annals of Arid Zone, 56(1&2): 23-36.

Mehra, K.L., Arora, R.K. 1982. Plant Genetic Resources of India- Their Diversity and Conservation. National Bureau of Plant Genetic Resources, New Delhi (NBPGR Science Monogram No. 4).

Mehta, K.C., Singh, S.M. 1960. Indian Journal of Horticulture. 17: 31-37.

Mishra, K.K. 2007. New karonda varieties from Pantnagar. Indian Horticulture, 52(6): 9-10.

Misra, K.K., Jaiswal, H.R. 1993. Effect of growth regulators on rooting of stool layers of karonda (*Carissa carandas* L.). Indian J Forestry 16(2):181-2.

Mishra, K.K., Jaisawal, H.R. 1999. Haryana Journal of Horticulture Sciences.,28: 176-178.

Mishra, K.K., Singh, R. 1990. Effect of growth regulators on rooting and air layers of karaunda (Carissa carandas L.). Annals of Agricultural Research11 (2): 208-210.

Muhammad, A., Kamal, M., Jawaid, T., Mohammad, K., Saini1, K.S., Kumar, A., Ahmad, M. 2016. Carissa carandas Linn. (Karonda): An exotic minor plant fruit with immense value in nutraceutical and pharmaceutical industries. Asian Journal of Biomedical and Pharmaceutical Sciences 6(58):14-19.

Naidu, C.V., Amritphale, D. 1993. Effect of light and storage in seed germination of *Caesuliaaxillaris*Roxb. Journal of Seed Research 22(1):58-61.

Nalawadi, U.G.,Jayasheela, N. 1975. Progressive Horticulture, 7: 37-38.

Patil, U.S., Athani, S.I., Naik, N.H., Jagadeesh, R.C., Allolli, T.B., Gopali, J.B. 2017. Quality and Yield Characterization of Karonda (*Carissa carandas* L.) Genotypes. International Journal of Pure Applied and Bioscience, 5(2): 374-378.

Panda, D., Panda, S., Pramanik, K., Sananda, M. 2014. Karonda (*Carissa* spp.). An underutilized minor fruit crop with therapeutic and medicinal use. International Journal of Economic Plants1(1):009-14.

Patil, R., Raut, V.U.,Wankhede, R.S. 2014. Sensory quality and economics of preparation of karonda candy. International Journal of Processing and Post Harvest Technology 5(2):169-172.

Rahmatullah, M., Noman, A., Hossan, M.S., Harun Or-Rashid, M., Rahman, T., Chowdhury, M.H., Jahan, R. 2009. A survey of Medicinal Plants in Two Areas of Dinajpur District, Bangladesh Including plants which can be used as Functional Foods. American-Eurasian Journal of Sustainable Agriculture 3(4): 862-876.

Rai, R. 2001. Standardization of micropropagation technique in *Carissa* Spp. Diss. Pantnagar: GovindBallabh Pant University of Agriculture and Technology.

Rai, R., Misra, K.K. 2005. Micropropagation of Karonda (Carissa carandas) through shoot multiplication. Scientia Horticulture 103(2):227-232.

Rajasekaran, A., Jeyasudha, V., Kalpana, B., Jayakar, B. 1999. Preliminary Phytochemical and Antipyretic Evaluation of Carissa carandas. Indian Journal of Natural Product 15(1):27-29.

Randle, A.B. 1925. In classification of flowering plants, University Press, Cambridge, 2: 468-474.

Reshu,V., Virmani, T.,Singh C.,Sorout, G., Gupta, J. 2017. Hidden Potential of Natural Herb *Carissa Carandas*L. (Karonda). Research in Pharmacy and Health Sciences 3(2):294-302.

Raut, U.A., Jadhav,G.G., Bhogave, A.F.,Deshmukh, M.S. 2015. Effect of different IBA levels on air layering of karonda (*Carissa carandas* L.). Research on Crops 16(3):537-541.

Sarma,J., Hussain, M.,Dey, S. 2020. A new rheophytic species of *Carissa* (Apocynaceae) from Northeast India. Journal of Asia-Pacific Biodiversity 13(2): 314-318.

Sawant, B.R., Desai, U.T., Ranpise, S.A., More, T.A., Sawant, S.V. 2002. Genotypic and phenotypic variability in Karonda (*Carissa carandas* L.). Journal of Maharashtra Agriculture University 27(3):266–268.

Saxena, D., Misra, K.K.,Rai, R. 2016. Studies on suitability of cultivars, picking dates and drying methods for the preparation of karonda (*Carissa carandus* L.) fruit powder. Indian Journal of Horticulture 73(2): 267-273.

Shafeeq, L., Athani, S.I., Madarakhandi, S., Mohammed, A.B. 2014. Studies on Fruit, Quality and Yield Parameters as Influenced by Different Seedling Progenies of Karonda (*Carissa carandas* L.). Trends in Biosciences 7(9):77-780.

Shaheel, S.K., Swami, D.V., Kumar, B.P., Krishna, K.U. 2015. Effect of Blending Of Karonda (*Carissa Carandas* L.) Juice With Guava, Papaya And Pineapple Juices On Its Qualityand Organoleptic Evaluation. Plant Archives 15(1): 187-192.

Shinde, V.B. 2014. Cultivation of Karonda. In: Production and Management of Fruit Crops in Arid/Drylands (Eds. Sontakke, M.B), Agrotech Publishing Academy, Udaipur, pp- 308-312.

Sharma, S.K. Banyal, S.K. 2010. Rehabilitation of marginal lands through karonda cultivation. Intensive Agriculture, Oct-Dec, pp.6-12.

Sharma, J.K.,Singh, H.B. 1997. Relative storability of seeds of some crop species under ambient conditions. Journal of Seed Research 25(1):37-40.

Shukla, S.K., Kumar, K. 2009. Underutilized Subtropical Fruits, International Book Distributing Co., Lucknow, p.184.

Singh, L.B. 1967. The aonla, mulberry and karonda in India. Farm Bulletin, ICAR, New Delhi.

Singh, I. S. 2001. Minor fruits and their uses. Indian Journal of Horticulture 58(1/2): 178–182.

Singh, H.B.,Arora,R.K. 1978. Wild edible plants of India, ICAR, New Delhi.

Singh, Y.,Bhatnagar, P. 2019. An Overview on Inherent Potential of Underutilized Fruits. International Journal of Pure Applied & Bioscience 7(3): 86-103

Singh, A.K., Bajpai, A.,Ravishankar, H. 2010. Biodiversity in underutilized fruits. Indian Horticulture 55(3): 53-56.

Singh, R., Chopra, D.P.,Gupta, A.K. 1982. Some Carissas for Western Rajasthan. Indian Horticulture, 27:10-11.

Singh, A.K.,Ravishankar, H. 2010. Prospects of production of quality planting material of underutilized fruits. In: National Conference on Production of Quality Seeds and Planting Material – Health Management in Horticultural Crops. 11-14 March, 2010, New Delhi, p. 149-150.

Singh, S., Singh, A.K., Apparao, V.V.,Bhargava, R. 2013. Thar Kamal: A new karonda variety. Indian Horticulture, 58(4): 9-10.

Singh, S., Singh, A.K., Meghwal, P.R., Singh, A.,Swamy, G.S.K. 2014. Karonda. In: Tropical and Sub tropical Fruit Crops: Crop Improvement and Varietal Wealth, pp-393-403.

Singh,A.K.,Singh,P. 1998. Power of significance of difference among fruit and seed size parameters of karonda (*Carrisacarandus* Linn.). Annals Agriculture Research, 19(1):66-71.

Srivastava A., Sarkar P.K.,Bishnoi S.K. 2017. Value addition in under-exploited fruits of Karonda (*Carissa carandus* L.): an earning opportunity for rural communities in India. Rashtriya Krishi,12 (2): 161-163.

Taylor, R.S.L., Hudson, J.B., Manandhar, N.P., Tower, G.H.N. 1996. Antiviral Activities of Medicinal Plant of Southern Nepal. Journal of Ethnopharmacology, 53(2):97- 104.

Tripathi P.C., Karumakaran G., Sankar V.,Senthilkumar R. 2014. Karonda : A Potential fresh fruit of Future. Technical Bulletin 7/2014, ICAR-IIHR, Central Horticultural experiment Station ,Chettalli, Kodagu, Karnataka, pp.-14.

Tyagi, S., Misra, K.K., Jaiswal, H.R. 1999. Effect of auxins on rooting of softwood stem cuttings of Carissa under mist. Scientia Horticulture 6:37-43.

Warrier, P.K., Nambiar, V.P.K., Ramankutty, C. 1993. Indian Medicinal Plants: A Compendium of 500 Species (Vol. I), Universities Press (India) Pvt. Ltd.

5

Variability in Nutmeg Species and Varieties

Pooja S. and Vijaykumar B. Narayanapur

Department of Plantation, Spices, Medicinal and Aromatic Crops
College of Horticulture, UHS, Bagalkot, Karnataka, India

1. Introduction

Spices are a high-value, low-volume commodity traded around the world. Spices have played a significant part in civilisation, exploration, and commerce throughout history. Since ancient times, India has been known as the spice capital of the world, with over a hundred different types of spices grown. India has long been a traditional producer, consumer, and exporter of spices, with nearly every state producing one or more of them. India has a total area under spice production of 39,27,430 ha, which produced 95,99,900 MT in 2017-18. (Spices Board, 2020a). Due to their rich aroma, texture, and flavour, Indian spices are in high demand around the world. India has the world's largest domestic spice market (Bhattan and Valasan, 2016). After consuming more than 70% of the spices produced domestically, India continues to be the leading spice exporter. Kerala, often known as India's spice paradise, is the country's biggest producer of tiny cardamom and nutmeg.

Important Spices of India

The important spices under commercial cultivation in India are black pepper, small cardamom, turmeric, ginger, coriander, fennel, fenugreek, cumin and garlic. India is a leading producer and exporter of black pepper in the world (Parthasarathy and Kandiannan, 2008). Among the wide range of spices cultivated in India, seventeen spices comes under the group 'tree spices'. Tree spices are considered as a unique group of spices. As the name indicates, they are tall plants with more canopy compared to other spices which can be utilized for intensive agriculture under mixed farming system along with other horticultural crops and are the best crops suitable for intercropping with coconut and arecanut (Haldankar et al., 2013).

Table. 1: Tree spices grown in India

S.No.	Botanical name	Family	Common name	Part
1.	*Myristica fragrans* Houtt.	Myristicaceae	Nutmeg	Kernel, Aril
2.	*Cinnamomum verum* Bercht & Presl.	Lauraceae	Cinnamon	Bark, leaf
3.	*Cinnamomum aromaticum* Nees	Lauraceae	Chinese cassia	Bark, leaf
4.	*Cinnamomum tamala* Nees	Lauraceae	Tejpat, Indian cassia	Leaf, bark
5.	*Syzigium aromaticum* (L.)	Myrtaceae	Clove	Flower bud
6.	*Garcina gummigutta* (L.) Robs.	Clusiaceae	Garcinia, cambogia	Pericarp of fruit
7.	*Garcina indica* (Thouars) Choisy	Clusiaceae	Garcinia, kokum	Pericarp of fruit
8.	*Pimenta dioica* (L.) Merr.	Myrtaceae	Allspice, Jamaica pepper	Immature fruit, leaf
9.	*Murraya koenigii* (L.) Sprengel	Rutaceae	Curry leaf	Leaf
10.	*Tamarindus indica* L.	Fabaceae	Tamarind	Fruit
11.	*Illicum verum* Hook	Illiciaceae	Star anise	Fruit
12.	*Juniperus communis* L.	Cupressaceae	Juniper	Fruit
13.	*Laurus nobillis* L.	Lauraceae	Bay leaf	Leaf
14.	*Capparis spinosa*	Capparidaceae	Caper	Flower buds
15.	*Punica granatum* (L.)	Punicaceae	Pomegranate	Dried seed (with flesh)
16.	*Averrhoa bilimbi* L.	Averrhoaceae	Bilimbi	Fruit
17.	*Averrhoa carambola* L.	Averrhoaceae	Carambola	Fruit

(Haldankar et al., 2013)

Of these tree spices, nutmeg, clove, cinnamon, garcinia and tamarind are the important tropical spices of India. Statistical reports show that area and production of nutmeg in India is showing an increasing trend whereas for clove and other tree spices it is almost static. As per Spices Board (2020b), during the year 2018-19, India exported 3,300 tons of nutmeg which accounts a value of 15,015 lakhs. Major largest importers of Indian nutmeg are U.A.E followed by Vietnam. With an area of 22,070 ha and production of 13,750 MT, Kerala is the largest producer of nutmeg in India.

In the early 2000s, a considerable number of small and medium spice cultivators in Kerala, particularly those in the Wayanad, Idukki, Ernakulam, and Kottayam districts, began cultivating Vanilla and Nutmeg as domestic prices of prominent spices such as black pepper and tiny cardamom fell. Nutmeg cultivation, which was originally limited to the Kalady, Koratty, and Kadukutty panchayats, as well as certain areas of the Kottayam district in Kerala, has now extended throughout the state, making Kerala the biggest producer of nutmeg in India (Thomas, 2009).

2. Family Myristicaceae

Under order Magnoliales, comes the primitive family Myristicaceae, which has about 21 genera and 500 species, out of which four genera namely, *Horsefieldia*, *Gymnocranthera*, *Knema* and *Myristica* are reported from India (Chandran and Mesta, 2001).Myristicaceae is composed of aromatic evergreen trees with their bark exuding a red sap. The flowers are small and dioecious, cymes, umbels or rarely in racemes. The female flowers have single celled, free and superior ovary. The fruit have a fleshy to leathery covering which splits into halves, rarely four on maturation exposing seed within a thin coloured and fleshy arillus either entire or lacerate covering the nut (Gamble, 1967; Parkinson, 1972).

3. Myristica Species

The name *Myristica* is derived from the Greek word 'Myron' which is a sweet liquid distilled from the plant (Everett, 1981). They are characterized by evergreen tree producing fragrant flowers. The genus *Myristica* is considered the most primitive and has more species than any other genera in the Myristicaceae (Sinclair, 1958) with nearly 175 species and out of these, five species are reported in South India (Banik et al., 2017). Most of the species of *Myristica* exist as wild and are considered endangered, endemic or threatened, available only in the vulnerable hotspots and are used either in medicine or as spice. Out of 175 species identified in the genus, fifteen are found in India namely, *M. malabarica* Lamk., *M. magnifica* Bedd., *M. beddomei* King, *M. fragrans* Houtt., *M. contorta* Warb., *M. longifolia* Wall., *M. amygdalina* Wall., *M. irya*Gaertn., *M. attenuata* Wall., *M. gibbosa* Hook.f. & Thomson, *M. glabra* Reinw. ex Blume, *M. glaucescens* Hook. f. & Thomson, *M. kingie, M. prainii* King and *M. andamanica* Hook (Nybe et al., 2006; Parthasarathy, 2010). Most of these are endemic to the Western ghat forests. Though they are not very much used for commercial purposes, earlier the mace was used as substitute for true mace from *M. fragrans*. *M. prainii* King and *M. andamanica* Hook were found endemic to Andaman and Nicobar Islands (Sharief, 2007).

3.1. Distribution Of Myristicaceae

The members of the family Myristicacae are distributed pantropical, mainly associated with the rainforests of Asia, Africa, Madagascar, South America and Polynesia. *Myristica* is distributed from India to Southeast Asia to New Guinea, northeast Australia, also in Fiji Islands (Banik et al., 2017). There are certain species endemic to India which is distributed along the evergreen forests of Andaman and Nicobar Islands, Meghalaya and the Western Ghats (Saji et al., 2019).The nutmeg of commerce, *M. fragrans*was brought to India by the Portuguese during the 16th century in the coasts of Kerala (Ravindran, 1998).

4. Origin of Nutmeg

Nutmeg has originated from the Banda Islands, known as the spice Island, in the Moluccas of Eastern Indonesia, so nutmeg is also referred as Banda nutmeg (Krishnamoorthy et al., 2010). It is associated with the hot and wet climate of the tropical rainforest. Being an introduced crop to India, the tree has a domestication history of about 200 years in the country and at present India is one of the major production centers of nutmeg in the world.

5. Myristica Swamps

Most of the *Myristica* species distributed in India are associated with the Western Ghats. The Western ghats is considered as one of the well-known global centres of biodiversity and are home for very special kind of ecosystem known as the myristica swamps. Myristica swamp is a unique, endangered fresh water swamp ecosystem found in few locations of South Western Ghats of India. It is a biodiversity hotspot for a number of plants and animals. It was described from Travancore and later from the valleys of Sendurney, Kulathuppuzha and Anchal ranges from Southern Kerala (Krishnamoorthy, 1960) and are distributed along Southern Kerala, Goa, Konkan coast, Uttara Karnataka and recently in North western ghats of Maharashtra (Sreedharan and Indulkar, 2018). As the name indicates, they are dominated by evergreen trees of the Myristicaceae, especially *Myristica* genus, forming 32 per cent of the total number of trees in the ecosystem (Chandran and Mesta, 2001). They are the only sites of occurrence of *Gymnacranthera canarica and Myristica fatua var. magnifica* of Myristicaceae and other species such as *M. dactyloides, M. malabarica* and *K. attenuate.*

Being an evergreen tree adapted to a waterlogged ecosystem, several physiological features have evolved enabling them to survive anaerobic conditions as they are endemic to the swamps. These trees possess two types of bizarre-looking roots known as knee roots and stilt roots. Adventitious roots termed as silt roots sprout from the main trunk providing mechanical support to trees growing in the soft, moist soil of the Myristica swamps. Such roots might become woody and flattened with age. Knee roots protruding from the ground helps the trees to exchange gases in waterlogged conditions. Such aerial roots are common among trees found in tropical mangrove swamps. Myristica swamps are one of the most threatened ecosystems of India and the decline of the swamps leads to decline of *Myristica fatua* var. *magnifica* and *Gymnacranthera canarica*. These swamps help in stabilizing ground water and efforts should be taken to the conservation of these ancient ecosystems.

5.1. Variability in *Myristica* Species

Variability can be defined as the state of being variable with respect to form, function, and other traits (Choudhary and Vijayakumar, 2012). An understanding of the variability existing in the crop is very essential for formulating crop improvement programmes. Genetic variability assessment of plant helps to improve agricultural production and productivity and understand the diversity within the crop. Wide variability is observed between the cultivated and the wild plants of *Myristica*. Morphologically, leaves of *M. fragrans* are relatively very small compared to the wild plants. The shape of fruits, nutmeg and mace are distinctly different between wild and cultivated species. The mace of *M. fragrans* is bright red due the lycopene present in it whereas it is not seen in *M. beddomeii*. Hence variability was observed between the wild and cultivated taxa of Myristicaceae with respect to morphology and primary and secondary metabolites (Maya, 2005).

Myristica malabarica Lamk. are seen in the evergreen forests of Western Ghats locally known as Kattujathi or Ponnampoo. It is an endangered tree endemic to Southern Western Ghats. It reaches to a height of 15 m and has slender flowering branches. Its ripe fruits called Bombay nutmeg and Bombay mace are extensively exploited for its aril locally as adulterants of *Myristica fragrans*. They have no odour and taste. The aril is reddish yellow, irregularly lobed, laciniate and extending to the apex of the seed (Gamble 1967; Jose and Pillai, 2016).

Myristica beddomei King are locally called Chithirapoovu or Pathiripoovu and are distributed in the evergreen forests, endemic to South India and Srilanka. The leaves are oblong or elliptic-lanceolate. Fruits are extensively exploited for its aril locally. The aril is orange-red and laciniate with their ends separate and had spherical seeds (Gamble 1967; Jose and Pillai, 2016). *Myristica fatua* Houtt. var. *magnifica*are endangered trees, characteristic species of Myristica swamps endemic to Southern Western Ghats. Tree is economically important for its wild nutmeg and mace which is used for both medicinal and industrial purposes. They produce large sized fruit and nut with triangular nut shape. *Myristica attenuata* Wall.are also endemic to Western Ghats, locally called Chorapathiri, Chorappayin due to its blood-coloured exudates from its bark. The mace fully covers the nut and seeds and are smooth surfaced without grooves. Tree is economically important for its wild nutmeg and mace which is used for both medicinal and industrial purposes (Jose and Pillai, 2016).

M. magnifica Bedd. produces oblong fruits and the arils are orange red in colour. A wide range of variation in shape of fruit from ovoid, elliptic to oblong were observed in *M. prainii* and the arils are red laciniated. Thin and glabrous pericarp with yellow arils lacerate at the tip,is the characteristic feature of *M. amygdalina* Wall. *M. andamanica* Hook. are endemic to Andaman and Nicobar Islands and they possessed stilt type of roots seen in species adapted to myristica swamps. The fruit resembles the size and shape of a hen's egg. Pericarp is thick and brown and the aril is having blood red colour (Parkinson, 1972; Skaria et al., 2004; Phulsagar et al., 2014).

A new nutmeg species, *Myristica trobogarii* wasreported from, Agasthyamala Biosphere Reserve of Western Ghats by Govind et al. (2020). The fruit is egg-shaped rather than round and were bigger. The uses are unknown. It is similar in many aspects to *Myristica beddomei*, *Myristica magnifica* and *Myristica malabarica*. The wild nutmegs are exploited widely for many purposes especially as medicines or root stock for *M. fragrans.* The fresh bark flakes decoction of *M. malabarica* Lamk.is used to prepare rice gruel for relieving body pain by tribes (Binu, 2011). They are considered as better than other rootstocks for productivity of *M. fragrans.* Samydurai et al. (2012) reported that the leaves and fruits of *M. dactyloides* King have sedative principle and are used for relief in bronchitis, cough and fever. The orange red-coloured arils of *M. magnifica* Bedd. are used for dyeing purpose and the seedscan be used for oilburning and candles making. *M. beddomei*ican also be used as a root stock for nutmeg (Phulsagar et al., 2014).

In a study conducted by Chetana and Ganesh (2013), in Kakachi, within the Kalakad Mundanthurai Tiger Reserve, Southern India, hundred fruits and seeds of wild nutmeg were collected, and their length, width and fresh mass were recorded and mean values were calculated. On comparing the fruit and seed dimensions of *M. dactyloides, M. malabarica and M. beddomei* high fruit weight and seed width was observed in *M. beddomei*, whereas seed size ratio (L/W) mm is least for the same (1.15 ± 0.02 mm), whose seeds are spherical compared to the other two species. Usually, the fruits of Myristicaceae on maturity dehisce on tree itself but in *M. beddomei* it did not dehisce on tree. On comparing the fruit and seed dimensions of *M. fatua, M. attenuata, M. malabarica and M. beddomei*, *M. fatua* shows large and heavy fruits compared to others (Jose and Pillai, 2016). Krishna and Somanathan (2016) observed that the seeds of *M. fatua* were large and heavy with 20.9 ± 2.93 g weight, 50.25 ± 4.2 mm length and 22.1 ± 3.5 mm width with little variation between trees.

*Myristica contorta.*Warb. is considered as a potential wild nutmeg species. One of the major problems with *M. fragrans* is its weak root system. Confining to the surface soil it makes the tree susceptible to falling in heavy winds as well as to pest and diseases. As opposed to this the wild nutmeg plant, *M. contorata* has a strong and extensive deeply going root system and is seldom attacked by pest & diseases. Hence this offers great potential as root stock for budding or grafting the nutmeg of commerce (Skaria et al., 2004). The nutmeg of commerce is a heavy fruit yielder producing 300- 3,000 of fruits per tree whereas *M. contorata* produce a moderate seed yield of 6,000- 11,000 per tree per year and dried mace of 70-80 kg per tree per year thus the seed and mace are used for adulteration.

Papua nutmeg (Myristica argentea Warb.) is another nutmeg species found in Indonesia, in addition to the well-known Banda nutmeg (Myristica fragans Houtt.). The plant's oval form and larger portions and fruits are its most noticeable morphological features. The shape of a tree's crown might be pyramidal, cylindrical, or spherical. After 8–9 years of age, Papua nutmeg trees begin to produce fruits, and a 10- to 12-year-old nutmeg tree can produce about 800–2,000 fruits per year from 2–3 harvests. Immature fruits fleshy pericarp is used in the production of jam and other products. Argenteane, a strong antioxidant equivalent to vitamin E, was extracted from the mace of M. argentea (Calliste *et al.,* 2010). The tree grows in the wild or in semi-cultivated areas, therefore productivity is low, yet it is used as a substitute for Banda nutmeg, fetching only about 60% of the price of Banda nutmeg and mace. Butter nutmeg is a fat obtained from the seed that is used in fragrance and soap making. Kino is a dye made from red bark sap that leaves a persistent brown stain. Because the tree is resistant to a root disease, it is considered a viable rootstock for Myristica fragrans (Ungirwalu *et al*., 2019).

Nutmeg requires large quantity of water for irrigation as it cannot withstand dry conditions. In water limited areas rootstock which can withstand drought or require less water would be ideal as it can be used for grafting. Krishnamurthy *et al.* (2008) compared *Myristica fragrans, M. beddomeii, M. malabarica, Knema andamanica and Gymnocranthera canerica* to identify drought tolerant rootstocks for grafting of nutmeg. On the fifth and tenth days after stress induction, membrane leakage percent and relative water content were determined in seedlings of various Myristica species. Control plants from each species were kept under sufficient watering. On the tenth day of observation, the results revealed that *M. malabarica* had the highest relative water content and least leakage, followed by *M. beddomeii,* and G. canerica had the lowest relative water content and highest leakage values. *M. malabarica* had a lower reduction in relative water content and increased leakage than the control. On water stress, *M. malabarica* took 17 days to wilt, but seedlings of *G. canerica* wilted extremely quickly in 11 days, followed by M. fragrans in 12 days. *M. malabarica* ranks best in terms of drought tolerance. *M. beddomeii* comes in second, along with *K. andamanica*, and *G. canerica* comes in third, tied with *M. fragrans*. As a result, high producing nutmeg types can be grafted onto *M. malabarica* rootstock to improve yield and drought tolerance in nutmeg.

6. Variability in Nutmeg (*Myristica Fragrans* Houtt.)

The only commercially important and cultivated kind of Myristica species is nutmeg (Myristica fragrans Houtt.) also known as Banda nutmeg. Nutmeg is a spreading, generally dioecious evergreen tree with dark green leaves, yellow flowers without petals, and enormous yellowish fruit, the nutmeg being the dry seed and the mace being the dried aril, according to Weiss (2002). Because it produces two distinct spices, nutmeg and mace, nutmeg is also known as twin spice. The dried kernel of the seed is nutmeg, and the dry aril surrounding it is mace (Krishnamoorthy et al., 2010). Although both spices have a similar flavour, nutmeg is slightly sweeter than mace and is more commonly used in cooking. Aside from nutmeg and mace, a variety of other items are important commercially. Nutmeg butter, oleoresins, essential oils are also derived from *M. fragrans* and they find varied use in the food, medicine and perfume industries.

Being cross pollinated, lot of variation exists in the seedling population in morphological and chemical parameters. Morphological variation has been observed in tree shape, tree size, leaf size, flowering characters, fruit shape, fruit size, fruit colour, size and thickness of aril characters (Rema et al. 1997; Haldankar et al., 2003; Shinde et al., 2006; Vikram, 2016). Variability study is considered as the basis for improvement in any crop. From the beginning of crop production, natural genetic variability has been exploited within crop species for farmer's and breeder's preferences. Observations on morphological and agronomic characters such as plant stature, roots, stems, leaves, flowers, fruit and seeds and use of molecular markers can be done in nutmeg for identification and characterization of genotypes (Purwiyanti, 2020).

The morphological differences are visible in the tree habit, leaf form and colour, flower, fruit, mace, and nut. Canopy design is a crucial feature in nutmeg, as it not only indicates the tree's bearing capacity but also aids in planting according to spacing and population density per unit area. Nutmeg has canopy shapes that are conical, pyramidal, rectangular, and spherical, according to Vikram (2016). The most prevalent canopy shape was pyramidal, whereas conical was uncommon. Conical trees grow tall with a narrow horizontal spread. This allows for a greater number of plants to be accommodated per unit area. As a result, they have added ornamental value and are well-suited to homesteads and high-density planting schemes. High density planting in nutmeg is an upcoming area of research due to reduced land availability. Nutmeg has two sorts of branching patterns: erect and spreading, with the latter being found in the majority of the population. The narrowly pyramidal canopy shape revealed the erect branching pattern. Obovate, elliptical, oblong, lanceolate, and ovate were the shapes of developed leaves. The narrowly pyramidal tree form is a distinctive feature in canopy shape that can be used in a high density planting strategy since it takes up less horizontal ground area. This tree's erect branching style encourages a narrow canopy spread (Priyanka, 2016).

Weiss (2002) observed wide range of variation in leaf shape from elliptic to oblong and lanceolate. Kakkappara-2 with narrow leaves and open canopy structure and Kakkappara -3 with broad leaves was spotted by Sasikumar et al. (2014). Priyanka (2016) reported small elliptic leaved nutmeg with dark green colour. The leaf area of this type is only half the area of a normal nutmeg leaf. Even though the leaves are small, the foliage density was abundant.

6.1. Variation in Sex Forms

Though nutmeg is usually a dioecious tree other sex forms occur sporadically. Sometimes male trees have been known to produce some female flowers and fruits and few of such trees were reported to bear only female flowers later on. The percentage of gynoecious and androecious trees formed from seeds is almost the same. Five sex forms in nutmeg namely, pure male, pure female, bisexual male, bisexual female and hermaphrodite trees were observed (Flach, 1966; Krishnamoorthy, 2000). Among this hermaphrodite were rarely observed. Existence of monoecious and dioecious sex forms in Kerala has been reported by Sasikumar et al. (2015); Rema et al. (2015) and Vikram (2016). An improved variety, Konkan Sugandha with bisexual flowers has been developed at Konkan Krishi Vidyapeeth, Dapoli, Maharashtra (Parthasarathy, 2010).

Sexual intermediates and sexual switching are common in dioecious species depending on location and season. Habitat degradation leading to reduction in population of native pollinators and climate change accelerate this. Breakdown of dioecy is the result of pollen limitation leading to males becoming co-sexual and thus self-reproducing and ensuring the species perpetuation. Under conditions of pollination limitation and stressful environment, population may shift rapidly towards autogamy by changing floral traits (Sasikumar and Aarthi, 2016). In monoecious type nutmeg, floral variety exposes three forms of flowers: pistillate, staminate, and hermaphrodite blooms. The androecium in hermaphrodite flowers has 1-4 anthers in the form of fused filament, free filament, or both, some anthers are united with the gynoecium, and the length of the anther varies from 3-6 mm, indicating intra flower variability. All flowers of *Myristica fragrans* consist of a three lobed perianth (Flach, 1966), where as in monoecious nutmeg, bilobed, tetralobed and pentalobed are observed rarely (Aarthi et al., 2018). The hermaphrodite flowers can be differentiated by the presence of staminodes and the developed fruit shows remnants of the stamen.

6.2. Variability in Flower, Fruit, Seed And Mace

Aside from these morphological characteristics, there has been a lot of variation documented in terms of growth rate, productivity, leaf size and shape, flower size and form, fruit set, and fruit size (Krishnamoorthy et al., 1996; Haldankar et al., 2004; Kumar et al., 2010)

According to Gamble (1967), females have fewer flowers per inflorescence than males, which range from three to five. Flowers in male genotypes were borne in clusters of three to seven, but flowers in female genotypes were borne either alone or in small groups, according to Haldankar et al. (2004). The number of female flowers per cluster varied, with one-two blooms per cluster being the most common, followed by one-three, one-four, and five or more (Vikram, 2016).

Earlier, fruit shapes such as ovoid, subglobose, and pyriform were seen (Gamble, 1967). Nitta discovered a unique pear-shaped and huge globoid fruit nutmeg in Moluccas Island (1993). Under Kerala circumstances, pyriform and round-shaped fruits were discovered (Thomas, 2008). Vikram (2016) found round, oval, ovoid, and pyriform shaped fruits, with round having the most frequency and ovoid having the lowest. The oval-shaped nuts came from the ovoid fruits. Fruit form is thought to be a hereditary feature in nutmeg and could be utilised as a morphological marker. Priyanka discovered an unusually formed oblong nutmeg with a fruit shape that differed from the standard nutmeg (2016). The nut and mace in this type were oblong shaped and had thin mace as compared to others, so it was very difficult to get flower grade quality mace from these fruits. But more number of fruits per tree was produced.

The average number of fruits each cluster is one or two. However, three to four fruits per cluster have been reported (Vikram, 2016). Although clustering of fruits is not a regular occurrence in nutmeg, it is a desirable trait. Eighty percent of the fruits on cluster fruited nutmegs were in three-fruit clusters, fifteen percent in two, and the rest in four to eight fruits per cluster. Fruit yield per tree was considerably higher in the cluster fruited accessions than in the other 23 accessions studied by Priyanka (2016). Priyanka and Miniraj (2016) described grape nutmeg as a novel fruit yielding. When compared to regular nutmeg, the tree grows a greater amount of little fruits, as the name suggests. As the name indicates, the tree bears a greater number of small fruits when compared to the normal nutmeg. As the fruit itself is small, fresh mace and nut weight were also significantly low, but can be exploited as ornamental types suitable for landscape horticulture. This can be considered as an ornamental tree with economic value.

The fresh colour of mace is a quality that attributes to the price for it. The yellow maced nutmeg is a desirable characteristic that is rarely spotted in the nutmeg family because mace sold as yellow flower gets highest market price. Two yellow accessions were reported by Priyanka (2016). The mace retained yellow colour even after drying. Red, deep red and orange coloured mace was reported by Vikram (2016) of which red and deep red fetches good market price. Nature of mace attached to nut varied as entire, slightly dissected, intermediate, and highly dissected. The entire mace is having higher mace weight. Even though thick mace is difficult to dry, maximum recovery of flower grade or first quality mace with high dry weight is obtained from this. But loose attachment of mace to nut helps in easy separation and recovery of unbroken mace, which again help to fetch higher market price. So both these are ideal trait in the selection of trees. Mace beak with probe was also observed in this study. The nut after removal of shell shows the kernel which vary in the appearance of grooves, as no grooves, shallow and deep in which smooth and shallow grooveness are positive traits as it aids in uniform drying and prevent the buildup of moulds during storage. Grooveness is a genetic trait which has definite bearing on final appearance of the produce. Colour of dried kernel is grey.

A unique nutmeg accession having normal fruit, but with rudimentary, sterile seed and mace with more in folding resembling the human brain was reported. Eighty per cent of fruits observed were with rudimentary sterile seed and remaining twenty per cent without rudimentary sterile seed (Priyanka, 2016; Sasikumar et al., 2017). The fruits were oblong in shape with yellow, netted pericarp and with an average fruit weight of 105 g. The seed or kernel was too minute or rudimentary. The shape of rudimentary seeds resembles banana seeds. The average fresh weight of mace is 9.3 to 12g and dry weight is 2 to 3g. Even though mutation is indicated to be the causative factor for female sterility in other plants, the exact reason behind this phenomenon in nutmeg is not yet clear (Saji et al., 2019).

Priyanka and Miniraj (2016) mention double seeded nutmeg and triangular mace nutmeg as other unusual nutmeg varieties. When the mace was joined to the nut, the triangular mace nutmeg was visible. The presence of few male flowers and an umbellate cyme inflorescence, typical of male trees, indicated that double seeded accessions were in transition of sex expression (Vikram, 2016). The number of fruits per tree was below normal, negating the benefit of being double seeded. On maturity, the fruits generally break into two, but in the case of double seeded nutmeg, three to four splits were seen. Ninety percent of the fruits were double seeded, and the remaining ten percent were slightly malformed. The double seededness and multiple splitting of fruits were generally associated with monoecious types (Miniraj et al. 2015).

Aside from these morphological characteristics, there has been a lot of variation in growth rate, productivity, leaf size and shape, flower size and form, fruit set, and fruit size (Krishnamoorthy et al., 1996; Haldankar et al., 2004; Kumar et al., 2010).

A study was conducted by Haldankar et al. (2004) on the growth and flowering pattern of nutmeg for assessing the extent of variability under Maharashtra conditions. 20-year-old nutmeg trees, raised from seedlings were used for the study. The majority of the genotypes in the study exhibited first vegetative flush during January-February and the second in May. Under Kerala conditions, Nazeem (1979) observed two peaks in nutmeg flushes during May-June and September. Considerable variation among the genotypes with respect to growth parameters was noticed and significant variations were noticed for duration of anthesis and fruit set which showed very low environmental coefficient of variation. These were least influenced by the environmental variation. The heritability in broad sense was very high for fruit set 94.44 per cent in comparison with the period required for anthesis as 77.93 per cent. Haldankar et al. (2009) studied thirty nutmeg genotypes under Maharashtra conditions and differentiated them into early with 225 days for maturation, mid with 250 days and late with 300 days duration. Nutmeg usually takes nine months for maturation. In this aspect, the early and mid-late genotypes can be recommended for early yield.

Variability and correlation for fruit characters in nutmeg was undertaken on seedlings under Maharashtra condition showed that there was good amount of variability for all fruit characters at phenotypic as well as genotypic level with least influence of environmental conditions. Being an obligatory cross-pollinated crop there exists wide variation not only with respect to vegetative characters but also the reproductive characters like fruit. Good variability for all fruit characters like fruit length, breadth, weight, nut length and breadth, fresh and dry nut weight, pericarp weight, fresh and dry mace weight, kernel and shell weight at phenotypic as well as genotypic level were observed (Shinde et al., 2006). The high heritability and genetic progress of the fruit character indicated that there was potential for exploiting variation through performance-based selection. Kumar et al. (2010) evaluated the fruit and yield characteristics of twenty-three nutmeg accessions grown in the high altitude region of Kodagu, Karnataka. Fresh seed yield per tree had the most variability, followed by dry seed yield per tree, fresh mace yield per tree, and dry mace yield per tree, with mace traits having medium to high variability. Leaf and seed characteristics have the least amount of variation. Aside from these few accessions, distinctive characteristics such as twin seeds, low pulp to seed ratio, triangular seed shape, deep red colour aril, rose colour aril, and high pulp seed ratio were observed.

Identification of the sex is one of the most difficulties in seedling raised nutmeg. Sex expression of nutmeg trees using the structure of flowers and fruit is time consuming, since it requires about 5-7 years to flower. Therefore, an attempt to differentiate male and female trees based on sex was made by Purwiyanti (2020). For this, 188 accessions of nutmeg were used as genetic material consisting of 104 female and 84 of male trees. The morphological characters observed were tree habits, leaves, flowers, fruit, seeds and mace. The relationship between sex expression and morphological characters were finally made. Based on the qualitative morphological characters in the vegetative phase, there are no characters that can distinguish the sex expression in nutmeg tree. The percentage of female and male trees having certain observed morphological characters was almost the same. There are no specific dominant characters which are associated with the male or female nutmeg trees. When using quantitative characters of the vegetative and generative growth phases, the width of the canopy and the circumference of the stem, the length of the corolla and the diameter of the corolla on female trees tend to be smaller than in male trees. The length of the pedicels and the diameter of corolla on female trees tend to be larger than in male trees. But it was proved that most of the morphological characters cannot be used to predict the expression of sex phenotypes in nutmeg. However, there is a correlation between the tree sizes to the sex phenotype since generally the male nutmeg trees are bigger than female ones, while the opposite is shown in the character of flower sizes.

7.3. Fruit Rind Constituents in Nutmeg

The thick and fleshy rind or pericarp of the nutmeg fruit is known as the pericarp. Fresh rind is astringent and aromatic, accounting for 80 to 85 percent of the total fruit weight. The use of rind for food is restricted due to these properties. Nutmeg rind's therapeutic properties, particularly its anti-oxidant, anti-microbial, and anti-diarrheal properties, have piqued interest. Based on the composition and quality of the rind, which is currently thrown as farm waste, it could be used to make value-added products such as jam, jelly, syrup, wine, and other products. For rind biochemical investigation, seventeen differently featured nutmeg accessions were chosen from core collections in central Kerala. The average moisture content in the recorded in the study was 88 per cent. The titrable acidity is an important parameter which affects the product quality and also acts as a protectant against the development of micro-organisms to a large extent. In the present study, acidity content ranged from 1.28 to 1.92 per cent. Low-acidity accessions could be advantageous in the processing industry. Nutmeg rind has a higher acidity than kernel and mace, according to Gopalakrishnan (1992). The amount of ascorbic acid in each 100g of food ranged from 4.5 to 9.5 mg. Ascorbic acid content is a good indicator of a material's antioxidant properties, therefore it's a popular feature that contributes to the rind's therapeutic worth. Vitamin C increases the nutritional value of the finished product. Tannins are responsible for the astringent flavour of nutmeg rind, and low levels of this element could be useful in developing value-added products from the rind. The most important carbohydrate in the human diet is starch, which is found in staple foods such as potato, rice, wheat, cassava, and corn. In nutmeg rind, starch content ranged from 0.30 to 1.23 g per 100g. Pectin is a valuable by-product that may be extracted from nutmeg rind. Fresh pericarp pectin concentration ranged from 0.21 to 1.08 percent. Because of its propensity to create gels, pectin is found in different levels in fruit cell walls and has both nutritional and industrial qualities. Pectin is used to make jams, jellies, marmalades, and preserves, as well as a thickening agent in sauces, ketchups, and flavour syrups. Furthermore, it has numerous applications in medicinal preparations such as pastes, cosmetics, and so on. The high pectin concentration of the rind could be regarded a good characteristic for processing selection (Miniraj et al., 2016). Low astringent nutmeg type with fruit rind containing lower acidity and tannin contents and higher pectin content was reported by Priyanka and Miniraj (2016). The rind acidity was less than 1.28-1.92 per cent and thus this type can be preferred for processing purpose.

7. Varieties of Nutmeg

Till now ten varieties are released in nutmeg through selection from evaluation through the nutmeg growing tracts and variability studies. From Dr. Balasaheb Sawant Konkan Krishi Vidyapeeth, Dapoli, Maharashtra, three varieties are released namely, Konkan Sugandha, Konkan Swad and Konkan Shrimanti. Konkan Sugandha is the only bisexual released variety in nutmeg which shows higher fruit set. The nut size is 5g, medium size nuts and they are moderate yielding with an yield of 500 fruits per tree per year. Konkan Swad is a female plant with higher per cent of fruit set and 5.8 g medium size nuts with an average yield of 700-750 fruits per tree per year. Konkan Shrimanti is also a female tree with higher per cent fruit set. It produces thick mace of 2.10 g and 10.0 g big size nut. From Indian Institute of Spices Research, IISR Viswashree was released during 2001. It is a clonal selection from elite trees. The tree produces dark red shining mace, black bold seed and high yield of 4,800 kg mace per ha per year and 31,220 kg nuts per ha. This variety is suitable for all nutmeg growing areas of Kerala (Haldankar et al., 2011). In 2013, IISR Keralashree was released (Mathew et al., 2016). This was derived from an open pollinated seedling progeny of a high yielding tree from Burliar, Nilgiris, Tamil Nadu. The tree is a pure female which flowers profusely and bears oblong shaped yellow fruits. The number of fruits, nut weight and mace weight of this variety was significantly more than IISR-Viswashree at three locations. Number of fruits, dry nut weight, and mace weight was 2455kg, 25.757 kg and 5.150 kg, respectively whereas, that for IISR-Viswashree was 1444kg, 13.150 kg and 1.889 kg, respectively. The dry recovery of both nut and mace are same. It is suitable for cultivation in Kerala and Tamil Nadu.

Five varieties of nutmeg are released in 2018 from Kerala Agriculture University namely, KAU Kochukody, KAU Mundathanam, KAU Poothara, KAU Punnathanam and KAU Pullan through selection from farmer's field. KAU Kochukody have a conical canopy shape and spreading growth habit. The nuts and mace are round. Single nut weight is 11.6g and single mace weight is 2.49 g. it is recommended for cultivation in Thrissur district. The nut yield is 20.88 kg per tree and mace yield is 4.48 kg per tree. KAU Mundathanam is having a pyramidal canopy shape and erect growth habit. It produces oval fruits, round nuts, dark red mace and dark brown nuts. Single nut weight is12.6 g and single mace weight is 2.49 g. It is recommended for cultivation in Palakkad district with a nut yieldof 19.7 kg per tree and mace yield of 3.89 kg per tree.

KAU Poothara also possess a pyramidal canopy shape but having spreading growth habit. They are characterized by round fruits and dark red mace. Single nut weight is10.0 g and single mace weight is 2.06 g. It is recommended for cultivation in Kottayam district. The nut yield is 22.0 kg per tree and mace yield is 4.53 kg per tree. KAU Punnathanam have a unique broadly pyramidal shaped canopy and have erect growth habit. They have oblong leaves, round fruits and attractive red coloured mace. The single nut weight is 13.85 g and single mace weight is 3.02 g with highest nut and mace weight among the KAU varieties. It is recommended for cultivation throughout the state. Nut yield is 19.39 kg per tree and mace yield is 4.23 kg per tree. KAU Pullan has a conical canopy shape with erect growth habit, oval shaped nuts and mace. The single nut weight is 10.85g and single mace weight is 1.36g. It is recommended for cultivation throughout the state and yields 22.79 kg nut per tree and 2.86 kg mace per tree. They showed highest nut yield compared to other varieties (KAU, 2020).

8. Biochemical Characterization

Nutmeg and mace are two of M. fragrans' main products, both of which are used as spices commercially. The dried kernel of the seed is nutmeg, while the dried aril is mace. The flavours of both spices are identical. Aside from nutmeg and mace, a variety of other items are important commercially. M. fragrans also yields oleoresins, nutmeg butter, and essential oils, which are utilised in the food, medical, and perfume sectors (Krishnamoorthy and Rema, 2001). The principal constituents of nutmeg are fixed oil, volatile oil and starch. It also contains proteins, cellulose, pentosans, resin and mineral elements. The percentage of constituents differs between the spices and this is a consequence of geographical origin, quality and duration of storage and even growing locations. Oleoresins, nutmeg butter and essential oils are derived from it.

8.1. Volatile Oil

The volatile oil in nutmeg is responsible for its flavour and medicinal properties. Nutmeg and mace are both used to extract volatile oil. It is usually obtained through hydro or steam distillation. Nutmeg oil is a colourless or yellow liquid with a distinct nutmeg aroma and flavour. The essential oil level of South Indian nutmeg ranged from 3.9 to 16.5 percent, while mace had a range of 6 to 26.1 percent (Maya et al., 2004). The oil's main components were sabinene, myristicin, elemicin, and safrole. Nutmeg oil contains hallucinogenic chemicals such as myristicin, elemicin, and safrole, while sabinene adds sweetness to the products. Thus, based on the low levels of myristicin, elemicin and safrole coupled with high sabinene content in the essential oil, selection can be made for future exploitation for food and pharmaceutical applications. Accessions with high oil yield in nut and mace, high butter content, high oleoresin in nut and mace, high myristicin and elemicin, low myristicin and elemicin, etc. have been identified from the germplasm collection.

The chief flavour contributing components, namely myristicin and elemicin, are present in low concentrations. Myristicin is reported to be a potent hepatoprotective principle in nutmeg. The lignan constituents possess anticarcinogenic property (Leela, 2008).

8.2. Fixed Oil

The fixed oil, often known as concrete or nutmeg butter, is an orange-colored aromatic semi-solid. Nutmeg contains 25–50% lipids in the form of fixed oil, primarily myristic, petroselinic, and palmitic acids (Purseglove et al., 1981). Essential oil content in crude fixed oil ranges from 10% to 12%. Trimyristin is the main component of fixed oil (Leela, 2008).

8.3. Oleoresin

Nutmeg oleoresin is obtained by solvent extraction from nut and mace. Oleoresins contain saturated volatile oil, fatty oil and other extractives soluble in the particular solvent employed. Nutmeg yields 31–37 per cent of oleoresins (Krishnamoorthy and Rema, 2001).

Fatty acids of nut and mace and leaf amino acids from *M. anadmanica, M. malabarica M. magnifica, M.amygdalina, M.beddomeii, M. prainii* and *Knema andamanica* were compared with *Myristica fragrans* to evaluate the variability among the species. Nut of *M. fragrans*, *M. beddomeii* and *M. prainii* are rich in myristic acid and mace of *M. fragrans* has high palmitic acid. *M. fragrans*and wild species except *Knema andamanica* leaf contains very high phenyl alanine. *K. andamanica* leaf is rich with aminoacids alanine and threonine. *M. prainii* is rich in amino acids the mace contains same pigment lycopene nutmeg.In nutmeg the fat content ranges from 28 to 34% and in mace it ranges from 23 to 26%. *M. prainii*possess almost the same content of butter as in *M. fragrans* in nutmeg while *M. malabarica* had the same mace fat as that of *M. fragrans* (Maya et al., 2006).

9. Germplasm Conservation Of Nutmeg

The germplasm collections of nutmeg are conserved at various research stations in their germplasm repositories. The important species conserved at IISR have 484 accessions which include *M. fragrans, M. fatua* var. *magnifica, M. malabarica, M. beddomei, M. andamanica, M. attenuata, M. prainii, M. amygdalina, Gymnacranthera canarica* and *Knema andamanica*. The unique types conserved include a tree bearing 1 to 4 seeds per fruit, an endangered species, *M. fatua* var. *magnifica*, nutmeg with bold nut and mace, nutmeg with high sabinene, yellow aril type nutmeg, monecious nutmeg and few elite lines (Saji et al., 2019). Germplasm of nutmeg is also conserved at NBPGR, Regional Station, Thrissur, Kerala, Horticultural Research Station, Pechiparai (28 accessions) and College of Agriculture, Dr. Balasaheb Sawant Konkan Krishi Vidyapeeth, Dapoli, Maharashtra (94 accessions).

10. Crop Improvement In Nutmeg

The easiest way to achieve quick results will probably be selection and vegetative propagation of highly productive females (Flack, 1966). In any program of selection it should be kept in mind that *Myristica fragrans* produces nutmeg and mace. The nutmegs are graded according to their size, the largest fetching the highest prices. The mace is sold by weight, on an average of twice the price of the same weight of nutmegs. Neither weight or size of seed and mace is a constant entity for any given tree, they vary with the number of fruits produced. Therefore, the variability study for the identification and characterization of the genotypes is really important.

11. IISR Vision 2050

Indian Institute of Spices Research have put forward a vision 2050 plan in which priority targets under genetic enhancement include, collection, conservation and characterization (molecular markers, bar coding) of germplasm and establishment of global gene bank of spices genetic resources and Novel nutmeg genotypes (bisexual) with altered mace colour and uniform seed size.The priority is given for development of bisexual and off-season types for increased yield, canopy architecture and development of early flowering, dwarf types for ultra-high density planting and urban horticulture, developing low myristicin types for food industry, isolating and utilization of myristicin and elemicin derivatives for cancer treatment (Krishnakumar, 2015).

12. Conclusion

Genetic variation is considered as the basis for crop improvement. Variability studies and selection within genera *Myristica*helps in relatively simple measures of selection for crop improvement. Collection of germplasm with special attributes is a prioritized area in the conservation and utilization of plant genetic resources (Priyanka, 2016).At present, there are many elite types of *Myristica* identified and popularized by progressive farmers in Kerala which excel the released varieties in yield as well as quality and such genotypes needs to be evaluated. Priority should be to given for identification of high yielding varieties with bold nut, thick and entire mace, to evolve high quality varieties for myristicin, elemicin, sabinine, safrole and to develop varieties tolerant to biotic stress such as leaf fall, anthracnose, die back, white thread blight, horse hair blight, gummosis and root rot and abiotic stress especially drought. Systematic hybridization programme is yet to gain momentum in this tree spice.

The variability in *Myristica* species need to be explored in terms of morphological and biochemical parameters. Among these, species such as *M. magnifica* is used as a dye and *M. attenuata* also produces red sap which can also beutilized, whereas *M. malabarica, M. dactyloides, M. fragrans* have high medicinal value. Myristicin, elemicin is considered as a negative trait because of its hallogenic properties and *M. Beddomeii* lacks these. So it can be marked for selection. *M. prainii* possess almost the same content of butter as in *M. fragrans* while *M. malabarica* had the same mace fat as that of *M. fragrans* and these species can also be used for butter extraction. The lignan types of the constituents in the nut of certain species are anticarcinogenic and those species need to be evaluated. Camphene present in the oil is used in the manufacture of camphor and related compounds and has strong antibacterial, antifungal and insecticidal properties. Dipentine is used in the manufacture of resins and is used as a wetting and dispersing agent. Studies in *Myristica* species can concentrate on this chemical component identification. Commercial and economical importance of *Myristica* species needs more exploration. Thus, the variability study in *Myristica* species needs to be strengthened for identification of elite species in terms of quantitative and qualitative characters so that they can be combined for further improvement in nutmeg. Further variability studies and identification of unique characters in Myristica species can help in utilization of the crop in various fields.

References

Aarthi, S., Rema, J., Sasikumar, B., Saji, K.V., Anandaraj, M. 2018. Diversity in floral characters of monoecious nutmeg (*Myristica fragrans* Houtt.). J. Spices and Aromat. Crops 27(1): 66-70.

Banik, D., Bora, P.P., Sampath Kumar, V., Bezbaruah, R.L. 2017. Conspectus on Indian *Gymnacranthera* and *Myristica*. Rheedea 27(1). 1-12.

Bhatt, A., Valasan, J. 2016. Spices Export from Kerala Current Trends & Opportunities Ahead. IRA- Int. J. Manag. Social Sci. 5(1), 54-65.

Binu, S., 2011. Medicinal plants used for treating body pain by tribals of Pathanamthitta district, Kerala, India. Indian J. Tradit. Know10(3), 547-549.

Calliste, C.A., Kozlowski, D., Duroux, J.L., Champavier, Y., Chulia, A.J., Trouillas, P., 2010. A new antioxidant from wild nutmeg. Food chemistry 118(3), 489-496.

Chandran, M.D.S., Mesta, D.K. 2001. On the conservation of the *Myristica* swamps of the Western Ghats. In: Shaanker U.R., Ganeshaiah, K.N., Bawa, K.S. (Eds) Forest genetic resources: status, threats, and conservation strategies, Oxford and IBH, New Delhi, pp.1-19.

Chetana, H.C., Ganesh, T. 2013. Reconciling natural history and species ecology: *Myristica beddomei* (Myristicaceae) in the Western Ghats, India. Trop. Conserv. Sci 6 (5), 663-673.

Choudhary, A.K., Vijayakumar, A.G. 2012. *Glossary of Plant Breeding: a Perspective*. Lambert Academic Publishing, pp.269.

Everett, T.H., 1981. *Myristica. Illustrated Encyclopedia of Horticulture.* Vol. 7. Garland Publishing Inc., New York, pp. 2264-2265.

Flach, M., 1966. *Nutmeg Cultivation and its Sex Problem* [e-book]. Dept. of Genetics and Tropical Crop Husbandry, Agricultural University, Wageningen, Netherlands. Available: https://edepot.wur.nl/294770 [20 Oct. 2020].

Gamble, J.S., 1967. *Flora of the Presidency of Madeas.* Botanical Survey of India, Calcutta, pp. 218-224.

Gopalakrishnan, M. 1992. Chemical composition of nutmeg in the Spice Islands. J. Spices Aromat. Crops1, 49-54.

Govind, M.G., Dan, M., Rameshkumar, K.B. 2020. *Myristica trobogarii* (Myristicaceae), a new species from southern Western Ghats, India. Phytotaxa 437(4), 206-212.

Haldankar, P.M., Joshi G.D., Patil, B.P.,Jamdagni, B.M., 2003. Repeatability of yield in nutmeg. J. Spices Aromat. Crops 12(1), 38-42.

Haldankar, P.M., Jamadagni, B.M., Haldavnekar, P.C., Shinde, V.V., Nagwekar, D.D., 2009. Studies on fruit growth pattern in nutmeg. J. Plant. Crops 37(3): 221-225.

Haldankar, P.M., Joshi, G.D., Patil, B.P., Haldavnekar, PC. 2004. Variability for growth, flowering and fruit set in seedling progenies of nutmeg (*Myristica fragrans* Houtt.). J. Spices Aromat. Crops 13(1), 28-33.

Haldankar, P.M., Khandekar, R.G., Haldavanekar, P.C. 2011. Genetic and varietal diversity and its conservation–Tree spices. In *Souvenir and Abstracts, National symposium on spices and aromatic crops (SYMSAC VI): Exploiting spices production potential of the Deccan region.* Indian Society for Spices, Kozhikode, pp. 58-62.

Haldankar, P.M., Lawande, K.E., Parulekar, Y.R. 2013. Tree spices – Problems and prospects. In: *Souvenir and Abstracts, National Symposium on Spices and Aromatic Crops (SYMSAC VII)* Post-harvest processing of spices and fruit crops, Indian Society for Spices, Kozhikode, Kerala, India, pp.143-147.

Jose, P. A., Pillai, P.C. 2016. Conservation through restoration of wild nutmeg tree populations of Western Ghats of Kerala. *Final Project Report No. 516.* Kerala Forest Research Institute, Peechi, Thrissur, pp. 82.

KAU [Kerala Agricultural University]. 2020. KAU home page [online]. Available: http://www.kau.in/basic-page/varieties-released [10 Nov. 2020].

Krishna, S., Somanathan, H. 2016. Spatiotemporal strategies that facilitate recruitment in a habitat specialist tree species. AoB plants, pp. 33.

Krishnakumar, N. K. 2015. Spices-way forward. In: *Symposium on spices, medicinal and aromatic crops, SYMSAC-VIII,* December 2015, Tamil Nadu Agricultural University, Coimbatore, pp. 3-14.

Krishnamoorthy, B. 2000. Sex conversion in nutmeg. Spice India. 13, 11-12.

Krishnamoorthy, B., Rema, J. 2001. Nutmeg and mace. In: *Handbook of herbs and spices.* Woodhead Publishing, New Delhi, pp. 238-248.

Krishnamoorthy, B., Rema, J., Leela, N.K., Jacob, T.K. 2010. Nutmeg. In: *Major Spices Production and Processing* (eds Parthasarathy, V. A. and Rajeev, P.), Indian Institute of Spices Research, Calicut, Kerala, pp. 274-288.

Krishnamoorthy, B., Sasikumar, B., Rema, J. 1996. Genetic variability and segregation of sex in nutmeg (*Myristica fragrans* Houtt.). J. Plant. Crops 24, 468-472.

Krishnamoorthy, K. 1960. Myristica swamps in the evergreen forests of Travancore. Indian Forester, 86(5), 314-315.

Krishnamurthy, K.S., Rema, J., Mathew, P.A., Krishnamoorthy, B. 2008. Identification of suitable *Myristica* species/related taxa as rootstock to combat drought in nutmeg. Indian J. Hort. 65(2), 204-208.

Kumar, R.S., Krishnamoorthy, B., Prasath, D., Venugopal, M.N., Ankegowda, S.J.,Biju, C.N. 2010. Variability in nutmeg (*Myristica fragrans* Houtt.) under high rainfall and high-altitude Kodagu region of Karnataka. Indian J. Plant Genet. Resour. 23(2), 191-194.

Leela, N.K. 2008. Nutmeg and Mace. In: Parthasarathy, V. A., Chempakam, B., Zachariah, T.J., (eds.), Chemistry of Spices. CABI international, pp. 175 189.

Mathew, P.A., Krishnamoorthy, B., Rema, J., Zachariah, T.J., Sasikumar, B. 2016. IISR Keralashree- a high yielding and high-quality nutmeg (*Myristica fragrans* Houtt.). J. Spices Aromat. Crops 25(1), 56-59.

Maya, K.M., 2004. Chemical composition of essential oil of nutmeg (*Myristica fragrans* Houtt.) accessions. J. Spices Aromat. Crops 13(2), 135-139.

Maya, K.M. 2005. Biochemical variability in nutmeg *Myristica fragrans* and related taxa. Ph.D. (Biochem.) thesis, University of Calicut, Kerala, 153p.

Maya, K.M., Zachariah, T.J., Krishnamurthy, K.S., Rema, J., Krishnamoorthy, B. 2006. Fatty acids and leaf amino acids in *Myristica fragrans* Houtt. and related taxa. Indian J. Hortic. 63(3), 316-318.

Miniraj, N., Vikram, H.C., Philip, M. 2015. Variability of nutmeg in Kerala. Indian J. Arecanut Spices Med. Plants 17(2), 6-14.

Miniraj, N., Vikram, H.C., Philip, M., Deepa, E.T. 2016. Fruit rind constituents in nutmeg (*Myristica fragrans*) morphotypes. J. Plant. Crops 44(3), 153-157.

Nair, N.C., Bahl, P.N. 1956. Vascular anatomy of the flower of *Myristica malabarica.* Phytomorphology 6, 127-134.

Nazeem,P.A. 1979. Studies on growth, flowering, fruit set and fruit development in nutmeg. M.Sc. (Hort.) thesis, Kerala Agricultural University, Thrissur, 106p.

Nazeem, P. A., Nair, P.C.S. 1981. Growth and flowering in nutmeg. Indian Cocoa Arecanut Spices J. 4, 81-83.

Nitta, A. 1993. Nutmeg found in Molucca Islands. J. Jpn. Bot. 68(1), 47-52.

Nybe, E.V., Miniraj, N., Thangaraj, T., Kennedy, R.R. 2006. Tree Spices In: *Advances in Spices Research* (eds: Ravindran, P.N.) Agrobios India, New Delhi, pp. 587-650.

Parkinson, C.E. 1972. *A Forest Flora of The Andaman Islands* - an account of trees, shrubs and principal climbers of the Islands. layyed Press, Ballimaran, New Delhi, pp. 223-224.

Parthasarathy, V.A. 2010. An overview of genetic resources of tree spices. In: *Proceedings of National Seminar on Tree Spices*; 5-7 March, 2010. Horticuture Research Station, Kanyakumari, pp. 5-21.

Parthasarathy, V.A., Kandiannan, K. 2008. *Spices and condiments.* India Institute of Spices Research, 171p.

Phulsagar, S., Dundi, M., Bhagwat, S., Girigaon, Y. 2014. An inside review of *Myristica fragrans* Houtt. – A potential medicinal plant of India. International J. Med. Sci. Clinical Inventions, 1(9), 500-513.

Priyanka, S. C. 2016.Collection and characterization of unique genotypes of nutmeg *(Myristica fragrans Houtt.)*. M.Sc. (Hort.) thesis, Kerala Agricultural University, Thrissur, 101p.

Priyanka, S.C., Miniraj, N. 2016. Morphological characterization of unique genotypes of nutmeg (*Myristica fragrans* Houtt.). J. Trop. Agric. 54(2), 120-128.

Purseglove, J.W., Brown, E.G., Green, C.L., Robbins, S.R.J. 1981. *Spices* (Vol. 1). Longman, London, pp. 174–228.

Purwiyanti, S., Wahyu, Y., Rostiana, O. 2020. Correlation between morphological characters and the sex phenotypes of *Myristica fragrans* Houtt Trees. In: *IOP Conference Series: Earth and Environmental Science,* IOP Publishing, 418(1), 6p.

Ravindran, P.N. 1998. Genetic diversity of major spices and their conservation in India. In: *Biotechnology Conservation and Utilization of Spices, Medicinal and Aromatic plants* (eds Sasikumar, B., Krishnamoorthy, B. Rema, J., Ravindran, P. N. and Peter, K. V.), Indian Institute of Spices Research, Calicut, pp. 16-44.

Rema, J., Krishnamoorthy, B., Mathew, P.A. 1997. Vegetative propagation of major tree spices – a review. J. Spices Arom. Crops. 87–106.

Rema, J., Saji, K.V., Sasikumar, B., Anandaraj, M.A., 2015. Monoecious nutmeg in Uttarkannada and Shimoga districts of Karnataka. Indian J. Arecanut Spices Med. Plants. 16(3), 3-5.

Saji, K.V., Sasikumar, B., Rema, J., Aravind, S., Babu, K.N., 2019. Spices Genetic Resources: Diversity, Distribution and Conservation. In: *Conservation and Utilization of Horticultural Genetic Resources*, Springer, Singapore, pp. 283-320.

Sammydurai, P., Jagatheeshkumar, S., Aravinthan, V., Thangapandian, V. 2012. Survey of wild aromatic ethnomedicinal plants of Velliangiri hills in the southern Western Ghats of Tamil Nadu, India. Int. J. Med. Arom. Plants 2(2), 229-234.

Sasikumar, B., Aarthi, S. 2016. Evolution of nutmeg towards monoecy. Spice India.pp. 14-16.

Sasikumar, B., Rema, J., Saji, K.V. 2014. In the trail of elite nutmeg. Spice India 27(9), 8-12.

Sasikumar, B., Saji, K.V., Rema, J. 2017. A note on a seed sterile nutmeg from the secondary center of domestication. J. Spices Aromat. Crops 26(1), 44-46.

Sasikumar, R., Rema, J., Saji, K.V.,Thomas, L. 2015. Scouting for elite nutmeg trees. Spice India 28(1): 24-25.

Sharief, M. U. 2007. Plants folk medicine of Negrito tribes of Bay Islands. *Ind. J. Tradl. Knowl.* 6, 468–476.

Shinde, S.S., Haldankar, P.M., Khandekar, R.G.,Haldavenkar, P.C. 2006. Variability and correlation for fruit characters in nutmeg (*Myristica fragrans* Houtt.). J. Plant. Crops 34(3), 152-154.

Sinclair, J. 1958. A review of the Malayan Myristicaceae. *Singapore Cargens Bull.*16. pp. 205-472.

Skaria, B.P., Joseph, R., Joy, P.P., Mathew, S., Mathew, G. 2004. *Myristica contorta* warb. – A potential wild nutmeg species for root stock and hybridization. In: *Proceedings of National Seminar on Research and Development in Production, Protection, Quality, Processing and Marketing of Medicinal and Aromatic Plants*, 27-29 February, 2004, 82p.

Spices Board. 2020a. Major spice/state wise area and production of spices [on-line]. Available: https://www.indianspices.com/ statistics.html [18 Oct. 2020].

Spices Board. 2020b. Major item wise export of spices [on-line]. Available: https://www.indianspices.com/export/major-itemwise-export.html [18 Oct. 2020].

Sreedharan, G., Indulkar, M. 2018. New distributional record of the northernmost Myristica swamp from the Western Ghats of Maharashtra. Curr. Sci. 115(8), 14-34.

Thomas, P. 2008. Characterization of dioecy in nutmeg (*Myristica fragrans* Houtt.). M.Sc. (Hort.) thesis, Kerala Agricultural University, Thrissur, 79p.

Thomas, T.P. 2009. Problems and prospects of spices trade in Kerala. Ph.D. (Commerce) thesis, Mahatma Gandhi University, Kottayam, 297p.

Ungirwalu, A., Awang, S., Maryudi, A., Priyono, S. 2019. Small Scale Ecology and society: Forest-Culture of Papua Nutmeg (*Myristica argantea* Warb.). J. Ilmu Kehutanan13, 114-126.

Vikram, H.C. 2016. Characterization and evaluation of nutmeg (*Myristica fragrans* Houtt.). Ph.D. (Hort.) thesis, Kerala Agricultural University, Thrissur, 323p.

Weiss, E.A., 2002. *Spice Crops*. CABI publishing, CAB International, Wallingford, UK, 411p.

6

Off- Season Leafy Coriander Production Under Shade Net Covered Tunnels: Tripling Farmer's Income

***Ravindra Singh*[1]*, Sharda Choudhary*[1]*, Ravi Y*[1]*, Madhu Choudhary*[2] *and Arvind Kumar Verma*[1]**

[1]*ICAR-National Research Centre on Seed Spices, Ajmer, Rajasthan, India*
[2]*ICAR-Central Soil Salinity Research Institute, Karnal, Haryana, India*

1. Introduction

Coriander (*Coriandrum sativum* L.) is a popular seed spice crop, used for dual purpose seed and leaves both. Since ancient times, coriander leaves have been used in the production of many popular Asian and East European foods. It improves the taste and flavor of vegetable, chicken, fish, and meat dishes when combined with other kitchen herbs and spices. Blending of green coriander leaves with other spice crops is used to make chatni (paste) which is a good appetizer, flavour enhancer and rich source of medicinal properties of coriander. The green leaves of coriander are a rich source of minerals like calcium, phosphorus and iron as well as vitamin A, C, riboflavin and folic acid. Coriander is one of the few plants which can be termed both as herb and spice, the seeds being used as spices (coriander) and leaves are widely used as herbs (cilantro). The green leaves of this herb contain 160 mg/100 g vitamin C, 12 mg/100 g vitamin A and 87.9% moisture, 3.3% protein, 0.6% fat, 6.5% carbohydrates and 1.7% mineral matter. Coriander has been shown to have antioxidant, anti-diabetic, antilipidemic, anti-mutagenic, and anti-spasmodic properties. This plant is also used to treat ailments such as digestive system problems, respiratory system disorders, urinary tract infections, kidney cleaning, and so on. For leaf purpose, coriander is grown all round the year in India. It can fairly tolerate high temperature to some extent and can be grown under ensured irrigation facilities during hot summers using protected structures.

Predominance of high temperature (38-46°C) and bright sun shine (850-1200 kilolux) with long day duration during summer season in Rajasthan and other parts of northern India is a common phenomenon. It is well known that extreme climate especially high temperature with bright sun shine having more UV rays in summer season is not conducive for the growth and development of coriander plant resulting into very less or sometimes no yield due to scorching of leaves. Under open field condition, it is very difficult to grow vegetables successfully in the summer season due to very high incidence of biotic and abiotic stresses. In extreme climates or at high temperature during the months of April to September the crop cannot survive in open conditions. High temperature and bright sun shine promotes early bolting (flowering) with stunted growth and pale colour in leafy coriander. Summer months are not ideal for germination of coriander as well as growth and yield of foliage. Bright sun shine with high temperature and light intensity are the limitations during summers, which not permit the farmers to grow leafy coriander in open condition. Although, during this hot season the demand for green leafy coriander is very high for use as garnishing and flavoring agent in various cuisines prepared in rural as well as urban areas of India. The pre-monsoon season is season of many social events and marriages in India and farmers/growers fetch high prices for their produce in the summer season. The prices of leafy coriander during off season (March- September) are 4 to 5 times higher ranging Rs 150-250 kg/ha than the rabi season leafy coriander production in Northern India specially in Rajasthan, sometimes prices may be 10 times higher than normal rabi season price due to very harsh climatic conditions, low harvest and high demand. Hence, protected cultivation in the summer days (off season) may be a possible alternative for farmers to control external climatic factors which may affect seed germination of coriander and growth of leaf and its production.

Tender leafy of coriander are used in the different form of spices but climatic condition and varieties affected the production and quality (Cowan et al., 2001). Some dual cultivars mostly grown by the farmers in India but recently ICAR-NRCSS, Ajmer has developed Ajmer Green Coriander especially for leafy purpose. Dual purpose varieties like CO-2 also gives an average green yield of 10 mt/ha (Singhet al., 2001). Ahlawat et al.(2000) reported that Panta Haritima is a good yielder of leaves (12.5-14.0 mt/ha) having the smooth leaves with good fragrance and attractive green color. A substantial increase in yield along with enhancement in yield attributing characters was recorded in protected cultivation as compared with open condition. Production of leafy coriander under shade net house has several benefits including minimum use of pesticides, off-season vegetable production, delayed bolting, increased productive period, improved quality and income per unit area is high. Controlling the microclimate allows for the development of higher-quality produce that is free of insect infestation, diseases, and chemical residue. As a result, increasing the output of leafy coriander through intense culture under shade net covered cultivation throughout the summer season is critical. By growing under protected structures/ shade nets, this crop can be produced easily for the green leaf purpose and one can fetch more income. Hence, protected cultivation in the summer days may be a possible alternative for farmers to control external climatic factors which may affect germination of coriander seeds and leaf yield. Protected cultivation of leafy coriander in peri-urban areas of Northern India provides the best way to increase the productivity and quality of the crop as well as better management of biotic and abiotic stress situation compared to open field condition. A substantial increase in yield along with enhancement in yield attributing characters has been found in protected cultivation compared to that of open. Green house, poly house and shade net house are some of the readily available options for off season production of the coriander, but these technologies are very costly and a few farmers can afford to adopt the technology due to high cost and requirement of higher skills and knowledge. Hence, efforts were made at ICAR-NRCSS, Tabiji, Ajmer to overcome the limitations of off-season cultivation of coriander. In this regard, shade net covered walk-in tunnels and high tunnels have been designed indigenously with available resources, tested and validated through field trials conducted at NRCSS Ajmer from 2014-15 to 2016-17 which is low cost, users friendly and adopted by the farmers with existing knowledge and skill. Walk-in tunnels and high tunnels are basically the low cost protected structure made from Pre-galvanized (GI) pipe and covered with UV stabilized shade nets for off season leafy coriander production. With the above in view the experiment has been carried out to optimize the shading intensity and colour of the shade net for production of leaf in protected situation during most demanding but off season (summers) for coriander. Among the different colour tested green (50, 75 and 90% shading) and black, white and red colour shade net (50 % shading) were tested and compared with open cultivation of the leafy coriander during off season. After standardizing the production technology of leafy coriander production during off season, the best methodology was validated and shade net covered walk-in tunnels and high tunnels technology were also demonstrated at farmer's field during 2017-2019 for off-season leafy coriander production during summer season for doubling the farmers income.

2. Challenges During Off Season Cultivation of Leafy Coriander

- Due to the high heat of the sun and heat storms in arid and semi-arid locations, only a few crops may be grown in the summer months, and crops grown must endure the burning heat of the sun and heat storms. The difficult situation occurs during the summer months due to the substantial diurnal temperature changes.
- Extreme radiation levels in the region impair the adaptation of new crops in these areas. High exposure with more UV rays raises the temperature and promotes water stress in many crops.
- Ground water available for irrigation is scarce and of varying quality. The ground water quality ranges from normal to saline and brackish, and it is currently underutilised in most areas due to a lack of adequate technologies.

- The region's soil stratum is noted for its poor fertility with very low levels of organic carbon (i.e. 0.2 percent only) and sandy loam to loamy structure with little water holding capacity, making crops less productive.
- The high temperatures and sparse foliage cover in these areas increase the likelihood of very frequent dust storms with strong wind velocity, causing severe crop damage.

3. Existing Protected Cultivation Technological Options

The kind of protective structures for crop production range from simple provisions such as rain shelters, shade houses, mulches, row covers, low tunnels, high tunnels, cloches to greenhouse structure with passive or climate control. While it would not be possible to describe each of the above detail, a brief description of some of these is given below (table 1).

3.1. Types of Tunnels for Off Season Production

In traditional agronomical practises, crops are grown / grown organically in the open field under natural climatic conditions, where the crops are more vulnerable to sudden changes in climate, such as heat, humidity, light intensity, photo period, and other conditions, which can affect and decrease crop quality and yield. Low tunnels, walk-in tunnels, and high tunnels are generally used to provide a favourable controlled environment such as light intensity, temperature, ventilation, humidity, irrigation, soil media, fertigation, disease control, and other agronomical practises throughout the season, regardless of the natural conditions outside. These tunnels have variety of applications, the majority being, advancing or delaying of the crops, growing of vegetables, floriculture, planting material acclimatization, fruit crop growing for export market. These low cost structures can be dissembled, moved throughout the farm, and work with the contours of the land. Similar to high tunnels, plastic covered low tunnels provide several degrees of protection against outer atmosphere to crops.

3.1.1. Low Tunnels (LT)

Low tunnels are flexible transparent plastic sheet / non-woven cloth covering that are installed over the rows or individual beds of vegetables to enhance the plant growth by increasing the air temperature in the microclimate of the plant as compared to open fields during the winter seasons. They also warm the soil by trapping the heat through convection and protect the plants from abiotic stress (hails, cold winds, frost injury etc) and biotic stress (insects). These tunnels help the farmers/ growers to advance the crop by 45-60 days than the normal season.These are seasonal structures covered with transparent plastic sheets or strips of synthetic, breathable fabric of various weights and mesh diameters, with supporting wire hoops, or over entire fields with rigid supports such as metal or plastic tubing, or electrical conduit. For preparing the LT, flexible galvanized iron hoops/ ribs are fixed manually on distance of 2.5 m to 3.0 m keeping the 0.75-

0.9 m width between two ends of hoops with a height of 0.6 - 0.75 m above the level of beds for covering the plastic sheet/ non woven cloth (crop covers) for making low tunnels. Low tunnels are typically 2 to 3 feet high and cover the width of a growing bed/ crop rows. Like row covers, they protect cold-hardy crops in winter and spring plantings from cold and wind. Cucumber, melons, watermelons, bitter gourds, squashes, and snake gourds, etc. can be grown in these tunnels. The crop can be advanced by 1.5 to 2.0 months in these low tunnels as compared to than open cultivation. The width and height of these low tunnels is hardly 3'. Soil preparation, spraying and picking is difficult in this tunnel, hence used only for a limited period to protect the crop from sudden climatic variations like low temperature or frost injury during winters.

3.1.2. Walk-In Tunnels (WIT)

Walk-in tunnel typically made from steel and covered in polythene/ shade net, usually semi-circular, square or elongated in shape. WT are lower in height (6-7' from centre) compared to high tunnels (10-12' from centre. Walk-in-tunnels provide higher yields than low tunnels because of better sun light and ventilation. The tunnel is suitable for growing tomatoes, cucumbers, sweet pepper and hot pepper. The interior heats up because the sun's incoming solar radiation warms plants, soil, and other objects inside the structure faster than heat can escape. The roof and walls keep warm air from hot internal surfaces inside the building. Temperature, humidity, and ventilation can be adjusted by leaving the end walls open or manually opening and closing vents all along side walls. Walk-in tunnels can be used to give a higher temperature and/or humidity than the environment but they can also protect crops from strong heat, bright sunlight, winds, hailstorm and cold waves. This permits fruits and vegetables to be cultivated at times when they would normally be deemed out of season; market gardeners frequently utilise walk-in tunnels to extend the season. Aside from seasonal extension, walk-in tunnels are used to allow cold-hardy crops to survive winter in areas where their hardiness is insufficient to thrive outdoors.

3.1.3. High Tunnels (HT)

By definition, the high tunnel is a walk-in, hoop or gothic-shaped, pipe frame structure that is covered with a single layer of plastic film during winters and shade net during summers. High tunnels are simple greenhouse-like structures over bare ground, without the elaborate heating and cooling systems of a greenhouse. They rely primarily on passive solar heating and passive ventilation. High tunnels generally have steel pipe frames set into the ground and are covered with one or two layers of greenhouse-grade plastic during winters or shade nets during summers. Steel tubing or galvanized iron pipes(GI pipes) or fence pipe is the standard material used for posts as well as the pre bended semi circular GI pipes as hoops. Posts are driven into the ground about 0.5 m deep to support the hoops. To get a higher ridge and better ventilation, the ground posts are sometimes extended

0.5 to 1.0 m above ground depending on the crops to be grown. The end wall remains open during summers for easy and fast ventilation. Open end or doors large enough to allow access for a small tractor in are desirable if the ground will be tilled for field operations. Keeping open end walls as open doors facilitate the ventilation in the tunnels which is more crucial during off season due to high temperature in Northern India. The use of HTs for the production of horticulture crops includes benefits such as protection from severe weather, faster warming of the soil in the winters and cooling in summers, and accurate application of water with the result of reduced disease incidence.

They are irrigated using drip systems, hand watering or small sprinklers. They may also have end-wall vents (louvers) or ridge vents. Crops grow in the ground, raised beds or containers. The ground may be bare, or it may be covered with landscape fabric, plastic mulch or organic mulch such as straw. High tunnels give maximum yield of crops and make easy soil preparation, picking, spraying due to its width and height. High tunnels are unheated, plastic-covered/shade net covered hoop-house structures primarily used for growing vegetables (tomatoes, cucumbers and sweet peppers), cut flowers or fruit brambles. High tunnels covered with shade net during summers offer more protection for crops than the open field, and serve to extend the growing season for many popular crops such as leafy greens (Spinach coriander, and fenugreek) and tomatoes. While similar to greenhouses in many ways, there are a few key differences between the two types of structures.

- High tunnels rely on covering materials like poly sheet during winters and shade net during summers to be used for maintaining the temperature inside the tunnel.
- High tunnel frames are usually anchored directly into the ground instead of being anchored on a foundation such as a concrete pad.
- To control the temperature, passive solar radiation for heat and natural ventilation for cooling needed.
- Application of shade nets of different colour with varying shading intensity used to break high solar radiation and bright sun shine during summers
- Seeds or seedlings are normally planted directly into the soil or in raised beds within the tunnel.

4. Low Cost Indigenous Design Developed for Installation of wit and HT

A new design of the low cost prototype of the walk-in tunnels and high tunnels were developed at ICAR-NRCSS during 2014-2016 for different purpose. These tunnels can be shifted from season to season as per requirement of the farmer.

- Plastic sheet covered walk-in and high tunnels during rabi season to protect the crops from frost injury

- Insect proof net covered walk-in and high tunnels during rabi season to protect the crops from frost injury
- Shade net covered walk-in and high tunnels during Off season (Kharif and summers) to protect the crops from frost injury
- Plastic sheet covered walk-in and high tunnels solar dryer for drying of kassuri methi leaves and rose petals

4.1. Tools for Installation of Shade Net Covered WIT and HT

Pipe bending machine, Step ladder, GI pipes of 3/4" and 1/2" OD, Hammer, Spade, Shade net, UV stabilised crop trailing thread, needle, Sledge hammer, Cutter and Grinder

5. Materials Needed for WIT or HT

To cover 100.0 m^2 area (4.0 m wide and 25.0 m length) by the tunnel, require the GI pipes as mentioned in table 2 and 3. For walk-in tunnels GI pipe of 3/4" OD, were cut in 0.6 m length with the help cuter and require 22 such pieces (11 piece in each side wall of the tunnel) to make posts for side wall of the tunnels. Whereas for high tunnels, the height of these pieces for side walls were kept 1.5 m. With the help pipe bending machine, 11 GI pipes of 1/2" OD with 6.0 m length were bended in semi circle shape for raising the hoops on the post. To give stability to the tunnel, two piece of GI pipes of 1/2" OD with 3.5 m length for walk-in tunnels and 4.5 m length for high tunnels also required. One end of these pipes were weld with 1.0" U shaped MS strip to grid the end wall pipes of the hoops and give extra support and stability to the end wall hoops.

6. Shade net Covered WIT and HT

Shade nets are perforated plastic materials used to cut down the solar radiation and prevent scorching or wilting of leaves caused by marked temperature and high amount of UV rays of bright sun shine increase within the leaf tissue from strong sunlight. These nets are available in different colours and having different shading intensities ranging from 25% to 90%. The application of different colures of shade nets as well as its shading intensity may vary for the different crops to be sown as per crop requirement and varying crop seasons. Leafy coriander have been grown under shade nets whose growth are significantly enhanced compared to non shaded plants when sunlight is strong during the Off season.

7. Steps for Installation of Shade net Covered WIT and HT

Major steps in the installation of walk in tunnel (WIT) and high tunnels (HT) are site selection and preparation, layout of the area for WIT and HT, drive in ground posts/holes in soil using hammer and rebar, installation of ground posts (side wall) with 3/4" GI pipes, raising hoops, trench digging and shade net attachment.

8. Agronomical Practices for Off Season Coriander Leafy Production

8.1. Variety

Select the late (delayed) bolting varieties available in the area having faster growth and profuse biomass production habit. We have selected AgCr-1 a green leafy coriander variety developed at NRCSS in the experiments as well as in demonstrations as it is having good aroma as well as delayed bolting (58$\pm$2 days) with higher biomass.

8.2. Seed Rate

As the purpose for raising the coriander is leaves and not the seed. hence higher see rate facilitate for plants per unit area which give competition within the plants and helps in more height and fast growth of the crop. Hence, 400 to 500 gm seed is sufficient for 100 m^2 area of the tunnel.

8.3. Date of Sowing

Coriander is sown from late March until early September during the off season. If the full plant uprooted, one cycle of the crop will take 50-55 days before harvesting. If cuttings will be taken crop will ready to harvest in 25-30 days after first cutting. But the keeping quality of the cut leaves is very less as fast lose of moisture due to harsh conditions and high temperature. But uprooted plants can be stored for longer period by sprinkling the water on the roots after uprooting. The maximum prices of leafy coriander in the market by harvesting the crop during mid May to June, hence plan the sowing date accordingly. To achieve a constant supply of leaves through the summer sow small amounts in batches every 3 weeks.

8.4. Sowing Method and Crop Durations

After preparation of the field in the tunnel, sowing is done by splitting the coriander seed in two halved by pressing the seed. Treat the seed with 1.0 g bavistan before sowing. Broadcast the seed uniformly in the tunnel and mix it in the soil. Broadcasted seed remain in the upper layer of the soil, germination will be faster. As the target is more foliage per unit area, broadcasting with higher seed rate (random sowing) give better results than line sowing as usually followed for seed production of the coriander.

8.5. Irrigations and Water Management

Install the 10 drip lines at 30 cm spacing with drippers at 30. cm interval with 2.1 litre/hr water discharge rate. After sowing the coriander seed, operate the drip irrigation till the full area saturate with the moisture. This will take 2.0 hrs for first irrigation in summers. After germination, give the irrigation after 4 days interval for 1/2 an hour will fulfil the water requirement of the crop

8.6. Nutrient Management

Standard management practices are followed throughout the growing period. In addition to 25-30 tonnes/ha well decomposed farmyard manure (organic manure), inorganic fertilizer was also applied. Nitrogen, Phosphorus and Potassium at the rate 60:40:20 kg/ ha (Irrigated condition). With one third of nitrogen along with full doses of phosphorus and potassium were applied as basal and rest nitrogen was top dressed twice after each cutting for boosting further vegetative growth. Irrigation was provided as per evaporative demand of atmosphere.

9. Pest Management

Incidence of thrips was observed at early stage of the crop growth (20-30 days after sowing) during the summers in the shade nets. Need based spray were made to control the pests.

10. Harvesting

The crop is ready to harvest within 45-60 days after sowing during summers depending on the variety. The crop should be harvested before bolting, otherwise quality will deteriorate. Crop may be harvested by cutting as well as uprooting of the plants. For enhance the keeping quality after harvest, uprooting is more advantageous than cutting. As crop is grown in shade nets, cutting will fasten the moisture loss from the leaves which leads to wilting symptoms in the harvested produce and reduce post harvest keeping (storage) quality.

11. Conclusions

Coriander is a strong smelling annual herb extensively grown inmany climates throughout the world. In India it is mainly grown during the *rabi* season and there no demand and supply gap during this season but when the temperature rises from February-March the supply of coriander leaves decreased and the price of coriander leaf in the market increased exponentially. The supply is reduced because to grow coriander leaf in open condition in the North India is not possible. So, by using shade net technology give the farmers new opportunities to get more prices of their produced. Under shade net technology two crops can be taken easily by the farmer during off-season. In the experimented conducted by the ICAR-NRCSS Ajmer, it was found that total production of 1st crop was produced 130 kg from 100 m^2 and sold in the market @ Rs 100.0/kg amounting the total revenue of Rs 13000, whereas the second crop was produced 105 kg and sold in the market @ Rs 85.0/ amounting the total revenue of Rs 8925. The cost of cultivation including the working cost of the tunnel were Rs 6750, hence net profit the farmer got 15175 from 100 m^2 are within 105 to 120 days during the off season when the prices of leafy coriander is very high. Hence it can be concluded that, production of leafy coriander under shade net house has several benefits including minimum use of pesticides, off-season vegetable production, delayed bolting, increased productive period, improved quality and income per unit area is high. Controlling the microclimate enables for the development of higher-quality produce that is free of insect assault, diseases, and chemical residue, thereby doubling or tripling the farmer's income.

References

Ahlawat, I.P.S., Prakash, O., Saini, G.S. 2000. Scientific Crop Production in India. Meerut: Aman

Chaulagain, R., Pant, S., Thapa, R., Sharma, M. 2013. Performance of coriander cultivars for green leaf production under late sowing condition. Journal of Agriculture and Environment 12, 67-73.

Cowan, A.K., Cripps, R.F., Richings E.W., Taylor, N.J. 2001. Fruit size: towards an understanding of the metabolic control of fruit growth using avocado as a model system. PhysiologiaPlantarum 111: 127–136.

Singh, D., Chaudhary, G.R., Singh S., Singhania, D.L. 2001. Production technology for seed spices. In: S. Agrawal, E.V. DivakaraSastry and R.K. Sharma (eds.), Seed Spices Production, Quality, Export. Jaipur (India): Pointer Publishers, pp. 89-94.

7

Cultivation of Underutilized Minor Fruit Plants in India

Sarvpriya Singh[1] and Sukhwinder Singh Aulakh[2]

[1]Krishi Vigyan Kendra, Punjab Agricultural University, Bathinda, Punjab, India
[2]Krishi Vigyan Kendra Punjab Agricultural University, Hoshiarpur, Punjab India

1. Introduction

There are around six hundred major and minor crop species but only two percent are cultivated commercially meeting the food (calorie) requirements. The most commonly cultivated food plants are even smaller counting to merely 20 crops (Wilkes, 1984). The huge number of underutilized crops may serve as potential fruit crops to meet the nutritional requirements of poor farming households in lesser common farming and subsistence agriculture. The adoption of underutilized crops on a commercial scale may not be possible without crop improvement, standardized agronomic and fertilizer management technologies, pre- and post-harvesting techniques and marketing (Bag Mal et al., 1997).

The wider adaptability and higher degree of tolerance against adverse climatic and soil situations make the plant species thrive under adverse conditions without proper management. These fruits species not only serve as source of protective food for people living in villages but also help in meeting the daily vitamins and minerals dietary requirements. The curative properties of these fruits have already been recognised in Ayurvedic and Unani system of medicine in India since time immemorial. Apart from medicinal and nutritive values, some of these underutilized fruits have excellent Aroma, taste and colour, that make them suitable for numerous processed products. Even then the cultivation of underutilized fruits is very restricted and grow wildly under barren lands and forests. Only few of the crop species has any specified variety, that is another major constraint to cultivation, production and productivity of these crops. The improved and high yielding varieties, advanced propagation methods, agronomic and fertilizer management interventions may be necessary for the sustainability of production of such crops. Some underutilized fruits such as jamun, mulberry, and phalsa are highly perishable, and impossible to be kept for more than 24 hours under ambient conditions. Some fruits are difficult to be used as a table fruit, e.g., bael, due to their hard shell, numerous seeds and sticky texture, thus is not popular as a desert fruit, even with huge nutritious value. Similarly, the strong astringent taste of fresh aonla fruit is not liked by many people. These fruits have great potential in the processed form and may be consumed worldwide. Some of the important indigenous underutilized fruits of India waiting future exploitation and commercialization are as follows.

Table 1: List of underutilized minor fruits grown in India

Common Name	Botanical Name	Family	Origin
Bael	*Aegle marmelos*	Rutaceae	India
Jamun	*Syzygium cumini*	Myrtaceae	India
Tamarind	*Tamarindus indica*	Fabaceae	Tropical Africa and Asia
Karonda	*Carrisa carandus*	Apocynaceae	India
Fig	*Ficus carica*	Moraceae	Mediterranean region
Phalsa	*Grewia subinaequalis*	Malvaceae	India
Chironji	*Buchanania lanzan*	Anacardiaceae	India
Mahua	*Madhuca indica*	Sapotaceae	Indo-China
Mulberry	*Morus alba*	Moraceae	China
Strawberry	*(Fragaria x ananassa)*	Rosaceae	
Rambhutan	*Nephelium lappaceum*	Sapindaceae	Malayan archipelago
Custard apple	*Annona reticulata*	Annonaceae	Mexico
Wood apple	*Feronia limonia*	Rutaceae	India

2. Use of Underutilized Fruits for Nutritional Security

Fruits, nuts and vegetables play a significant role in human nutrition (Table 2), especially as sources of vitamins [ascorbic acid (C), carotenoids (A), thiamine (B1), riboflavin (B2), niacin (B3), pyridoxine (B6), and folacin], minerals (Table 3), fat, protein, and dietary fiber (Craig and Beck, 1999; Wargovich, 2000). Processing and value addition of underutilized tribal fruits into more useful, digestive, tasteful, and convenient products can improve the economic value of underutilized commodities, increase their reach to masses and create nutritional security.

Table 2: Nutritional content in some underutilized fruits

Common name	Protein (mg/100 g pulp)	Fat (g/100 g pulp)	Carbohydrate (g/100 g pulp)	Energy (Cal./100 g pulp)
Bael	1.80	0.39	31.8	137
Jamun	0.53-0.65	0.10	-	-
Tamarind	3.1	0.1	67.4	284
Karonda	2.3	9.6	67.1	-
Phalsa	1.3	0.9	14.7	-
Khirni	0.48	2.40	27.7	134
Mahua	-	1.60	22.7	112
Custard apple	1.6	0.3	23.5	104
Wood apple	7.10	3.70	17.0	-

Source: Mitra et al. (2010)

Table 3: Vitamin and mineral composition of unutilized minor fruits

Common name	Vitamin A (µg/100 g)	Vitamin B1 (mg/100 g)	Vitamin B2 (mg/100 g)	Vitamin B3 (mg/100 g)	Vitamin C (mg/100 g)	Calcium (mg/100g)	Phosphorus (mg/100g)	Iron (mg/100g)
Bael	55	0.07	1.19	1.10	8	85	50	0.7
Jamun	73-100	-	-	-	30.3-40.7	0.02	0.01	0.1
Tamarind	60	0.44	0.07	0.7	3	170	110	17
Karonda	-	-	-	-	10.26-17.94	_	_	4.15
Fig	162	-	-	0.60	-	80	30	1.0
Phalsa	419	-	-	0.3	22	129	39	3.1
Khirni	495	0.13	0.14	0.70	15.67	83	17	0.92
Mulberry	-	-	0.13	0.50	-	70	30	2.3
Custard apple	-	0.07	0.17	-	42	17	47	4.31
Wood apple	61	0.04	0.17	0.80	3	130	110	0.48

Source: Gopalan et al. (2004)

3. Medicinal/ Nutraceutical Value of Underutilized Minor Fruits

Underutilized fruits are major source of raw material for drugs and traditional medicines. A lare number of underutilized fruits are potential source of remedies for diseases like cancer, diabetes, jaundice, asthama, and nutritional deficiency. In India, fruits of aonla, bahera, and harad are the most common entering in to 219 patented drugs. About 60 patented drugs are derived from Bael. Aonla is the main ingredient of ayurvedic dietry supplement- chyavanprash. Similarly, ber fruit is used in joshanda. Jamun seeds are useful for diabetic patients, and black mulberry in docking of AIDS virus. (Quebedeaux and Bliss, 1988; Quebedeaux and Eisa, 1990; Produce for Better Health Foundation, 1999; Prior and Cao, 2000 and Wargovich, 2000. The stem and bark of *Khejri* are used for curing boils, leprosy, dysentery, asthma, piles, and tumors. Daily use of *aloe* juice helps in significant improvement in glycemic control, lipid profile, and BMI of Type-2 diabetes. The medicinal properties of various fruits are given in Table 4.

Table 4: Therapeutic uses of underutilized minor fruits

Fruit	Therapeutic use
Bael	appetizer, stomachic, cooling, restore vitality
Jamun	stomachic, anemia, improves hemoglobin in blood, diabetes
Phalsa	blood purification, anemia, curing hepatitis
Karonda	antiscorbutic, anemia
Wood apple	diarrhea, dysentery, liver and cardiac tonic, antivenom activity
Tamarind	useful in heart care and against stone formation in urinary system
Mahua	cough, cold, bronchitis
Kiwi	reduced platelet aggregation and blood triglyceride level
Pistachio nut	cure lung cancer, cardiovascular diseases and diabetes

Source: Goyal and Sharma (2009), Pareek et al. (1993) and Sen (2003)

4. Underutilized Minor Fruits as Source of Sustainable Living and Income

Underutilized fruits are the source of sustainable income in tribal areas. Many plant parts (root, leaves, fruits, gum, etc.) have high economic value in market. Many economically useful fruits like, custard apple, anola, tamarind, ber, jamun, bael, ker (*Capparis apphyulla*), lasoda (*Cordia mixa*), sangari (*Prosopis cineraria*), karonda, dates, tumba (*Citrullus colocynthis*), kumtia (*Acacia senegal*), kachari (*Cucumis melo*), timru, etc. are either propagated or naturally available in the tribal areas of Rajasthan (Anonymous, 2011). The tribal (women and children) and weaker sections of the society collect these fruits and useful parts of the plant. A part of their collection is retained for their own consumption as raw or in dried form and surplus is marketed (Jain, 2003; Anonymous, 2011). According to one survey these minor fruits can support triabal livelihoods by supplementing their income as given in Table 5. The market of underutilized fruit in-turn is influenced by a number of factors like consumer preference, processing ability, value addition, export potentail, domestic consumption which again is based on our knowledge about the health promoting qualities and nutritional value of the that particular fruit (Chundawat, 2003). The global forum on Agriculture Research (GFAR) in 1999 also emphasized the role of underutilized species in raising income of the rural poor.

Table 5: Underutilized minor fruits as a source of income for tribals

Fruit plants	Plants/ha	Average yield of fruits/pl.	Average total yield (Qt./ha)	Average market rate (Rs./Kg.)	Expected total income (Rs./ha)
Fig	100	25	25	15	37,500
Mulberry	277	20	55	20	1,10,000
Lasoda	100	60	60	15	90,000
Bael	100	50	50	15	75,000
Jamun	100	50	25	25	1,25,000
Phalsa	625	2	12	20	1,20,000
Karonda	400	20	80	15	25,000
Custard apple	277	30	83	10	83,000
Tamarind	100	80	80	12	96,000
Khirni	100	20	20	25	50,000
Mahua	100	70	70	30	2,10,000

Source: Field survey in tribal areas of Rajasthan by Nandal & Bhardwaj during 2008 to 2011

5. Value Addition in Minor Fruits

Although local polulations and tribals earn a good and sustainable earning by selling the raw produce, their income can be increased many folds by processing and marketing value added products from minor fruits. There are many reasons for processing fruits besides high returns, like reduction in post harvest losses, quality enhancement, maintaining the nutritional value, year around availablity, ready to use products etc. There is huge potential in processing and value addition of underutilized fruits and converting into products such as jam, pickle, fruit drinks, jelly, preserve, dried products, candy, confectionery and bakery products, etc. For instance, mature fruits of bael are used for preserve, candy and canning purose, whereas its ripe fruits are used for beverages (RTS, squash, cider etc.), mixed jam with mango, powder, toffee etc. Similary, candy can be prepared from red type karonda. Dried flakes of karonda are used as substitute for cherry in pastry and other foods. Lemon sherbet fortified with 'Lalima', a 'natural colourant cum nutraceuticals-supplement' is derived from karonda genotype CIAH Selection-1 (Krishna et al. 2014). Chironji oil is substitute for almond oil and used in the confectionary industry and chironji kernels are used as dried and roasted form and are valued for nutritious kernels rich in calcium - 279mg, phosphorus - 528mg, sodium & potassium - 436mg. The list of some value-added products is given in Table 6.

Table 6: Value added products of minor fruits

Products	Fruits suitable for value added products
Jam	Jamun, Custard Apple, Bael, Wood Apple
Jelly	Tamarind, Jamun, Karonda, Wood Apple
Preserved	Bael, Karonda
Candy	Karonda, Datepalm, Tamarind
Glazed fruits	Tamarind
Juice/squash	Bael, Jamun, Phalsa, Mulberry
Chutney	Wood Apple, Mahua, Tamarind, Karonda
Wine	Karonda, Wood Apple, Tamarind
Sauce	Wood Apple, Tamarind, Karonda
Pickle	Karonda, Tamarind
Dehydration	Bael, Karonda, Phalsa, Tamarind, Custard Apple
Frozen puree	Tamarind, Bael, Karonda, phalsa, Tamarind, Custard Apple
Canning	Jamun
Confectionary	Tamarind
Dry fruits	Pistachio, Pecan, Hazelnut, Chironjee

Source: Goyal and Sharma (2009), Singh et al. (2008) and Sen (2003)

5.1. Bael

Bael *(Aegle marmelos)* belongs to family Rutaceae and it grows wild in sub-Himalayan tracts of the Central India, Sri Lanka, Pakistan, Bangladesh, Thailand and most of the South East Asian countries. Due to nutraceutical properties of this fruit, they are commonly used in medicines. This fruit is rich source of Vitamin especially riboflavin (Vitamin B2) and minerals. The mature fruit is a stimulant, curative, purgative, good to cure digestive and stomachic, heart and brain related problems. At processing point of view, it is mainly used in RTS, nectar, squash candy, toffees and dry form of powder. Fruit comprises 20–25 per cent of peel from total fruit weight which also depends upon the varieties/genotype. The fruit peel is very hard, which cannot be consumed by animals whereas dry form of this fruit, fed to the dairy animals (Kaushik and Dhawan, 1996). Bael tree can be grown under poor & neglected soil and adverse climatic conditions up to an altitude of 1200 m, not much affected by lower temperature. Most of the bael trees planted in India are of seedling type leading to variability. A little genotypes have been made in parts of Uttar Pradesh, Bihar and West Bengal (Srivastava et al., 1998) and extensive exploration conducted in various states indicate considerable variability in physico-chemical characters like: fruit shape (spherical, semi-spherical, flattened-round, elleptical, oval, oblong, pyriform); length (6- 26 cm); width (4–20 cm); weight (0.4–3.2 kg); fibre, more to less, muscilage, less to high; seeds, few to numerous (10–300); pulp (40–60%); fibre content (1.4–7.6%); acidity (0.36–0.80%); vitamin C (14.0–20 mg/100 g); TSS (20–46%); flesh colour (creamy yellow, whitish-yellow, deep yellow); peel thickness (1.0–4.0 mm); stem end cavity, shallow to deep. The different type of genotype has diverse vegetative, reproductive phases and mature in February-September. Commercially fruits come into bears after 7–8 years from seedlings while the asexual propagated plants start bearing in 4-5 years. The flowering occurs in month of May-June and the fruit matures in next April-May. The average yield of bael tree is about 400 fruit/ plants after 10-15 years. The treatment of ethephon @ 1000-1500 ppm helps for uniform ripening, normally it takes two to three months. Maturity of bael fruits occurs in 18–24 days by artificially methods (Mitra, 2010).

- **Varieties**: NB-4, NB-7, NB-16, NB-17, NB-5, NB-9, CISH B-1, CISH B-2, Goma Yashi, CHES B-5, Pant Aparna, Pant Shivani, Pant Sujata, Pusa Urvashi
- **Distribution**: Uttar Pradesh, Bihar, Jharkhand, Madhya Pradesh, West Bengal, Orissa, Chhattisgarh, Rajasthan, Uttarakhand
- Bael Goma Yashi: Released by CIAH in 2010
- Dwarf, Fruit weight 1.25 kg, yield 50-60 kg per tree in 7th year, suitable for high density planting

5.2. Jamun

Jamun (*Syzygium cuminii*) is evergreen fruit tree, indigenous to India belongs to Myrtaceae family. It is mostly used as windbreak and boundary plantations. The fruits are nutritious and contain iron, calcium and phosphorus, protein, fat, fibre and carbohydrates. The fruit is mainly used for curing the stomachie, carminative, diarrhoea diuretic, diabetes and dysentery related problems. The different studies have been noticed that jamun fruit helps to lowers down the blood pressure whereas the powdered form of jamun seeds helps to reduce the sugar level in urine. The lotion prepared from jamun powder cures the ringworm infection. In processing industry, there are so many products made from ripe fruit of jamun likewise: juice, squashes, vinegar, jellies and wine. (Roy, 1997). Worldwide, jamun is mainly grown in Thailand, Philippines, Israel, Algeria, Madagascar, West Indies and in some other countries. Due to hardy in nature, this tree has wider adaptability under waste and marshy land as compare to other fruit plants.

The improved genotype is not available but in In North India, 'Ra Jamun' is the common variety. Its fruits are big, oblong, deep purple to bluish black in colour and juicy in nature with very less stone which matures in June – July. It is distributed throughout the tropical and subtropical regions India mainly in Rajasthan, Gujarat, Utter Pradesh and Tamil Naidu. Jamun is mainly grown under sandy loam soils without water logging condition whereas dry spells in jamun stimulate the flowering and fruiting. It is propagating by sexually whereas vegetatively, it can be propagated by layering, inarching, veneer grafting and budding (shield, patch, forekert). The plantation should be done in square system at 10–12 m spacing but cultivar Goma Priyanka can be panted at 5m x 5 m distance with modified leader system and pruning of annual extension growth 25 % may be followed for canopy management. The planting is done in the month of February-March and July-August. The trees are not usually pruned; however, at the later years the dry twigs and cross branches should be removed. Jamun is a cross pollinated tree. The pollination is done by honeybees, houseflies and wind. Maximum fruit set (32.6 to 36.0%) occurred at one day after anthesis. The problem of fruit drop (60-65 per cent) in jamun can be resolved by two foliar applications of GA_3 at 60 mg^{-1}, firstly at the time of flowering and secondly at fruit setting (15-day interval). If the fruit is grown from seedling, it takes 8 to 10 years come into bearing commercially while through vegetative propagated means, it comes into bearing after 6 to 7 years. The average yield of tree varies from 10 to 80 kg/plant. Recentlly, ICAR funded National Network Project on Underutilized Fruits (NNPUF) intensive work on jamun was carried out at CISH, Lucknow. Some of the imroved varieties of Jamun are given in Table 7

Table 7: Improved varieties of Jamun

Name of variety	Institute that developed the variety
Goma Priyanka	CIAH, Bikaner, Rajasthan
CISH J-37, CISH J-35, CISH J-42	CISH, Lucknow, Utter Pradesh
CHES J-19	Godhra, Gujarat
Jamun GJ-2, Jamun GJ-8	Anand, Gujarat
Konkan Bahadoli	RFRS, Vengurla, Maharashtra
Rajamun	Local genotype (Commonly grown in the north India)

Source: CIAH, Singh et al. 2008, Dhillon 2013

5.3. Tamarind

Tamarind *(Tamarindus indica)* a drought tolerant tree and grown under rainfed conditions therefore, suitable for poor farmers of dry tracts. The tree is a heavy bearer (produces 2-3q pods/tree) with 60-70 years of longevity. Tamarind is used as dried fruits; hence there is no problem of marketing. Fruits are rich in sugar (55.8%) and minerals. It is a major ingredient in the preparation of Rasam-a special drink before dinner in southern India. Thailand has done fairly well in tapping potential of this fruit tree. This tree can be found in Tamil Naidu, Andhra Pradesh, Maharashtra, Karnataka, Kerala, Orissa, Madhya Pradesh, Bihar, Rajasthan, Gujarat and West Bengal. There are few selected varieties developed by different Institutes in India (Table 8).

Table 8: Improved varieties of Tamarind

Name of variety	Institute that developed the variety
Goma Prateek (Sweet type with Long pods)	CIAH, Bikaner, Rajasthan
Tamarind T-10	CHES, Godhra, Gujarat
PKM- 1 (Sour type)	Horticultural Research Station, Periyakulam, TN
Pratisthan (sweet type with short bulged pods)	Fruit Research Station, Aurangabad, MAU, Parbhani, Maharashtra
Yogeshwari	Marathwada Agricultural University, Parbhani, Maharashtra
DTS-1, DTS-2	University of Agricultural Sciences, Dharwad, Maharashtra

Source: Singh, 2012

5.4. Karonda

The karonda *(Carissa carandus)* is mainly grown under dry spell area where rainfall intensity is very less. Due to bushy in nature, less care and maintenance is required after planting. This fruit is native of India but now grown in South Africa, Australia, India, Malaysia, Sri Lanka, Bangladesh and Myanmar. Matured fruits are less acidic to sweetened in taste depends upon the variety. There are so many products prepared from karonda such as jelly, squash and chutney whereas the dried fruits may be substitute for raisins and candied like cherry. The mature fruits also utilized for wine making which contains about 14–15% alcohol. The milky latex of raw fruit used for chewing gum and rubber. Fruits of karonda also used in dyeing and tanning industries. The extracts isolated from karonda roots helps to cure lumbago, chest complains and venereal diseases. Karonda is an evergreen, branched, spinous shrub or small tree belongs to Apocynaceae family which is widely used for hedge because it has spinous nature that why it protects from animals. It is classified, according to fruit colour as green-fruited, pink-fruited and white-fruited. It can be widely grown under semi-arid and arid regions. It is mainly propagated by seeds, however, asexually propagated by cutting, layering, grafting (inarching, softwood) and budding (shield). The spacing should be done at 2 m x 2 m and require pruning regularly. The flowers of karonda comes into bearing in the month of March-April to November depending upon altitudes. The fruit matures in 120-130 days from fruit set whereas flowers are borne on terminally and axillary and produces 5-10 kg fruits/plant. The fruits can be stored week at 13°C and 95% relative humidity for 7 days where as under ambient condition, the keeping quality is very less . Some of the improved varieties of karonda are given in Table 9.

Table 9: Improved varieties of Karonda

Name of variety	Institute that developed the variety
CHES K-2 CHES K-3	Godhra, Gujarat
Konkan Bold	Konkan Krishi Vidyapeeth, Ratnagiri, Maharashtra
Pant Manohar Pant Sudarshan Pant Suvarna	G.B. Pant University of Agricultural and Technology, Pantnagar, Uttarakhand
Karonda Accession-3	Regional Fruit Research Station, Vengurla, Maharashtra

Source: CIAH, Salvi et al. (2006)

5.5. Fig

Fig (*Ficus carica*) also known as *Anjeer* and having nutty texture. Its fruits are mainly used for various value addition products and consumed as fresh, dried, preserved, candied, canned and jam preparation. Fig is a deciduous plant which have capacity to with stand under adverse climatic condition, if temperature is extreme in summer which caused scorching in terminal leaves. It performs well under clayey, sandy loam and light sandy soils with better drainage system. This plant is found in Maharashtra, Gujarat, Andhra Pradesh, Karnataka, Tamil Nadu, Utter Pradesh, Himachal Pradesh and Punjab. Punjab Aricultural University, Ludhiana has recommended a variety Black Fig 1 which matures second fortnight of June to July and gives average fruit yield of about 13 kg per tree. Another variety from PAU i.e Brown Turkey which is a heavy bearer cultivar and average yields about 53 kg fruit/tree that matures from May to June. Fig plants are propagated using 25-30 cm hardwood cuttings with 3-4 buds during January. Planting is done at of 6m×6m spacing during January to February. It is trained by modified leader system. The training is done in first 3-4 years of planting. Plant bear fruits in the leaf axils and on current growth. Pruning of new wood stimulates the growth and yielding parameters. Every three years, there is required to heavy pruning to stimulate the tree growth and balancing the C:N ratio. Fruits are picked when they become fully coloured but are still firm. Some of the improved varieties of fig are given in Table 10.

Table 10: Improved varieties of Fig

Name of Variety	Institute that developed the variety
Poona	
Brown Turkey	Punjab Aricultural University, Ludhiana, Punjab
Black Fig 1	Punjab Aricultural University, Ludhiana, Punjab
Black Ischia	
Excel	
Deanna	
Dinkar	Marathawada Krishi Vidyapeeth, Parbhani, Maharashtra

Source: Singh (2012), Rattanpal et al. (2011)

5.6. Phalsa

Phalsa (*Grewia subinaequalis*) native to India and belongs to the family Tiliaceae. This can be survived under water stress condition i. e why grown under dry region of Punjab, Haryana, Rajasthan, Uttar Pradesh, Madhya Pradesh, Maharashtra, Gujarat, Andhra Pradesh, Bihar, West Bengal are the main growing areas for this fruit. Phalsa fruits are enriched with vitamin (A &C) and minerals (phosphorus and iron). It is attractive in colour from crimson-red to dark purple. The fruits of phalsa are used several beverages, viz., nectar, squash and crushed phalsa packed in glass bottles remain acceptable for up to 8 months if stored properly. Baskets prepared from phalsa shoots are mainly used by the vegetable vendors.

Asexually, it is propagated by hardwood cutting; however, air-layering, stooling and softwood grafting can also be done. Pruning of shoots is mandatory to maintain better fruit quality. In Phalsa fruit plant, flowering is mainly occurring in February-March to till May. It is double sigmoid type of growth curve of fruit and their respiration curve is non-climacteric type. Fruits starts mature (60–65 days of fruit set) in last week of April. Fruit should be harvest in early morning or evening hours due to its perishable nature and marketed as soon as possible (Mitra, 2010). The average yield mature is about 6-10 kg. Some of the improved varieties of fig are given in Table 11.

Table 11: Improved varieties of Phalsa

Name of variety	Institute that developed the variety
CHES P-1 CHES P-2 CHES P-3	Godhra, Gujarat
Sharbati	Local selection HAU, Haryana

In addition to above there are some other minor fruits which are of low importance at present but have potential in the future. List of such fruits, their varieties and areas where they are grown is given is given in Table 12.

Table 12: Important Varieties and saleint characteritics of other minor fruits

Fruit/ Varieties	Distribution and Important Characteristics
Chironji CHES C-3 CHES C-7 CHES C-9 CHES C-10	- North, west and central India - Grows well on yellow sandy-loam soil - Fruit ripens in April - Decidous tree and propagation though soft wood grafting in July - CHES C-7 was found promising cultivar with 22% TSS, 1.20 g fruit weight, 0.12g kernel weight and 30.0 % kernel protein
Khirni CHES K-1 CHES K-10	- Rajasthan, Gujarat, Madhya Pradesh, Maharashtra, Bihar - CHES K-10: Fruit weight 6.00g, TSS 21° Brix

Mahua MH-10 MH-14 MH-35 MH-63	- Madhya Pradesh, Rajasthan, Gujarat, Maharashtra, Uttar Pradesh, Chhattisgarh, Jharkhand, Bihar and Andhra Pradesh - MH-63: Flowers recorded 2.50 g weight, 68.00 % juice, 24.37 % TSS and 23.50g fruit weight.
Mulberry CISH Mulberry-2 CISH Mulberry-8	- Throughout India under subtropical, tropical and temperate regions.
Rambutan CHES – 31 CHES – 14 CHES – 27 CHES – 26	- Tamil Naidu, Kerala, Karnataka - It has unusual appearance and sweetness - Potential source of an antihyperglycemic agent. - Yellow (CHES-31, CHES- 14); Red (CHES-26, CHES-27)
Custard apple -Arka Sahan (Hybrid) -Balanagar (Selection) -Washington 107005 (Introduction)	- Andhra Pradesh, Tamil Naidu, Assam, Chhattisgarh, Madhya Pradesh, Bihar, Karnataka, Maharashtra, Orissa, Rajasthan, Gujarat, Uttar Pradesh. - Potential of Custard Apples Under Rainfed Conditions - Promising genotype of custard apple with high pulp content and less number of seeds.
Woodapple CHES-1 CHES-2 CHES-3 CHES-4 HB-10	- Maharashtra, Madhya Pradesh, Utter Pradesh, Chhattisgarh, Bihar, Jharkhand and Rajasthan - CHES-4 is promising selection with Fruit Weight: 627.33g, Pulp: 319.00g, TSS: 9.60°Brix, Acidity %: 2.08

So far, we have discussed about underutilized indiginous fruit plant but recently in last few years the demand for exotic/imported fruits in Indian market is steadily building up and there is potential of growing these fruits in India and capture the market demand with local production instead of importing these and spending valuable foreign exchange. These are some exotic fruits (Table 13) that can be successfuly grown in various parts of the country.

Table 13: Important Varieties and distribution of exotic fruits in India

Fruit	Varieties	Distribution
PASSION FRUIT *(Passiflora edulis)* Passifloraceae Southern Brazil	Kaveri (Purple x Yellow) Purple cultivar Yellow cultivar	Tamil Naidu, Kerala, Karnataka, Nagaland, Manipur, Mizoram, Meghalaya
HAZELNUT *(Corylus avellana)* Betulaceae Black sea coast areas	Tonda Giffani Tonda Romaana Kashmir Selection	Jammu & Kashmir, Himachal Pradesh, Uttarakhand
PECANNUT *(Carya illinoinensis)* Juglandaceae North America	Burkett Nellies Mahan Schley Cheyenne	Himachal Pradesh, Uttarakhand, Jammu & Kashmir
PISTACHIO *(Pistacia vera)* Anacardiaceae Western Asia	Peter Kerman Chiko Red Allepo Joley	Jammu & Kashmir
STRAWBERRY *(Fragaria x ananassa)*	Chandler Tioga Torrey Selva Pajaro Camarosa Winter Down Festival Sweet Charlie	Himachal Pradesh, Uttarakhand, Maharashtra, Haryana, Jammu & Kashmir, Rajasthan, West Bengal, Meghalaya
KIWI (*Actinidia deliciosa*) Actinidiaceae China	Abbot Bruno Monty Allison Hayward Tomuri	Arunachal Pradesh, Sikkim, Himachal Pradesh, Mizoram, Uttarakhand, Nagaland, Meghalaya
PERSIMMON (*Diospyros kaki*) Ebenaceae China	Fuyu Jiro Hyakume Hachiya	Jammu and Kashmir, Himachal Pradesh

Source: Singh (2012)

Other indigenous under utilized fruits of India.

A. Chironji, **B.** Khirni, C.Mahua, **D.** Rambutan, **E&F.** Mulberry, **G.** Custard apple, **G.** Wood apple

Potential exotic fruits of India.
A. Passion fruiti, **B.** Hazelnut, **C.** Pecanut, **D.** Pistachionut, **E.** Strawberry, **F.** Kiwi, **G.**

6. Constraints in Cultivation of Minor Fruits

- Unavailability of planting materials/ Lack of high yielding commercial varieties
- Lack of awareness of nutritional, medicinal and economic aspects of minor fruits
- Lack of processing industry and absence of scientific resources for testing valuation and post-harvest management.
- Deforestation, forest fragmentation and hill cutting is responsible for the loss of nutrient rich minor fruits
- Non-availability of statistical data on minor fruits in general and marketing in particular.
- Very small scale and scattered production accompanied by inconsistent supply and demand due to seasonality and perishable nature of produce makes this activity a risky venture.

7. Research Focus and Future Thrust Areas

There is urgent need to identify early and late germplasm and collect, evaluate and conserve the local germplasm. Introduction of exotic germplasm for selection of widely adapted superior varieties is another area that need immediate attention. Supply of good quality planting material and development of improved production practices should be priorty areas. In addition, creation of infrastructure like proper marketing facilities, processing units and standarlization of various products are needed. These are main research institutes (Table 14) which work on the different minor fruits in India for commercialization on the national level.

Table 14: Research Institutes working on minor fruits in India

Research Institute	Works on minor fruits
IIHR, Bangalore (Karnataka)	Custard apple, Mulberry, Jamun, Phalsa, Bael
CHES, Godhra (Gujarat)	Bael, Karonda, Phalsa, khirni, Chironjee, Jamun, Tamarind
CISH, Lucknow (U.P.)	Bael, Jamun
CIAH, Bikaner (Rajasthan)	Bael, Karonda, Phalsa, khirni, Chironjee, Jamun, Tamarind
CAZRI, Jodhpur (Rajasthan)	Bael, Karonda, Phalsa, wood apple
NDUAT, Faizabad (U.P.)	Bael, Karonda
RFRS, Vengurla (Maharashtra)	Tamarind, Jamun
GBPUAT, Pantnagar (Uttarakhand)	Bael, Phalsa, Karonda
YSPUHF, Solan (Himachal Pradesh)	Pecan nut, Strawberry, Kiwi
CITH, Srinagar (Jammu and Kashmir)	Pecan nut, Pistachio, Strawberry

Indegineous underutilized fruits of India. (a) Bael, (b) Jamun, (c) Tamarind, (d) Karonda (e) Fig and F Phalsa

8. Conclusion

In India, diversification of different underutilized fruits is available in varied agroclimatic zone. These types of fruits have great potential to enhance the income of local people as well as help us to conserve the germplasm of the particular area. These fruit species have a wide adaptability to survive under adverse climatic condition. In our country, wide range of germplasm available for underutilized fruits but no standard variety has been developed so far. Many of these fruits have great nutraceutical properties. These fruits crops provide sustainable agriculture particularly for small farmers by augmenting their income with the least risk. So there is need to develop suitable varieties/genotypes for commercialization of these fruit crops at large scale.

References

Anon., 2011. Baseline Survey About Utilization and Future Prospects of Underutilized Fruits Available in Sirohi district of Rajasthan. Krishi Vigyan Kendra, Sirohi.

Bhag, M., Pardoda, R.S., Kochar, S. 1997. Underutilized crops and their implication in farming systems in India. p. 30–45. In: J. Smart and N. Haq (eds.), Domestication, Production and Utilization of New Crops. ICUC, UK.

Chundawat, B.S. 2003. Constraints in the cultivation of underexploited horticultural crops. In: Proceeding of Winter School on Exploitation of Underutilized Horticultural Crops at MPUAT, Udaipur, pp. 233–234.

Craig, W., Beck, L. 1999. Phytochemicals: health protective effects. Can. J. Diet. Pract. Res. 60, 78–84.

Dhillon, W.S. 2013. Fruit Production in India. Narendra Publishing House, Delhi-110006, India.

Gopalan, C., Ramashastri, B.V., Balasubramanian, S.C., Rao, N.B.S., Deosthale, Y.G., Pant K.C. 2004. Nutritive Value of Indian Foods, National Institute of Nutrition, ICMR Hyderabad.

Goyal, M., Sharma, S.K. 2009. Traditional wisdom and value addition prospects of arid foods of desert region of North West India. Indian J. Tradit. Know. 8(4), 581–585.

Jain, N.K. 2003. Post harvest technology of underutilized fruits, vegetables and medicinal plants. In: Proceeding of winter school on Exploitation of underutilized horticultural crops at MPUAT, Udaipur. pp. 170–177.

Kaushik, R.A., Dhawan, S.S. 1996. Popularizing nutritive potential of bael. Indian Hort. 41(3), 56.

Krishna, H., Chauhan, N., Sharma, S.K. 2014. A natural colourant cum nutraceuticals-supplement derived from karonda. ICAR-Central Institute for Arid Horticulture, Bikaner.

Mitra, S.K., Pathak, P.K., Chakraborty, I. 2010. Potential Underutilized Tropical Fruits of India. Proc. III IS on Trop. and Subtrop. Fruits Eds.:M. Souza and R. Drew. Acta Hort. 864.

Nandal, U., Bhardwaj, R.L. 2014. The Role of Underutilized Fruits in Nutritional and Economic Security of Tribals: A Review Critical Reviews in Food Science and Nutrition, 54, 880–890.

Pareek, OP., Sharma, S. 1995. Underexploited fruits for advance ecological conditions. In: Impact of Modern Agriculture and Environment-2, pp. 111–232.

Prior, R.L., Cao, G. 2000. Antioxidant phytochemicals in fruits and vegetables: Diet and health implications. Hort. Sci. 35, 588–592.

Quebedeaux, B., Bliss, F.A. 1988. Horticulture and human health. Contributions of fruits and vegetables. In: Proc. 1st Intl. Symp. Hort. And Human Health., Prentice hall, Engelwood, NJ.

Quebedeaux, B., Eisa, H.M. 1990. Horticulture and human health. Contributions of fruits a n d vegetables. In: Proc. 2nd Intl. Symp. Hort. And Human Health. Hort Science 25, 1473–1532.

Rattanpal, H.S., Kahlon, G.S., Jawandha, S.K. 2011. Performance of exotic varieties of fig (*Ficus carica*) under the agro climate of Punjab. Crop improvement 38, 87.

Roy, S.K. 1997. Potentiality of processing of underutilized fruits of India. p. 264–69. In: J. Smart and N. Haw (eds.), Domestication, Production and utilization of New Crops. ICUC, UK.

Salvi, B.R., Malshe, K.V., Bendale, V.W., Kshirsagar, P.J., Jagaonkar, V.N. 2006. Book of abstracts, Nat. Symp. On underutilized horticultural crops held at IIHR, Bangalore on June 8-9, 2006, pp.31.

Sen, N.L. 2003. Underutilised horticulture crops: Trends, challenge and opportunities in the 21st century. In: Proceeding of winter school on Exploitation of underutilized horticultural crops at MPUAT, Udaipur, pp. 224–231.

Singh, G. 2012. Checklist of commercial varieties of fruits. Year of Horticulture – 2012, Published by Department of Agriculture & Cooperation, Ministry of Agriculture, Krishi Bhawan, New Delhi, India.

Singh, D., Wangshu, L., Prahalad, V.C. 2008. Processing and Marketing Feasibility of Underutilized Fruits Species of Rajasthan. Contributed paper presented at IAMO Forum.

Srivastava, A.K., Singh, H.K., Singh, I.S. 1998. Bael (*Aegle marmelos*). Indian Hort. 42(4).

Wargovich, M.J. 2000. Anticancer properties of fruits and vegetables. Hort. Sci. 35, 573–575.

Wilkes, G. 1984. Germplasm conservation towards the year 2000. In: C. Yeatman, D. Kefton and G. Wilkes (eds.), Plant Genetic Resources: A conservation imperative. Amer. Assoc. for the Advancement of Science, Washington, DC, USA.

8

Doubling Farmers Income Through Latest Technology of Horticulture Intervention

Mahaveer Suman and Ashok Kumar

College of Horticulture and Forestry, Jhalawar Campus
Agriculture University Kota, Jhalrapatan, Rajasthan, India

1. Introduction

India's population is likely to reach 1.7 billion by 2050, making it as the thickly populated nation in the world. To feed the ever rising population, food production need augment by 70%. This augment is very serious since Indian agriculture is facing numerous challenges due to disintegration of holdings, ill brunt of climate change, sustaining investment, soaring fluctuation in prices and production, unorganized markets, inconsistent market performance, renovation of agricultural land intended for non-farm purposes and enlarged cost of production (Gangaiah, 2018). The current conditions of the Indian agriculture sector reflect that the farmers are facing severe deficit of income which is to an imperative issue and as a result dissuading the rural youth to take up agriculture as livelihood operation. The Doubling farmer's income has become a watchword for every agricultural organization and is formulated by bridging the efficiency gaps, employing most recent production technologies, advanced dispensation technologies and improved marketing. Horticulture has emerged as one of the prospective agricultural endeavor in accelerating the growth of economy in India. It offers not only a wide assortment of options to the farmers for crop diversification, but also provides abundant scope for sustaining great number of Agro-industries which produce huge employment opportunities. Horticulture has been evolving in form of horticulture industry in our country owing to its significant role in nutritional safety, poverty mitigation and employment invention. The production of fruits, vegetables and flowers has acquired much importance in recent times due to their rising command. India has a great diversity of climate and soil on which horticultural crops such as, fruits; vegetables, ornamentals, plantation crops and spices, medicinal and aromatic plants are cultivated. Horticultural crops for which the Indian geography and agro climates are well matching is an ideal way of achieving sustainability of small holdings, rising employment, civilizing environment, providing a huge export prospective and above all achieving dietary protection.

2. Horticulture in India- Present Status

The production of horticulture produce has been around 293.98 Million tons (fruits, vegetables and flowers) in the year 2020. The present estimated area under fruit cultivation in India is about 6.76 million ha with a production of 102.08 million tons and productivity of 15.1 MT/ha while in case of vegetable crops it is about 10.30 million hectares with a production of 188.90 million tons and productivity of 9.3 MT/ha, year 2020. Presently estimated area under flowers have grown in around 0.32 million hectares with the production of 3.00 MT along with 9.3 MT/ha of productivity (2020).

However, during the last decade, horticulture sector was a whole has thus, registered a compound growth of over 6 per cent as seen from the trend of horticulture growth in India. The increase in the production has been highly significant in fruits (80%), vegetable crops (91%) in last decade while in ornamentals has registered 99.19 % increase in case of loose flowers and 76% increase in cut flowers in last five years.

3. Interventions for Doubling the Farmers' Income Through Latest Technologies

3.1. Quality Planting Material

The planting material forms the support of the orchard, hi-tech or farm produce and is the key to excellence produce. Consequently, the unadulteratedvalue in terms of genetic purity, vigorous and disease and pest free plant material is it seed or asexually propagated material requirements to be guaranteed.Appreciation of nursery, perform of plant propagation and promote hardening at nursery stage are important considerations. Resources of planting material *i.e.* mother plant, growing media and procedures followed for the similar needs to homogeneous. Nursery decontamination including tools for tools and equipments too needs to be observed to make possible proper operations and ease the harboring of pests by weeds. However, another important parts of nursery management are record maintenance of different tricks like sowing, planting, procedure or activity schedules, products targets, labeling, germination, observations, disease and pest control actions *etc.* Maximum precedence should be given towards expansion of sound systematic nursery in the country. Various problems in ensuring production of quality planting materials are; (i) lack of genetically standardized rootstocks, (ii) be short of variability for rootstocks, (iii) not much information on rootstock-scion relations (iv) throughout the year production by specialized production structures, and (v) lack of micro propagated protocols for the majority of crops and be deficient of asexually propagation technique for seed propagated crops. There is a dire require to establish rootstock-breeding selection programme. Scions with low vigour should also be a priority area of research. With multi-species and multi-varietal orchards becoming one of the components to orchard efficiency, production of quality planting material are assuming greater importance.

3.1.1. Important Considerations For Development of QPM are as Mentioned Below

- Popularization of hybrids varieties might be proved as timely effort for escalating the income. The hybrid skill have ability to revolutionize the production of vegetable crops and demand for hybrid seeds is constantly rising. Maximum production, earliness, best quality, consistent produce and tolerance to biotic and abiotic stresses are the main benefits of F_1 hybrids varieties. Acceptance of hybrid varieties may be increase 1.5-3.0 times more yields which means additional profits from increased output from similar area (Kumar and Tripathi, 2020).
- Grafting in vegetables should be followed for mitigating biotic stresses under protected cultivation and thus minimizing use of plant protection chemicals.
- In citrus, micro-grafted/ micro-budded indexed plants should be made available on the pattern of Nagpur mandarin being produced from ICAR-CCRI, Nagpur. Similarly micro-propagated banana, pomegranate, strawberry and potato planting material should also be made available to growers. PPS interventions are required in bringing the cost down of TC date palm and oil palm plants of the benefit of the growers.
- Dwarfing varieties of many fruit crops such as mango, banana, papaya, sapota and dwarfing rootstocks in apple, pear, plum, peach, mango, guava, citrus fruits and berare highly suitable for high density planting system and ultra high density planting system for accepting more plants per unit area and getting more output and income. Rootstock technologies have ability to double the yield and also helpful to grow fruit crops under adverse soil as well as climatic condition.
- Promotion of plug plant production in vegetables and flower crops through Hi-tech nursery raising technology. Pro-tray nursery preparation skill by using coco-peat, vermiculite and perlite without soil by using soilless media should be adopted for raising off-season seedlings of different horticultural crops, particularly vegetable crops.
- Enhancing the capacity of existing micro-propagation units in PPP mode. Establishing Zonal primary and secondary hardening facilities to minimize cost of TC plantlets.

3.2. Enhanced Nutrient Use Efficiency in Horticultural Crops

Improving nutrient use efficiency (NUE) is a fundamental challenge faced by the fertilizer industry in general and horticulture in particular. Bio-fertilizers such as biological N fixers, VAM fungi and other beneficial microbial agents play a central role in effective nutrient use efficiency. Fertigation which is the application of water soluble fertilizer along with irrigation water,mainly with drip irrigation method, could be a money-spinning method of fertilizer application to attain healthier distribution uniformity and fertilizer use efficiency in horticultural crops. The availability of ready to mix solid fertilizers and liquid fertilizers are promoting the application of fertigation. The use of efficient and automatic fertigation equipment enhances ease of application of measured quantity of fertilizers. Application of nitrification inhibitor is another novel approach and it has been found that meliacins with ammonium sulphate and urea resulted in a significant increase in uptake of N, K and micronutrients from the rhizosphere of Kinnow. Hence, efficacy of neem-coated urea will be increased by coating urea with only fraction *of* meliacins from neem oil.

3.3. Plastic Culture

Plastic culture includes plastic film mulches, drip irrigation, row covers, low tunnels, high tunnels, silage bags, hay bale wraps, and plastic trays and pots used in transplant and bedding plant production. Use of plastic in horticulture crop production has increased dramatically in the last two decades.

- Plastic reservoirs and irrigation systems-Drip lines, sprinklers, etc
- Plastic for raising planting material, Floating row cover, multi-celled plastic plug-trays
- Protected cultivation: Polyhouse and. Low Tunnel polyhouse shade net house
- Postharvest handling- boxes, crates for crop collecting, handling, packaging and transportation

3.4. Technology Driven Farming and Farm-Mechanization

There is dire need in development of technology for small land holding farmers with booming entrepreneurial models. Creation of self-help groups and provision for credit providing agencies for establishing efficient mechanized farms needs to be developed. In perennial horticultural crops, mainly in fruit crops, the brown space available in the pre-bearing age of the orchard can be effectively utilized to grow short duration horticultural crops as intercropping or for inter spaced planting and cultivation to gain higher income. Farm mechanization facilitates timely, precise and scientific farm operations, increasing farm input and labor use efficiency also helpful to doubling the farmer's income (De and Singh, 2017).

3.4.1. Advantages of Farm Mechanization

- Increase in crop intensity and yield thus ensuring better returns to the farmer
- Reduction of weather risk and risk of non-availability of labour thus minimizing post harvest wastages
- Improved working conditions and enhanced safety for the farmer
- Conversion of uncultivable land to agricultural land through advanced tilling technologies
- Increased rural employment due to substitute of animal labour and stipulation of management of farm machines.

3.5. Advanced Irrigation Systems

Water is a vital component for crop management. The quality and amount of irrigation water is critical and is based on crops grown, weather conditions, time of year and the environment control system. Alkalinity, pH and soluble salts along with presence of heavy metals is needed to be considered. Assuring quality of water, the mode of irrigation is also important. Drip and sub-irrigation systems are the most efficient means of irrigation over overhead sprinklers and hand watering. Adoption of such systems not only results in enhanced water use efficiency but also minimizes the environmental pollution a common problem under conventional irrigation practices. Though, its potential benefits are well established at research as well as in farmer fields and although Government is promoting it through subsidies, its adoption rate is very low by the end users. Facilitation of micro-irrigation in horticulture sector in rain-fed areas. Among various methods tried drip irrigation has proved successful in exhibiting high water productivity by saving irrigation water from 25 to 60 per cent in various orchard crops and vegetables with a 10 to 60 per cent increase in yield as compared to the conventional method of irrigation. It is one of the latest methods of irrigation which is becoming popular in areas with water scarcity and salt problems. The impact of micro-irrigation on resource conservation (saving in input costs to the farmer) is estimated in the range of 20 to 40 per cent in case of horticultural crops.

3.6. New Intercession in Greenhouse Technology

Greenhouse technology modifies/controls partially or fully microclimate surrounding the plant as per the requirements of the plant species grown during their period of growth, with the added advantage of assured remuneration on account of quality produce and in some cases off-season production, early production and prolonged production- a desirable trait in fruits, vegetables and flowers production (Saikia and Bordoloi, 2019). There have been several novelties and innovations under protected horticulture world over with having potential of doubling farmers' income as discussed herewith. (i) Soilless cultivation adopted for raising plant nurseries, micro-greens, vegetables, flowers and certain fruits like strawberry, (ii)

Vertical horticulture production system (leafy vegetables, fruit crops like strawberry and some of the annual flowers) where plants are fed hydroponically and natural or LED light helps in photosynthesis, (iii) Edible seedlings/microgreens - new generation smart food young edible greens, (iv) Plasticulture or use of plastics in horticulture for irrigation, micro-irrigation, fertigation, mulching, claddings to protective structures, nets, containers- innovative plastics-plastics with shelf-life enhancers, crop covers etc. economize cost of production of horticulture crops leading to better returns to farmers, (v) Tissue culture of horticulture plants- Banana, carnation, gerbera, orchids, chrysanthemum and other herbaceous crops.

3.7. Rejuvenation of Old Orchards

Old, unproductive and un-economic fruit orchards are the major hurdles for enhancement of fruit productivity in India. Rejuvenation is thus gaining popularity to make such orchards productive. Rejuvenation technology has helped in restoring the productivity and profitability of several fruit crops like mango, guava, aonla, litchi, cashew in the shortest possible period besides reducing the tree height to a manageable limit. Rejuvenation studies on 45-year-old trees of mango cv. Dashehari suggested initially higher cumulative fruit yield in fourth order pruned trees (330.92 kg/tree), but, long term studies indicated that though there was no fruiting in the initial two years in severely pruned trees (first and second order), while from third year onwards, there was gradual increase in fruit yield in these treatments. In moderate and light pruning treatments (third to fifth order pruned trees), there was continuous fruiting from first year onwards with an increasing trend for initial 3-4 years, there after inconsistent yields were obtained. Now, the technology has spread in more than 11,000 ha area in the country. The impact studies have shown the success of rejuvenation ranging from 63.6 to 99.7%. Centre opening, thinning out of few centrally located branches or a portion of them, in case of mid-age mango orchards has also resulted in enhanced light penetration, reduction of tree height, reduced insect pest infestation and improved fruit yield and quality. As a result of centre opening, reduction in tree height was achieved by 7.5-10%, besides enhanced light availability in the inner canopy by 55-100%.

3.8. Precision Farming

Precision farming is a management strategy that employs detailed, site-specific information to precisely manage production inputs. Maximization of quality yield by means of mechanization of horticulture farm, application of microprocessor and computers, new emerging scope of modern precision farming with the upcoming latest technologies, and having more emphasis on sustainable farming.

3.9. Biotechnological Interventions in Horticulture

Molecular markers is a very important technique and can be used in a breeding programme for recognizing desirable parents, enhancing assortment of best alleles at loci controlling important characters, pyramiding favourable alleles at multiple loci affecting either a single trait or multiple traits and finger printing cultivars for identification and intellectual property protection. Marker locus-trait combinations can be used as selection criteria to complement parent selection and that of desirable plants in segregating populations. Molecular markers linked with reproductive traits such as sex forms (cucurbits and papaya), cytoplasmic male sterility (sweet pepper, cole crops and carrot), self-incompatibility (cole crops) facilitates rapid conversion of promising lines into desirable sex form for use in hybrid breeding. The linked molecular markers to genes for resistance in tomato (leaf curl, bacterial wilt, late blight, root knot nematode), cauliflower (black rot, down mildew), cucumber (papaya ring spot virus, *Fusarium* wilt, downy mildew), banana (*Fusarium* wilt, browning nematode), potato (cyst & golden nematodes), grape (powdery mildew), apple (powdery mildew, scab), rose (black spot) and coconut (mites and lethal yellowing), and quality traits such as carotenoids and anthocyanins (carrot and cauliflower); lycopene (papaya and watermelon); pungency &Sulphur assimilation (onion), glucosinolates (Cole crops) and reduction of antinuritional factors such as tannins (broad bean); fruit acidity (citrus); and seedlessness (grape and cucumber) will have great promise in marker-assisted selection. For abiotic stresses like salinity, drought, temperature extremes and heavy metals, breeding resistant genotypes continue to be a challenge due to genetic complexities of the traits and difficulties in sequential phenotyping conventionally. Regulation of flowering and vernalisation requirement in fruits and biennial winter vegetables can be explored by studying regulatory pathway mechanisms using molecular markers. Efforts to bring marker assisted breeding in to wide adoption for horticultural crops improvement will benefit from targeted utilization of synteny among related crops and information from model plants to develop markers for use in standard breeding methods.

3.10. Production of Seed Spices Through Hi-Tech Interventions

Seed spices play a very important role in national economy as India is the largest producer, consumer and exporter of spices and seed spices. Seed spices are low nutrient and water requiring crops and giving higher returns compared to cereals, pulses and oilseeds. At present average productivity of major seed spices (cumin, coriander, fennel and fenugreek) is only 9.8 q/ha, which need to be enhanced up to 15.20 q/ha by 2050 to meet domestic as well as export demand. Bridging the wide gap between present and targeted productivity of seed spices is great challenge to researchers due to their susceptibly to insect pests and diseases under ambient climatic condition. Protected cultivation technologies, automatic fertigation, raised-bed planting techniques, plug-tray seedling production and transplanting technologies, insect proof net covered walk in tunnel, plastic mulching and shade-net covered walk in tunnel for off-season cultivation of green leafy seed spices during peak summer season are important hi-tech intervention for raising productivity of seed spices. Application of water and nutrient through fertigation and sowing of crops with raised-bed technique results in enhancing 30- 35% yield, with saving of 25-30% nutrient and 45-50% water, which ultimately enhance profitability in seed spice cultivation.

3.11. Pesticides Residue Management Strategies

In recent years, occurrence of pesticide residues in horticultural produces has posed menace in international trade. As per the evaluations made by the European Food Safety Authority (EFSA), fruits and vegetables, followed by nuts and processed horticultural products, account for the major number of rapid alerts (RASFF) related to pesticide residues in Europe. A comparative statistical data reflect that the majority of these non-compliances are reported for the products originated from various Asian countries, including China, Thailand and India. The main cause for such food safety related non-compliances occurs owing to the lack of good agricultural practices recommendation and irregular applications of pesticides for the management of insect insects and diseases incidence. Mainly in the importing countries, the maximum residue limit (MRLs) is being set at the default level of 0.01 mg/ kg. The treatments receiving crop nutrients through biological options and crop protection using locally available low-cost bio-pesticides yielded comparable to conventional agriculture treatment, Rupela et al. (2005). Thus, the crop at harvest might contain residues above the MRLs, and therefore, become unfit for export. The sale of such non-compliant commodities is also not legally permitted within the country, considering the Food Safety Standards Authority of India (FSSAI) implementation of the new MRLs including default value of 0.01 mg/ kg. An effective residue monitoring program requires the support of scientifically established methods for the sampling of produce and analysis of contaminants. Since the MRLs are mostly set at ppb levels, the methods for their analytical determination should ensure enough sensitivity to detect the contaminants at trace levels with confirmation about identity. The food safety traceability program implemented in case of table grapes is an excellent example which is being replicated to resolve the issues of contaminants in other horticultural produces. A comprehensive residue monitoring of pesticides, therefore, promotes consumer acceptance within and across the countries as well as ensures higher return.

3.12. Organic Horticulture

Development of GAP certified farms need support of state governments and the state agricultural universities / national research centers under ICAR. This work can be facilitated by adopting a cluster development approach by setting up model GAP compliant farms. Thereafter, we need to motivate farmers and educate the consumer to demand GAP certified produce. The farmer must be able to see merit in GAP implementation. This requires the development of packages of practice (POP) for each horticulture or agricultural crop. IPM would be the best way to implement GAP. Implementation of this GAP would increase awareness among the farmers as well as the consumers about the need for consumption of good quality and safe produce. This would facilitate compliance with better standards for our citizens. It would meet the objectives of the government to provide safe food to the consumers (Purohit and Gehlot, 2006). It would also contribute to improvement in the environment and soil fertility. Needless to add, this would also support employment generation. The requirement of organic horticultural products mainly fruits and vegetables are increasing at a quick pace. Such horticultural produce grown through organic means is nutritionally better and free from the harmful and dangerous fertilizers and pesticide. Use of resistant varieties, management of soil, use of vermi-compost, FYM, and bio-fertilizersand management of disease and pests using bio-pesticides are worked out. Thus organic farming is not simply a replacement of bag fertilizers by similar quantity of nutrients applied through compost which is indeed a scarce item. Thus this type of crop production and protection system cannot be a threat to food security of any nation as feared by Chhonkar (2003).

3.13. Post-Harvest Loss Management Strategies

There left overs a large gap between the gross production and net accessibility of fruits and vegetables owed to serious post-harvest losses in case of horticultural produce. It is estimated that India incurs post-harvest fruits and vegetable losses worth over two lakh cores each year largely owing to the absence of modern cold storage facilities and lack of proper food processing units. More emphasis needs to be given to post-harvest management of fruits and vegetables. Besides storage and processing, the cold-chain is seen as a value adding activity as it allows farmers to capture greater value. Harvesting, sorting/ grading, transportation, storage at wholesaler and retailer levels are the main operations and channels where losses were found to be high. To reduce the harvest and post-harvest losses the foremost work needed is to launch a massive drive to sensitize and train the farmers for following good harvesting practices (GHP). Skill development in all stakeholders for harvesting and handling, infrastructures development such as construction of multi-commodity cold storage in rural areas and mandies; CIPHET-evaporatively cooled storage structures at farm and retail levels; gap filling of broken cold chain, and service sector development such as introduction of long distance transport using horti-train of CIPHET-NHB are recommended. Further, establishment of horticulture crop processing training-cum-incubation centres, Horti-processing centre (HPC), Horti-parlour as start-up company will not only reduce post-harvest losses, but will also generate employment and reduce migration of unemployed youth from rural to urban India. Further, nanotechnologies incorporating 'nano matrix' and 'nano-film'with smart packaging of fruitsare underway to scale up the technologies with smart packaging that can be highly exploited to minimize the post harvest losses in fruits, vegetables and flowers.

4. Conclusion

India is either first or second in the world in terms of overall production of many agricultural commodities, but it is far behind in terms of productivity, that is the yield per unit input. Current agricultural practices are neither economically nor environmentally sustainable in India. There is a huge potential for technological innovations and startups as the farmers are facing lots of challenges such as traditional agricultural practices, financial crisis, lack of labourers, scarcity of water, extreme weather events, and unavailability of fair market besides other problems. In present scenario, adoptions of new technologies and skill development have been the major driving force for increasing agricultural productivity and promoting overall development. In this regards, precision farming, mechanization of farm operations using smart machines, devices and tools, use of nanotechnology, storage technologies, cleaning, sorting and processing at the farm level, e-commerce, renewable power based machines for all operations are some important measures to achieve the goal smoothly. The successful achievement of above target required huge technical staff coupled with emerging technologies. Hence there is urgent need to focus skill development programmme technology transformation for the farmer and youth of the country.

Reference

Chhonkar, P.K. 2003. Organic farming: science and belief. Journal of the Indian Society of Soil Science 51(4), 365-377.

De, L.C., Singh, D.R. 2017. Natural resources in North East region of India. International Journal of Agricultural Science and Research. 7(5), 51-66.

Gangaiah, B. 2018.Challenges and opportunity for Indian Agriculture.3rd International Conference on Agriculture and Horticulture. 2168.

Horticulture Statistics Division, Ministry of Agriculture and Farmers Welfare, Govt. of India. (Final Data, Year–2020)

Purohit, S.S., Gehlot, D. 2006. In 'Trends in Organic Farming', Agrobios India, Jodhpur Rajasthan.

Rupela, O.P., Gowda, C.L.L., Wani, S.P., Ranga, Rao, G.V. 2005. Lessons from non-chemical input treatments based on scientific and traditional knowledge in a long-term experiment. Pages 184-196 in the Agricultural Heritage of Asia: Proceedings of the International Conference (YL Nene ed.). 6-8 December 2004, Asian Agri-History Foundation, Secunderabad-500 009, AP, India.

Saikia, D., Bordoloi, R. 2019. Strategic interventions on horticultural sector for doubling farmers income in Assam by 2022-23. Acta Scientific Agriculture 3 (1), 56-58.

Sanjeev, K., Tripathi, V.K. 2020. Doubling Farmers' Income through Horticulture Interventions. Just agriculture 1(1), 271-278.

9

Biofortification: A Modern Approach for Doubling Farmers' Income and Fight Against Malnutrition

***Praveen Kumar Singh Gurjar*[1], *Nikhil Parihar*[1], *Ravi Y.*[2] *and Abhijeet Ghatak*[3]**

[1]*RVS Krishi Vishwa Vidyalaya, Gwalior, Madhya Pradesh, India*
[2]*ICAR-National Research Centre on Seed Spices, Ajmer, Rajasthan, India*
[3]*Bihar Agricultural University, Sabour, Bihar India*

1. Introduction

Total population of India is estimated to cross 1.64 billion in 2050. Increasing population, decreasing economic status of Farmer, improper diet and nutrition, fixed farm land, degradation of cultivable land and malnutrition in Indian population is major key problem in 2 to 3 decades. More than 600000 children below 5 year age die due to improper nutrition each year in developing countries (White and Broadley, 1998). All the data provide a scary image about future. Almost 60% of the malnutrition children are iron deficient. Iron deficiency is very common among the girls; 30% each are deficient in zinc and iodine. The incomplete diet and malnutrition can lead to several disorder and degrade the human health (Shuman, 1998).

Fruits and vegetables are the vital addition to the human diet because they supply crucial minerals, vitamins, and fibre that are necessary for good health. India is the world's second largest producer of fruits and vegetables followed by China. Due to insufficient preservation facilities, a large amount of produce is squandered, as just 1.5 to 2% of total harvest is processed in the country. One of the primary cornerstones of the food sector is the preservation of fruits and vegetables. The food preservation and processing sector has evolved from a luxury to a necessity (Gregorio et al., 2000).

The process of enhancing a crop's nutritional wealth is known as biofortification. It states to agricultural nutrient supplementation to combat the adversarial economic and well-being effects of vitamin and mineral deficits in humans. VAD (Vitamin A deficiency) is a serious public health issue in poor nations, especially among offspring and women of reproductive age (Cavagnaro, 1998) with an estimated 600,000 fatalities per year among children under the age of five. According to data supplied to the World Health Organization (WHO) by the Government of India, 62% of the children in preschool have VAD. Also, deficiencies in iron (Fe), zinc (Zn), and selenium (Se) are critical civic health concerns as well as significant soil limits to agricultural productivity, particularly in poor nations.Table 1 shows biofortified target crops and countries release program.

2. Methods of Biofortification

2.1. Biofortification by Agronomic Methods

Fertilizers are used to boost the micronutrient content of edible components. The suppleness of mineral elements into the soil and in the plant is related to the degree of agronomic biofortification success. Zinc (foliar applications of $ZnSo_4$), Iodine (soil solicitation of iodide or iodate), and Selenium are the most acceptable micronutrients for agronomic bio fortification (as selenate). Application of inorganic Se fertilizers ensued in a more than 10-fold rise in Se meditations. In China and Thailand, the application of inorganic Fe and Zn had an influence on plant enrichment on a national scale. Due to the alteration of Fe+3, which are inaccessible to plant roots, Fe ($FeSO_4$) has a poor mobility in soil. Foliar application of micronutrients (Fe, Zn, Cu, etc.) in plants is a rapid and straightforward form of fertilizer application. Mycorrhizal interactions have been observed to enhance Fe, Se, Zn, and Cu contents in crop plants in several investigations. Micronutrients such as Zn, Cu, and Fe are better absorbed and used by arbuscular mycorrhizal fungi (Hakala et al., 2003).

2.2. Biofortification Through Classical Plant Breeding

Traditional breeding has mostly focused on yield attributes and resistance breeding for the past four decades, with nutritional elements receiving less attention, and existing cultivars have a reduced nutritional status. Minerals including Fe, Zn, Cu, and Mg have dropped in mean dry matter content in a range of plant-based diets. The fortification of key vitamins, antioxidants, and micronutrients has become a focus of recent advances in conventional plant breeding. If enough genetic variability in concentrations of -carotene, other bioactive carotenoids, iron, zinc, and other minerals exists across cultivars, traditional breeding can enhance the micronutrient density of staple foods, allowing for the selection of nutritionally suitable breeding materials (Hakala et al., 2003).

2.3. Biofortification by Use of Genetic Engineering

When there is insufficient distinction among genotypes for the chosen characters/ traits within a species, or when the crop is not appropriate for conventional method of plant breeding (due to absence of sexuality; e.g. banana), genetic engineering is a viable option for aggregate the concentration and bioavailability of micro nutrients in palatable crop tissues. One of the major issues is the so-called 'gene flow' environmental issue, which refers to the risk of foreign genes being transferred to non-target species. Transgenes could be used to re-distribute micronutrients among tissues, improve the efficiency of biochemical activities in edible tissues, or even reconstruct specific pathways. 'Antinutrients' were removed from some of the approaches (Hakala et al., 2003).

3. Application of Biofortification in Horticultural Crops

3.1. Nutri-Rich Banana

Because commercial banana/plantain (Musa) types are sterile triploids(3X), breeding them is difficult. There might be a lot of cross incompatibility among the productive groupings. In addition, the Musa crop cycle is lengthy. Genetically modified (GM) bananas can be grown alongside non-GM bananas in the same field because most edible bananas are vegetative propagated and transgene outflow is small. Furthermore, because the GM bananas are sterile, there will be no heritable mixing of GM and non-GM cultivars in nature, and the current banana diversity in India would be unaffected. Unfortified bananas have 0.4 mg Fe per 100 g, but fortified bananas have 2.6 mg Fe per 100 g. (Yadava et al., 2017)

3.2. Orange Sweet Potato (OSP)

The International Potato Centre (CIP) in South Africa and Uganda (HarvestPlus) and National Agriculture Research and Extension System (NARES) began a study in 2002-2007 to enhance the goal level of 30 ppm of provitamin A in sweet potatoes in the first variety released in 2002. This cultivar has the potential to retain more than 80% of provitamin A after baking or steaming and at least 75% after sunlight or sun drying, as well as being high yielding and drought tolerant. OSP was supplied to over 24,000 homes in Uganda and Mozambique by HarvestPlus and its partners. Biofortified cultivars are already being introduced in several parts of Africa and South America, as well as China. CIP launched the Sweet Potato for Profit and Health Initiative (SPHI) in 2009 (Yadava et al., 2017).

3.3. Nutri-Rich Cowpea

GB Pant University of Agriculture and Technology, Pantnagar, India, has initiated pioneering research on biofortification of cow pea. Pant Lobia-1 (82ppm Fe and 40ppm Zn) and Pant Lobia-2 (100ppm Fe and 37ppm Zn) are two early maturing high iron and zinc fortified kinds released in 2008 and 2010. Pant Lobia-3 (67 ppm Fe and 38 ppm Zn) and Pant Lobia-4 (51 ppm Fe and 36 ppm Zn) were released in 2013 and 2014. Brazil also released three high-iron cowpea varieties manufactured by Embrapa, as well as bio availability, in 2008 and 2009. (Yadava et al., 2017).

3.4. Potato

In 2009, the CIP (International Centre for Potato, Lima, Peru) began a study to generate Fe-rich potato varieties using traditional bio fortification methods, and the varieties would be released in 2017(Yadava et al., 2017).

4. Bio-Fortified Varieties Released by ICAR

- Cauliflower
  - Pusa Beta Kesari 1 (β-carotene rich variety)
- Sweet potato
  - BhuSona (β-carotene rich variety)
  - BhuKrishna (Anthocyanin rich variety)
- Pomegranate
  - SolapurLal (Iron, Zinc & Vitamin-C rich variety)

1. Pusa Beta Kesari 1
   - Country's first biofortified cauliflower variety
   - Pusa Beta Kesari 1 contains high β-carotene content (8.0-10.0 ppm) incomparison to slight β-carotene content in popular varieties
   - Curd yield: 40.0-50.0 t/ha
2. Bhu-Sona (β-carotene rich variety)
   - Bhu-Sona is rich in Beta carotene and contains about (14 mg/100 gm) which is very high as compared to commercial varieties.
   - Average yield is about 19.8 t/ha
   - Suitable for nutritious food processing industry
   - Salinity stress tolerant variety
   - Suitable for sowing in Orissa
3. Bhu-Krishna (Anthocyanin rich variety)
   - Rich in anthocyanin (90 mg/100 g) content in contrast to common varieties with minor anthocyanin content
   - Bhu-sona have tuber yield of 18 t/ ha
   - It is also tolerant to salinity stress
   - It contains starch of about 19.5 %
4. SolapurLal
   - Bio-fortified variety superior to Bhagwa variety famous in Maharashtra
   - Contains enhanced iron & Zinc contents
   - Vitamin C content 19.4 to 19.8 mg/ 100 g
   - Average fruit yield is 23-27 t/ ha
   - Ideal for semi-arid region

Table 1: Biofortified target crops and countries-release schedule

Crop	Nutrient	Targeted country	Leading institutions	First release year
Banana/ Plantain	Provitamin A Carotenoids	Nigeria, Burundi, Ivory Coast, Cameroon, DR Congo	IITA, Bioversity	Unknown
	Provitamin A, Iron Carotenoids	Uganda	Queensland University of Technology, NARO	2019
Bean	Iron(Zinc)	Rawanda, DRCongo	CIAT, RAB, INERA	2012
		Brazil	Embrapa	2008
Cassava	ProvitaminA, Carotenoids	DR Congo, Nigeria, Brazil	CIAT	Unknown
	Provitamin A, Carotenoids and Iron	Kenya, Nigeria	Donald Danforth Plant Science Centre	2017
Cowpea	Zinc, Iron	India	GB Pant University	2008, 2010
	Zinc	Brazil	Embrapa	2008, 2009
Irish potato	Iron	Ethiopia, Rwanda	CIP	Unknown
Pumpkin	Carotenois, Provitamin A	Brazil	Embrapa	2015
Sweet potato	Carotenoids, Provitamin A	Uganda	CIP, NaCCRI	2007
		Mozambique	CIP	2002
		BraziL	Embrapa	2009
		China	Institute of Sweet Potato, CAAS	2010

Source: Susana et al., 2013

5. Conclusion

Biofortified crops developed either by conventional breeding methods or by modern biotechnological tools are not the solutions for malnourishment. The ultimate aim in global nutrition remains a sufficient and diverse diet for the world's population. However, biofortified crops can complement existing micronutrients interventions; can have a significant impact on the lives and health of millions of people, especially those most in need. Now a day, it's an important approach to provide maximum nutrition through biofortified farming produce and flush out the problem of malnutrition in children, adolescent girls and pregnant ladies. So many varieties have been bred through different methods in India and other countries for providing nutritive and quality foods to those people who are highly vulnerable to malnutrition and other nutrition-based health issues. Farmers have a good opportunity for getting good price of their fortified produces in the present scenario and there has a huge scope toproviding quality fortified produce to Indiaas well as other countries of the world. Hence, there are so many malnutrition vulnerable counties are looking towards India for buying these products to minimize their problems, which result in economic upliftment of Indian farmers for having good opportunities of earning ample amount from selling such products which play a vital role in making the doubleincome of the farmer.

References

Cavagnaro, T.R., 2008. The role of arbuscular mycorrhizas in improving plant zinc nutrition under low soil zinc concentrations: Areview. Plant and Soil 304, 315-325.

Directorate of Knowledge Management in Agriculture, ICAR. 1, 16-19

Gregorio, G.B., Senadhira, D., Htut, H., Graham, R.D. 2000. Breeding for trace mineral density in rice. Food Nutr. Bull. 21, 382-386.

Hakala, M., Lapveteläinen, A., Houpalahti, R., Kallio, H., Tahvonen, R. 2003. Effects of varieties and cultivation conditions on the composition of strawberries. J. Food Copost. Anal.16, 67-80.

Shuman, L.M. 1998. Micronutrient fertilizers. J. Crop Prod.1, 165-195.

Susana, M.P., Carvalho, Marta, Vasconcelos, W. 2013.Producing more with less: strategies and novel technologies for plant-based food biofortification. Food Res. Int.54, 961-971.

White, P.J., Broadley, M.R. 2003. Calcium in plants. Ann. Bot. 92, 487-511.

WHO.2009.GlobalprevalenceofvitaminAdeficiencyinpopulationsatrisk1995-2005.WHO Global Database on Vitamin A Deficiency. Geneva.

Yadava, D.K., Choudhury, P.R., Firoz, H., Dinesh, K. 2017. Biofortified varieties: sustainable way to alleviate malnutrition. ICAR, New Delhi.

10

Agronomic Bio-Fortification Through Zinc to Enhance Nutrition in Food Crops

Vajinder Pal Kalra

School of Organic Farming, Punjab Agricultural University, Ludhiana, Punjab India

1. Introduction

Zinc is an important micro-nutrient for human population and its deficiency can cause major diseases, mortality and thus becomes a major challenge in public health care systems of the developing nations. It was estimated that a major part of population is affected with deficiency of zinc (Combs, 2001) which causes lower growth, abnormal or decrease in infant birth, lowers the energy of body, decreases the power to lower the deadly infections, increase the mortality rate in children etc. (Gibson 2012; Krebs et al., 2014). Zinc also catalyzes various enzymes such as RNA polymerase, superoxidase dismutase and helps in working efficiency of cellular proteins. Zinc being an important part of nearly 3000 proteins, which constitutes 10% of the total proteins in human body actively taking part in various neurological and physiological functioning of body (Andreini et al., 2006). According to World health organisation (WHO) more than 8 lakh of people annually affected by the deficiency of zinc, out of which half of the mortality comes from the children under the age of five. Therefore, zinc has becomes the top leading causes of death and diseases in developing nations (Prasad, 2006). The response to deficiency of zinc varies according to the number of factors like amount of zinc present in the body, age etc. However, several problems are directly influenced by the low intake of zinc are diarrhoea, skin diseases, low immunity to infectious diseases, dwarfing in children, low birth outcomes in pregnant and adverse ulcers etc. (Prasad et al., 2014; Hambidge, 1997). Although the exact mechanism of working of zinc is not known yet in several diseases such diarrhoea as well as pregnancy in women, the intake of zinc is highly prescribed. Therefore the huge importance of zinc in human body, scientists across the worlds has developed an interest in studying the different component relating to the intake of zinc. There are numerous strategies being made to increase the intake and bioavailabilty of zinc in human body (Moretti et al., 2013; Udechukwu et al., 2016; Saha et al., 2017). There is also a correlation between the presences of high deficiency of zinc in major population consuming cereals in developing countries as cereals possesess low source of nutrients and high amounts of phytate; therefore lowering the zinc availability to end users (Sharma et al., 2013).

2. Zinc in Soils

The deficiency of zinc in soil causing the loss in yield and as well as lowering the nutritional quality of produces. It is estimated that half of the world's soil is affected by the zinc deficiency. Major cereal growing countries including India also faces a lower amount of zinc concentration in the cereals as it is grown on the zinc deficient soils further lowering the amount in humans. The analysis of soil samples from northern plains of India revealed that more than 45 per cent of soils are deficient in zinc however, nearly half of the total soils of India are deficient in zinc (Singh and Yadav, 2006; Narwal et al., 2010). This lower amount of zinc in soils can lead to lowers the crop yield and in severe cases can cause total crop failure in the field. The crop raised on these types of soils also results in lower nutritive crops (De Valença et al., 2017; Liu et al., 2017). The advances in numerous technologies like mutation or improvement in breeding approaches introduces the several varieties capable of high nutrient uptake, however not successful in fortifying the zinc content in soils having low amount of zinc (Ortiz-Monasterio et al., 2011). Different physio-chemical properties of soil affects the concentration of different micronutrients in the soils and arid as well as high problematic soils induces zinc deficiency in several parts of the world. The high incidence of zinc deficiency in Indian soils might be due to the intensive cropping intensity, lower availability of organic manure, high use of chemical fertilizers and higher use of phosphorus fertilizer in the field etc. There are chances of increasing the zinc deficiency in the fields in future also. It might be due to the more and more non-cropped area is being going to use under the intensive crop cultivation area.

3. Possible Solution: Zinc Biofortification

There are numerous options available to fortify the food we consumed. It may be through industrial fortification, supplementation and biofortification. First, industrial fortification requires a huge amount of infrastructure, high purchasing power of the consumers and easy access to these products from the markets. Secondly, supplementation requires high initial cost, higher access to urban people as compare to non-urban areas and ready to available forms of zinc for easy bio-availability in the human body. Third and the most important form of biofortification is genetic and agronomic biofortification. It involves developing the new varieties with genetic potential to accumulate higher amount of zinc as compared to traditional varieties. This approach is highly costly and requires tremendous amount of initial infrastructure and cost reaches the varieties in the farmer's field. The last and most cost-effective approach is agronomic biofortification of crops. It includes different method of applying the zinc fertilizers to the crop so that the nutritional status of the plant must be enriched with zinc. There are high chances of increasing the nutritional status of plant through genetic approaches in the long run but agronomic approaches are cost effective and successful in fastest route to improve the zinc status in the plant. Cereal grain should contain nearly 40-60 mg of zinc/kg however in present scenario the availability is 10-30 mg of zinc/kg (Cakmak 2008). Among agronomic biofortification techniques, foliar application of fertilizers is considered as the most cost effective and easy approach to enhance the nutrient load in the reproductive part of the plant. In zinc, lower rate of translocation from root to shoot and diffusion are the two most important factors responsible for the inhibiting the zinc concentration in plant cells.

4. Agronomic Biofortification Mechanisms

Zinc can be applied through different methods to crops. It includes soil application, foliar application, seed priming or combining any of two or three combinations. All the methods follow the different mechanisms to enhance the zinc content in the reproductive parts of the plant.

4.1. Soil Application

Soil application of fertilizer is the most common and easiest way to boost the nutritional content of the plant. However, the results of applied fertilizer truly depend upon the different physical and chemical properties of the soil. Zinc is relatively available in acidic soil as compared to other soil types. Its solubility decreases 100 times with a unit increase in pH (Lindsay and Mortvedt, 2018). If liming will be done in acidic soils it will lower the zinc availability to the crops (Prasad et al., 2014). In alkaline soil, zinc availability is also low to the crops. Similarly in calcareous soils, zinc deficiency can be due to the formation of insoluble calcium zinc (Prasad, 2007). Soils physical properties such as lower soil moisture and organic matter also decrease the diffusion process in the rhizosphere thus lowering the zinc availability to the crops (Cakmak, 2008; Cakmak and Kutman, 2018; Rengel, 2015; Cakmak et al., 1996). Therefore, soils with arid and semi-arid conditions are more prone to the lower availability of zinc as compared to irrigated conditions. Zinc availability is also affected by the presence of other nutrients. It has positive interaction with nitrogen (Erenoglu et al., 2011; Kutman et al., 2010) while having negative interaction with phosphorus (Mousavi, 2011; Prasad et al., 2016) iron, zinc and copper (Prasad et al., 2014). These interactions can also be varied according to the soil type and amount of nutrient present in the soil. The soil biological characteristics also help in increasing or decreasing the zinc content in the plants. Plant growth promoting rhizobacteria and Arbuscular mycorrhiza increases the zinc availability in the soil (Vejan et al., 2016; Cakmakçi et al., 2006; Cavagnaro, 2008). The optimum dose of zinc fertilizer and type of crops decides the actual dose of the zinc fertilizers.

4.2. Foliar Application

Foliar application of nutrients has numerous advantages as compared to other methods. In foliar application, low amount of fertilizer is needed as compared to soil applied methods. Zinc applied through foliar tissues will easily penetrate the outer cell wall and enter the phloem tissue (Cakmak et al., 2018) and then can be transferred to vegetative tissue as well to the developing grains. Foliar application has been found to be the most efficient method for improving the nutrient content in different crops like wheat, chickpea etc. Pal et al (2021) reported that foliar application is also the most economical way to boost the nutrient content of the crop. Nutrient application at the later stages of crop growth has higher chances of increasing the nutrient load of grain as compared to the foliar application at the early stages of crop growth (Malesh et al., 2016). It might be due to the presence of more active photosynthates as at the reproductive stage and higher mobilization of these photosynthates to the reproductive parts of the plant which are the final sink organs of the plant. Higher mobility of phloem tissue is also seen at reproductive stage of the plant (Haslett et al., 2001) which helps in retranslation of zinc from vegetative part to reproductive parts of the plant. In chickpea, foliar application of the zinc at pod formation stage has led to the highest accumulation of zinc in sink tissue (Pal et al., 2019). In multi-location experiments in seven different countries reported that application of zinc fertilizer at heading and milking stage resulted in the highest zinc accumulation in wheat grain (Zou et al., 2012). The efficacy of applied fertilizer is also depends upon the timing of fertilizer application (Cakmak, 2008). Foliar applied fertilizer in the evening hours of the day has the highest tissue permeability (Alshaal and El-Ramady, 2017). Foliar application also helps in triggering the defence mechanism for oxidative damage caused by the stress and thus increases the final yield of the crop (Karim et al., 2012; Cakmak, 2000). Foliar application of zinc does not face any type of interaction with other nutrients thus not forming any type of fixation.

4.3. Seed Priming

Seed priming unlike foliar application of fertilizer has been not observed as the best method to increase the zinc content in grain (Zaman et al., 2018) unless applied on the low fertile and resource less conditions in the field. Zinc transportation is mainly done through diffusion process from the soil and then transported through phloem tissues to the upper parts of the plants. Therefore, soil primed seed will not be equally effective in low moisture conditions. Under rainfed conditions, seed priming is not considered as the best method to improve the zinc content of the crop. Seed priming has increased the zinc content in the plant to the tune of 12%, 29% and 19% in wheat, chickpea and maize (Harris et al., 2008).

4.5 Importance of Biofortication in Present Scenario

There is tremendous pressure on world agriculture to produce more food per unit of area to feed the nutritional demands of ever increasing population. The inclusion of high yielding cereal varieties along with application of higher doses of micronutrients has led to depletion of micronutrients like zinc in the soils (Imtiaz et al., 2010). Therefore, our crop production system should be aimed at producing nutritious crops (bio-fortified) along with higher production to fulfil the needs of ever growing population (Jha and Warkentin, 2020). Most of our staple food is low in micronutrients and cannot fulfill the daily demand of body's micronutrients (Prom-u-thai et al., 2020). Different scientific approaches were used to raise the crop plants enriched with Zn for easily adoption by the mass population having higher risk of malnutrition (hidden hunger). To cure these deficiencies in modern health care system, developed as well as developing nations has to spend a huge amount of money and thus faces a deeper economic crunch (Stein, 2014). These deficiencies is mainly due to lower intake rate of micronutrients (Zn etc.) in diet, higher consumption of cereal based food and less use of non-vegetarian products in daily lifestyle. The modern intensive crop production system is not planned to supply nutritious crop plants along with higher production. This might be the reason for ubiquitous presence of hidden hunger in every country of the world (Welch et al., 2013). This approach will help the farmers as nutrient rich seed will fetch higher price in the market as compare to the traditional practices yield. The farmers will be benefitted by lowering the amount spent on micronutrient fertilizers in the coming years by sowing the seeds with zinc content. It will also help in decreasing the micronutrient deficiencies in the human who consume them.

5. Limitations and Benefits of Agronomic Biofortification

The main drawback in biofortification is the extra costs involved in the zinc fertilization. It might not have the best returns until the premium price is not fetched for the produce from the zinc fertilization. But, severity of the zinc in soil zinc application through foliar application has considerable effect on the crop yield and quality. Scientific reports have reported that there is marginal increase in cost of produce as compared to yield benefits achieved through it (Shivay et al., 2008; Joy et al., 2016). Joy et al. (2015) reported that zinc fertilization through foliar sprays is cost efficient strategy to increase the grain zinc in cereals, as the cost is same as achieved through zinc fortification in flour. The major cost involved through zinc fertilization is the cost of application of zinc however it can be avoided through the coinciding the timing of application with the pesticide treatments of the crops (Ram et al., 2016). There are no published reports available which suggests the compatibility issue of pesticides with zinc fertilization. It can also be mixed with other nutrient in same tank mixture to avoid the extra cost involved in spraying treatments (Pal et al., 2021). There should be farmer awareness camp regarding the positive impact of the tank mix application of zinc nutrient with other nutrient as well as to the pesticide application. Zinc helps in increase in yield along with benefitting through activating the stress tolerance mechanisms of the plant. Is also helps in increasing the germination, vigour, tolerance to different diseases (Cakmak, 2008). There is indirect benefit of zinc fertilization are recorded by different scientists across the world. These include higher performance of zinc rich seeds in field as compared to the low zinc content seeds. The seedling with higher zinc content performed better in the field, having better germination and establishment rates thus lowers the seed rate of the field and provide the better economic returns to the farmers (Yilmaz et al., 1998; Braun, 1999). Zinc is reported to affect the several fungal diseases in wheat caused by fusarium, rhizocotina sp. etc. it helps in lowering the disease infestation by increasing the disease strengthens mechanism (Braun, 1999). One of the main health benefits is through competing with cadmium which is heavy metal, very toxic in nature and not needed in human health. The main cereals like rice and wheat are reported to increase the cadmium in dietary intake of humans (Cakmak, 2009). The crops grown of cadmium rich soils lead to have higher uptake of cadmium and thus have high grain cadmium concentration (Harris & Taylor, 2013). Zinc and cadmium having the similar chemistry thus follows the same transporting mechanisms. The zinc and iron regulating transporters such as ZRT, IRT and heavy metal transporters also transport the cadmium from root to the upper portion of the plants (Cun et al., 2014). The genetic modifications that effect the gene coding for transport mechanisms of heavy metal including cadmium uptake also affects the zinc accumulation in grain. Detterbeck et al., (2016) reported that in barley, high zinc accumulation was correlated with high cadmium content in grain. In this context, agronomic biofortification has attained a considerable affect in optimizing the cadmium content in grain. Agronomic interventions with zinc will help in decreasing the uptake, root to shoot translocation, retranslation as well deposition of cadmium in grain tissues (Cakmak et al., 2000; Jiao et al., 2004). However, apart from this zinc also possess the property of heavy metal and its high amount can considerably create the health hazards in the humans. Zinc toxicity is very rare and usually limited to area affected with mining and smelting activities, affected mostly with polluted water, industrial waste and accumulation of sewage sludge. The plant tissue shows toxicity at 300 micrograms per gram of dry weight of plant (Marschner, 2012). Even after attaining high content in plant tissue or soil rhizosphere zinc toxicity is less toxic as compared copper or cadmium in plant tissue (Alloway, 2008).

6. Zinc Doses in Field

Zinc fertilization is based on crop, soil and climate specific. However, it also varies on the age and type of the crop. The higher rates in soil are associated with problematic soils such as alkaline as well calcareous soils higher rates (Alloway, 2008). In Indian context, blanket application of 5 to 25 kg $ZnSO_4.7H_2O$ is recommended in every soil type for treating the soil deficiency of zinc. In the absence of zinc sulphate heptahydrate $ZnSO_4.H_2O$ is recommended to supply nearly 5 kg zinc per hectare in every year or in alternate years as a soil application. In foliar, spray generally 0.5% spray of $ZnSO_4$ is recommended in every soil type. In certain soil types zinc residual can last upto 10 years depends upon the type or crop and intensity of crop cultivation (Singh, 2008). In foliar application of zinc sulphate, it contains 2-5 g of $ZnSO_4 \cdot 7H_2O$ per litre which generally supplies the zinc at the rate of 1 kg per hectare or even less. Therefore it is completely safe for the soil-plant ecosystem as compared to soil application of the zinc fertlization.

7. Limitations in Genetic Biofortification

The other important approach of bio-fortification is using genetic engineering to develop varieties which will have more sink size for zinc accumulation or can uptake more zinc from the soils. The varieties develop through genetic engineering will serve as the long term strategy to alleviate the malnutrition in humans.

8. Process in Genetic Engineering

The genetic approach provides the long term strategy to alleviate the micronutrient deficiencies by providing the new varieties enriched with higher amount of zinc in reproductive or edible parts of the plant (White and Broadley., 2005; Zouet et al., 2012). The initial investment is very high and it takes a very long time to develop a new variety but once developed it does not have any additional cost to bear with the farmers. The main cost is remained with the breeding agencies/scientists who find the possible germplasm. The first and foremost step is finding the germplasm suitable for genetic variation followed by selection of the plants. The next few steps includes the maximum times in this strategy are crossing and back crossing then stabilization of the traits in different environments followed by improved agronomic practices of different environments are tested (Cakmak, 2008). This breeding strategy of enriching the food is created by welch and graham (Welch and Graham., 2004). They stated that while developing the new genotype for enriched nutrient it must maintained its nutrients content and yield across different environments and it should be eaten by the end consumers in the traditional way of eating.

Among the different approaches to enrich the staple food systems particularly rice-wheat; genetic and agronomic bio fortification approaches were commonly used (Cakmak and Kutman., 2018). Even during the green revolution era focus was to develop the high yielding varieties though. But, no efforts were made to enrich the new genotypes to be nutritionally rich (*i.e* bio fortification) along with higher production. Today, there are very few (almost none) Fe and Zn-biofortified varieties released across the globe for use in the general cultivation. While increasing the yield there was decrease in the nutritional quality of the new varieties due to the dilution effect (Davis 2009; Garvin et al., 2006).

9. Conclusion

From this review, it can be concluded that agronomic biofortification approach has huge potential to dwindle the deficiency of micronutrients in humans. The main aim of agronomic biofortification is to enhance the concentration of nutrients in the staple food crops which should be quickly adopted by the large masses without any compromise in dietary pattern. The ease of availability at lower price is another reason for its wider adaptability across the globe. Apart from this, strict policies must be implemented by the government to enhance the dissemination of such technologies to concur the malnutrition problem in the humans in under developing countries.

References

Alloway, B.J. 2008. Zinc in Soils and Crop Nutrition, 2nd edn. International Zinc Association and International Fertilizer Industry Association, Brussels.

Alshaal, T., El-Ramady, H. 2017. Foliar application: From plant nutrition to biofortification. Environment Biodiversity & Soil Security 1, 71-83.

Andreini, C., Banci, L., Rosato, A. 2006. Zinc through the three domains of life. Journal of Proteome Research, 5, 3173-3178.

Braun, H.J. 1999. Prospects of Turkey's wheat industry, breeding and biotechnology. In: *Hububat Sempozyum* (ed. H. Ekiz), pp. 1–74. International Winter Cereal Research Center, Kenya.

Cakmak, I., Kalayci, M., Kaya, Y., Torun, A.A., Aydin, N., Wang, Y. 2010. Biofortification and localization of zinc in wheat grain. Journal of Agricultural & Food Chemistry. 58, 9092-9102.

Cakmak, I., Kutman, U.B. 2018. Agronomic Biofortification of Cereals with Zinc: A Review. European Journal of Soil Science 69,172-180.

Cakmak, I. 2010. Biofortification of cereals with zinc and iron through fertilization strategy. In: Proceedings of the 19th World Congress of Soil Science, Soil Solutions for a Changing World. 4-6. 1–6 August 2010, Brisbane, Australia published on DVD.

Cakmak, I. 2008. Enrichment of cereal grains with zinc: agronomic or genetic biofortification? Plant and Soil.302, 1-7.

Cakmak, I. 2009. Enrichment of fertilizers with zinc: an excellent investment for humanity and crop production in India. Journal of Trace Element in Medicine & Biology, 23, 281–289.

Cakmak, I. 2000.Tansley Review No. 111: Possible roles of zinc in protecting plant cells from damage by reactive oxygen species. New Phytologist. 146, 185-205.

Cakmak, I., Ozkan, H., Braun, H.J., Welch, R.M., Romheld, V. 2000. Zinc and iron concentrations in seeds of wild, primitive and modern wheats. Food & Nutrition Bulletin, 21, 401-403.

Cakmak, I., Kutman, U.B. 2018. Agronomic biofortification of cereals with zinc: A review. European Journal of Soil Science 69, 172-180.

Cakmak, I., Yilmaz, A., Ekiz, H., Torun, B., Erenoglu, B., Braun, H.J.1996. Zinc deficiency as a critical nutritional problem in wheat production in Central Anatolia. Plant and Soil, 180, 165-172.

Cakmakçi, R., Dönmez, F., Aydın, A., ¸Sahin, F. 2006. Growth promotion of plants by plant growth-promoting rhizobacteria under greenhouse and two different field soil conditions. Soil Biology and Biochemistry 38, 1482-1487.

Cavagnaro, T.R. 2008. The role of arbuscular mycorrhizas in improving plant zinc nutrition under low soil zinc concentrations: A review. Plant and Soil, 304, 315-325.

Combs, G.F. 2001. Selenium in global food systems. British Journal of Nutrition 85, 517-547.

Cun, P., Sarrobert, C., Richaud, P., Chevalier, A., Soreau, P., Auroy, P. 2014. Modulation of Zn/Cd P (1B2)–ATPase activities in Arabidopsis impacts differently on Zn and Cd contents in shoots and seeds. Metallomics, 6, 2109-2116.

Darton-Hill, I., Webb, P., Hunt, J.M., Dalmiya, N., Chopra, M., Ball, M.J. 2005. Micronutrient deficiencies and gender: social and economic costs. The American Journal of Clinical Nutrition, 71, 1198S-1205S.

Davis, D.R. 2009. Declining Fruit and Vegetable Nutrient Composition: What Is the Evidence? HortScience. 44, 15-19.

De Valença, A., Bake, A., Brouwer, I., Giller, K. 2017. Agronomic biofortification of crops to fight hidden hunger in sub-Saharan Africa. Glob. Food Security, 12, 8-14.

Detterbeck, A., Pongrac, P., Rensch, S., Reuscher, S., Pecovnik, M., Vavpetic, P. 2016. Spatially resolved analysis of variation in barley (*Hordeum vulgare*) grain micronutrient accumulation. New Phytologist, 211, 1244-1251.

Erenoglu, E.B., Kutman, U.B., Ceylan, Y., Yildiz, B., Cakmak, I. 2011. Improved nitrogen nutrition enhances root uptake, root-to-shoot translocation and remobilization of zinc (65 Zn) in wheat. New Phytologist 189, 438-448.

Garvin, D.F., Welch, R.M., Finley, J.W. 2006. Historical shifts in the seed mineral micronutrient concentration of US hard red winter wheat germplasm. Journal of food science and Agriculture, 86, 2213-2220.

Gibson, R.S. 2012. Zinc deficiency and human health: Etiology, health consequences, and future solutions. Plant Soil 36,291-299.

Hambidge, K.M. 1997. Zinc deficiency in young children. American journal of clinical nutrition 65, 160-161.

Harris, D., Rashid, A., Miraj, G., Arif, M., Yunas, M. 2008. 'On-farm' seed priming with zinc in chickpea and wheat in Pakistan. Plant and Soil, 306, 3-10.

Harris, N.S., Taylor, G.J. 2013. Cadmium uptake and partitioning in durum wheat during grain filling. BMC Plant Biology, 13, 103-119.

Haslett, B.S., Reid, R.J., Rengel, Z. 2001. Zinc Mobility in Wheat: Uptake and Distribution of Zinc Applied to Leaves or Roots. Annals of botany, 87, 379-386.

Imtiaz, M., Rashid. A., Khan. P., Memon, M.Y., Aslam, M. 2010. The role of micronutrients in crop production and human health. Pakistan Journal of Botany 42:2565-2578.

Jha, A.B., Warkentin, T.D. 2020. Biofortification of Pulse Crops: Status and Future Perspectives. Plants 9:1-29.

Jiao, Y., Grant, C.A., Bailey, L.D. 2004. Effects of phosphorus and zinc fertilizer on cadmium uptake and distribution in flax and durum wheat. Journal of the Science of Food & Agriculture, 84, 777–785.

Joy, E.J.M., Ahmad,W., Zia, M.H., Kumssa, D.B., Young, S.D., Ander, E.L. 2016. Valuing increased zinc (Zn) fertilizer-use in Pakistan. Plant and Soil, 411, 139-150.

Karim, R., Zhang, Y.Q., Zhao, R.R., Chen, X.P., Zhang, F.S., Zou, C.Q. 2012. Alleviation of drought stress in winter wheat by late foliar application of zinc, boron, and manganese. Journal of Plant Nutrition and Soil Science. 175, 142-151.

Krebs, N.F., Miller, L.V., Hambridge, K.M. 2014. Zinc deficiency in infants and children: A review of its complex and synergistic interactions. Paediatr Int Child Health 34, 279-288.

Krezel, A., Maret, W. 2016. The biological inorganic chemistry of zinc ions. Archives of Biochemistry & Biophysics. 611, 3-19.

Kutman, U.B., Yildiz, B., Ozturk, L., Cakmak, I. 2010. Biofortification of durum wheat with zinc through soil and foliar applications of nitrogen. Cereal Chemistry. 87, 1-9.

Lindsay,W.L., Mortvedt, J.J. 2018. Inorganic Equilibria Affecting Micronutrients in Soils. Micronutrients Agriculture and Nutrition 4, 89-112.

Liu, J., Yang, M., Li, H., Li, D., Shi, X., Zhang, Y. 2017. Genetic processes of iron and zinc accumulation in edible portion of crops and their agro-biofortification: a review. American Journal of agriculture and Forestry. 5, 65-72.

Malesh, A.A., Mengistu, D.K., Aberra, D.A. 2016. Linking agriculture with health through genetic and agronomic biofortification. Agriculture Sciences. 7, 295-307.

Marschner, P. 2012. Marschner's Mineral Nutrition of Higher Plants, 3rd edn. Academic Press, Elsevier, San Diego, CA.

Moretti, D., Biebinger, R., Bruins, M.J., Hoeft, B., Kraemer, K. 2013. Bioavailability of iron, zinc, folic acid, and vitamin A from fortified maize. Annals of New York Academy of Sciences. 1312, 54-65.

Mousavi, S.R. 2011. Zinc in crop production and interaction with phosphorus. Australian Journal of Basic and Applied Sciences, 5, 1503-1509.

Narwal,R.P., Malik, R.S., Dahiya, R.R. 2010. Addressing variations in a few nutritionally important micro-nutrients in wheat crop. Proceedings of 19th World Congress of Soil Science. Published on DVD.1-6 A Brisbane, Australia.

Ortiz-Monasterio, I., Cardenas, M.E., Cakmak, I. 2015. Zinc biofortification in wheat through foliar fertilization combined with pesticides. *4th International Zinc Symposium: Improving Crop Production and Human Health*, Sao Paulo [WWW document]. URLhttp://www.fertilizer.org/en/doc_library/Events/2015_International_Zn_Symposium_Abstract_Catalogue_Oct_2015.pdf.

Prasad, R., Shivay, Y.S., Kumar, D. 2014. Agronomic biofortification of cereal grains with iron and zinc. Advances of Agronomy, 125, 55-91.

Prasad, R. Crop Nutrition-Principles and Practices; New Vishal Publications: Delhi, India, 2007, pp. 272.

Prasad, R., Shivay, Y.S., Kumar, D. 2014. Agronomic Biofortification of Cereal Grains with Iron and Zinc. Advances in Agronomy. 125, 55-91.

Prasad, R., Shivay, Y.S., Kumar, D. 2016. Interactions of zinc with other nutrients in soils and plants-A Review. Indian Journal of Fertilizer2016, 12, 16-26.

Prom-u-thai,C., Rashid, A., Ram, H., Zou, C., Guilherme, L.R.G., Corguinha, A.P.B., Guo, S., Kaur, C., Naeem, A., Yamuangmorn, S., Ashraf, M.Y., Sohu, V.S., Zhang, Y., Martins, F.A.D., Jumrus, S., Tutus, Y., Yazici, M.A., Cakmak, I. 2020. Simultaneous Biofortification of Rice With Zinc, Iodine, Iron and Selenium Through Foliar Treatment of a Micronutrient Cocktail in Five Countries. Front Plant Sci 11:1-12.

Pal, V., Singh, G., Dhaliwal, S.S. 2019. Agronomic biofortification of chickpea with zinc and iron through application of zinc and urea. Communications in Soil Science and Plant Analysis 50, 1864-1877.

Pal, V., Singh, G., Dhaliwal, S.S. 2021. A New Approach in Agronomic Biofortification for Improving Zinc and Iron Content in Chickpea (*Cicer arietinum* L.) Grain with Simultaneous Foliar Application of Zinc Sulphate, Ferrous Sulphate and Urea. Journal of Soil Science and Plant Nutrition 21,883-896.

Ram, H., Rashid, A., Zhang, W., Duarte, A.P., Phattarakul, N., Simunji, S. 2016. Biofortification of wheat, rice and common bean by applying foliar zinc fertilizer along with pesticides in seven countries. Plant and Soil, 403, 389-401.

Rengel, Z., Batten, G.D., Crowley, D.E. 1999. Agronomic approaches forimproving the micronutrient density in edible portions of field crops. Field Crops Res.60, 27-40.

Rengel, Z. 2015. Availability of Mn, Zn and Fe in the rhizosphere. Journal of Soil Science and Plant Nutrition.15, 397-409.

Roohani, N., Hurrell, R., Kelishadi, R., Schulin, R. 2013. Zinc and its importance for human health: An integrative review. Journal of Research in Medical Sciences. 18, 144-157.

Saha, S., Chakraborty, M., Padhan, D., Saha, B., Murmu, S., Batabyal, K., Seth, A., Hazra, G.C., Mandal, B., Bell, R.W. 2017. Agro-nomic biofortification of zinc in rice: Influence of cultivars and zinc application methods on grain yield and zinc bioavailability. Field Crops Res. 210, 52-60.

Sharma, A., Patni, B., Shankhdhar, D., Shankhdhar, S.C. 2013.Zinc-An indispensable micronutrient. Physiology and Molecular Biology of Plants, 19, 11-20.

Shivay, Y.S., Kumar, D., Prasad, R. 2008. Effect of zinc-enriched urea on productivity, zinc uptake and efficiency of an aromatic rice–wheat cropping system. Nutrient Cycling in Agroecosystems, 81, 229-243.

Singh, M.V. 2008. Micronutrient deficiencies in soils and crops of India. In: Alloway BJ (ed) Micronutrient deficiencies in global crop production. Springer Sciences Business Media, Dordrecht, pp 93-125.

Singh, R., Yadav, D.S. 2006. Effect of rice (*Oryza sativa)* residue and nitrogen on performance of wheat (*Triticum aestivum* L.) under rice-wheat cropping system. Indian Journal of Agronomy. 51, 247-250.

Singh, M.V. 2008. Micronutrient deficiencies in crops and soils in India. In: Micronutrient Deficiencies in Global Crop Production (ed. A. Brown), pp. 93-125. Springer, New York.

Stein, A.J. 2014. Rethinking the measurement of undernutrition in a broader health context: Should we look at possible causes or actual effects? Global Food Security 3, 193-199.

Udechukwu, M.C., Collins, S.A., Udenigwe, C.C. 2016. Prospects of enhancing dietary zinc bioavailability with food-derived zincchelating peptides. Food Funct. 7, 4137-4144.

Vejan, P., Abdullah, R., Khadiran, T., Ismail, S., Nasrulhaq Boyce, A. 2016. Role of plant growth promoting rhizobacteria in agricultural sustainability-A Review. Molecules, 21, 573.

Welch, R.M., Graham, R.D., Cakmak, I. 2013. "Linking agricultural production practices to improving human nutrition and health, expert paper written for ICN2," in Second International Conference on Nutrition Preparatory Technical Meeting, 13-15 November, Rome.

Welch, R.M., Graham, R.D. 2004. Breeding for micronutrients in staple food crops from a human nutrition perspective. Journal of Experimental Botany, 55, 353-364.

White, P.J., Broadley, M. 2005. Biofortifying crops with essential mineral elements. Trends in Plant Sciences. 10, 586-593.

Yilmaz, A., Ekiz, H., Gultekin, I., Torun, B., Barut, H., Karanlik, S. 1998. Effect of seed zinc content on grain yield and zinc concentrations of wheat grown in zinc–deficient calcareous soils. Journal of Plant Nutrition, 21, 2257-2264.

Zaman, Q.U., Aslam, Z., Yaseen, M., Ihsan, M.Z., Khaliq, A., Fahad, S., Bashir, S., amzani, P.M.A., Naeem, M. 2018. Zinc biofortification in rice: Leveraging agriculture to moderate hidden hunger in developing countries. Archieves of Agronomy and Soil Science 64,147-161

Zou, C.Q., Zhang, Y.Q., Rashid, A., Ram, H., Savasli, E., Arisoy, R.Z., Ortiz-Monasterio, I.; Simunji, S.;Wang, Z.H.; Sohu, V.S. 2012. Biofortification of wheat with zinc through zinc fertilization in seven countries. Plant and Soil, 361, 119-130.

11

Approaches of Precision Farming in Horticulture

Pranava Pandey[1], N.K. Meena[2] and Arvind Kumar Verma[2]

[1]V.K.S. College of Agriculture, Buxar, Bihar Agricultural University, Sabour Bhagalpur, Bihar, India
[2]ICAR- National Research Centre on Seed Spices, Ajmer, Rajasthan, India

1. Introduction

Global population is expected to reach 9.7 billion by 2050 (UN report, 2019) and food requirement will betwofold of the current levels to feed the growing population. Asia and the Pacific regionaccounts for> 70% of the global population but has only 30% of the world's arable land (FAO, 2021). An increase in yield of these regions over the years has largely been due to excessive and indiscriminate use of inputs like irrigation, seeds, fertilizers, pesticides, agrochemicals etc. and at considerable expense of its natural resources. Most of the pesticide recommendations are also region specific, which are on a few random observations of pest density. Because of low quality of agrochemicals often due to adulteration and increased pest resistance, indiscriminate application is now common especially in high valued crops. Hefty subsidies on power for pumping water in countries like India have led to misuse of ground water and misuse on agricultural farms leading to water logging and salinity. Complete and unswerving information about land use/cover, soils (extent of waste lands and degraded lands), agricultural crops, water resources (both surface and underground), natural hazards/ calamities like drought and flood and agro meteorology is essential. Seasonal information on crops, acreage, strength and production allows countries to take appropriate steps to address shortages and implement appropriate support and procurement policies.

India is typically a country of small agricultural farms, where approximately 80% of total land holdings in the country are less than 2 ha (5 acres) with 30% irrigated land only. India produces wide variety of agricultural products because of its varied agro-climatic regions but low farm productivity is major concern, which is around 33% of the best agricultural farms world over. Indian farmers can get more remuneration from the same piece of land with fewer inputs, once the productivity gets increased. Agriculture sector alone contributes about 14.5% towards gross domestic product (GDP) of India. Food grain production has experienced a considerable jump from about 51 million tonne during 1950 to 291.95 million tonnes in 2019-20 as per second advance estimates. The production of horticulture products has doubled over the past quarter century and the value of global tradein horticulture crops now exceeds that of cereals. For the first time, horticulture production surpassed food production in India during 2013-14 and continues to excel currently.

With increasing population, urbanization and over exploitation of natural resources, there has to be a paradigm shift in farmers' awareness from production towards productivity and profitability. In current scenario, agriculture is facing foremost challenges due to shrinking land holding, depleting irrigation water and other related resources. There is an urgent need for adopting farmer friendly location-specific production and management strategies in a concerted manner to achieve vertical growth in horticulture production with ensured quality of produce and judicious use of natural resources for better return per unit of area. In this context, precision farming has potential of utilizing resources per unit of time and area efficiently for achieving targeted production in horticultural crops.

Indian farmers are growing their crops in traditional way on large scale. The traditional farming system practices are area specific. There is a general set of crops cultivated throughout an area. Every farmer in that particular area follows a set of particular procedures of sowing, nutritional requirement, intercultural operations, irrigation and harvesting of the crops. This sequence of steps leads to impulsivity, over consumption of natural resources, costly labour and the production of unmanaged waste. Before using the technique in agriculture, farmers had a better chance of getting good produce than tossing a coin and wanting a head. The farmers did not have information about their farm, so there was no way to know the cause of the crop failure. This practice has driven farmers into losses and balance due. Precision agriculture is the introduction of precision practices that use techniques to meet the needs of individual plots and crops. You can use big data analysis software like CropIn or robots like drones to get more information about plots, soil types, proper plants, irrigation, and fertilizer requirement. The information obtained will be used to adjust the very accurate selection of plants, the amount of fertilizer and the need for watering. Precision agriculture helps farmers lead a debt-free life by reducing production costs and losses and minimizing their overall environmental impact.

Therefore, precision farming is a new advanced way for farmers to provide optimized inputs such as water and fertilizers to improve productivity, quality and yield of the crops (Gebbers and Adamchuk, 2010). It requires a vast amount of information on growing plant conditions and plant health with high spatial resolution. Regardless of the data source, precision farming main goal is to allow farmers to monitor their farms. There are many ways to support this, but ultimately it usually reduces the resources required. Hence, we can say the precision agriculture is the science of getting better crop yields and assisting management decisions using advanced technology sensor and analysis tools (Mulla, 2013). Precision agriculture (PA) can be defined as management of spatial (space-related/ geographic) and temporal (time related) variability in fields using information and communications technologies (ICT) (Blackmore et al., 2003). Temporal changes within or between years have been addressed in good agricultural practice (GAP) by means of laboratory analyses of example spots, while spatial patterns of plant growth, which have also been known for a long time have been quantified in large scale with the assistance of PA (McBratney et al., 2005).

2. Concept of Precision Farming

The term precision farming(PF) means a sequence of strategies and equipment that allow farmers to optimise and enhance soil quality and yield putting in place a series of targeted key interventions, a result that can be accomplished thanks to the introduction of increasingly advanced technologies.It is referred to as "precision" because thanks to the state-of-the-art tools used, it is possible to perform the right intervention, in the right place, at the right time, responding to the specific demands of particular crop and particular areas of land with superior levels of precision.Technologies are used, first and foremost, to collect the data and information required to make decisions on how to boost production and secondly to put in place the necessary corrective actions to achieve this goal. Nowadays there is increasing talk about Agriculture 4.0, which is the evolution of the precision farming concept: this term refers to all the tools and strategies that use cutting-edge technologies in an interconnected manner starting with the use of data to get better and optimise production

2.1. Advantages of Precision Farming

Producing more with less resource while maintaining the superior quality standards the ultimately goal of precision farming.

This type of approach generates a two-fold advantage:

- **For farming enterprises**, which can optimise efforts and resources, decrease consumption and waste and boost land productivity. The work also becomes more profitable for farmers and contractors as the processes are managed more rapidly and effectively leading to a fall in hourly costs. Last but not least, they boost upthe performance while reducing human resource fatigue.
- **For the environment**, given that there is a reduction in waste of fertilisers and herbicides, emissions and soil compaction thanks to a more rational use of resources. It is necessary to remember that precision farming is a practice that is becoming increasingly imperativebecause of the universal increase in food demands, following the exponential raise in the population we have

witnessed in recent decades.The increase in the production of food is the main priority without, however, generating an increase in the exploitation of vital resources such as water and fertiliser. It is also crucial to achieve these results without compromising the physical and chemical qualities of the soil: this is the only way to implement a sustainable food production model.

2.2. Main Advantages of Precision Farming

It is sophisticated agricultural practices and selection of crop based on soil suitability

- Increased productivity
- Prevention of soil degradation on cultivated land
- Annulment of volatility and uncertainty
- Proper utilisation of farm waste
- Reduced production costs at farm
- Reduced environmental impact on crop
- Judicious use of fertilizers and pesticides
- Better water management

Precision Agriculture (PA) also known as precision farming, prescription farming, variable rate technology (VRT),site-specific farming (SSF), site-specific management (SSM), site-specific crop management (SSCM), is painstaking as the effervescent farming system of the 21st century, as it symbolizes an enhanced balance between trust on traditional knowledge as well as information and management thorough technologies. Professor Pierre C. Robert, who is considered as the father of precision farming defined precision farming as precision agriculture is not just the injection of new technologies but it is rather an information revolution, made possible by new technologies that result in a higher level, a more precise arm management system. It basically teaches the judicious utilization of natural resources at the right time and the right location within a field. Precision agriculture (PA) ("precision farming", "site-specific farming", "farming by the foot", "spatially variable crop production", "grid farming"), is an innovative conception of agricultural production based on information technologies in crop production.

US House of Representatives, 1997 have given the following two definitions of precision agriculture:

Precision Agriculture (PA) is integrated information and production based farming system that is designed to enhance long term, site specific and whole farm production efficiency, productivity and profitability, while minimizing unintended impacts on wildlife and the environment.

2.3. Site-Specific Crop Management in Precision Farming

It is a form of precision agriculture whereby decisions on resource application and agronomic practices are improved to fit well to the requirements of crop and soil based on heterogeneity of the field. The United States Department of Agriculture (USDA) called this kind of farming "as needed farming" and define it as "a management system that is information and technology based, site specific and uses one or more of different sources of data like soil types, crops, nutrients, pests, moisture or yield for optimum profitability, sustainability and protection of the environment".

In simply term, precision agriculture can be defined as information and communication technology (ICT) based agricultural management system, by placing all crop production practices in the right places, at the right time in the field. Identify, analyze, and manage. The methods implemented for optimal profitability, sustainability and conservation of natural resources. In India, the average land holdings are very small, even with large and progressive farmers. The more suitable definition for precision farming in the context of Indian farming scenario could be- precise application of agricultural inputs based on soil, weather and crop requirement to maximize sustainable productivity, quality and profitability *i.e.* minimum input-maximum output approach.

3. Rationale of Precision Farming

The "Green revolution" of 1960's has made our country self-sufficient in food production attributable to High Yielding Varieties (HYVs), fertilization, irrigation, pesticides and increase in cropping intensity and lastly automation of agriculture. However, Green Revolution Technologies (GRTs) has also led to some challenges, which needs to be addressed for enhancing productivity of crops and simultaneously safeguarding environmental interest.

3.1. Green Revolution Fatigue

India has certainly seen remarkable growth in agriculture, but the productivity of many major crops is still below the expectations. India has not even reached the lowest potential productivity levels of India's high-yielding varieties, but the world's most productive countries have yields well above the potential of India's HYVs.

3.2. Depletion of Natural Resources

The green revolution has been linked with some pessimistic factors leading to ecological inequality. The Indian environment statistics shows that about 182 million ha of the country's total geographical area of 328.7 million ha is affected by land degradation, of which large proportion of 141.33 million ha is due to water erosion, 11.50 million ha due to wind erosion, and 12.63 and 13.24 million ha are due to water logging and chemical deterioration (salinity and loss of nutrients), respectively. On the other hand, India shares 17% of world's population, 1% of gross world product, 4% of world carbon emission, 3.6% of CO_2 emission intensity and 2% of world forest area. Therefore, there is an urgent need to transform this green revolution into an evergreen revolution with an agricultural system approach that helps produce more from available land, water and labor resources without environmental or social imbalances.

Table 1: Difference between traditional farming and precision farming

Traditional farming	Precision farming
Arable site is considered as a homogenous for applying treatment at field level.	Arable site is regarded as different from one point to another and considered as heterogeneous at field level.
Nutrient management is based on representative samples taken from a field.	Nutrient management is based on point like (site specific) samplings taken at specific sites and GPS.
Plant protection applications for diseases and insect-pests are generally made based on average survey done for plant damage.	Plant protection applications are made based on GPS and point (specific) like plant survey.
Selection and planting of plant species and variety is limited to overall arable site.	Plant species and variety are selected as per the suitability to a specific site/location.
Farm operations are adhered to same machinery in the arable site.	Farm operations by machinery are adjusted specifically to the arable site.
Plant stocks are minimally organized into homogeneous blocks at arable sites hence scope for better management is less.	Unified plant stocks are organized into homogeneous blocks at arable sites for better management.
Few data are available influencing decision making at arable site.	A lot of data are available to make conclusive decisions for better growth and development of a crop.

The critical differences between traditional and precision faming clearly define the prospects of precision farming for sustainable development of agriculture/ horticulture in India.

4. Precision Agriculture Goals

4.1. Enhanced Productivity, Profitability and Sustainability

The agriculture fields can give highest profit by optimizing the exact application of inputs such as crop varieties, seeds, fertilizers, herbicides, pesticides, etc., depending on the requirements of the crop affected by the weather and soil characteristics and management of all the inputs in holistic manner

4.2. Most Advantageous Production Efficiency

Identifying potential yield variability can provide an opportunity to optimize production levels within each site or zone by using different approaches under specific field conditions.

4.3. Optimizing Product Superiority

Optimization of product quality by using sensors for the detection of quality attributes of a crop, which helps in taking decision on input application as per the target.

4.4. Increasing Effectiveness of Inputs

Efficient use of fertilizers, seeds and other inputs, depending on the potential of soil that affect the crop yield at a particular location.

4.5. Effective and Efficient Pest Control

To minimize the cost of inputs for crop production is one of the important objectives of precision farming, which can result in getting higher return and better environmental services. Site-specific variable rate application recommends the application of chemicals *i.e.* herbicides, pesticides at the area of problem with targeted approach in comparison to conventional farming methods.

4.6. Conservation of Energy, Water and Soil

A comprehensive approach to precision agriculture begins with crop planning and an assessment of field variability. This includes cultivation practices that preserve or minimize soil destruction. In addition, based on the principle of "increasing the yield per drop", we use water efficiently by methods such as drip irrigation. All of these precision applications use very little energy, which also saves energy.

4.7. Protection of Surface and Ground Water

Efficiently use inputs such as fertilizers and chemicals to protect the environment and prevent leaching and spillage into groundwater.

4.8. Minimizing Environmental Impact

Better management decisions in precision agriculture are made to modify the inputs to meet production needs, which make the inputs to the environment missing or negligible.

4.9. Minimizing Risk

Risk management is a common strategy adopted by most of the farmers today and is expected from an income and environmental perspective. In production systems, farmers often practice risk management by misjudging additional inputs while estimating the unit price of a particular input as "low". Therefore, farmers can apply additional sprays add additional fertilizers, buy additional machines, hire additional labours to produce / harvest / sell products on time and guarantee the return on investment (Singh et al., 2021).

5. Outlook for Precision Agriculture

The effectiveness of precision farming is highly dependent on two factors:

1) Existence and measurement of inconsistency within the production system
2) Producer / farmer ability to identify and apply the best management techniques for sub-areas in each area based on variability assessment. Proper use of the spatial data provided by precision agriculture can realize the following potential benefits:

5.1. Better Profits Through Increased Efficiency

A producer can use agricultural inputs more efficiently as per the variability present for those components in specific sites of a field by applying more or less of them depending upon the specific requirements.

5.2. Reduced Agronomic Inputs

Producers/farmers can reduce their overall use of agricultural inputs like fertilizers, lime, chemicals etc. by making adjustment to a field's fertility level based on assessment, thus the amount of inputs applied to a specific site can specifically be applied.

5.3. Better Record Keeping

Large amounts of data on spatial records of inputs and outputs for agricultural fields generated by precision farming can help to develop more accurate management plans and strategies.

5.4. Improved Production Decisions

Precision farming can effectively be used to make decisions about land use. Profit maps of a farm showing the spatial distribution of a field's profitability can help a producer to make decisions about adoption of cropping systems at the best.

5.5 On-Farm Research

The ability to quantify the spatial performance of crops allows conducting more comparison trials within fields of a location. For instance, it is relatively simple to compare the performance of different varieties on different soil types, when practicing precision farming.

5.6. Reduced Environmental Impact

Production inputs are applied and used more efficiently under precision farming, so very low or minimum fraction of such inputs will leave the field through surface water and ground water. Decreased environmental impact will certainly have impact on the acceptance of farm produce in near future.

5.7. Property Advantages

Many landlords are giving first choice to farmers who can create yield maps and other files of spatial data for the fields under production practice. This spatial history may also enhance the value of cropland.

5.8. More Area Farmed

The records generated under precision farming can allow supervision more cropland effectively than the effectiveness observed in past.

6. Elements/Components of Precision Farming

6.1 Information or Data Base

- Soil: Texture, structure, physical condition, moisture, nutrients, etc.
- Crop: Plant population, plant tissue nutritional analysis status, plant stress, weed patch (weed type and intensity), insect or fungal invasion (type and intensity), Crop yield etc.
- Climate: Temperature, humidity, precipitation, solar radiation, wind speed, etc.

6.2. Technology

Precision farming deals primarily with understanding the variability of field and its management over time and space. Precision farming includes a variety of tools of hardware, software and equipments like Computer system, Global Positioning System (GPS), Differential Global Positioning System (DGPS), Geographic information systems (GIS), Remote sensing, Sensors, Variable Rate Applicator, yield mapping technology/yield monitors etc. for variability assessment and its management in the field.

6.2.1. Computer System

Precision farming requires the acquisition, management, analysis and output of large amount of spatial and temporal data generated for a specific field. Computer software has now become better with time and the knowledge needed for managing variability and decision making on the field can be generated via such software to execute management strategies.

6.2.2. Global Positioning System (GPS)

GPS provides continuous real-time location information, while in motion. Positioning system works with the help of different constellations of satellites and developments of such different positioning systems are the main technological interventions bringing precision farming concept into reality (Singh et al., 2021). GPS receivers can directly be carried to the field or mounted on implements that allow users to return to specific locations to sample or treat such areas and helps to allow mapping of soil and crop measurements at any time. GPS provides an accurate positioning system for field implementation of variable rate technology. It helps to manage application of inputs by equipments and identify the precise location of farm equipment within specific positions of the field for precision application of fertilizers and pesticides to the soil plant characteristics. GPS receiver with electronic yield monitors is generally used to collect yield data across the land in precise way. Global positioning systems (GPS) are widely used for mapping yields (GPS + combine yield monitor), variable rate planting (GPS + variable rate planting system), variable rate lime and fertilizer application (GPS + variable rate controller), field mapping for records and insurance purposes (GPS + mapping software) and parallel swathing (GPS + navigation tool).

6.2.3. Differential Global Positioning System (DGPS)

The Differential Global Positioning System is a GPS accuracy improvement technique that uses pseudo-distance errors measured at known locations to improve the measurements of other GPS receivers within the same geographic area. DGPS uses a series of satellites to locate agricultural machinery within a few inches of the actual location of the site. Knowing the exact location helps to compare the location of the soil sample with the results of the lab on a soil map, a fertilizer that closely matches soil characteristics such as clay and organic content, and soil conditions (relief and drainage). It helps to prescribe pesticides and other inputs. One can use the information generated by this system to make various tillage settings. The system can also monitor and record yield data as just walk through the fields.

6.2.4. Geographic Information System (GIS)

Geographic information systems (GIS) are computer hardware and software with featured attributes and location data to produce maps. GIS includes organized collection of computer hardware, software, geographic data and personal designed to capture, store, update, manipulate, analyze and display all forms of geographically referenced information efficiently. Agricultural GIS stores a variety of information such as yields, soil survey maps, remote sensing data, crop exploration reports, and soil nutrient levels. GIS includes basemaps such as terrain, soil type, nutrient levels, soil moisture, pH, fertility, weeds, pest intensities, etc., integrating all types of information and recommending levels of nutrients or pesticides to apply (Grisso et. al., 2009, Singh et al., 2021). It helps in precision farming in two ways:

- One is to link and integrate GIS data (soil, harvest, weather, field history, etc.) into the simulation model.
- Another is the support of engineering components for the design of attachments and GPS guided machines.

6.2.5. Remote Sensing

Remote sensing is a tool for collecting, processing and analyzing data to obtain information from the surface of the earth without any physical contact with the earth. It shows the potential of precision agriculture and may be able to monitor spatial fluctuations over time with high resolution. Rapid improvement in the accurate monitoring of agricultural growth and its health assessment is important for the wise use of agricultural resources and the management of crop yields (Ryu et. al., 2011). These challenges can be addressed by implementing remote sensing (RS) systems such as hyperspectral imaging to create accurate biophysical index maps across different cycles of plant growth. Remote sensing data captures stress associated with crop health and health concerns related to water, nutrients, soil compaction, crop disease, and other crop health concerns commonly and easily identified in overhead images. Remote sensing can identify and analyze seasonal fluctuations that affect crop yields and help make timely management decisions to improve the profitability of current crops. Remotely sensed data obtained either by aircraft or satellite containing electromagnetic emittance and reflectance data of a crop can provide information useful for soil condition, plant growth, weed infestation *etc*. This type of information is cost effective and can be very useful for site-specific crop management programmes (Singh et al., 2021).

6.2.6. Sensors

Remote sensorsare generally categorized as aerial or satellite sensors that can provide instant maps of field characteristics. An aerial photograph is optical sensors that can show variations in field colour corresponding to the changes in soil type, crop development, field boundaries, roads, water *etc*. Both aerial and satellite imagery can be processed to provide vegetative indices reflecting plant health also. Sensors are used to determine crop stress, soil properties, pest incidence *etc*. These can also be used to measure soil and crop properties as the tractor passes over the field, as a scout goes over the field on foot or as an airplane or satellite photographs of the field from the sky. Yield monitors are the primary sensing system that makes measurements. Commercial sensing systems are available that claim to measure soil properties and make changes in application rate.

Sensors can also be carried by a scout to the field and used to sense the health of plants and soil properties. These use light reflectance on the leaf to determine chlorophyll levels, which directly correlate nitrogen levels in the plant proportionally to the chlorophyll production. Several soil chemistry mini kits sensors are also available to measure soil pH, nitrogen, potassium, phosphorus *etc*. directly in the field.

6.2.7. Variable Rate Applicator

The variable rate applicator has three components eg. computer controller, locator and actuator. A computer system attached to a variable rate applicator uses a prescription and a GPS receiver to control a product delivery system that changes the quantity and type of products according to the prescription *e.g.* combine harvesters with yield monitors. Two methods are generally used for variable rate applications *viz.*, first one is map based on historical data (previous or present year) that help the process control technologies to draw information from the GIS (prescription maps) and adjust fertilizer application, seeding and pesticide selection and application rate, thus providing proper management of the inputs. The second method uses sensors that can adjust the applications rates of equipments by detecting some characteristics of the crop or soil.

6.2.8 Yield Mapping Technology

Yield is ultimate indicator of variations of different agronomic/horticultural parameters in different parts within a field. Therefore, mapping yields and their interpretations and correlating this map with the spatial and temporal variability of various crop parameters will help develop strategies for managing crops in consecutive seasons. Present yield monitors measure the volume or mass flow rate to generate periodical record of quantity of harvested crop for that period and time periodic yield data is then synchronized with location obtained from onboard GPS system to create most common colour coded thematic map. Yield mapping can be carried out easily in mechanized crops, for example, loading cells weigh the crop passing on a conveying belt or an array of sonic beam mounted over the grape discharge chute to estimate the volume and the tonnage of fruit harvested. Radio-frequency identification (RFID) or bar code tags on the bins have also been used for yield mapping in which a weighing machine is combined with a tag reader and a GPS to record the weight and the place of each bin. The data collected are used to produce yield maps of the orchard (Singh et al., 2021).

7. Applications of Precision Farming

7.1. Water Management

Drip irrigation, sprinkler irrigation and fertigation are the important applications in precision farming for water management (Singh et al., 2021).

7.1.1. Drip Irrigation

It is a regulated or slow application of irrigation water through emitters at frequent interval around crop rhizosphere over a period of time. It saves water by 50-70%, reduces labour and energy cost and has almost negligible weeds problem.

7.1.2. Sprinkler Irrigation

Water is sprayed into the air and allowed to fall on the ground surface somewhat resembling to rainfall. The spray is developed by the flow of water pressure through small nozzles.

7.1.3. Fertigation

Application of fertilizers with irrigation water to the crop, which provide nutrients directly to the active root zone, less likely to be lost thus minimizing the wastage of expensive nutrients and ultimately helps in improving productivity and quality of produce. Fertigation through microirrigation system is the most advanced and efficientpractice of fertilizer application. It combines the two main and essential componentsof plant growth development, namely, water and nutrients. The suitable combinationwater and nutrient is the key of high yield and quality (Singh et al., 2021).

7.2. Surface Covered Cultivation

7.2.1. Mulching

It is used to reduce or increase the temperature, suppress weed growth and conserve soil moisture.

7.2.2. Soil Solarization

It is a method of heating the soil by covering it with transparent polythene sheets during hot periods to control soil-borne pathogens.

7.3. Controlled Environmental Condition

It includes greenhouse, polyhouse, grow tunnels, shade net houses *etc*., which are framed structures covered with a transparent or translucent material providing protection and creating congenial environment for growth and development of crops.

7.4. Organic Farming

It uses natural, biodegradable inputs and intentionally avoids the use of synthetic fertilizers and chemicals. This mainly includes the use of vermicompost, fertilizers, organic fertilizers, livestock, green manure, biological management, crop rotation, etc.

7.5. Precise Space Utilization

7.5.1. High Density Planting (HDP)

HDP is a system of planting more number of plants than the optimum or normal through manipulation of tree architecture. Different methods like genetically dwarf scion cultivars, dwarfing rootstock and interstock, training and pruning, use of growth retardants are generally adopted to manage the size of plants to fit well under HDP.

7.5.2. Meadow Orcharding

This system is famous as an ultra-high density planting system and commonly practices in fruits crops. By using this system of fruit planting 20000-100000 plants / ha can be accommodated. To keep the shape of the tree, a strict top cut is made, similar to cutting grass.

7.6. Micro Propagation

This refers to the production of plants from very small plant parts such as tissues, or cells that grow aseptically in test tubes or vessels under controlled condition.

7.7. Integrated Pest Management (IPM)

Pests control strategies including various components such as use of resistant varieties, management of natural enemies, adoption of cultural practices, physical, mechanical, biological and therapeutic methods. By using precision techniques, wise application of pesticides used to minimize the pests below the threshold level.

7.8. Integrated Nutrient Management (INM)

It involves application of nutrients from organic as well as inorganic sources with the aim of improving physio-chemical and biological properties of the soil.

8. Applications of Precision Farming in Horticulture

Its application started over the last twenty-five years, when global positioning systems (GPS) and yield monitors were made available in field crops. Harvesting was mechanized and sensors were placed on harvesting machines to measure the spatial distribution of yield continuously. Applications started in cereals using impact or γ-ray grain flow sensors. When first yield monitors were developed and yield maps were created, it was shown that yield and soil properties varied highly within the field. This fact marked the development of modern PA. However, applications in fruit and vegetables did not start until the 1990's and were published even later. This delay may be attributed, at least in part, to lack of appropriate technology to record yield or quality data of the crop in an automated way.

8.1. Yield Mapping

Examples of Yield monitor for handpicked horticultural crops (Zude-Sasse et al., 2016)

- *Citrus*- Weighing pallet bins using load cells from neighbouring trees on tractor platforms. Estimating yield by tree canopy (ultrasonic sensor, Lidar, multi-spectral camera)
- *Apples / Pears / Olives*- Weighing bins of handpicked fruits of neighbouring trees, geo-referenced using DGPS
- *Palm / Plum / Pear / Cranberry*- Numbering each tree before harvest and measuring the mass of fruits picked manually. Topographic model or local referencing
- *Peaches / Kiwis*- RFID or barcodes on the bins together with a weighing machine, RFID or barcode reader and DGPS.

- *Potatoes*- Load cells under the conveying chains. 2-D vision system above the conveying belt.
- *Pecan / Broccoli*- Load cells and GPS to weigh the volume and position of the platforms transferring the crop in the field on the go.
- *Onions / Watermelons*- Dividing the field into block and weighing the platforms carrying the fruits per block.

8.2. Remote Sensing Applications

Remote sensing is a group of techniques that can collect field data without being in contact to the object (plant or soil) using reflectance or emission of light from plant or soil. Light reflectance (sun or artificial) has been used in PA to calculate vegetation indices. The most frequently used vegetation index is the normalized difference vegetation index (NDVI) that is feasible in low-chlorophyll fruits and canopy imaging. NDVI is, therefore, correlated to vigour of plants and has strong interaction with yield and sometimes quality. The photochemical reflectance index (PRI) is a normalized difference index used as stress index providing an effective indicator of, e.g., photo-inhibition and water stress in plants.

Table 2: Spectral photometric methods available as portable systems for in situ analysis of fruit(Zude-Sasse et al., 2016).

Domain	Measuring principle	Features
Frequency	Hyper- and multispectral spectroscopy in the visible range.	Anthocyanins, carotenoids, chlorophylls
	Near infrared spectroscope	Dry matter, soluble solids content
	Hyper- and multispectral imaging	Same as visible
	Photogrammetry	Size, shape, colour, biospeckle
	Fluorescence	Chlorophyll, phenols
Time	Distribution of time of flight	Anthocyanins, carotenoids, chlorophylls, effective path length
Space	Spatially-resolved hyper- and multispectral Imaging	Wavelength-dependent, same as NIR or Vis, scattering properties

8.3. Variable Rate Applications

VR is an important component for PAs and is becoming more common among large land owners. In VR, a collection of field variable information and other input data helps define the proper set of chemical inputs required for a field.

- In citrus orchards in Florida, tree canopy measured by ultrasonic or laser scanner was correlated to yield (Zaman et al., 2005). This property was used to vary fertilizer application.
- In olive trees, created sitespecific fertilization maps based on leaf nutrient spatial variability (Lopez-Granados et al., 2004).
- In apples, used soil analysis data and nutrient removal from the soil by the crop to prepare prescription maps for fertilizer application (Aggelopoulou et al., 2010).

- VR irrigation is also of importance due to shortage of water reserves and necessity of irrigated crops for food security. Irrigation systems for perennial crops have to be designed from the beginning to achieve VR irrigation.

9. Limitations of Precision Farming Under Indian Conditions

- Small sized farms/ small land holdings.
- Presence of heterogeneous cropping systems.
- Precision farming technology demands high initial capital investment.
- High cost of obtaining and analyzing site specific data.
- Complexity of precision tools and techniques requiring new technical skills.
- Precision farming as a new technology needs to be narrated and demonstrated to farmers with its on-farm and off-farm impacts on yields for better adoption.
- Lack of local technical expertise.
- Uncertainty on returns from investments on new equipments and information management system.
- Culture and attitude including reluctance of farmers towards adoption of new technologies and lack of awareness about environmental problems.
- Knowledge and technological gaps.

References

Anil Kumar Singh.Precision Farming.http://apps.iasri.res.in/ebook/EBADAT/6-Other%20 Useful%20Techniques/14-Precision%20Farming%20Lecture.pdf

Abdullahi, H.S., Mahieddine, F., Sheriff, R.E. 2015. Technology Impact on Agricultural Productivity: A Review of Precision Agriculture using Unmanned Aerial Vehicles. DOI: 10.1007/978-3-319-25479-1_29.

Aggelopoulou, K.D., Wulfsohn, D., Fountas, S., Gemtos, T.A., Nanos, G.D., Blackmore, S. 2010. Spatial variation in yield and quality in a small apple orchard. Precision Agriculture 11: 538-556.

Baidya, A.S. 2017. Precision farming: A new vista for Indian horticulture. RashtriyaKrishi 12(2), 59-63.

Blackmore, S., Godwin, R., Fountas, S. 2003. The analysis of spatial and temporal trends in yield map data over six years.Biosystems Engineering 84,55-466.

Bradley, K.,Khosla, R. 2003. The Role of precision agriculture incropping systems. Journal of Crop Production 9(1-2), 361-381.

Brahma Singh, 2019.Precision Farming and Protected Cultivation. New India PublishingAgency. New Delhi. 270p.

Das, U., Pathak, P., Meena, M.K.,Mallikarjun, N. 2018. Precision Farming a PromisingTechnology in Horticulture: A Review, *International Journalof Pureand Applied Bioscience* 6(1),, 1596-1606.

Gebbers, R., Adamchuk.2010.Precision Agriculture and Food Security. Scienc.327, 828 - 831.

Grisso, R.B., Alley, M., Wysor, W.G., 2009. Gordon Groover. Precision Farming Tools: GPS Navigation. Virginia Tech Publication. 442-501.

FAO Agricultural Outlook 2021-2030 © OECD/FAO 2021

Hakkim, A.V.M., Joseph A.E., Gokul, A.J., Mufeedha, K. 2016. Precision Farming: The future of Indian agriculture. Journal of Applied *Biology and Biotechnology* 4(06): 068-072.

http://www.fao.org/india/fao-in-india/india-at-a-glance/en/.

http://www.drishtiias.com/upsc-exam-gs-resources-Precision-farming1of914052018

http://www.ncpahindia.com/application.php?appl_code=9.

https://aicrp.icar.gov.in/pet/.

https://www.ncpahindia.com/precision-farming.

https://www.prsindia.org/report-summaries/economic-survey-2018-19.

https://www.researchgate.net/publication/283641594_Technology_Impact_on_Agricultural_Productivity_A_Review_of_Precision_Agriculture_Using_Unmanned_Aerial_Vehicles?enrichId=rgreq-cf6f2080dd47fc4bff95e8a375d4d8ac-XXX&enrichSource=Y292ZXJQYWdlOzI4MzY0MTU5NDtBUzoyOTQyNDg5NDg3NDgyOTNAMTQ0NzE2NTgyNTEyMg%3D%3D&el=1_x_2&_esc=publicationCoverPdf

Liaghat, S.,Balasundram, S.K. 2010. A Review: The role of remote sensing in precision agriculture. American Journal of Agricultural and Biological Sciences 5(1), 50-55.

Lopez-Granados, F., Jurado-Exposito, M., Alammo, S., GarciaTorres, L. 2004.Leaf nutrient spatial variability and site-specific fertilization maps within olive (*Oleaeuropaea* L.) orchards. European Journal of Agronomy 2, 209-222.

Mandal, D.,Ghosh, S.K. 2000.Precision farming -The emerging concept of agriculture for today and tomorrow.Current Science 79(12), 1644-1647.

Mandal, S.K.,Maity, A. 2013. Precision Farming for small agricultural farm: Indian scenario. American Journal of Experimental Agriculture 3(1), 200-217.

McBratney, A., Whelan, Ancev, B.T., Bouma,J. 2005. Future directions of precision agriculture. Precis. Agric. 6, 7–23

Mishra, A., Sundaramoorthi, K., Chidambara R.P., Balaji, D. 2003. Operationalization of Precision Farming in India. In: Map India Conference 2003. © GISdevelopment.net.

Mondal, P.,Basu, M. 2009. Adoption of precision agriculture technologies in India and in some developing countries: scope, present status and strategies. Progress in Natural Science 19, 659-666.

Mulla, D.J. 2013. Twenty five years of remote sensing in precision agriculture: Key advances and remaining knowledge gaps. Biosys. Eng. 114, 358–371.

Pandey, P., Pandey, A.K., Kumar, S. 2021. Horticulture-Basic Principles and Recent Approaches, ISBN: 978-93-91383-96-1, Publisher: New India Publishing Agency (NIPA), New Delhi-34,

Patil, S.S.,Bhalerao, SA. 2013. Precision farming: The most scientific and modern approach to sustainable agriculture. International Research Journal of Science and Engineering 1(2), 21-30.

Rains, G.C., Thomas, D.L. 2011. Precision Farming- An Introduction.University of Georgia Research and Extension Personnel. https://www.researchgate.net/publication/277791290

Ryu C, Suguri M, Iida M, Umeda M, Lee C. 2011. Integrating remote sensing and GIS for prediction of rice protein contents. Precision Agric.12, 378-394.

Sahoo, R.N. 2003. Precision farming: Concept and approaches. In: Proceeding of in DOS funded 12th Winter School on Remote Sensing Application in Agriculture with special emphasis on Precision Farming 1-10.

Singh, A., Dhankhar, S.S., Dashiya, K.K. 2015. Protected cultivation of horticultural crops. Doi: 10.13140/RG.2.218172.95364

Singh, R., Choudhary, S., Verma, A.K. 2021. Precision farming in horticultural crops. In Book: "Current Horticulture: Improvement, Production, Plant Health Management and Value-Addition" (Editors; Balraj Singh, A.K. Singh, B.S. Tomar, J.K. Ranjan and Som Dutt). Brillion Publishing, New Delhi– 110005, India. pp. 385-399. ISBN: 978-93-90757-42-8

Srinivasan, A. Relevance of precision farming technologies to sustainable agriculture in Asia and the Pacific (http://www.sristi.org/mtsa/s_7_4.htm)

Stafford, J.V. 2000.Implementing Precision Agriculture in the 21st Century.Journal of *Agricultural Engineering Research.*76, 267-275..

Sudduth, K.A. Engineering Technologies for Precision Farming.https://www.semanticscholar.org/paper/Engineering-Technologies-for-Precision-Farming-Sudduth/c81f739bf0208f20a85bc58446c8119311b7d8cb#citing-papers.

Tran, D.V., Nguyen, N.V. The concept and implementation of precision farming and rice integrated crop management systems for sustainable production in the twenty-first century.https://pdfs.semanticscholar.org/f67d/28db59c04514336f8b16860c7e25dd4e5e5f.pdf?_ga=2.228042840.454061046.1590062449-444517366.1512281757. http://agropedia.iitk.ac.in/content/precision-agriculture-glance.

Wagh, R.M. 2017. Precision farming, remote sensing, geographical information system: A new paradigm for agricultural production in India. International Archive of Applied Sciences and Technology8(4), 04-09. Doi: .10.15515/iaast.0976-4828.8.4.49.

Zaman, Q.U., Schumann, A.W., and Miller, W.M. (2005).Variable rate nitrogen application in Florida citrus based on ultrasonically-sensed tree size. Applied Engineering in Agriculture 21: 331-335.

Zhang, C. and Kovacs, J.M. 2012. The application of small unmanned aerial systems for precision agriculture: a review. Precision Agric: An International Journal on Advances in Precision Agriculture 13, 693-712.

Zude-Sasse, M., Fountas, S., Gemtos, T.A. Abu-Khalaf, N. 2016. Applications of precision agriculture in horticultural crops: a review. Eur. J. Hortic. Sci. 81(2), 78–90.

12

Commercial Vegetable Nurseries Types, Establishment, Operations Management and Economics

Sukhwinder Singh Aulakh[1], Sarvpriya Singh[2] and Navjot Singh Brar[3]

[1]*Krishi Vigyan Kendra, Punjab Agricultural University, Hoshiarpur Punjab, India*
[2]*Krishi Vigyan Kendra, Punjab Agricultural University, Bathinda Punjab, India*
[3]*Vegetable Research Farm, Punjab Agricultural University Khanaura, Hoshiarpur, Punjab, India*

1. Introduction

Nursery raising is as old as agriculture and has many advantages compared to direct seeding in special situations like vegetable cultivation and other high value plants. There are several types of nurseries to cater different crop species like ornamental plants, fruit plants, forest plants and vegetables, although the fundamental concept of all these nurseries is the same. With the advent of drip system, soilless media and plasticulture the nursery system has also become more modern and automated. A nursery is a area or place for growing and looking after young seedlings/plants until they are ready for planting in the main field. A nursery is place, where seedlings, cuttings, layers, composite plantlets etc. are grown under favourable conditions until they are ready for final planting. Vegetable nursery is usually needed for solanaceae (tomato, chilli, capsicum, brinjal etc.), cruciferous (cauliflower, cabbage etc.), cucurbitaceous (bottle gourd, bitter gourd, sponge gourd etc.) and other small seeded vegetabes like onion, lettuce, asparagus etc. The demand for quality vegetable seedlings is always high and command premium. For instance, the cumulative demand for seedlings of major vegetables in India was around 1000 billion seedlings (Resource book on horticulture 2012, NAIP, ICAR). Therefore, nursery establishment and production of quality vegetables seedlings is going to be profitable venture. With required technical knowlegde, guindance and business skills this remunerative activity can be successfully adopted by farmers as a means of full time source of income and livlihood.

In this chapter the fundamental knowlegde required for setting up of small to large nursery is discussed. Discussion regarding selection of growing medium, sterlization techniques, type of protrays, protected structure, and plant protection is attempted. The financial outlay of a medium sized nursery is also presented for the benefit of the farmers who are willing to venture into this activity.

2. Advantages and Disadvantages

2.1. Advantages of Nursery Raising

There are certain advantanges and some disadvantages also of raising nursery instead of direct sowing of seeds as mentioned below.

1. It is convenient to take care of small seedlings in nursery bed.
2. In some vegetables, seeds are of very small size, making it almost impossible to do direct sowing properly.
3. The hybrid vegetable seeds are costly and nursery raising can decrease cost by using less seed rate compared to direct sowing.
4. In direct seeded crop land is occupied for longer duration, whereas if nursery is raised then it can be used to grow some other crop during nursery raising.
5. Nursery being a small area is easy and economical to manage and follow plant protection.
6. Week and diseased seedlings can be discarded during transplanting.
7. Nursery raising generally takes 3-5 weeks which is sufficient to prepare main field for transplanting.
8. Waste utilization – e.g., coconut coir pith.
9. Mechanization is possible, hence reduction in drudgery of farm workers.
10. Proper seed germination, uniform growth, less seedling mortality.

2.2. Disadvantages of Nursery Raising

1. Transplantation shock is common in nursery raised plants but if the plants are grown in protrays then it can be reduced to a great extent.
2. Seedlings are expensive than seeds which increases the cost of production.
3. Similarly, labour is required for transplanting, which further adds up the cost.

3. Types of Nurseries

3.1. Conventional Nursery

a. It consists of elevated beds of well ploughed and pulverized fine tilth soils which are 3m long (or any length) × 1m wide × 0.15-0.20m high (Fig1)

b. In these beds 20 to 25 kg well rotten FYM and 200 g SSP is incorporated followed by treatment with 15-20 g fungicide like Dithane M-45/Bavistan.

c. Beds are treated with formaldehyde (25 ml/litre of water) after which these are covered with air tight polythene for about 72 hours. After removing the polythene the soil is turned for two weeks to remove the formalin fumes. This step will get rid off many soil borne diseases.

d. Before sowing seed is treated with Captan or Bavistin (2 -3 g/kg of seed) to check damping off disease of nursery.

e. Sowing is done in lines which are about 5 cm apart and seed is placed at 0.5-1.0 cm depth (depending upon seed size). The seeds are covered with light friable material like well rotten FYM, sand or cocopeat mix.

f. Nursery beds are mulched with polythene or dry grass. Water is applied to keep beds moist till the seedlings emerge out.

g. After seedling emergence, mulching is removed and beds are drenched with Bavistin (0.1%). When seedlings are 8-10 cm tall, 0.3% urea is sprayed.

h. After about 4-6 weeks the seedlings are ready for transplanting (Fig 1C)

i. Before transplanting hardening is done to reduce transplanting shock by withholding irrigation for 3-4 days. On the transplanting day, water is applied in the nursery bed to remove seedlings easily without causing any damage to roots.

j. Seedlings are transplanted in the main field during cooler periods/evenings for good crop establishment.

Fig. 1: Upper panel - Conventional nursery. Raised nursery beds **(A)** sowing of seeds **(B)** and brinjal plants grown on raised bed **(C)**. Lower panel- Soilless media nursery. Black sheet to act as soil barrier, mulch and temperature regulator **(D)** sowing of seeds in protrays **(E)** and protrays covered with clear polysheet for faster germination of seeds **(F)**

3.2. Modifications in Conventional Nursery

a. **Covering with polythene sheets:** To fasten the germination, transparent polythene sheet of 150 micron thickness can be used. In order to do this, after completion of sowing, beds are irrigated and then covered with polythene. The polythene edges are covered with the soil and it is removed after germination of seeds. The remaining steps are similar to conventional nursery (Fig. 1C-D).

b. **Poly tunnels for normal weather:** In this method, after seed sowing and irrigation is done, the beds are covered with the tunnels. Tunnels of 3 m length (or any desired length) 1.5m width and central height of 1.0m made from G.I. wire are used for this. The semi circular structures are covered with 200 micron UV-stabilized polythene sheet with 75 per cent transmittance. Both the openings can be closed during winters.

c. **Sunken nursery for weather extremes:** In this method, a trench of suitable length, 1.2m wide and 0.5m deep is dug. Inside the trench, a raised bed of about 1.0m wide and 5-10 cm height is prepared. Addition of FYM, fertilizers and soil treatment is as mentioned in conventional method. After sowing the seeds, trench is covered with white, transparent polythene sheet, and is tapered on both sides and made air tight from all sides. After seed emergence (in about 10 days), irrigation may be scheduled, as required till the 4 leaf stage is achieved.

d. **Poly bags for cucurbits:** For raising early crop, seeds can be sown in alkathene bags and germinated under low tunnel/ polyhouse. The seedlings are transplanted when they have 2-true-leaves. Normally, cucurbits do not with-stand transplantation shock beyond this stage due to injury to tap root. It also saves seeds and nearly 50-60% percent seeds are saved in this method compared to direct seeding. This practice is very common in Punjab, for raising muskmelon and ashgourd.

3.3. HI-Tech Nursery

a. **Low cost naturally ventilated polyhouse:** For commercial nursery growing, naturally ventilated polyhouses can be used. In a polyhouse of one kanal (500 m^2) area, about 200,000 seedlings can be raised in one batch and in one year a total of five such nursery batches can be produced.

b. **Net House-partially controlled environment:** In low cost naturally ventilated polyhouse, environmental factors can not be controlled, whereas hitech polyhouse are too expensive. Therefore, nursery grower prefers to have manually or semiautomatic control arrangement. The structure for this type of polyhouse is constructed using galvanized iron (G.I) pipes. The cover is attached with structure with the help of screws. Whole structure is firmly fixed with the ground to withstand the wind pressure. To regulate the temperature and humidity levels exhaust fans with thermostat, evaporative cooling pads, misting instruments etc. are installed. The limitation of this type of nursery is that, they require a lot of attention and care, and it is very difficult to maintain uniform environment throughout the cropping period and therefore these types of nurseries may suitable for dry and composite climatic zones.

c. **Hi-tech Nursery-fully controlled environment:** The nursery establishments where the devices that monitor and control the environment function automatically are classified as hitech nurseries. For year-round nursery operations and for mass multiplication through tissue culture, hi-tech nursery is required. The primary reason for achieving optimum growth in hi-tech nursery is due to high relative humidity, temperature control, adequate day length and light intensity. The optimum humidity level for vegetable production is around 60% to 65% and often polyhouse have higher humidities. Therefore by providing vents around the polyhouse excessive humid air can be expelled.

4. Establishment of Model Nursery

4.1. Selection of Site

When selecting an area for establishment of a nursery, following points should be considered.

1. The land should be fertile, well drained and rich in organic matter. But for setting up of hitech nursery soil type and fertility are not important as seedlings are to be grown in soilless media in germination trays. Infact, non-cultivable land is a good choice for nursery.
2. Areas with intensive fruit or vegetable cultivation should be avoided for nursery establishment to prevent disease and insect infestations (e.g. mites, thrips, whitefly, or blight).
3. The area should be free from the water logging. A gradient of 1-2% is preferable.
4. It should always be away from shade to get desired sunlight.
5. The nursery area should be close to water source and fenced to protect from pets and wild animals.

4.2. Layout of Model Nursery

Many commercial nursery operation starts as small backyard units, without any systemic planning for future expansion. Different features of nursery vary with the type of nursery and specific production requirements. It is a good idea to have a layout for model nursery production for effective utilization of inputs and to do things in efficient and proper way. Layout varies as per production requirement and available land size. Most modern nurseries have dedicated areas where garden tools, soil mix, pots, fertilizers and small implements etc. are also offered for sale along with planting material. Additional structures may consist of propagation structures (for fruit nursery), hardening space (for fruit nursery), store and office etc.

4.3. Establishment of Hi-Tech Nursery

Hi-tech structures are environmentally controlled and have various devices/gadgets to control temperature, humidity, light intensity, day length, and concentration of CO2 etc. There is also variation in cladding material (covering, see Table1) and shade nets (Table 2) depending upon purpose. Additionally, hitech nursery uses nursery tables, soilless growing medium and protray for seedling production. In some cases some or most of production process is also mechanized and automated (Fig2A-C)

Fig. 2: Hitech Nursery. Top panel **(A-C)**, Overview of nursery tables. Middle panel **(D-F)**, seedling of tomato **(D)**, cabbage **(E)** and brinjal **(F)** in protarys. Lower Panel, close p view of capsicum **(G)** and cabbage **(H)** seedling in protarys

Table 1: Type of structure coverings for nursery

Sr.No.	Type of Covering	Use/purpose/advantageous/disadvantageous
1	Transparent polysheet	Economical, 3-5 year life, more light penetration, high humidity build up
2	Translucent polysheet	Economical, 3-5 year life, gives diffused light, high humidity build up
3	Yellow polysheet	Economical, 3-5 year life, less light penetration, high humidity build up
4	Polycarbonate/fiberglass	expensive, long life, light weight
5	Glass	expensive, long life, heavy

Table 2: Type of shade nets

Sr.No.	Type of shade nets	Use/purpose/advantageous/disadvantageous
1	White	Used for vegetable nursery and produces compact seedlings
2	Silver	Used for vegetable seedlings
3	Green	Good for ornamentals. Not suitable for vegetables as it produces lanky seedlings
4	Red	Used for capsicum cultivation, improves fruit color and quality
5	Black	Suitable for ornamental nursery

4.4. Seedling Production Using Pro-Trays

Protrays contains different number and size of cells (Fig 2D-H). Selection of tray cell size (number and volume) to grow seedlings depends on many factors, including crop, seed size, economics, seedling growth rate, and customer demands. Normally tray cell number varies from 72 counts to 800 cells per tray. Cell size can vary by both the diameter (round cells) or size-length (square cells) and depth of the cell. Small size of cell can create problems with moisture retention, aeration and increase in soluble salt. Cell size controls the amount of media used as well as water retention. With smaller cell size, more number of plants can be produced in per unit area and time. Smaller cells also need lesser medium and transport costs are lower. On the other hand, seedlings grown in larger cells are vigorous, have earlier yields and grow rapidly after transplanting.

4.5. Germination and Growing Media

Successful commercial nursery production depends on the chemical and physical properties of the rooting media (Table 3) and its uniformity among batches. An ideal rooting medium is free from weeds and disease spores, well decomposed, light weight, porous, have high cation exchange capacity, well drained but have high water holding capacity. Other parameters to consider include low cost, easy availability, consistency and stability between batches. A growing media must be sterile and may or may not be inert.

Table 3: Physical and chemical properties of various media components

Sr.No.	Component of medium	Function	Properties
1	Sand	Improves aeration and drainage	Particle diameter 0.05 to 2.0mm
2	Coco peat	High water holding capacity (retains 5x water by volume)	Coconut coir fibre, contains potassium, high C:N ratio
3	Perlite	Improves CEC and aeration	Aluminium silicate, light weight
4	Vermiculite	Increases pore size and provide nutrients , Mg (8-12%), K (4-8%) and Iron	Hydrated Mg Iron silicate, light weight, absorb large amount of water
5	Sphagnum moss	Retains moisture	Dehydrate residue of *Sphagnum* sp., acidic pH

4.6. Nursery Media Treatment

Nursery growing media may be harbouring many disease causing organisms, pests like nematodes and insects and weed seeds, so it IS important to sterilize the growing media. Some selective sterilization methods are presented in Table 4.

Table 4: Different types of treatment to sterilize/disinfect growing media

Treatment	Name of chemical and method of application	Advantages
Soil solarization	Deep ploughing in summer	Kills most stages of insects, soil born fungus, open soil for bird herbivory, cheap
Fumigation	Formalin (37%) @20ml/lit. of water, drench with 2lit/ft3, cover with polythene for 2-3 days	Kills most of pathogens if done properly and effectively
Fungicide	Captan, Thiram, Ridomil @ 5-6g/m3	Kills soil borne fungus like *Pythium, Phytophothora* and *Fusarium*
Insecticide	Drenching 6" top soil/media with Chloropyriphos @2-3ml/lit. water	Kills all stages of most insects in soil/media
Herbicide	Gramoxone @5-10 ml/lit. water	Kill most of plant life. Non-selective herbicide
Biopesticide	Neem cake @ 400g/m2	Control nematodes
Biological agents	Trichoderma @ 5-8g/m2	Control various soil fungi

5. Nursery Operations and Management

5.1. Seed Treatment and Priming

Seed treatment is the most common method adopted for healthy nursery production and this is very effective and economic method. In this method seeds can be treated with *Trichoderma viride* 4 g or *Pseudomonas fluorescens* 10 g or Carbendazim 2 g per kg of seeds 24 hours before sowing. Also, just before sowing, treat the seeds with *Azospirillum* @ 40 g / 400 g of seeds. Many vegetable seeds are of irregular shape and size and need to be pelleted for uniform sowing, especailly in mechanized operations. Commonly pelleted vegetable seeds include tomato, brinjal, chilli, onion, lettuce and celery. Sometimes seed coating is done to make seeds more flowable in the seeders. This practice is very common in cucumber, watermelon, squash and melon

5.2. Media filling, sowing and seed germination

The seedling trays (pro-trays) are filled with the growing medium. Mix about 300 kg clean and sterilized cocopeat with 5 kg neem cake along with 1 kg each of Azospirillum and Phosphobacteria. Approximately 1.2 kg of cocopeat is required for filling one protray. Small depressions (0.5 cm) are made with fingertip or some tools in the center of each cell and one seed per cell is sown at appropriate depth and covered with medium. Well hydrated moist cocopeat is used and hence no irrigation is required immediately after sowing. In large scale operations, a seeder assembly can be used for automated tray filling, sowing, covering and watering having capacity of about 100 trays/hour. After sowing transfer the sown trays to a darkened germination room / closed chamber that is kept at a mild temperature to conserve moisture for germination. Many growers use bottom heat or root zone heating to provide warm, even temperature. A black polyethylene is placed on the top of the bench to help spread the heat with skirts on the side to help contain the heat. For small nurseries a corner of nursery can be used for this purpose. Broad spectrum fungicide is sprayed 7-10 days after seed germination. Nutrients are provided by liquid fertilizer mix. Apart from the N, P, K, Ca, Mg and S, micronutrients such as Fe, Mn, Zn, Cu, B, Mo and Cl are also necessary for maintaining the normal health and metabolism of plants. The seedlings would be ready in about 3-6 weeks for transplanting. Before transplanting seedlings are hardened by withholding irrigation and reducing the shade.

5.3. Plant protection

The most common nursery insect pests are white flies, thrips, leaf minors and aphids (Table 5). Sticky traps are very useful in monitoring and controlling these pests. Yellow sticky trap attract white flies, winged aphids and leaf minors. Blue traps are good for attracting thrips. For controlling diseases, general sanitation and use of sterile media and seed treatment are important.

Table 5: Common nursery insect pests and diseases and their management

Insect pest/Disease	Control
Thrips, Whiteflies, leaf minor	Thiomethaxam @0.3g/lit. of water
Damping off, blight	Cymoxanil + Mancozeb@ 2-3 ml/lit. of water
Collar rot	Drenching with Cymoxil 8% + Mancozeb 64%

6. Economics of Vegetable Nursery

We have discussed about various types of vegetable nurseries ranging from small traditional nursery to hitech fully automatic commercial unit.

Table 6: Economics of 500 sq.m polyhouse nursery unit (annual production capacity of 200000 seedlings)

Sr.No.	Particulars	Rate (Rs.)	Quantity	Amount (Rs.)		
				Year I	Year II	Year III
A Fixed Cost						
1	Polyhouse civil work (including polysheet)	1045/sq. m	500m2	522500		
2	Shade Net	20/sq. m	500m2	10000		
3	Irrigation system (drip, fogger etc)	150/sq. m	500m2	75000		
4	Land preparation/leveling	50/sq. m	500m2	25000		
5	Small implements/tools	Lumpsum		10000		
	Total fixed cost (FC)			642500		
B Variable Cost						
1	Growing media (cocopeat, neem cake etc.)	30/kg	2400 kg	72000	72000	72000
2	Protrays	25/tray	2000 trays	50000	50000	50000
3	Planting material (Hybrid seed) Crop 1- Capsicum/ Tomato	100/g	1.2 kg	120000	120000	120000
4	Planting material (Hybrid seed) Crop 2-Cucumber/ brinjal	50/g	1.0 kg	50000	50000	50000
5	Fertilizers and Plant Protection	Lumpsum		12000	12000	12000
6	Electricity	Lumpsum		3000	3000	3000
7	Labour	325/day	40 mandays	13000	13000	13000
8	Transportation	Lumpsum		2000	2000	2000
9	Depreciation on structure and machinary	10%		61750	55575	50017.5
10	Interest on fixed cost	10%		64250	56925	51233
11	Total variable cost (VC)			448000	434500	423251
12	Total cost of project (FC+VC)		1090500	434500	423251	
C Returns			Based on 80% capacity			
1	Sale of seedlings (crop 1)	4	156800	627200	627200	627200
2	Sale of seedlings (crop 2)	3	156800	470400	470400	470400
3	Total Income per year (I)		1097600	1097600	1097600	
D Profit (I- (FC+VC))			7100	663100	674350	

Assumptions

1 Rates are indicative and approximate. Actual figures may vary based on location,quality and other factors

2 Cost of land is not included

3 Cost of borewell and electricity connection are not included

4 Depreciation is calculated assuming useful life of 10 years

5 Useful life of polysheet is assumed to be 4 years.

6 A third and fourth nursery crop can be taken (not included in calculations) during one year period.

7. Conclusion

The advent of different nursery growing media, plasticulture and covering materials and modification of environmental factors has opened the new vistas for growing vegetable crops in any month of the year. Such innovative techniques are facilitating the nursery growers to produce and market quality seedlings and that inturn are helping production of off season vegetables for fetching remunerative prices. Setting up of hitech nursery enterprise is a skilled business activity that requires training, skills, guidance market study and capital. However, with technical and financial support from various government and some private agencies, easier capital access from commercial banks, hardwork and vision, this business model could be a profitable venture for farmers and young enterpreneurs.

References

Bose, T.K., Som, M.G. 1986. Vegetable crops in India. Naya Prokash, Bidhan Sarani, Calcutta. 128p.

Joshi. 2007. Production technology of spices in Himachal Pradesh. Seed Technology and Production centre, UHF, Nauni, Solan. 19-20.

Lawwa, R, Balraj, S. Centre for Protected Cultivation Technology, Indian Agricultural Research Institute, New Delhi, India.

Nandpuri, K.S., Surjan, S. 1986. Vegetable growing in Punjab. PAU, Communication centre, Ludhiana. Pp 141.

13

Ornamental Plants Nursery An Evergreen Enterprise

Simrat Singh, R.K. Dubey and Nepu Rana

Department of Floriculture and Landscaping, Punjab Agricultural University Ludhiana, Punjab, India

1. Introduction

Nursery is a place, where seedlings and saplings of different ornamental plants (trees, shrubs, climbers, annuals, potted plants etc.) are propagated and maintained until they are out-planted in a prepared bed or in pots. Establishment of an ornamental nursery is emerging as a profitable venture, especially around cities. Rapid urbanization has led to the development of community parks and various sites for people to engage in active and passive recreation. This has led to surge in demand for quality planting material, which ensures rapid establishment and beautification of the landscape site. The following chapter deals with the basic principles and practices of establishment of ornamental plant nursery, with reference to its effective management to enhance the income and reap better profit.

2. Scope and Importance of Ornamental Plant Nurseries

Nursery production of ornamentals is no longer restricted to development of large community parks and gardens, but this segment of horticulture industry has entered to cater the demand of ornamentals for beautification of multi-storey apartments, roof tops, commercial complexes, offices, industrial areas, hospitals, hotels, residential backyards, roadsides in cities etc. In India, around 50,000 ornamental plant nurseries situated in suburban areas propagates different ornamental plants (trees, shrubs, climbers, pot plants, succulents etc.). There is an urgent need to plan and establish plant nurseries in villages to extensively beautify the surroundings and to ensure the safety of our ecology and environment (Bhardwaj and Sarolia, 2011).

Raising nursery of native ornamental plants is a remunerative enterprise for farmers, provided they own or lease the required land at appropriate locations near the cities. Nursery business offers an opportunity to get self-employed and also provide employment opportunities to both skilled and un-skilled local youth, with a reasonable capital investment (Dumroese, 2008). Importance of ornamental plants nursery can be well acknowledged as follows:

- Ornamentals plants nursery add diversity in prevalent cropping system.
- Offers highest returns per unit area.
- Better returns during different occasions, festivals and ceremonies.
- Plants can be traded in bulk, maintained and sold later at higher price.
- To mass propagate the tree saplings for avenue plantation.
- To display and sell garden accessories/pots and garden ornaments.
- Nursery enterprise can extend its network to provide consultancy landscape services and maintenance contracts.

3. What is a Nursery?

Nursery is an enterprise of ornamental horticulture sector where seedlings and saplings of plants (including ornamental trees, shrubs, climbers, potted plants, forest tree species, vegetables, medicinal & aromatic plants and planting materials are nourished and raised for planting on site or for sale. Nursery is an important component of ornamental or fruit crop production. The relevance of having a well-planned nursery can be attributed to fulfill the following objectives:

- Production of healthy, vigorous plants devoid of any abiotic and biotic disorders.
- Seedlings of same size and age will ensure uniformity in planting.
- Mortality of plants is minimized, if the plants are well nourished in nursery before and after the transit
- Tender saplings can be hardened in nursery itself before planting and exposing to outside environment.
- Exotic plants can be kept in nursery for acclimatization.
- Plants can be grouped together during harsh weather to create a unique micro-climate.

3.1. Types of Nurseries

The ornamental nursery can be broadly classified as follows:

3.1.1. On the Basis of Size

i) **Backyard nursery**: It comprises a small area in the backyard of a residential garden to propagate and raise new plants for meeting the demands of grower's own garden. The main objective is to engage whole family in gardening and meet nutritional, aesthetic and functional requirements of family. It also gives an opportunity to children to get engaged with nature and fulfill one's creative pursuits.

ii) **Commercial Nursery**: Commercial Nurseries are established in a relatively larger area to propagate the plants for sale to earn profit. Such nurseries hire professionals with technical expertise and are managed so as to utilize the inputs and resources in judicious manner for production of plants. Further, such nurseries can be localized either in rural or urban areas.

a) **Rural nursery**: These nurseries are located in a village on a common piece of land (Panchyat land/ religious places/govt. schools etc.) and may be owned either individually or by group of persons upon a mutual agreement. The sole purpose is to raise sufficient number of plants for mass plantation campaigns or for beautification of village premises.

b) **Urban Nursery**: This type of nursery is located in an urban/sub-urban areas to provide quality planting material for beautification of community parks and also sale exotic plants to city residents. The size of urban nursery is usually small due to costly and limited availability of land.

3.1.2. On the Basis of Business/Turnover

i) **Wholesale Nursery**: Such nurseries involve mass multiplication of plants for sale to retail nursery outlets. Usually, such nurseries specialize in propagation of particular plant species that is in great demand and such plants are maintained as mother stocks to obtain propagating material (seeds/cuttings/suckers etc.). These nurseries are usually located in rural areas and employ several skilled and unskilled rural youth in different cultural operations.

ii) **Retail Nursery**: The retail nurseries purchase plants from wholesale nurseries in bulk and usually maintain and sell the plants at 5-10 times the procurement cost. Such nurseries are located near a town or a city and may also have a separate section for display and selling of garden tools, implements, ornaments, fertilizers, pots, seeds etc.

iii) **Landscape Nursery**: Such nurseries are a property of Municipal Corporation's or city council, with an objective to raise and provide perennial plants (trees and shrubs) for roadside planting and for beautification of community parks. The city residents may also procure plants at govt. approved rates.

iv) **Mail Order or e-Nursery**: Such nurseries, either in wholesale or retail business sell plants and other garden accessories on online portal through well managed and update websites. The delivery of their products could be PAN India (present across nation) or in localized regions. Such nurseries make their presence and advertise themselves on social networking sites to attract the customers and garden amateurs. Customers order their products online, that are delivered either by courier or privately owned messengers.

v) **Agency/trading nursery:** These nurseries sell their stock in bulk at faraway places through agents or sales representatives. Such nurseries are highly specialized and are usually few in number, but have significant annual turnover.

3.2. Components of A Nursery

An ornamental plant nursery is an industry in itself, which employs several skill and unskilled people at different capacities to channelize the work and achieve the prescribed mandate. A reputed nursery propagating and maintaining a stock of quality planting material has several independent units, which operate simultaneously and are managed by skilled workforce. Following are the major components that are required for successful operation of an ornamental plant nursery. Their material, construction design and maintaince issues may vary depending upon the scale of the nursery, extent of horticultural machinery and tools & implements in use, number of workforce employed and most important aspect is the annual turnover, that is likely to sustain the expenditure of maintaince of different components of nursery.

1. **Water reservoir**: A reservoir of suitable dimension is required to hold sufficient water, which can be pumped and delivered to nursery stock. A tank of 12 x 10 x 5ft capacity can hold approx. 17000 liters of water. Water reservoir should be sheltered with shade net or woven nylon net to prevent falling of leaves, which may choke the pipe, decompose and make the water turbid.

2. **Store-house**: A store house is necessary for storing the agri-inputs such as fertilizers, pesticides, garden tools & implements and other nursery stocks such as polybags seeds, bulbs, corms, cuttings etc.

3. **Multipurpose Shed**: An outdoor shed is constructed under the canopy of trees for the purpose repotting of plants with sufficient space for stocking of required soil and organic manure for filling of polybags and pots. The shed ceiling should be fabricated with durable metal anchoured securely with hinges to prevent impact of rain or hot weather. The dimensions of shed may vary depend upon the available space and volume of work done.

4. **Nursery beds**: These are beds made 1.5-2.0 feet deep and filled with sieved sand for propagation of different ornamental plants. The bottom of the beds is preferably kept bare earth or may be laid with bricks. Drainage of excess water should be ensured from the bottom. The bed width is kept preferably 60 cm with length depending upon the availability of space.

5. **Mist Chamber**: Mist chambers are an important component of nursery for propagation of leafy soft-wood and hard to root cuttings, under high humidity conditions. The chamber are installed with foggers and misters that are operated for few seconds to raise the humidity (65-70%). The ideal temperature range of 22-35°C is maintained. The chambers have

propagation benches raised 4 feet from ground to ensure comfortable working and facilitate the placement of pro-trays and small pots.

6. **Shade net-houses**: A shade net house fabricated with 50% green shade net is required to place the potted foliage plants, to protect these from scorching sun-rays and hot winds. A shade net house has the provision of foggers and misters to raise the ambient humidity and create a congenial micro-climate.

3.3. Planning and Layout of Ornamental Plant Nursery

An ideal nursery should be able to provide an ideal 'propagation environment' for the plants and therefore, the entire facility and its different components should be arranged in proper orientation and proportion (Table 1), so as to withstand the climatic conditions round the year. Following considerations may help to plan and layout a nursery:

- The site should preferably be near the link road or on highway to facilitate easy transport of inputs and nursery plants.
- Soil and water analysis of the selected site must be ensured to determine the soil and water pH and EC properties, since most of the ornamental plants are susceptible to higher salinity and alkalinity levels. A soil pH of 6.0 to 7.0 is most suitable for propagating ornamental plants.
- The soil should preferably be well drained with medium to clayey texture, fertile and had good capacity to hold moisture.
- The land terrain should be plain with a gentle slope of 1-1.5% to facilitate effective drainage of rainwater.
- A good barbed wire fence must be erected all around the nursery to prevent the trespass of animals and theft. A thorny bush planted at a closer spacing also serves the purpose.
- The nursery entrance should have a provision of adequate parking space and beautified with well-maintained lawn.
- The display area for saleable ornamentals and sale counter should be attractive and have a sufficient space and seating arrangement for the customers.
- The boundary should be planted with wind breaks (north-west direction) to reduce the impact of wind.
- Old trees should be taken advantage of shade and providing micro-climate.
- A compost yard of suitable dimension with 5 feet depth should be made to decompose fallen leaf debris and other green waste for *in-situ* preparation of organic manure.
- Mother stock area of woody perennials must be kept in a sunny site to yield healthy cuttings.

- Outdoor sheds should be oriented in a way to receive adequate sunlight and ventilation for the workers.
- The sand beds or trenches should be oriented in east-west orientation to gather maximum light intensity, preferably be built under sparse foliage trees.
- The paths should be carefully planned for easy access to different components of nursery. The path should be raised 20 cm from the ground and of suitable width to facilitate movement of workers and customers.
- Irrigation system with drip laterals must be carefully planned to ensure maximum coverage of area.

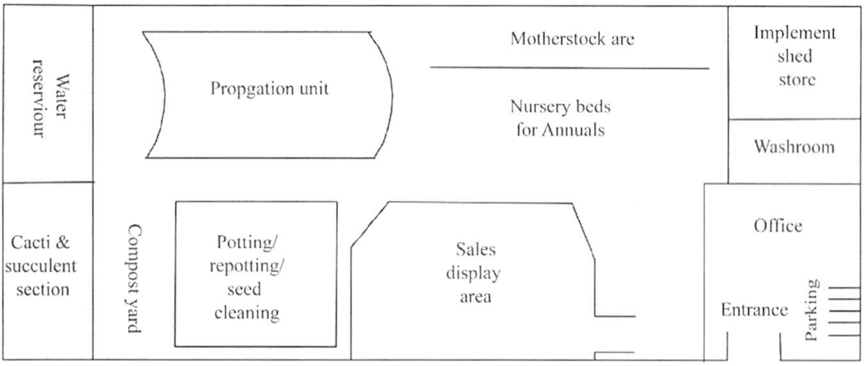

Different components of ornamental plant nursery

Table1: Approx. size allotted to different components of nursery [for 0.5 acres (2000 sq m)]

Component/facility	Purpose	Approx. Size (sq m)	Percent proportion of the total area (%)
Open field for raising annuals/Rose bushes/ Mother stock for perennials	Seed production of annual flowers/to get bud wood and cuttings for propagation	560	28.0
Raised nursery beds	For raising seedlings of annual flowers, and saplings of trees, shrubs and climbers	550	27.5
Beds/trenches	For placement of polybags with planted with rooted cuttings / planting of hardwood cuttings	350	17.5
Area under shade net	Display of potted foliage and flowering plants	200	10.0
Propagation units (Mist chambers/ shade nets/naturally ventilated polyhouse etc.)	For mass multiplication of foliage plants/ bulbous ornamentals	36	2.0

Component/facility	Purpose	Approx. Size (sq m)	Percent proportion of the total area (%)
Shed	Filling of polybags/pots/repotting/ formulating growing media etc.	27	1.5
Store/office and other paved structures	For safe storage of implements/ Office for record keeping of skilled and general purpose workforce/room for customer-owner interaction etc.	27	1.5
Total area		1750	-
Extra allotted area (15%)	For passage/drainage/boundary plantation etc.	260	15
Total area		**2010**	

3.4 Propagation Environment

Raising the tender nursery plants is just like nourishing a child in a comfortable environment that is not limiting for proper growth and development. There are several environmental and edaphic factors that limit the plant growth, provided these factors appear in sub-optimal state for the plants. Limiting factors in the soil include water and mineral nutrients whereas, temperature, light, carbon dioxide, and humidity can be limiting factors in the atmosphere. The best possible congenial environment for propagation of tender nursery plants is provided by erecting suitable protected structures, to safeguard these from the unfavourable environmental conditions.

Whether building a new nursery or modifying an existing one, it is critical to analyze the limiting factors present on the site. It is usually expected that the costs of nursery operations will increase with erection of protected structures, however, it should be well understood that, it is much less expensive to operate an environmentally controlled nursery, than that which is exposed to the natural conditions, in terms of quality of produce and fear of loss of nursery due to harsh weather conditions. Several protected structures offer an enclosed micro-climate that is congenial for the young plants, however, the design specifications, covering or glazing material, framework of structure depends upon regional annual climatic variation, topography and estimated budget. Following protected structures are commonly fabricated in ornamental nurseries.

3.4.1. Shade net-houses

These are low cost and are fabricated either with metal GI pipes or bamboo framework covered with green or white coloured shade nets to intercept the direct sunlight depending upon their density of woven nylon per unit area. Shade net houses have been very successful in tropical and subtropical regions, offering many advantages such as protection from harsh winds during winters and summers, regulating the micro-climate by operating overhead micro sprinklers or misters to reduce the impact of water droplets over tender plants.

3.4.2. Naturally ventilated poly Houses

These are fabricated structures constructed with metal GI pipes with plastic film of 200 micron thickness as roofing material. In northern latitudes an East West orientation is preferred, as sunlight can enter through the sides (with maximum surface exposure) rather than through the ends of the greenhouse. These protected structures should be protected from strong winds, for which windbreaks (seedling Mango and Jamun) are planted on the North and West sides of the greenhouse

3.4.3. Intermittent-Mist chambers

These propagation environments are engineered to provide high levels of humidity for rooting of hard to root cuttings or for off-season production of nursery plants. Rooting of cuttings is much easier in these environments as frequent misting maintains a thin layer of moisture over plant tissue and the rooting media that help the cuttings remain turgid and in active state of growth. Mist systems require delivery of water supplied through PVC pipes, under high pressure.

4. Tools and Implements

Several routine cultural operations are carried in ornamental plant nursery which includes formulation of growing media, filing of polybags, preparation of field for planting of mother stock plants, weeding, sowing of nursery plants, manuring, irrigation and following spray schedules for different plant protection measure against insect-pests and diseases. All these operations require various types of garden tools and implements (such as bill hook, spade, wheel barrow, tree pruner, grass shear, shovel, trovel, hand cultivator, electric chain saw, lawn mower, knapsack sprayer, brush cutter etc.) in a good working condition.

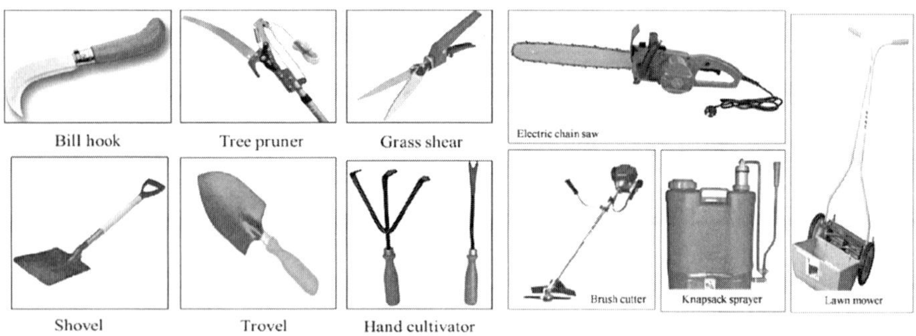

4.1 Precautions to be Taken when using Tools and Equipment

i. Keep all tools and equipment out of children's reach.
ii. Handle them with care and follow the instructions regarding their operation and maintaince.
iii. Ensure periodic oiling of all the tools wherever desired to prevent rusting.
iv. Sterilization of secateurs and blades with 2% household bleach (Sodium hypochlorite) or Isopropyl alcohol (>70%) is essential to check the spread of viral diseases.
v. Spray of insecticides, weedicides and fungicides should be done following safety measures such as wearing the mask and gloves to avoid the drift of the spray.

4.2. Care and Maintenance of Garden Tools and Equipments

i. All garden tools and equipments should be cleaned before and after use to avoid rusting.
ii. Machinery, garden tools and other equipments should be safely placed in a room with a paved floor, windows for receiving adequate bright light and ventilation.
iii. The sprayers and the nozzles should be properly flushed with clean water before and after use.
iv. Periodic monitoring and repair of garden tools and equipments must be ensured.
v. Power operated equipments and garden tools should be greased and oiled to prevent rusting and ensure their smooth operation.
vi. Sharpen the blades of harrow, cultivators, hoes and cutters regularly.
vii. Always ensure cleaning and grinding of cutting blades to check rusting during storage.

5. Growing Media

Successful propagation of ornamental plants in a nursery is largely dependent on the prevailing climate, quality of water and physico-chemical properties of the rooting or the growing media. The rooting medium should be free from weed seeds and disease inoculum. The media, either used individually or in formulation should be properly sieved, free from any physical impurities, light weight, porous enough to facilitate drainage of excess water and also acquire high CEC. Other considerations such as local availability, cost, consistency between batches and stability of the substrates properties over time are also crucial factors for success in ornamental nursery business (Bunt, 1988). Following table (Table 2) highlights the growing and rooting media used in nursery production with their respective utility and associated properties:

Table 2: General media for raising saplings

S.N.	Media	Use/Role	Comments
1	Sand	Improve drainage	Basic component of rooting substrate with particle size ranging from 0.05 to 2.0 mm dia
2	Coco peat	Enhanced moisture retention	Eco-friendly biodegradable, soilless, sterile growing media, rich in potash but poor in nitrogen.
3	Perlite	For adequate aeration	White aluminum silicate mineral of volcanic origin, very light in weight, requires a spray of fine mist of water before mixing to avoid suspension of dust in air
4	Vermiculite	Enhance porosity and availability of trace elements Mg (9-12%), K (5-8%) and iron etc.	Hydrated magnesium iron silicate is light in weight and absorbs sufficient moisture that keeps the propagating media sufficiently moist
5	Sphagnum moss	For enhancing the water holding capacity	Dried residue of *Sphagnum papillosum, S. capillaceum and S. palustre* can be effectively utilized in propagation of ornamentals through air layering

5.1. Nursery Media Treatment

It is essential to treat and disinfect the nursery growing medium, which may contain harmful disease causing organisms, nematodes, insects and weed seeds that may affect the growth and development of nursery plants. Following table mentions the source, method of application, and recommended practices to be followed to effectively disinfect the nursery media and make it suitable for propagating the nursery saplings/seeds.

Table 3: Nursery growing media treatment

Agent	Method	Recommendation	Reference
Soil solarization	Summer deep ploughing, cover beds with 200 gauge white polythene	Kill insect-pests, soil born fungus due to expose in open sun light and heat	Agropedia.iitk.ac.in/ content/nursery management-vegetable
Formalin	20 ml/l of water (37%)	Apply 2litre/cu.ft. cover for 14 to 36 hr and aerated for at least 14 days	Landis et al., 1994
Captan, Thiram, Ridomil-72WP	5-6 g/m^2	Kill Pythium, Fusarium, Rhizoctonia and Phytophthora like soil born fungus	Central Insecticide Board and Registration Committee (http:// cibrc. nic.in/)
Chloropyriphos	2ml/l at 15-20cm depth of soil	Kill insect's egg, larval stage	http://npic@ace. orst.edu/ factshees/ chlorpgen.html

Agent	Method	Recommendation	Reference
Glyphosate	90ml/15l	Systemic non selective herbicide	Rana&Rana, 2015
Quizalfop ethyl 5% EC	30ml/15l	Against all grassy weeds	
Tricoderma	5-8g/m^2	Against various soil fungus	Riad et al., 2010
Neem cake	400g/m^2	Against nematodes as repelient	Mbora et al., 2008

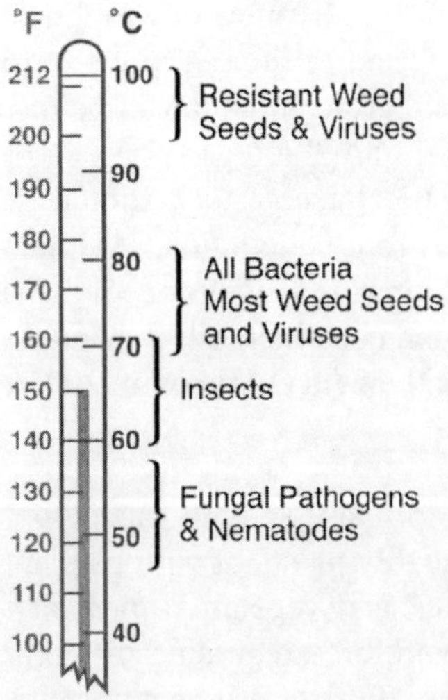

Required temperatures for heat pasteurization need to be held for 30 minutes in the target range to be effective on that pest (Wilkinson et al., 2014)

6. Selection of Containers

Ornamental nursery has diverse varieties of plant species varying in size, growth habit, which require selection of appropriate containers to anchor the plants besides being light in weight to facilitate manual handling and transport from one place to another. The selection of containers can be made depending upon the growth habit of plants, on root system morphology, target plant criteria (seedling/ sapling), intended display, aesthetics, durability of the material, availability and cost economics. The dimensions of container affects root development, determines the total volume of pot, placement of number container over benches or as per the layout (Landis, 1995). In general, a nursery grower should follow these guidelines regarding the selection of containers-

- Seed of annuals can be raised in pro trays with cavities (72 / 98 / 104 / 221) to save the expensive seed and ensuring space for the seedling to develop to 2 pair leaves for transplanting.

- Shallow rooted plants with fibrous root systems (annuals/herbs/succulents/ foliage plants etc.) need to be planted in shorter containers/small polybags (6 x 4 inches).
- Tap-rooted plants such as woody perennials grow better in taller containers or bigger polybags of size 8 x 3 or 9 x 7 inches.
- Plants with multiple, thick and fleshy roots should preferably be raised in wide containers such as bonsai shallow pots or air prune pots for development of fibrous root system.

Often, one can visualize ground penetration of roots, especially of woody saplings, if these are made to rest at one location for longer period of time. Any effort to pull out such plants disturb their root system and often results in wilting. To avoid such a situation, old nursery stock must be replanted into 2" bigger size bag or container, after necessary pruning of the old taproots. Nursery stock should be rotated periodically to new place to avoid in-ground root penetration. It is also advisable to selectively prune the roots while transplanting the saplings in polybags (Landis et al., 1999) or raise the seedlings in air prune containers till they attain required age and appropriate size for transplanting in the field (Whitcomb, 2003).

7. Propagation of Ornamental Plants

Rapid and mass multiplication of ornamentals is the primary activity carried out in a nursery with an objective to raise disease free quality plants of various plant species. Ornamental plants are propagated by sexual and vegetative means. It is essential for the owner to know and guide the workforce regarding different methods and techniques employed in the propagation of diverse ornamental plants. The different methods of propagation have been broadly classified as follows:

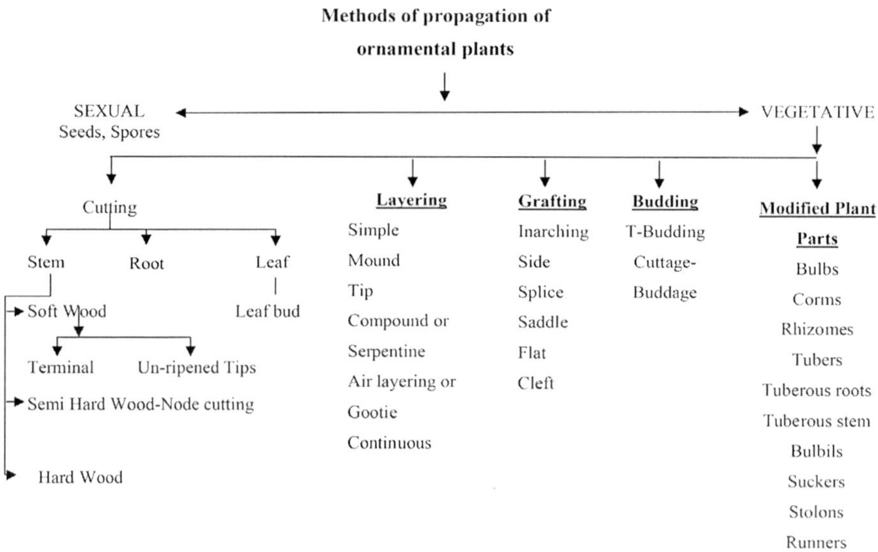

8. Nursery Management

Production of ornamental plants in nursery is a continuous process interlinked with various other cultural operations, which are indispensible for production of quality planting material. Nursery plants being very tender and succulent are susceptible to disease and insect infestation during their early stages of growth. Delay in timely control measures can cause total devastation of progeny and may impair the quality of survived plants. Presence of undesirable vegetation (weeds) in nursery also host array of disease causing pathogen, besides completing with the tender saplings for nutrients, space and moisture. It is therefore essential to under an integrated approach to check the spread of disease, insects and weed infestation for production of disease free and quality planting material. The following strategies will help nursery grower to raise good quality plants with judicious use of resources with minimum investment.

8.1. Integrated Approach for Diseases & Insect-Pest Management

The integrated approach to check the spread of disease and insect-pests on tender nursery plants involves following of cultural, chemical and mechanical measures at right time, using appropriate dose of chemicals and spray equipment (Gunjal Surya, 2012). The cultural methods include deep ploughing of nursery beds for effective soil solarization during May-June. Growing media for filling of polybags must also be sterilized to prevent soil borne pathogens that can rot the tender seedlings. Following good agricultural practices (GAPs) involving clean cultivation and removal of undesirable crop residues also helps to check the harbouring of pathogens (Jaenicke, 1999; Mbora et al., 2008). Periodic change of nursery growing media every 2-3 years, maintaining appropriate plant density, use of trap crops, installation of adhesive lures for trapping aphids, jassids and thrips and avoiding excessive irrigation and nitrogenous fertilization will help to check the spread of disease causing pathogens. Following integrated approaches are very essential to successfully operate an ornamental nursery for production of quality planting material.

- Selection of healthy seeds/ propagules for seedling production
- Seed treatment with 0.2% Carbendazim/Benomyl/ Thiram, prior to seed storage of annuals.
- Soil and growing media sterilization by deep ploughing and sun exposure during hot summer months.
- Sterilization of budding knife, secateurs and scissors with 70 % isopropyl alcohol/hydrogen
- peroxide/ chlorine bleach.
- Transplanting of seedling after root dip treatment in 0.02% Carbendazim solution prior to transplanting.

- Ensuring proper sunlight for development of vigorous and disease free planting material
- Frequent examination of nursery stock and culling of disease infected or weak seedling
- Foliar spray of 0.2% Carbendazim/Dithane M-45 at monthly intervals or as per need

Table 4: Management of insect, pest and diseases in nursery (Sarolia et al., 2018)

S.N.	Insects /Disease causing pathogens	Management
Insect-Pests		
1	White grub, caterpillars, worms	Methyl Parathion -50EC, Phosalone 35EC, Quinolphos 25EC -1.5% powder @ 25 kg ha^{-1} and Phorate 10G 10kg ha^{-1}
2	Aphid, Jassid, White fly, Thrips & other sucking insects	Systemic insecticides Dimethoate, Metasystox, Acephate, Profenofos 50EC @ 1-1.5ml $litre^{-1}$ or Imidacloprid 17.85L @ 1 ml per 3 litre water
3	Leaf minor, leaf rollers	Sanitation, Neem cake or NSKE @ 5% and insecticide Imidacloprid 17.85L @ 1 ml per 3 litre water
4	Butter fly (lemon, pomegranate)	Nuvacron or Rogar 1-1.5ml $litre^{-1}$
5	Inderbela or bark eating caterpillar	Clean the site and apply Dichlorvos 100EC in kerosene and pour in hole then cover it
6	Termite	Regular irrigate & apply Chloropyriphos-25 EC, Fipronil 5%SC @1-1.5ml $litre^{-1}$
7	Nematode	Soil treatment prior to seedling raising with Carbofuran@ 5-7 g/sqm or neem cake 400g/sqm. Grow trap crop (Marigold)
8	Mite	Propargite 75% EC, Ethione, Abamectin 1ml/l
Diseases		
9	Damping off (*Pythium* and *Phytopthora* spp.)	Soil treatment with Formalin 2ml/l and seed treatment with Captan-50WP or Thiram @2-3 kg^{-1} seed
10	Wither tip/dieback (*Colletotricum* spp.)	Cut infected portion and apply Bordeaux paste (1.0%) or Blitox @ 0.3-0.4%. Ensure proper drainage
11	Malformation (Complex syndrom)	Use healthy scion apply NAA@200 ppm
12	Wilt (*Fusarium* or *Rhizoctonia* spp.)	Proper drainage, crop rotation and improve soil condition through amendments (lime) and Carbandazim-50WP apply as foliar (0.1%) and drenching (0.2%) method
13	Powdery mildew (*Oidium spp.*or *Uncinula spp.* or *Erysiphe* spp.)	Sulfex or Dinocap 80% EC (Kerathane) spray @0.1%
14	Downey mildew (*Peronospora spp.*)	Bordeaux mixture (0.8%) &Dithane Z-78 @0.1%
15	Die-back, leaf blight	Dithane M-45, Copper oxy chloride @2g/l
16	Anthracnose	Tebucanzole 26% (0.05%)
17	Blight (*Alternaria* spp.)	Difolatan and Mencozeb @ 0.2%
18	Virus	Control vectors and use disease free scion

8.2. Weed Management

Undesirable plants growing in and around nursery seed beds, mother block area, alongside water channels and other open spaces of nursery provide a habitat for the disease causing pathogens that perpetuate in form of spores or fungal mycelia. Therefore, such undesirable plants must be roughed out, preferably manually as spraying of herbicides may cause drift on nursery plants and may affect their leaf appearance. Mulching with 25 micron polysheet or with locally available straw residue/stones/chips etc. will suppress the weed growth and make the nursery colourful and weed free.

8.3. Record Management

For management of employed human resource, sowing time, stock sales, wages, repairs, loan repayment and financial statement of all the inputs, it is essential to maintain a record in writing to carry out the annual feasibility analysis of operating a nursery. It is essential to maintain the records in terms of stock inventory of all purchased inputs (seeds, planting material, agro-chemicals, shade nets, polybags, media *etc.*). Observations of field data (dates of sowing, germination, germination percentage, growth and floral parameters, pest and disease outbreaks *etc.*) must be maintained separately on a register. Muster rolls for labor attendance, engagement in various duties, wages and movement register should be maintained up to date. Various records of expenditure and income are recorded in special log books such as sale-purchase book, ledger, cash book, dispatch register (Ratha Krishnan et al., 2014).

8.4. Marketing

It is essential to follow certain strategies prior to effective marketing and sales of nursery plants for getting good returns per unit of the investment made. A haphazardly managed nursery will likely confuse the customer and distract its attention. To ensure pleasant environment for the customers, it is essential to beautify the display area and arrange the plants properly, preferably in a vertical tier system, so that all the plants are visible and customers may handle these efficiently. Safe and hurdle free access to different section of display area, provision of luggage trolley will likely encourage the customer to spend time and purchase more number of items. Nurserymen should account following points to make an attractive display of the products for the customers.

- The entrance to nursery should preferably be beautified with placement of seasonal flowering annuals displayed on pot stands to attract the customers.
- Only the best quality graded nursery plants be displayed with tags citing the name of plant in vernacular language.
- Plants to be displayed must be of uniform size and age.
- Each of the products in retail sale must be visible to the customers and none of the products get visibly concealed.

- Plants in polybags should preferably be watered during morning or evening hours to avoid disturbance to root system due to repeated handling and touching by the customers.
- A colourful attractive catalogue or a pamphlet containing the detailed information about what is available at nursery should be made available to customers.
- A small model of a residential landscape at a nursery will likely attract the customers and provides them a choice for selecting suitable planting material for different landscape use.

8.5. Disposal of Nursery Plants and Planting Materials

Year round ornamental nursery operations likely generate lot of green biomass as a result of pruning of hedges, shrubs and trees, mowing of grass, annual leaf fall of deciduous trees etc. This green waste must be collected in a heap and shade dried for few days to release excess moisture and reduce the bulk volume. Shredding the green waste (using chopper) in to small pieces will increase its surface area for decomposition and also reduce the bulk volume. Thereafter, the green waste should be effectively decomposed in pits (dug 5-6 feet deep) spread in layers sandwiched between well decomposed organic manures. Depending upon the weather, it may take around 6-9 months for effective decomposition and turning in to a nutrient rich organic manure, free of cost.

9. Recent Trends in Ornamental Plant Nurseries

The trend of buying plants online is certainly picking up in India, In the last five years, nearly 50 big and small online plant nurseries have come up in the country. These e-nurseries have qualified professionals in different capacities to provide services to their clients. The services include preparation and execution of landscape design for residential, industrial, commercial plazas, hotels & hospitality sector, railways, National highways authority of India (NHAI) and several other govt. and public sectors. In addition, several of these e-nurseries also offer maintaince contracts with the clients for year round beautification of their premises.

Besides, these e-nurseries provide employment opportunities to several local people in different capacities at their regional headquarters and sub-centres, and also offer several opportunities for university graduates to work as paid intern trainees for few months to one year, or as per the contract. as inter, Needs. Several maintenance services like periodic sweeping, irrigation, pruning, fertilizer application, mowing and decomposing of garden waste are offered to the clients as per their contract. Some of the online retail platforms present across Nation (PAN- India) and at select locations selling live plants and garden accessories are as follows. The trend of buying plants online is certainly picking up in India, Most advertise only through Facebook and Instagram and a few, like Nurserylive, Ugaoo, myBageecha, gaamla.com and Nurturing Green, enjoy followers in hundreds of thousands on these social media platforms.

Table 5: E-retail nurseries delivering PAN India coverage and localized regions

Sr. No.	Name of e-nursery/ website	Head office/ Sub-offices	Products/Accessories sold	Delivery order
1.	Nurserylive.com	Hadapsar, Pune Maharashtra	Live plants	PAN India coverage
2.	LivingGifts.co.in	Tilak Nagar, New Delhi	Garden ornaments, gift wraps	PAN India coverage
3.	Nurturinggreen.in	Noida, U.P.	Live plants	PAN India coverage
4.	GangaNursery.com	Kolkata, W.B.	Garden ornaments, gift wraps	PAN India coverage
5.	Mysunny Balcony	Bangalore	Garden ornaments, gift wraps	PAN India coverage
6.	Gamlaa.com	Hyderabad	Garden tools and accessories	Hyderabad
7.	Nurseryplants.in	Mumbai	Live plants	Mumbai
8.	Gardenworld.in	Bengaluru	Live plants, plants as gift	Bengaluru
9.	Hariththarang	Chennai	Live plants, plants as gift	Chennai
10	Javitree.com	Delhi NCR	Live plants, plants as gift	Delhi NCR
11	Givegreen.in	Delhi NCR	Garden tools and accessories	Delhi NCR
12	chhajedgarden.com	Pune	Garden ornaments, gift wraps, succulents, growing media, woody species	Pune

Table 6: Year round production operations in an ornamental plant nursery January

Ornamental plants	Cultural operation undertaken
Rose	Planting of rootstock hardwood cuttings in sand beds
Chrysanthemum	Heading back of plants 3-4 inches above the ground for development of lateral shoots. Continue removing dried flowers in pots for extending their display
Shrubs	Pruning of deciduous shrubs in nursery mother block and preparing of hardwood cuttings for rooting of deciduous shrubs *(Dombeyas pectabilis, Barleria cristata, Erythrina crista-galli, Lagerstroemia indica, Vitex negundo, Jatropha spp.)*
Climbers	Planting of hard wood cuttings of climbers eg:*Adenocalymmaalliaceum, Beaumontia grandiflora, Thunbergia grandiflora, Wisteria sinensis, Pyrostegiave nusta*etc.
Marigold	Raising of marigold nursery under polysheet for summer season planting.
Dahlia	Staking of plants with bamboo sticks to keep these erect
Routine work	Addition of well rotten farm yard manure in all woody plantations followed by irrigation, Protect the tender plants by erecting straw *kullies* to protect from frost

February

Ornamental plants	Cultural operation undertaken
Chrysanthemum	Division of the plants and planting in raised beds for production of suckers. Empty the pots in which flowers have dried and no longer presentable.
Marigold	Transplanting of marigold in end February in field as summer season crop.
Shrubs	Pruning of dormant shrubs can be continued till new growth starts.
Rose	The T- budding operations over well rooted stock plant continues upto15th February
Annuals	Cleaning of seeds of summer annuals and preparation of nursery beds
Routine work	Stocking of well rotten manure and soil for pot and polybag filling

March

Ornamental plants	Cultural operation undertaken
Annuals	Nursery raising of summer season annuals through seeding over nursery bedseg: Zinnia, Kochia, Gomphrena, Portulaca, Cosmos, Gaillardia etc.
Shrubs	Planting of hard wood cuttings of *Poinsettia pulcherrima, Bougainvillea, Hibiscus rosa sinensis*, *Lantana spp,Cestrum spp.*, *Nerrium indicum, Punica granatum, Thunbergia erecta*etc., Seed sowing of shrubs (*Cassia glaucca, Tecomastans* etc.)
Indoor plants	Planting of cuttings (Dracaena,Pedilanthus, Sanseveriaetc.)/ divisions of foliage pot plants (Alocassia, Chlorophytum, Aglaonema, Peace lily, Syngonium, Potosetc) for mass multiplication
Routine work	Preparation and mixing of growing media for filling in pots for transplanting summer season annuals, Weeding, potting and repotting of old potted plants

April

Ornamental plants	Cultural operation undertaken
Gladiolus	Lifting of corms and cormels of gladiolus followed by cleaning and shade drying and their treatment with captan (0.1%)prior to cold storage
Chrysanthemum	Encourage the branching of root suckers transplanted in the last month by pinching and fertilization and regular watering in nursery stock plants.
Routine work	Weeding, potting and repotting of old potted plants, periodic watering and shading potted foliage plants from afternoon sun rays.

May

Ornamental plants	Cultural operation undertaken
Chrysanthemum	As mentioned in April, regular watering, fertilization and remove off type undesirable plants.
Bougainvillea	Pruning of potted Bougainvillea to confine its canopy

Media	
Annuals	Collection of seeds of winter annuals as and when they ripe
Dahlia	Harvest and lift the tubers, shade drying followed by fungicidal treatment with Bavistin 0.1%.
Routine work	Soil and growing media solarization in hot afternoon sun for disinfestation, -same as in April-

June

Ornamental plants	Cultural operation undertaken
Annuals	Nursery raising of rainy season annuals eg: Celosia, Cock's comb, Balsam, Tithonia, Amaranthus.
Chrysanthemum	Planting of terminal cuttings (4–6 cm) taken from the mother plants in third week. Keep cuttings under shade and moist.
Shrubs	Seed sowing for raising nursery of *Barlariacristata, Bauhinia accuminata, Cassia glauca, Ceasalpiniapulcherima, Dodoneaviscosa, Nyctanthusarbortristis, Tecomastans, Thevetiaperuviana, Vincarosea.* Propagate Jasmine by ground layering.
Indoor Plants	Propagate Rubber Plant and Crotons though air layering
Climbers& Trees	Seed sowing of *Aristolochia elegans* (Duck Climber), *Bignonia gracilis* (Cat Claw), *Hiptage benghalensis* (Madhavi Lata), seed sowing of Tree species in polybags for raising saplings.
Marigold	Nursery raising of marigold in end June for rainy season planting
Routine work	Same as in May

July

Ornamental plants	Cultural operation undertaken
Chrysanthemum	Continue propagation of chrysanthemum, and plant rooted cuttings in prepared beds/pots when these are rooted.
Trees	Seed sowing of trees for raising nursery. e.g.: *Araucaria cookii, Azadirachta indica, Bauhinia varigata, Butea monosperma, Chorisia speciosa, Melia azedarch, Saraca ashoka*etc.
Shrubs	Planting of soft wood and semi-hardwood cuttings eg: *Acalypha, Beloperone, Duranta, Exocaria bicolor, , Mussaenda, Plumbagocapensis, Poinsettia pulcherrima, Russelia juncea, Sanchezia, Tecoma capensis*
Climbers	Planting of soft wood cuttings of *Ficus repens, Ipomea palmata, Jasminum* sp. *Vernoniaela egnifolia*
Jasmine	Ground layering of tender shoots for multiplication
Indoor plants	Propagation of indoor foliage plants eg: *Aglaonema, Chlorophytum, Philodendron, Portulacaria and Tradescantia, Asparagus, Aspidistra* etc.
Annuals	Transplanting of rainy season annuals in pots
Routine work	Periodic watering of plants, collection of plant material and media for Bonsai making, Suitable time for potting and repotting of new and old plants in new pots.

August

Ornamental plants	Cultural operation undertaken
Chrysanthemum	Ensure transplanting of all rooted cuttings in field or pots by mid of August.
Trees	Seed sowing of tree species continues. e.g. *Azadirachta indica, Melia azidarch, Saraca ashoka*, plant root suckers of *Millingtonia hortensis*, Time to plant terminal cuttings of *Ficus benghalensis, F. benjamina, F. elastica, F.retusa* etc.
Shrubs	As mentioned in July
Climbers	Preparation of soft wood cuttings as mentioned in July and time to plant root suckers of *Clerodendron splendens*.
Marigold	Transplanting of marigold in field first fortnight of August
Routine work	Periodic removal of weeds and other undesirable plants. Suitable time for potting and repotting of new and old plants in new pots.

September

Ornamental plants	Cultural operation undertaken
Chrysanthemum	Disbudding of standard type and pinching of spray type varieties. Time to put net on spray types to give them support, insert stakes tied with jute thread alongside the pots to contain the plant canopy
Bulbous plants	Treat corms of Gladiolus with 0.2% captan, shade drying.
Annuals	Sowing of seeds of winter annuals over well prepared nursery beds e.g. Petunia, Calendula, Phlox, Stock, Nemesia etc. in end September.
Dahlia	Remove tubers from cold store and place these on moist sand beds for sprouting for getting healthy terminal cuttings.
Trees	Seed sowing of *Grevillea robusta, Magnolia grandiflora* and other tree species continues.
Routine work	Same as in August

October

Ornamental plants	Cultural operation undertaken
Rose	Pruning of roses in second fortnight followed by application of organic manures, and covering of cut ends with Blitox + linseed oil. Suitable time to plant new roses in field/pots, The digging of stock plants for budding or the previous year budded plants begins end of the month.
Annuals	Nursery raising of winter annuals continues
Chrysanthemum	As mentioned in September.
Marigold	Nursery raising over raised beds for winter season crop
Dahlia	Excise 3-4 cm cuttings from tubers and plant them in sand beds
Routine work	Field preparation for sowing nursery seedlings of winter season annuals

November

Ornamental plants	Cultural operation undertaken
Annuals	Transplanting of winter season annuals on prepared beds/pots.
Rose	Budding of roses on compatible rootstocks.
Chrysanthemum	Stake of the plants continues and pot are painted for exhibition display
Dahlia	Plant rooted cuttings in field or pots.
Routine work	Periodic removal of weeds from newly transplanted winter annuals.

December

Ornamental plants	Cultural operation undertaken
Chrysanthemum	Selection and labeling of best plants for maintaining purity in germplasm
Shrubs	Time to prune winter deciduous shrubs begins
Routine work	Occasional watering to regulate soil temperature and to prevent frost injury to shrubs and other plant material, stocking of well rotten organic manures.

References

Bhardwaj, R.L. Sarolia, D.K. 2011. Modern nursery management, Agrobios (India).pp, 1-403.

Bunt, A.C. 1988. Media and mixes for container grown plants. London, United Kingdom: Unwin Hyman. 309 p.

Dumroese, R.K., Luna, T., Landis, T.D. 2008. Nursery manual for native plants: volume 1, a guide for tribal nurseries. Agriculture Handbook 730. Washington, DC: U.S. Department of Agriculture, Forest Service. 302 p.

Gunjal, Surya. (COPI, NAIP e- publication-2012) Resource book on horticulture nursery management, NAIP, ICAR: 1-268. www.fdcm. nic.in/PDF/ Horticulture %20plant%20 nursery.pdf

Jaenicke, H. 1999. Good tree nursery practices: practical guidelines for research nurseries. International Centre for Research in Agroforestry. Nairobi, Kenya: Majestic Printing Works. 93 p.

Landis, T. D., Tinus, R. W., McDonald, S E., Barnett, J. P. 1994. Nursery planning, development and management. Vol. 1, The container tree nursery manual. Agriculture Handbook 674. Washington, DC, USA: US Department of Agriculture, Forest Service: 188.

Landis, T.D. 1995. Improving polybag culture for sustainable nurseries.Forest Nursery Notes. (July 1995), 6–7.

Landis, T.D., Tinus, R.W., Barnett, J.P. 1999. The container tree nursery manual: seedling propagation. Agriculture Handbook, 674.Washington, DC: U.S. Department of Agriculture, Forest Service, p 167.

Mbora, A., Lilleso, J. P. B, Jamnadass, R. 2008. Good nursery practices: A simple guide (ISBN: 978-92- 9059-235-8). Nairobi. The World Agroforestry Centre: 36.

Rana, S. S., Rana, M.C. 2015. Advances in weed management. Department of Agronomy, College of Agriculture, CSK Himachal Pradesh Krishi Vishvavidyalaya, Palampur 55p.

Ratha, Krishnan. Rajwant,P., Kalia, K., Tewari, J. C., Roy, M. M. 2014. Plant nursery management: principles and practices. Central Arid Zone Research Institute, Jodhpur (Raj.): 1-40.

Riad, S.R., El-Mohamedy, Ziedan, E. H., Abdalla, A.M. 2010. Biological soil treatment with *Trichoderma harzianum* to control root rot disease of grapevine (*Vitis vinifera*L.) in newly reclaimed lands in Nobaria province. Archives of Phytopathology and Plant Protection 43 (1): 73–87.

Sarolia, D.K., Samadia, D.K., Choudhary, B.R., Singh, D. 2018.Production of quality seed and planting materials. CIAH/Tech./Pub. No. 75. ICAR-Central Institute for Arid Horticulture, Bikaner, Rajasthan (Indian), 38

Whitcomb, C.E. 2003. Plant production in containers II. Stillwater, OK: Lacebark Publications. 1,129 p.

Wilkinson, K.M., Landis, T.D., Haase, D.L., Daley, B.F., Dumroese, R.K. 2014. Tropical Nursery Manual.A Guide to Starting and Operating a Nursery for Native and Traditional Plants.U.S. Department of Agriculture, Forest Service Agriculture Handbook 732, 378p.

14

Plasticulture: A Hi-Tech Approach for Betterment of Horticulture

Rajesh Kumar Sharma[1], Bhagchand Yadav[2], Dhuni Lal Yadav[1] and Sanjay Kumar[3]

[1]Agriculture University, Kota, Rajasthan, India
[2]College of Agriculture, Swami Keshwanand Rajasthan Agriculture University, Bikaner, Rajasthan, India
[3]ICAR- National Research Centre on Seed Spices, Ajmer, Rajasthan India

1. Introduction

The word 'PLASTIC' is a Greek word which involves two words 'PLASSIEN' and 'PLASTIKOS'. Denotation of word is to mould or shape into a soft substance which can be permanent or temporary. The use of plastics in agriculture as well as in horticulture has made substantial progress during the last decade. We all know that plastics contribute not only in planting but also in post-harvest handling and processing of many agri-horticulture crops. At every stage of horticultural crop life cycle plastics are used *i.e.* in seeds and fruits packaging, planting and propagation, mulching, irrigation, harvesting and preservation etc. Plastics application in agriculture and horticulture sector is commonly known as "plasti-culture". In India, about 5 million tons of plastics produce every year but only about 0.35 million tons of plastics are used in agri-horticulture. Today plastics are used in different ways like greenhouses to promote growth, production and productivity, mulching to maintain soil temperature, moisture and suppress weeds as well as plastic containers for seedling and soil-solarization to reduce incidence of pest and diseases. Characteristic properties of plastic are colored, melted, shaped, squashed, cheapness, light weight, impermeable to moisture and gases, made into fibers or rolled into sheets etc. Because of such flexible properties these are very useful and can be used in any shape, size and thickness. Now a day available plastic have benefits over traditional materials like greater thermal insulation, tremendous corrosion resistance, higher strength/weight ratio, more flexibility, resistance to harmful chemicals, higher moisture barrier properties, permeability of desirable gases, smooth surface causing in decrease in friction losses, higher transmissibility of light, enhance shelf-life, hygiene and better visibility of the produce etc.

Uses of plastics in agriculture, smart horticulture, irrigation water management and related different areas represent "Plascticulture". Different final products of plastic like plastic films/sheets, pipes, nets, moulded products and many more are used in Plasti-culture. Micro irrigations systems such as sprinkler & drip, lining of farm ponds/reservoirs, colour plastic mulching, polyhouse/greenhouse, plastic low and high tunnels, shade net houses, different protection nets such as insect-proof nets, bird nets, hail protection nets, fencing nets and modern nursery management through use of plastic pro-trays, bags & pots, hanging baskets, etc. are the significant role of plastic playing in agriculture and allied sectors. Increase earliness and total crop productivity in tomato crops by using balck coloured mulch polythene was the initial driving force to use of plastics in horticulture (Baki et al., 1992).

2. Why Plasticulture?

Multitude benefits of plasti-culture techniques are considered as most significant indirect agricultural inputs, which consequence in moisture maintenance and conservation, precise water and nutrients application, economically viable controlled environment agriculture, reduction in insects and disease incidence with the help of nets and helps in increasing shelf-life, storage and transportation of fruits & vegetables through innovative packaging solutions..

3. Different Approaches of Plasticulture

3.1. Plastic in Plant Propagation and Nursery Management

In vegetable crops, plug tray seedling production technique is working as a well-established nursery industry in several developed countries. In plug tray nursery technology, seedling are grown in plastic multi-celled plug trays (60 cm x 30 cm size or any other suitable size) in soil-less media or in poly bags in particularly designed polyhouse/greenhouses or other protected structures. Virus free healthy quality seedlings of different vegetable crops can be grown in pro-trays in small area of green house for main season as well as for their off-season cultivation and generate additional employment in agricultural sector. Normally three ingredients; coco peat, vermiculite and perlite (3:1:1) are used as rooting medium for raising/ growing seedling in nurseries. These trays are made of polypropylene and re-usable. Shelf life of the tray depends on the handling of the seedling trays. To produce quality healthy seedlings, grafts and plants, plastic bags & pots, pro-trays/ plug trays, hand/ knapsack sprayers; hanging baskets are used for easy handling and transportation. Following are some reasons why plug-tray vegetable seedlings are needed;

- It promotes better germination and root development of transplants with less mortality.

- Provide better drainage and aeration.
- Provides independent area for each seed to develop where is no any competition for place.
- Minimise mortality rate due to proper care and less infestation of disease insect and pest.
- Root damage during transplanting is almost avoided. This results in uniform crop with early maturity.
- Maintain uniform and healthy seedlings because accurate distribution of nutrient.
- Pro-trays are easy to handle and required minimum space.
- To raise healthy and quality seedlings for round the year crop cultivation.
- Economical in transportation of seedling.

Fig. 1: Plants grow in plastic pots/bags/pro-tray

3.2. Pond/Reservoir Lining

Management of rain water is the inexpensive and cleanest source of water. Harvesting of the rain water in the form of different structures like pond, lakes, wells, tanks and reservoirs helps to conserve this water so that it can be used as life-saving or supplementary irrigation later on during dry spells and insufficient rains. Water management through plastic lined pond, canal and reservoirs are the most effective because these provide check against seepage as well as avoiding depletion of stored/harvested water used for irrigation and portable use. This water may be also used for fisheries, livestock etc. Main advantages of these reservoirs are that they increases water availability for extended period of time thus reducing water leakage/seepage, prevents soil erosion, removes water logging and avoids upward imposition of salts into stored water, extremely useful in porous soils where water holding capacity is very low. Pond/reservoir linings or sealings are commonly required in areas where natural ground is rocky, difficult to obtain fine clay, short supply of water, existing of underground faults, toxicity in ground water and where a small pond is to be installed.

Fig. 2: Plastic lined farm ponds

3.3. Drip Irrigation System

Application of water and nutrients in precise and regulated quantity at low pressure and frequent intervals as drops through a closed network of plastic pipes fitted with drippers/emitters directly into the root growing area of the plant is refers as drip irrigation. In present situation, drip irrigation has become compatible with plastics that are durable and simply moulded and complexity of shapes requisite for pipe and emitters. In this system water discharge is very low (@ less than 12 lph) which supports in water saving and enhanced crop production and productivity. Embracing of recent irrigation techniques is needed to be emphasized to escalation water use efficiency and covering more area under cultivation. Drip irrigation along with fertigation system is the most efficient way to supply water as well as nutrients to the plant for increasing yield of fruits and vegetable crops (Tiwari et al., 2003). Clogging of drippers, precipitation of chemical and accumulation of salt at wetting front are the major issues related to drip irrigation system.

Fig. 3: Drip irrigation installed on field

Advantages

- Reduces soil erosion and controls disease/pests.
- Water saving by 40 % to 70 %.

- Saves 30 % to 50 % fertilizer as compared to conventional method.
- Prolongs crop maturity duration.
- Improves quality & quantity of produce and allows early maturity.
- Saves 10 % to 15 % labour cost on weeding as compared to conventional methods.

3.3.1. Sprinkler and Micro Sprinkler Irrigation System

Application of water under high pressure in the form of spraying water in the air and allow falling on the ground surface is referred as sprinkler irrigation. High pressure is generally achieved with the help of a pump. The water is discharged similar to rainfall through the orifice of the nozzle placed in the pipes. These systems are suited in moist of the soils due to its wide range of discharge capacity. Darko et al. (2016) also suggested that sprinkler system of irrigation with deprived uniformity results in reduced yields due to water stress as well as water logging which added to increase in the cost of irrigation and other associated issues. Uniformity in irrigation is the central goal of the sprinkler system design. Properly and uniformly spaced sprinklers in the field will respond to a quite uniform application of water over the irrigated land. Irrigation uniformity may have a different perception which depends on whether it is observed by an engineer or by an agronomist. With the perception of engineer, it will be geared headed for the design and management of the sprinkler systems. With agronomist, perfect uniformity may be observed as each part of the rising crops obtain uniform application of water. Sprinkler discharge capacity must be similar to the infiltration rate of the soil in the field. Water will run off the field or relocate within the field, if the discharge rate surpasses the soil intake rate which consequence in over and under watered areas. So, selection of nozzle sizes, operating pressure and sprinkler spacing is atmost.

Fig. 4 : Sprinkler irrigation system installed on field

Advantages

- Requires no land levelling and reduces soil erosion.
- Application of fertilizer into the soil can be regulated.

- Water saving up to 30-50% with no soil compaction.
- Provides protection against the frost.
- Moderates micro climates.
- No interference with cultural practices.

3.3.2. Soil Solarization

Soil solarization is a non-pesticidal technique to control weeds, soil borne pathogens and pests by covering moist soil of field with transparent polyethylene film exposed to sunlight during periods of high temperature. This is generally done during summer months (May-June) when the air temperature more than 35°C. However, climate and weather conditions of that area also affect the energy released by sun on soil temperature. This technique can rises the soil temperature to a level that kills many soil borne pathogens, nematodes, weed seed and seedlings etc. Solarized soil not only improve soil physical and chemical properties like soil structure, colour, temperature, moisture content and nutrient availability but also enhances plant growth, development, harvestable yield and crop quality. Effect of soil solarization may last for more than one growing season. Nowadays light plastic films (LPFs), which are able to increase soil temperature more than 20°C above air temperature, are commonly used particularly in open and greenhouse vegetables, flowers crop cultivations and orchards. Application of LPFs as soil solarization is an environmentally-friendly alternative with low cost and low environmental impact. Soil solarization is a crop protection technique that considerably convey on locally existing resources and farmers did not depend on agrochemical producers. Furthermore, in achieving Indian food security and to fill the yield gap, soil solarization may play an important role.

Fig. 5: Soil solarization on field

3.3.3. Plastic Mulching

Mulching is a technique of covering the soil surface around the plant with plastic film to create more favorable microclimates for better plant growth, development and yield. Word mulch means 'covering of soil'. Natural mulches (organic mulch) have been used for centuries but introduction of synthetic plastic materials has changed the techniques and advantages of mulching. Plastic mulches are completely impermeable to water which acts as a physical barrier for direct evapotranspiration of moisture from the soil surface and conserve water by preventing the water losses as well as avoids the rise of water containing salts. Besides this, plastic mulches also prevent weeds growth, enhances root growth, nutrient uptake, early fruit ripening, improved fruit yield and quality, protect crops from soil borne pathogens, prevent wind and water erosion and hail damage. Better soil aeration, moisture and temperature condition favour, higher beneficial microbial biomass in the soil and ensuring added complete nitrification. Plastic mulch keeps ripening fruits above the soil surface and reduced contact with the soil and water. This decreases fruit rot as well as keeps the fruit and vegetables clean (Deschamps *et al.* 2019). Plastic films are available in different thickness and colours as per requirement but for medium duration crops generally 25 to 50 μ and for longer duration crop 50 to 100 μ thickness is best use. Disposal of the plastic film after its use, which causes an environmental pollution, is a major constraint for its commercialization as mulching material. However, this limitation has led to development of biodegradable and photodegradable mulches. Double cropping, which allows growing two or more crops on the same mulch is another alternative to reduce plastic mulch waste. Ecological conditions and other aspects of mulching also affect the selection of mulches which are as under;

Type of Mulch	Particulars
Thicker mulch	Orchard and plantation
Thinner film	Early germination
Thin transparent film	Soil solarization
Perforated mulch	Rainy season
White colour film	Summer cropped land
Black colour film	Sandy soil and Saline water use
Silver colour film	Insect repellent

Fig. 6: Use of plastic mulch on field

3.3.4. Greenhouses / Polyhouses

Greenhouses are inflated or framed structures covered with transparent/translucent materials which are more suitable to grow crops under controlled or partially controlled environmental conditions to get optimal plant growth and productivity. This technology provides better space utilization and maintain conducive environment for crops against vagaries of climate such as heavy rainfall, snowfall, high wind velocity etc. Besides, from protection to adverse climatic condition, vegetables grown under protected production gives higher yield and better quality in terms of shape, size and colors. Enclosed structure and changed micro climate inside the poly house leads to lower incidence of insects and pests. Consequently, there is minimum use of chemical insecticides. Early and offseason cultivation of vegetables and fruits fetches higher market prices (Nguyen *et al.*, 2009). Green houses are also used for hardening of tissue culture plants, production of genetically superior transplants, maintenance of stock plants, grafted and micro-propagated plant-lets. In cold climates like Ladakh region greenhouses are being constructed for extending the growing season of vegetables from 3 to 8 months. Several changes in the type of houses have been recommended to obtain higher energy efficiency and economic use of construction materials. Multiple span houses with higher roofs are supposed to be more appropriate for the continental climate than the single span greenhouses. Other advance techniques of crop cultivation like hydroponic, aeroponic and nutrient film techniques (NFT) are possible only under protected structures.

Fig.7: Use of plastic material in greenhouse structure

3.3.5. Plastic Low and Walking Tunnels

Plastic low and walking tunnels are miniature/small frame structures to cover crop canopy and creating greenhouse like effect. These are generally 4 to 6 feet wide and 2 to 4 feet tall in height. Generally, these tunnels are made in arc shape from PVC and covered with plastic films. Entrapped outer CO_2 inside the tunnel leads to increase photosynthetic rate of the plant that help to obtain higher crop yield. Nowadays, these tunnels are being used for production of healthy and quality seedling of high value crops. These tunnels not only provide protection from high wind, rain, frost and snow but also protect the crops from cold wave by increasing inside temperature during winter season. These are most effective for cultivation of melons, cucumber, tomato, peppers, strawberries, radish, beans, bitter gourds, squashes etc. For beginners this technology is a very cost effective investment option. Vegetable cultivation under plastic low tunnels is a steadily growing agricultural sector but concern developed over environmental sustainability of this farming system.

Fig. 8: Use of plastic tunnels on field

Advantages

- Provides microclimate for seedling and grafts.
- Creates favorable condition to enhance photosynthetic activity for better nutrients uptake by the plants.
- Ease of construction and dismantling as well as mobility.
- Disease free and healthy saplings for nursery are developed.
- Substitute for crop protection and season extension.

3.3.6. Plant Protection Nets

Plant protection nets are widely used in agri-horticulture sector to safeguard the vegetables and fruits crops. Nowadays plastic nets are used as insect proof net, shade net, screen vents in greenhouses, soil coverings and post-harvest packaging etc. Insects proof nets (IP Nets) are eco-friendly substitute to pesticides and widely used in organic cultivation. For virus free crop production, single /double layer IP nets are used as covering in screen houses. IP screening not only checks the entry of insects and avoid feeding of disease causing vectors but also avoid the escape of pollination insects from the greenhouse. Therefore, colour and size of mesh are the most important factors. Thus, IP nets are considered to have very low porosity and permeability. With respect to colour, white or transparent fibers are used to reduce solar radiation transmittance.

In order to reduce air temperature and light level inside the greenhouses shade nets are widely used. The effectiveness of shading depends on the size, colour and shade factor of the net. Depending upon the use, shade nets are available in different mesh size and shade factors between 15 to 90 %. Presently shade nets are available in different colours also like white, red, blue, yellow, black and green having response on crop/plant yield attributes. It offers partially controlled environment by reducing the light & temperature which is harmful for the crop. In order to reduce heat losses (convection & thermal radiative) from the green houses during winter season thermal screen are used. Plastic nets are also used against the climatic hazards (wind, hail, snow, frost and rain) to protect crop plants. Windbreak nets are installed to avoid possibility of breaking of branches, flowers, reduce evapo-transpiration, increase pollination activities, protect crop from dust and sand. Reduction in wind load through windbreaks depends on the height of the frame, porosity of net and distance from crop plantation.

Fig. 9: Use of plant protection nets on field

Advantages

- These are flexible and low weight.
- Protects crop/plants from birds, insect, pest, hail etc.
- Can be placed upon simple structure.
- UV stabilized and resistant to bacteria, chemical agents etc.

3.3.7. Plastic Vermi-Beds

Vermi-culture/vermi-composting through earthworm is economically viable and easy technology to transform all bio-degradable waste into a beneficial soil amendment and good quality organic manure. Plastic vermi-beds offer strength and flexibility as well as portability besides ease of installation. Nowadays, these plastic vermi beds are available with different size and capacity according to the requirement.

Fig. 10: Use of plastic for preparation of vermi-beds

Advantages

- Maintain appropriate aeration to increase the population of earthworms, thereby increasing the vermicomposting process.
- Ease of handling, assembling and transportation.
- Customized design for collection of vermiwash.
- Reduction in labour cost.
- The vermi-wash produced is used foliar spray.
- Increases physical and chemical quality of soil, aeration and water holding capacity and prevents soil erosion.

3.3.8. Sub-Surface Drainage

Sub-surface drainage is a system of removing excess water and salts from the root zone of the plants either naturally or artificially through linkage of pipes lay underneath the ground surface. The piped network is designed at lower water table to allow optimum root development preventing capacity movement of salts into the root zone of the plant. Therefore, increase crop yields and rural income and living standard. Manual and full mechanized installation techniques are there to cover small farm holdings and thousands of hectares, respectively. The cost of installation depends on physical and economic conditions and installation methods adopted. Introduction of advance installation equipment and materials has made large scale implementation feasible.

Advantages

- Regulates water table and ground water flows.
- Removes excess water from the soil to prevent water logging.
- Most effective method to controlled soil salinity.

- Increases soil environment for vegetative growth and reduction in soil and water erosion.
- Controls surface runoff, thereby reduces loss of nutrients.

3.3.9. Innovative Packaging Solutions

Distribution and marketing of agricultural produce is one of the most critical aspects wherein packaging of agricultural produce and handling plays a vital role. A variety of agro-horticultural products either raw or processed need to be packed and stores. Plastic product due to its inevitable properties such as flexibility, light weight, safe & hygienic, cost effective with better display options makes irreplaceable contribution for the processing, storing, preserving and transportation of agri-horticultural produces. More than 30 % of farm produce is lost between the channel of farm and consumer thereby affecting the quantity and returns to the growers. Therefore transportation, climate and handling of the farm produce demands well designed and sound packaging products to arrest these post-harvest losses. Therefore, different type of plastic materials like plastic field boxes, pallet boxes, HDPE crates, plastic films, polyethylene waxing etc. are used. Modified atmospheric packaging (MAP) with thin plastic films, designer films and coatings are an advance and smart packaging system which interact with internal gas composition by limits air movement and water losses. Thus, extend the shelf life of fruits and vegetables with better retention of produce quality for longer duration. Shrink-wrap packaging is also a new packaging technique for safe handling of fruits and vegetables. Cryovac (9 μm), LDPE (25 μm) and polyolefin (13 μm) are some example of heat shrinkable plastics. Heat shrinking is done at high temperature (120 °C) with short period of time (10 seconds).

Fig. 11: Packaging of agricultural produce in different plastic containers

4. Agencies Involved in Promoting the Plasticulture

4.1. GOI Subsidy Scheme

Ministry of Agriculture & Farmer's Welfare, Government of India (GoI) is providing financial assistance to the beneficiaries for adopting the above Plasticulture applications with a view to popularize these technologies at PAN INDIA through a well-structured mission management with State implementing agencies. The subsidy is channelized through State Directorates of Agriculture/ Horticulture.

4.2. National Committee on Plasticulture Applications in Horticulture (NCPAH)

NCPAH, Ministry of Agriculture & Farmer's Welfare, Government of India, has been promoting use of various Plasticulture applications in agriculture and horticulture. NCPAH under its mandate and major activities provides suitable policy measures and technical support on Plasticulture & Precision farming for improving productivity and quality of the produce, facilitate, monitor and supervise applied research at 22 Precision Farming Development Center's (PFDCs) to generate data base on plasticulture, suggest ways & flagship scheme covering various plasticulture applications and precision farming components, develops and update quality standards for various plasticulture applications and precision farming components.

5. Disposal of the used Plastics from the Plasticulture System

This is the one burning question which is always queried by people who are concerned in or are using plastics to modify the cropping micro as well as macro environment. "What do I do with the plastics when I am done using them?" It is definitely a problem that is a concern worldwide in its scope. There have been continuously to be many attempts at solving this problem, such as usage of plastics several times which are photodegradable or biodegradable, reducing the weight of films (less material to deal with), recycling, and incineration. Recycling of drip irrigation tapes and mulching materials are the hardest because since after a season in the field, these materials have a lot of plant debris, dirt and moisture associated with them, which makes it difficult. Incineration of these hard materials is the alternate option to recover the considerable fuel value (1 kb of plastics has as much potential energy as an equivalent amount of fuel oil). Burning of these used plastics in energy plants creates one problem is that plastics burn hotter than the surrounding waste; thus, "hot spots" are created in the waste stream. A group of researchers at the Penn-sylvania State University, University Park, are testing "Plastofuel," a "fuel nugget" made from used agricultural plastics such as plastic mulches, drip irrigation tape and tubing, row covers, pots, trays, flats, silage bags, and hay wraps. To supplement to coal or other waste products or by itself plastofuel can be burned in a burner unit developed in South Korea, which is to be tested in the near future to heat a wide variety of structures (Garthe *et al.,* 2003). In addition to the emission testing to be conducted on the burner unit, research proposal to be prepared on the appropriate collection methods, preparation and transportation of the plastics from the crop production site to the point where it is processed further (Plastofuel) and then to the site of incineration.

6. Conclusion

Plasti-culture has been showed effective tools for increasing agricultural production throughout the world. In broadest concept, Plasti-culture employs plastics in agriculture sector. In India, it is attaining importance. Still, at farmer's level, the actual benefits of the plasti-culture are not being recognized due to lack of latest information, non-standardized designs and package of practices (PoP), indigenous level services and non-availability of inexpensive tools and technologies. In plasticulture, plastic materials are used in plasti-culture for effective utilization of the available resources through numerous applications which results in boosted productivity and sustainability. For increasing horticultural production and quality, plastics may be used for several purposes i.e. propagation nurseries, mulching, covering materials for protective structures, protected nets, drip and sprinkler irrigation system, drainage system, soil solarization, plastic traps and packaging. Nowadays, plastics have developed tantamount with modern life style nevertheless it is difficult to dispose off and have become a significant source of environmental pollution. But with the development of biodegradable and photodegradable plastic it is possible to easy decomposition of plastics by soil microorganisms and persists for shorter period of time in nature.

References

Baki, A., Spence, C., Hoover, R. 1992. Black polyethylene mulch doubled yield of fresh-market field tomatoes. Hort Science. 27(7), 787-789.

Darko R.O., Yuan S, Hong L., Lui J., Yan H. 2016. Irrigation, a productive tool for food security- a review. Acta Agriculturae Scandinavica, Section B- Soil & Plant Science 66(3), 191–206.

Deschamps, S.S. Vance M.Whitaker, V. M. and Agehara, S. 2019. White-striped plastic mulch reduces root-zone temperatures during establishment and increases early season yields of annual winter strawberry. Scientia Horticulturae 243, 602-608.

Ennis, R.S. 1987. Plastigone TM a new, time controlled photodegradable plastic mulch film. *Proceedings, 20th National Agricultural Plastics Congress*, pp. 83-90.

Garthe, J.W., Miller, B.G., Wasco, R.S., Lamont, W.J., Orzolek, M.D. 2003. Used agricultural plastic as a coal fuel supplement. Proc. 31st Natl. Agr. Plastics Congr. Pp.53–57.

Hanafi, A. 2003. Evaluation of different types of insect screens for the exclusion of white flies and natural enemies. Proceedings of the IOBC/WPRS Working Group *Integrated Control in Protected Crops, Mediterranean Climate*, 26, 43-47.

Kaddi, G., Tomar, B.S., Singh, B., Kumar, S. 2014. Effect of growing conditions on seed yield and quality of cucumber (*Cucumis sativus*) hybrid. Indian Journal of Agricultural Sciences 84(5), 624-627.

Natwick, E.T., Durazo, A. 1985. Polyester covers protect vegetables from white flied and virus disease. California Agriculture 39(7+8), 21-22.

Nguyen, T.H.N., Borgemeister, C., Max, J., Poehling, H. M. 2009. Manipulation of ultraviolet light affects immigration behaviour of Ceratothripoides claratris (Thysanoptera: Thripidae). Journal of Economy Entomology 102(4), 1559-1566.

Redditt WM. Factors affecting sprinkler uniformity. Sprinkler Irrigation Engineering Manual. Hawaiian Sugar Planters Association, Honolulu, Hawaii, 1965.

Teerling, C.R., Murphy, G. and Lenteren, J.C. 1999. Experiences with insect exclusion screening of greenhouse vents in Ontario, Canada. IOBC/WPRS Working Group Integrated Control in Glasshouses 22: 247-250.

Tiwari, K.N., Singh, A., Mal, P.K. 2003. Effect of drip irrigation on yield of cabbage (*Brassica oleracea* L. var. *capitata*) under mulch and non-mulch conditions. Agricultural Water Management 58(1),19-28.

Wells, O.S. 1991. High tunnels shelter early crops. American Vegetable Grower 39(2), 44, 46-47.

15

Fruit Drop
Causes and Management in Fruit Crops

Manu Tyagi

Krishi Vigyan Kendra, Pathankot, Punjab Agricultural University Punjab, India

1. Introduction

India has diverse agro-climatic conditions which, widely promote the cultivation of different types of fruit crop. A wide variety of fruit crops are grown across the country such as banana, mango, citrus, papaya, guava, grapes, pomegranate, apple etc. Globally, we rank second in fruit production after China. The total area under fruit crops is 66.64 lakh ha with a production of 99.06 million Mt (Anon., 2018). However, the average productivity (13.03 t/ha) in fruit crops is quite low than other countries. Further, the per capita fruits availability in our country is 207.9 g/ day which is below than the recommended quantity of 230 g/capita/day (Anon., 2018). Therefore, the increase in fruit production is necessary for internal as well as global market. The fruit production both in terms of yield as well as quality is hampered by various factors such as non-availability of quality planting material, unfavourable climatic conditions, lack of irrigation, inadequate nutrition, poor disease and insect pest management. The combined influence of such factors often promotes fruit drop.

According to Marcelis et al., (2004) "Fruit drop is a natural phenomenon in which the number of fruits on a tree are reduced by lack of photosynthates caused by auto regulation". The per cent of natural shedding varies with crops (Table 1). In general, the fruit crops like apple abscise about 95% of blooms and developing fruits, while the small fruit crops like blueberry shed only 20-30% of blooms and young fruits (Westwood, 1967). But, the abnormal shedding of flowers and fruits is a serious concern which, begins soon after flowering and continues till harvesting. In most of the fruit crops despite of profuse flowering and adequate fruit set, only a few fruits reach to maturity. In case of mango, individual panicles exhibit hundreds of ovaries bearing flowers but, a meagre percent of these form fruits and attain maturity. Similarly in Kinnow, less than 1 per cent of flowers develop as fruits. Likewise in Aonla, the fruit set varies from 58.94 % to 13.75 % while, the fruit retention ranges between 12.37% to 3.42% (Hiwale, 2015). The excessive fruit drop is thus, a significant factor limiting fruit production. It is not a continuous phenomenon, rather, it occurs during varied phases of growth and development of fruits. A proper understanding of these stages is imperative to comprehend the physiology of fruit drop.

Table 1: Fruit setting and retention percentage in few important fruit crops (Stephenson, 1981)

S.No.	Fruit Crop	Fruit setting (%)	Fruit retention (%)
1.	Mango	13-28	<0.1-0.4
2.	Lemon	52	7.0
3.	Sweet Orange	34.9-61.5	0.2-1.0
4.	Pear	39-83	3-7
5.	Peach	-	16.4-25.9
5.	Macadamia nut	15-50	0-4

2. Stages of Growth and Development in Fruit Crops

A typical fruit development is characterised of pollination, fertilization, fruit set, fruit growth, ripening and senescence. These phases are associated with three different growth stages *viz.* cell division, cell enlargement and cell maturation. According to Gundewadi et al. (2018), a majority of fruit species such as banana, mango, citrus, apple, pomegranate, avocado etc. constitute a single sigmoidal growth curve wherein the cell division stage (or stage I) involves slow growth but high cell division. This stage begins with the opening of flowers (also known as anthesis) followed by pollination and fertilization (Tadeo et al., 2020). It is succeeded by cell enlargement stage (or stage II) which is marked by rapid increase in fruit size owing to resource accumulation in seeds and pericarp. Due to rapid augmentation in fruit size, this stage is often referred as the period of exponential growth. The final stage or cell maturation stage (stage III) is characterised by a decline in fruit growth and increase in metabolic activities which, eventually lead to fruit ripening. The developing fruit incurs a series of physiological and biochemical changes during this stage i.e. softening of pericarp, breakdown of stored carbohydrates into sugars, decrease in astringency etc.

Another group of fruit crops like grapes, papaya, guava, sapota, olive, ber, stone fruits etc. exhibit a double sigmoid type fruit growth curve. In such crops, the initial rapid growth phase involves cell division and augmentation in seed size, nucellus and integuments. The stage II or the slow growth phase is characterised of stone hardening and embryo enlargement followed by stage III or the second rapid growth phase. This phase is associated with increase in the size of ovary leading to fruit ripening.

3. Stages of Fruit Drop

The fruit drop occurs prior to stage II, in case of crops with single sigmoid fruit growth while, the fruit crops with double sigmoidal growth experiences drop before stage III (Stephenson, 1981). But, in general, three waves of fruit drop are observed in majority of the fruit plants.

3.1. Post Bloom Drop (First Wave)

The first wave or post bloom drop is also known as cleaning fruit drop. It begins soon after flowering and lead to shedding of very small fruits. This drop is ascribed to heavy flowering and can be considered as a natural phenomenon adopted by plants to reduce the excess crop load. The rate of post bloom drop depends on the rate of pollination and fertilisation. Poor fertilisation leads to higher rate of post bloom drop (Racsko et al., 2004). Yuan and Huang (1988) also found that the initial wave of fruit drop is associated with absence of fertilisation. In case of stone fruits like peach, plum, cherry etc. few other factors such as belated pruning, poor fertilization and lack of soil moisture may also contribute to this drop (Racsko et al., 2004). According to Singh and Singh (2020), elevated status of growth inhibitors and suppression of growth promoters is responsible for post bloom drop in mango. The extend of first wave of fruit drop may vary depending on the fruit crop or variety (Table 2). This fruit drop is a little matter of concern to the orchardists due to negligible economic losses.

Table 2: Duration of first wave drop in different fruit crops

S.No.	Fruit crop	Phase of Ensuing First Wave Drop	References
1.	Apricot	1 or 2 weeks after bloom	Racsko et al., 2006
2.	Sweet cherry	1.5 -2.5 weeks after bloom	Racsko et al., 2006
3.	Pecan nut	14-45 days after pollination	Racsko et al., 2006
4.	Litchi	Upto 7 days after full bloom	Yuan and Huang, 1988
5.	Aonla	Upto 21 days of flowering	Bajpai, 1968

3.2. Summer or June Drop (Second Wave)

The second wave of fruit drop begins 5-6 weeks after flowering. It is characterized by the shedding of immature fruits or fruitlets (Arseneault, 2016). Stephenson (1981) mentioned that this fruit drop occurs before the beginning of cell division in embryo but after the completion of cell division in endosperm, during the stage I of fruit growth. Singh and Singh (2020) reported that in case of mango, the second wave of fruit drop commences from 28-35 days following pollination and fertilization. A study conducted by Yuan and Huang (1988) revealed the second fruit drop in litchi occurs after 21 days of full bloom prior to the full development of liquid endosperm. In case of citrus, second wave or June drop is initiated one to two months after bloom. The fruits abscised during this drop have a diameter of approximately 12 cm (i.e. marble size). A number of factors are ascribed to second wave of drop. Firstly, the increasing competition among the developing fruitlets for photosynthates. Secondly, as the name suggests "June drop" is associated with high temperatures during June. Lack of irrigation under such conditions lead to excessive fruit drop (Rattanpal et al., 2019). Further, Singh and Singh (2020) mentioned that nutrition imbalances and growing incidence of fungal diseases viz. *Collectotrichum gloeosporiodes* may be responsible for the second wave of fruit drop in citrus.

3.3. Pre-Harvest or Pre-Mature Drop (Third Wave)

This fruit drop occurs in the beginning of ripening phase and comprises physiologically mature to harvestable fruits. The occurrence of pre-harvest fruit drop corresponds with poor levels of auxin in the fruit (Luckwill, 1953; Wright, 1956). It may be associated with reduced carbohydrate reserves, inadequate nutrition, hormonal imbalances, diseases and insect pest incidence (Mollapur et al., 2016). This drop is initiated at the junction of peduncle and calyx resulting in the shedding of fruits without peduncle attachment (Rattanpal et al., 2019). The pre-harvest fruit drop causes severe economic losses to the growers due to poor yields at harvest. The yield losses due to pre-harvest drop in apple are about 30% (Byers, 1997; Schupp and Greene, 2004). The intensity of pre-harvest drop may dependupon the cultivars as few varieties are prone to excessive fruit drop while, some are tolerant. A study conducted by Yuan and Huang (1988) indicated the occurrence of pre-harvest fruit drop in litchi cultivars (with aborted seeds) during 2-3 weeks prior to horticultural maturity which, resulted in approximately 42.7% yield loss to growers. The pre-harvest fruit drop in Aonla (begins from second fortnight of August and continues till October) is mainly found to be associated with embryological and physiological factors (Bajpai, 1968).

4. Factors Influencing Fruit Drop and Their Management

The fruit drop is influenced by a series of internal and external factors. A detailed description of these factors along with their management is given below.

4.1. Internal Factors

The internal factors such as lack of pollination and fertilization due to anomalies in floral biology *viz*., defective sex organs, poor stigmatic receptivity, non-viable pollen, sterility, incompatibility etc. or absence of suitable variety for cross pollination may severely affect the fruit retention. Both pollination and fertilization are crucial for normal fruit growth and development in majority of fruit crops. Absence of pollination and self-fertilization may gradually lead to abscission of fruits.

4.1.1. Management

The pollination failures can be managed by growing pollinizer varieties with the main variety. This promotes cross pollination and increase in fruit set and retention.

Haldhar et al. (2018) reported promotion in fruit set (68 per cent) and improvement in fruit quality due to cross pollination in pomegranate. Similarly, the pollination problem in plum due to prevalence of heterostyly (variable style length) can be overcome by growing pollinizers (Racsko et al., 2006). For effective pollination and consistent good yields, pollinizer varieties must possess following attributes:

- The flowering period of the pollinizers must coincide with the flowering period of main variety.
- Pollinizer should be compatible with the main variety and produce a copious amount of viable pollen for successful pollination.
- Pollinizer variety should possess commercial value.
- It should be well adapted to climatic and soil conditions.

The pollinizer varieties of some important fruit crops are mentioned in the Table 3.

- Pollination can also be augmented through introduction of bee hives in the orchard. Good produce can be harvested by keeping one honey bee colony per acre. The best time for placing bee hives in orchard is during flower emergence. Uniform placement of hives in the orchard ensures sufficient bee activity for pollination particularly, during unfavourable climatic conditions.

In case of aonla, introduction of pollinators *viz.* honey bees in the orchard and plantation of 10 per cent pollinizers is effective for better fruit set and good yield. (Allemullah and Ram, 1990). Pollination can further be enhanced by placing blooming shoots (large branches with dehiscing and unopened flowers) of other cultivars also known as bouquets in orchards. These can be kept in vessels containing water.

Development of self-fertile cultivars through breeding and selection is crucial in fruit crops to attain good fruit set and final yield.

Table 3: Pollinizers for important fruit crops (Din et al., 2019)

S.No.	Cultivars	Pollinizer
Apple		
1.	Irish Peach, Early Shanburry, Benoni, Mollie's Delicious, and Early Red One.	Tydeman's Early Worcester, Summer Queen.
2.	Gala Mast, Oregon Spur, Imperial Gala, Vance Delicious, Well spur, Firdous, Cooper IV, Red chief, Red Fuji, Red Spur, Shireen and Top	Sparten, Red Gold, Ginger Gold, McIntosh and Lord Lambourne.
3.	Ambri, Royal Delicious, Rich-a-Red, Red Delicious,Lal Ambri and Akbar.	Granny Smith, Golden Spur, More Spur Gold, Sunhari, Golden Delicious

S.No.	Cultivars	Pollinizer
Pear		
1.	Bartlett	Fertility, Bosc, d'Anjou, Clapps Favourite, Comice and Conference.
2.	Bosc	Comice, Bartlett, d' Anjou, Seckel
3.	Comice	Bosc, Bartlett, d' Anjou, Seckel.
4.	Seckel	Comice, Bosc
5.	Conference	Docteur Jules Guyot,William, Beurre Bosc and Anjou
Peach		
1.	Earlihale, J.H. Hale, Candoka, Hal-Berta and July Elberta of peaches.	Other varieties of peach can pollinate these self-unfruitful varieties
Plum		
1.	Burbank	Satsuma, Shiro, Santa Rosa.
2.	Satsuma	Shiro, Santa Rosa, Burbank.
Almond		
1.	Makhdoom	Shalimar and Waris
2.	Shalimar	Makhdoom and Waris
3.	Waris	Makhdoom And Shalimar
4.	Peerless and Price	Non-Pariel
5.	Parbat	Makhdoom
Aonla		
1.	NA-7	NA-6, Chakaiya
2.	Banarasi	Chakaiya

4.2. External Factors

4.2.1. Physiological Factors

The extent of fruit drop varies in orchards as per the climatic and agrological conditions. The crop load in a particular orchard differs between consecutive years (Basak and Buczek, 2010; Robinson et al., 2010). The climatic and agrological conditions such as wind, temperature, nutrients, water, sunlight etc. may vary between years. Under unfavourable growth conditions, the plant is subjected to stress which, may also lead to fruit drop. The main physiological factors associated with fruit drop are discussed below:

4.2.2. Unfavourable Climatic Conditions

High temperatures during growing season severely affect the fruit retention and final crop load. In fruit crops like apple, the duration of fruit growth and development period is drastically reduced under hot growing season (Hoffman, 1940). As a result, ripening is advanced. The coexistence of ripening with high temperatures causes severe and early pre-harvest fruit drop. In line with this, Walsh (1977) also found that prevalence of high temperature during fruit ripening (or initiation of ethylene production in fruits) leads to advancement in ethylene rise and subsequent fruit drop. In aonla, pre-mature fruit drop was augmented with the sudden rise in temperature (Hiwale, 2015).

Further, the occurrence of frosts, low temperature or winds during flowering may also damage the blossoms and affect the fruit set by hampering the pollinator activity in apple and peach (Murneek, 1937). The fruit drop is aggravated in pear variety 'Hardy' during windy days (Byers, 1997). High wind velocities (20-25 miles per hour), temperatures lower than 50°F or continuous rains severely affect the insect flight and visits which influence pollination, fertilization, fruit set and retention. Bees cover short distances under such conditions. In addition to this, cloudy weather with low light conditions induces June drop in apple /cherry(Byer, 1997).According to Wani et al., (2010)prevalence of low atmospheric humidity may induce post blossom drop due to drying of stigmatic fluid.

Water deficit or lack of irrigation during cropping season is unfavourable for good crop. The water deficit conditions often induce stress in plants which, augments the synthesis of growth retarding hormone i.e. abscissic acid. Higher concentrations of abscissic acid advances fruit abscission (Léchaudel and Joas, 2007). According to Roemer et al., (2008) the IAA export out of fruits is retarded in plants under stress which, advances the abscission processes, gradually leading to fruit drop.

4.2.2.1. Management

Regular irrigation at weekly intervals is crucial during active growth for higher fruit retention. Adoption of micro-irrigation techniques like drip (surface or sub-surface), sprinkler irrigation etc. favourably affects the fruit retention and crop yields besides, saving water and controlling weeds. Sharma et al. (2003) recorded minimum fruit drop in apple under micro-irrigation at 80% ET. Use of organic or inorganic mulches helps in creating favourable environmental conditions for growing plants. Mulch application is thus, an important cultural practice which, may improve fruit retention through moisture conservation and regulation of extreme temperatures (Bakshi et al., 2015). It also checks weed competition for resources which, improve the crop load. In line with this Pande et al., (2005) found that use of dry grass as mulch significantly improved the fruit retention in apple. On tree fruit bagging using brown paper or PP non-woven bags may be beneficial in overcoming biotic stresses in fruit crops. A study by Sharma et al. (2020) indicated that fruit bagging in guava fruits using polypropylene non-woven bags significantly reduced the fruit abscission. The natural perforation in such bags promote free air circulation and prevents acceleration in temperature and relative humidity. Significant reduction in fruit drop was recorded in litchi using perforated transparent polythene bags at 14 days before harvest (Pal et al., 2016). Similar results were also obtained by Gaikwad et al. (2004). Establishment of windbreak checks fruit drop by reducing the mechanical injury caused due to abrasion of shoots, leaves, flower or fruits. Besides, it also conserves moisture by checking evaporation losses and provides protection against sun burn (Norton, 1988).

4.2.3. Inadequate Nutrition

Inadequate nutrition is another major factor associated with pre-harvest fruit drop. A heavy fruit set often induces nutrient shortage due to competition among growing fruitlets for nutrition which, ultimately lead to fruit drop (Racsko, 2006). The deficiency of macro as well as micro nutrients is a critical stress factor for quality yield in bearing trees. A study by Addicott (1982) indicated that deficiency of boron in fruit crops severely affects the carbohydrate transport leading to cell deterioration and fruit abscission. Likewise, Wooldridge (2002) revealed severe fruit drop in apple due to boron deficiency. Excessive nutrient application may also accentuate fruit drop. In a study by Hoffman (1940), a heavy pre-harvest fruit drop was observed in apple *cv.* McIntosh which, can be attributed to the high N status of orchard.

4.2.3.1. Management

Effective nutrient management through timely application of nutrients helps in increasing female flower per cent and checks fruit drop (Hiwale, 2015).The application of nitrogenous fertilizers about a few days prior to flower emergence favours good fruit set and yields. Foliar sprays of recommended micro-nutrients is helpful in correcting nutrient deficiencies and improving crop load. A study by Ghosh et al. (2009) found that foliar application of borax increases fruit retention in aonla by improving transport of photosynthates and various other physiological processes such as N metabolism, active absorption of salts, hormonal metabolism, fat metabolism etc. within the plant. The micronutrients like boron, zinc etc. plays a significant role in reducing fruit drop through maintenance of cell membrane integrity and prevention of abscission layer formation (Zoremtluangi et al., 2019). Ashraf et al. (2012) found that foliar sprays of nutrients *viz.* potassium and zinc are beneficial in improving fruit retention in kinnow. Correcting Mg deficiency through foliar sprays improves pre-harvest fruit drop in apple (Arseneault and Cline, 2016).The application of different micronutrients for regulating fruit drop and subsequently, fruit retention is mentioned in Table 4. Thinning can be also practised to improve fruit retention in different crops. It reduces competition for nutrition between growing fruitlets. Wani et al. (2010) mentioned that early fruit thinning in pear by retaining one fruit per cluster improved the fruit retention.

Table 4: Effect of various nutrients on fruit drop per cent of different fruit crops

Sl.No.	Fruit crop	Nutrient& Effective dose	Response	References
1.	Guava	$ZnSO_4$ (0.6%) + B (0.6 %)	Fruit drop per cent reduced i.e. 34.40 %	Yadav et al. (2014)
2.	Guava	Mg (0.4 %)	Maximum fruit retention (46.27%)	Singh et al. (1982)
3.	Aonla	B (0.6 %)	Fruit drop was reduced (66.21 %) and fruit retention was improved (29.85 %)	Tripathi et al. (2018)
4.	Mango	B (125g) + Potash (K) 800g	Reduced fruit drop (81.95 %)	Gupta et al. (2016)
5.	Kinnow Mandarin	Zinc and Boron soil application (250 g each) and foliar spray (3g+2g)	Reduced fruit drop per cent (28.48 %)	Davinder et al. (2017)
6.	Kinnow Mandarin	$MgSO_4$ (0.6 %)	Checked fruit drop and Increased fruit retention	Kaur et al. (2016)
7.	Khasi Mandarin	Zn (0.5 %) + Cu (0.4 %) + B (0.1 %)	Increased fruit retention (65.83 %)	Zoremtluangi et al. (2019)

4.2.4. Hormonal Imbalance

Plant hormones are natural compounds that act at low concentrations for proper growth and development of fruits. Among the phytohormones, auxin and ethylene primarily play a crucial role in fruit development and maturity (Racsko et al., 2006). The endogenous level of these hormones directly influence abscission. Auxin regulate ethylene synthesis and fruit abscission. In addition to this, Yuan and Huang (1988) revealed a positive correlation between endogenous ABA concentrations and fruit abscission. They found that fruits subject to abscission exhibit higher ABA status as compared to fruits retained on tree. Likewise Singh and Singh (2020), indicated that a high status of growth inhibitors and low status of growth promoters play a major role during first and second waves of fruit drop in apple.

4.2.4.1. Management

Appropriate use of plant growth regulators with or without macro- and micronutrients may aid in regulating excessive fruit drop in horticultural crops.

The plant growth regulators recommended for checking fruit drop in different fruit crops are mentioned in Table 5.

Table 5: Use of plant growth regulators (PGRs) for fruit drop regulation (Singh and Singh (2020) and Bons and Kaur (2020))

Sl. No.	Fruit crop	Growth regulator	Concentration	Time of application
1.	Litchi	NAA or 2, 4-D	10-15ppm +1% ZnSO4.	During fruit development
2.	Apple	NAA or 2, 4-D or 2,4,5- T	10 ppm 20-50 ppm	Soon after the petal fall.
3.	Grapes	Gibberellic acid GA3 or PCPA	100 ppm	Before harvesting
4.	Mango	NAA or 2, 4-D	20-30 ppm	Fourth week of April or as the fruits attain marble size.
5.	Citrus	2, 4-D or GA3	8-10 ppm 50 ppm	During the growth period of young fruits, two foliar sprays may be given.
6.	Bael	NAA	20 ppm	Spray after 7 days of initiation of growth &II spray in last week of June)
7.	Sapota	NAA	200 ppm	Two sprays (1 month before flowering and at pea stage)
8.	Custard apple	GA3	75 ppm	Flowering and fruit set stage (twice)
9.	Ber	NAA	30 ppm	Two sprays (third week of Oct and third week of Nov)

4.2.5. Diseases

The pathological fruit drop may be induced by different plant pathogens. *Colletotrichum gloeosporioides*, *Diplodia natalensis* and *Alternaria citri* are the common pathogens causing fruit drop in citrus (Rattanpal et al., 2019). The abscised fruits are unmarketable and cause serious yield as well as economic losses to orchardists.

4.2.5.1. Management

Timely removal of dead, decaying or disease infested shoots,whichact as the primary source of disease inoculum, through pruning. Apply Bordeaux paste on the cut ends of pruned shoots or spray copper based fungicides to check any further infection. Collect and bury fallen mummified fruits in soil to check secondary disease infection.

4.2.6. Insect Pests

Pre-harvest drop may also be induced by severe incidence of insects such as mites, leaf miners, fruit fly and other insect species. Heavy insect infestations may suppress photosynthetic activity of leaves resulting in limited availability of photosynthates to the fruits. Akter et al. (2020) recorded a significant drop in mango yield due to fruit fly damage. Excessive pre-mature fruit drop was observed in citrus fruits like kinnow, lemon, sweet orange, grapefruit etc. due to damage by fruit fly and fruit sucking moth (Rattanpal et al., 2019). Similarly Karar et al. (2009) indicated that severe infestation of mealy bug aggravates fruit drop in mango.

The coincidence of insect damage with nutrient imbalances or water stress can further aggravate the proportion of drop in various fruit crops.

4.2.6.1. Management

- Orchard floor should be kept clean by collecting and destroying the damaged fruits and plant debris.
- Ploughing may be practiced in orchards.It minimizes the insect incidence by exposing hibernating insectsor egg-mass.
- The host crops like wild plants, weeds, bushes etc. growing in and around the orchard should be uprooted to control the insect pest population.
- Practise bagging in fruit crops. This technique creates a physical separation between developing fruit and environment resulting in better protection against insect pests (Akter et al., 2020; Sharma et al., 2020).
- Install fruit fly traps in orchard @ 16 traps per acre to control the adult female fruit flies (Rattanpal et al., 2019).
- Integrated pest management using sticky bands, conservation of natural enemies, spray of botanicals like neem seed extract, garlic oil etc. may aid in suppressing the entomological factors.

5. Conclusion

Successful fruit production depends on final crop yield which is a function of fruit set and retention. Fruit drop is an undesirable event which causes severe economic losses to growers. There is a comprehensive list of factors associated with this problem. Based on studies, the adoption of different techniques has been recommended for managing fruit drop. The use of plant growth regulators and micronutrients is quite extensively studied and adopted by the orchardists. However, integrated approaches encompassing cultural, mechanical as well as chemical methods may be more beneficial for better and sustainable results. More research studies are required in this context.

References

Addicott, F., 1982. Abscission, 1st ed. University of California Press, Berkeley. 1–369.

Akter, M. M., Islam, M. T., Akter, N., Amin, M. F., Bari, M. A., Uddin, M. S. 2020. Pre-harvest fruit bagging enhanced quality and shelf-life of mango (*Mangifera indica* L.) cv. Amrapali. Asian J. Agricl.Horti. Res. 45-54.

Anonymous. 2018. National horticulture board. Retrieved from http://www.nhb.gov.in

Arseneault, M.H., Cline, J.A. 2016. A review of apple preharvest fruit drop and practices for horticultural management. Sci. Horti. 211, 40-52.

Ashraf, M. Y., Yaqub, M., Akhtar, J., Khan, M. A., Ali-Khan, M., Ebert, G. 2012. Control of excessive fruit drop and improvement in yield and juice quality of Kinnow (Citrus deliciosa x Citrus nobilis) through nutrient management. Pak. J. Bot., 44, 259-265.

Bajpai, P. N. 1968. Studies on flowering and fruit development in Aonla (*Emblica officinalis* Gaertn.). Hortic Adv VII. 38–67

Bakshi, P., Wali, V. K., Iqbal, M., Jasrotia, A., Kour, K., Ahmed, R., Bakshi, M. 2015. Sustainable fruit production by soil moisture conservation with different mulches: A review. African J. Agricl. Res. 10(52), 4718-4729.

Bons, H.K. Kaur, M. 2020. Role of plant growth regulators in improving fruit set, quality and yield of fruit crops: a review. J. Horticl. Sci. Biotech. 95(2), 137-146.

Byers, R.E. 1997. Effects of amino ethoxy vinyl glycine (AVG) on preharvest fruit drop, maturity, and cracking of several apple cultivars. J. Tree Fruit Prod. 2, 77–97.

Davinder, A.M., Kumar, A., Singh, R., Pratap, S., Singh, B. 2017. Impact of zinc and boron on growth, yield and quality of Kinnow (Citrus deliciosa x Citrus nobilis) in sub-tropical conditions of Punjab. J. PureApp. Microbio. 11(2), 1135-1139.

Din, S., Wani, R.A., Ab Waheed Wani, F.N., Farwah, S., Rizvi, S., Nisar, T.F.W.S. 2019. enhancement in temperate fruit crops. J. Pharmacog. Phytochem. 8, 2, 1203-1216.

Gaikwad, S. C., Ingle, H. V., Panchbhai, D.M., Ingle, S.H. 2004. Effect of different types of mulches on soil moisture, soil temperature and fruit drop in Nagpur Mandarin. Agric. Sci. Dig. 24(1), 71-72.

Gupta, V., Das, R., Bhattacharyya, D., Hazarika, B. 2016. Effect of Boron and Potash on Control of Fruit Splitting and Fruit Drop in Mango (*Mangifera indica* L.) cv. Amrapali. Advances. 2093.

Haldhar, S. M., Kumar, R., Samadia, D. K., Singh, B., Singh, H. 2018. Role of insect pollinators and pollinizers in arid and semi-arid horticultural crops. J. Agri. Eco. 5, 1-25.

Hiwale, S. 2015. Aonla (Indian gooseberry). In Sustainable Horticulture in Semiarid Dry Lands (pp. 83-96). Springer, New Delhi.

Hoffman, M.B. 1940. The pre-harvest drop of mature McIntosh apples as influenced by applications of nitrogen carrying fertilizers. Proc. Amer. Soc. Hort. Sci. 37, 438-442.

Jackson, J.K. 2003. Biology of apples and pears. Cambridge University press, UK. pp :268-340.

Karar, H., Arif, M. J., Sayyed, H. A., Saeed, S., Abbas, G. and Arshad, M. 2009. Integrated pest management of mango mealybug (*Drosicha mangiferae*) in mango orchards. Int. J. Agri. and Bio. 11, 81-84

Kaur, R., Kaur, N., Rattanpal, H.S. 2016. Effect of mineral nutrients and the growth regulators in management of fruit drop and improvement of fruit quality in Kinnow mandarin. The Bioscan. 11, 1, 589-596.

Léchaudel, M., Joas, J. 2007. An overview of preharvest factors influencing mango fruit growth, quality and postharvest behaviour. Brazilian J. Plant Physio. 19, 287-298.

Luckwill, L.C. 1953. Studies on fruit development in relation to plant harmony. I. Hormone production by the developing apple seed in relation to fruit drop. J. Hort. Sci. 28, 14-24.

Marcelis, L.F.M., Heuvelink, E., Baan Hofman-Eijer L.R., Den Bakker J., Xue L.B. 2004. Flower and fruit abortion in sweet pepper in relation to source and sink strength. J. Exp. Bot. 55, 2261-2268

Mollapur, Y., Miri, S. M., Hadavi, E. 2016. Comparison of foliar fertilizers and growth regulators on pre-harvest drop and fruit quality of Thompson Navel orange. Open Agri. 1, 1.

Murneek, A. E. 1937. Pollination and fruit setting. University of Missouri, College of Agriculture, Agricultural Experiment Station.

Norton, R. L. 1988. Windbreaks: Benefits to orchard and vineyard crops. Agriculture, Ecosys. & Envi. 22, 205-213.

Pal, M., Lal, R.L.,Nautiyal, P. and Joshi, P. 2016. Effect of chemicals and physical means on harvesting span, yield and quality of litchi (*Litchi chinensis* Sonn.) *cv.* Rose Scented. J. Appl. Hortic. 18, 71–75.

Pande, K.K., Dimri, D.C., Kamboj, P. 2005. Effect of various mulches on growth, yield and quality attributes of apple. Ind.J.Horti. 62(2), 145-147.

Racsko, J., Nagy, J., Soltész, M., Nyeki, J., Szabo, Z. 2006. Fruit drop: I. Specific characteristics and varietal properties of fruit drop. Int. J. Horti. Sci. 12, 2, 59-67.

Rattanpal, H.S., Singh, G., Arora A., Singh, S. 2019. Management of Fruit drop in citrus. Punjab Agricultural University, Ludhiana.

Roemer, M.G., Hegele, M., Wünsche, J.N., Huong, P.T. 2008. Possible physiological mechanisms of premature fruit drop in mango (*Mangifera indica* L.) in Northern Vietnam. In IX International Symposium on Integrating Canopy, Rootstock and Environmental Physiology in Orchard Systems 903, 999-1006.

Singh, S., Singh, A. 2020. Significance of Fruit Drop in Horticulture. Agriculture and Food: e Newsletter. 4, 2, 738-741.

Singh, S.P., Singh, B.P., Moti, R. 1982. Effect of magnesium and season on growth, flowering, fruiting and yield of guava (*Psidium guajava* L.) *cv*. Allahabad safeda, Haryana J. Hort. Sci. 11(3-4), 188-191.

Schupp, J.R., Greene, D.W. 2004. Effect of amino ethoxy vinyl glycine (AVG) on preharvest drop, fruit quality, and maturation of 'McIntosh' apples. I. Concentration and timing of dilute applications of AVG. Hort Sci. 39, 1030–1035

Sharma, R.R., Nagaraja, A., Goswami, A.K., Thakre, M., Kumar, R., Varghese, E. 2020. Influence of on-the-tree fruit bagging on biotic stresses and postharvest quality of rainy-season crop of 'Allahabad Safeda' guava (*Psidium guajava* L.). Crop Prot. 135, 105216.

Sharma, L.K., Mankotia, M.S., Chauhan, P.S., Sud, A. 2003. Studies on the effect of micro-irrigation levels on growth, yield, fruit quality and nutrient assimilation of delicious apple. In VII International Symposium on Temperate Zone Fruits in the Tropics and Subtropics-Part Two. 696, 193-196

Stephenson, A. G. 1981. Flower and fruit abortion: proximate causes and ultimate functions. Annual review of ecology and systematics. 12(1), 253-279.

Tripathi, V.K., Pandey, S.S., Kumar, A.J. A.Y., Dubey, V., Tiwari, P. 2018. Influence of foliar application of gibberellic acid, calcium and boron on fruit drop, yield and quality attributes of aonla (*Emblica officinalis*) cv. NA 7. Ind.J.Agricl. Sci. 88,(11), 1784-88.

Walsh, C.S. 1977. The relationship between endogenous ethylene and abscission of mature apple fruit. J. Amer. Soc. Hort. Sci. 102, 615-619.

Wani, I.A., Bhat, M.Y., Lone, A.A., Mir, M.Y. 2010. Unfruitfulness in fruit crops: Causes and remedies. African J. Agricl. Res. 5(25), 3581-3589.

Wright, S.T.C. 1956. Studies on fruit development in relation to plant hormone. III Auxin in relation to fruit morphogenesis and fruit drop in the black currant, Ribes nigrum. J. Hort. Sci. 31, 196-211.

Yadav, R.K., Ram, R.B., Kumar, V., Meena, M.L., Singh, H.D. 2014. Impact of micronutrients on fruit set and fruit drop of winter season guava (*Psidium guajava* L.) cv. Allahabad Safeda. Ind. J. Sci. Tech. 7(9), 1451-1453.

Yuan, R., Huang, H. 1988. Litchi fruit abscission: its patterns, effect of shading and relation to endogenous abscisic acid. Sci.Horti. 36(3-4), 281-292.

Zoremtluangi, J., Saipari, E., Mandal, D. 2019. Foliar application of zinc, manganese, copper and boron influenced the fruit growth, development and quality of Khasi Mandarin (*Citrus reticulata* Blanco). J. Pharma. Phytochem. 8(3), 3324-3327.

16

Forestry Based Lac and NTFP for Sustainable Livelihood

Praveen Kumar Singh Gurjar[1], Nikhil Parihar[1], Ravi Y[2] G.L. Kumawat[3] and Abhijeet Ghatak[4]

[1]RVS Krishi Vishwa Vidyalaya, Gwalior, Madhya Pradesh, India
[2]ICAR-National Research Centre on Seed Spices, Ajmer, Rajasthan, India
[3]Sri Karan Narendra Agriculture University, Jobner, Rajasthan, India
[4]Bihar Agricultural University, Sabour, Bihar India

1. Introduction

Lac and other non-timber forest products (NTFPs) provide vital sustenance, employment, and financial revenue to producers and collectors as well as serving a raw material for a range of businesses. They are also important for ecological balance and long-term growth. Weather and price fluctuations have a significant impact on the production of such natural items. Food, fodder, and medication are all ingested straight from these plants. The Indian subcontinent is a significant centre of animal and plant biodiversity. In tropical climates, a variety of forest products play an important role in social and economic life. Wood and Non-Wood Forest Products are the two types of forest products (NWFPs). NWFPs are "products of biological origin other than wood, obtained from forests, shrublands, and tree plantations," according to FAO (1999).

Many NWFPs have economic value and make a significant contribution to the local and national economies. The number of commercially important non-wood plant species worldwide is estimated to be between 4000 and 6000 (FAO, 2001). Natural resins, gums, exudates, leaves (kendu or tendu), pine turpentine, and fragrance oils from roots, stumps, and fruits of numerous tree species are all examples of NWFPs. The majority of NWFPs are export-oriented, and many of them are well-suited to local small-scale enterprises. Natural resins, gums, and tamarind seeds are more lucrative than timber in tribal and poor district regions of India. Despite the importance of NWFPs, forest management has historically been dominated by wood output (Yogi et al., 2016).

India has surpassed China as the world's top producer, user, and dealer of non-wood forest products (NWFPs). This quick evolution has resulted in significant changes in the worldwide image of NWFPs commerce, as well as increased awareness of future demand and supply on a global scale. India has a population of 1.252 billion people and is a tropical (dry and wet) country. About 21.23 percent of the land is covered with forest (FSI, 2013). Killman (2003) reported that 25 percent of the world's 6.2 billion people rely on forest resources to some degree for their life, and 350 million people live in or near thick forest regions and rely heavily on them for sustenance or livelihood. For their basic health and nutritional needs, around 80% of people in developing nations rely on non-wood forest products such as fruits and herbs (FAO, 2008).

2. Global Scenario

Lac is mostly manufactured in India, with modest production in Indonesia, China, Myanmar, the Philippines, Vietnam, Cambodia, and a few other countries. India is the world's largest lac producer, and Indian jujube (*Ziziphus mauritiana*) is one of the key contributors. Sticklac was exclusively available in Switzerland. Pakistan (45%) was the leading exporter of sticklac, followed by the United States (34%), and Germany (34%).The top three export destinations for Indian shellac were Bangladesh (33%), Iraq (14%), and Egypt (9%). Aleuritic acid, a high-value lac product, has been sold to a variety of nations, with Switzerland (82%) and China (18%) taking the lead in exports this year. The United States (84%) and the United Kingdom (14%) were discovered to be the top importers of bleached lac from India (Jaiswal et al., 2012).

3. National Scenario

The average lac output in India in the 1950s was around 42320 tonnes, with a downward trend in lac production in the 1970s (52 percent), 1980s (19 percent), and 1990s (4 percent). However, during the resurgence phase of the 2000s, there was a 24 percent growth. Efforts in the lac sector in terms of policy, research, and development percolated at the grassroots level throughout this decade, and stakeholders' interests received critical support. Only around 5% of lac host trees are currently being cultivated in a systematic manner.A comprehensive detail of lac production contributed by different states of India is presented in Tale 1.

Table 1: Lac productionin India during 2018-19 (in tonnes)*

Name of states	Name of lac crop/strain						Total (% share)
	Baisakhi	Katki	Rangeeni	Jethwi	Aghani	Kusmi	
Jharkhand	634	286	920	4640	4991	9631	10551 (56.92)
Chhattisgarh	760	505	1265	550	1635	2185	3450 (18.61)
Madhya Pradesh	1222	826	2048	42	46	88	2136 (11.52)
West Bengal	69	610	679	80	203	283	962 (5.19)
Maharashtra	295	405	700	7	15	22	722 (3.90)
Odisha	12.5	9	21.5	189	436	625	646.5 (3.49)
Assam	6	15	21	0	0	0	21 (0.11)
Uttar Pradesh	15	1	16	0	0	0	16 (0.09)
Gujarat	5	1	6	4	6	10	16 (0.09)
Andhra Pradesh	6	2	8	1	1	2	10 (0.05)
Meghalaya	5	1	6	0	0	0	6 (0.03)
Total	3029.5 (16.34)	2661 (14.36)	5690.5 (30.70)	5513 (29.74)	7333 (39.56)	12846 (69.30)	18536.5 (100.00)

**Source*: Indian Institute of Natural Resins and Gums, Ranchi

4. Potential of Lac Production

India is the world's biggest supplier of NWFPs based raw materials for the food, paint, varnish, cosmetic, and pharmaceutical sectors. According to a survey conducted in the Jharkhand districts of Ranchi and West Singhbhum, the percentage of lac host plants used was quite low. The majority of lac producers have 26-50 palas and ber trees, as well as roughly 5 kusum host plants. At the farmer level, the proportion of palas, ber, and kusum used for lac cultivation was 28 percent, 54 percent, and 17 percent, respectively, of the total accessible lac host trees. The maximum usage of accessible lac host trees was clearly noticed for ber (Killman, 2013).

5. Institutional Arrangement

Tribal people rely on the forest for 60 percent of their food and medicine, as well as more than half of their revenue. According to the Act of 2006, "Minor Forest Produce" comprises all NTFP of plant origin, such as bamboo, brushwood, stumps, cane, tassar, cocoons, honey, wax, lac, tendu leaves, medicinal plants and herbs, roots, tubers, and so on. State governments of Jharkhand, Chhattisgarh, Madhya Pradesh, and Andhra Pradesh including Telangana, Maharashtra, and Gujarat are active in coordinating lac-related activities such as production, processing, marketing, training, and research and development. Many non-governmental organisations, including VikasBharati, Torpa Rural Development Society (TRDS), Ramakrishna Mission Ashram (RKMA, Ranchi), PRADAN, Kadamba Charitable Foundation (KCF), and others, are actively working to promote NTFP-related initiatives.

6. Brood-Lac Farm

There are around 50 large broodlac farms dispersed across lac-growing states, with a large number (more than 2.0 lakh) of lac host trees that have the capacity to produce enough broodlac if maintained properly. Ranchi, Khunti, Gumla, Saraikela, Hazaribag, Palamu, Dhanbad, Bokaro, Godda, and Chatra are among the districts in Jharkhand where broodlac farms can be found. West Bengal (Purulia), Chhattisgarh (Kanker, Ambikapur, Bilaspur, Korba, Raigarh, and Bastar), Odisha (Sundergarh), Madhya Pradesh (Seoni and Balaghat), and Maharashtra (Seoni and Balaghat) are home to the remaining farms (Gondia and Garhchiroli). Many non-governmental organisations, such as VikasBharati in Gumla, Torpa Rural Development Society (TRDS) in Khunti, Ramkrishna Mission Ashrama (RKMA) in Ranchi, Jhalda Lac Growers Association (JLGA) in Jhalda, and so on (Jaiswal et al., 2011).

7. Lac, Lac Insect and Lac Crops

Lac is a natural resin produced by the lac bug, a small insect. Lac is a member of the Homoptera order's Coccid group. Lac insects are found in India in 19 species, with three of them, *Kerria lacca*, *Kerria chinensis*, and *Kerria sharda*, being the most commonly employed for lac production. *Kerria lacca* is the most significant lac bug for commercial use (Kerr.). Rangeeni and kusmi, two strains of *K. lacca*, generate two harvests per year (bivoltine) and contribute considerably to lac production. The kusmi bug thrives on kusum (*Schleichera oleosa*), whereas the rangeeni strain thrives on palas (*Butea monosperma* Taub.). In a year, the Rangeeni strain produces two crops: katki (rainy season) and baisakhi (summer season).

In India, lac host plants such as palas (*B. monosperma*), ber (*Ziziphus mauritiana*), and kusum (*S. oleosa*) are commercially grown. Palas is used to grow rangeeni lac, kusum is used to grow kusmi lac, and ber is used to grow both forms of lac. Lac insects thrive in mild climates. Lace farming may benefit from annual rainfall of 1000 to 1500 mm and typical temperatures of 24°-27°C. Rainfall distribution is critical for optimal lac production. During key periods of lac growth, consecutive weeks with more than 50 mm rainfall have an impact on lac yield. Lac producers harvest two crops every year, one from the rangeeni strain (October-November and June-July) and the other from the kusmi strain (January-February and June-July).

8. *Z. mauritiana* for Commercial use

The Rhamnaceae family includes *Z. mauritiana*. *Z. jujube* is found in tropical Africa, Burma, Barbados, Jamaica, Guadeloupe, Iran, Martinique, Sri Lanka, Syria, and areas of the Mediterranean as natural vegetation. It has a significant commercial significance in India and the Chinese subcontinent. *Ziziphus mauritiana* is a prickly, evergreen shrub or small tree with a trunk diameter of 40 cm or more and a mature height of up to 15 m.

More unique characteristics of the tree include a spreading crown, stipular spines, and many drooping limbs. It produces fruit in a variety of shapes and sizes. Depending on the variety, this might be oval, obovate, oblong, or round, with a length of 2.5-6.25 cm (1.0-2.5 inch). The Indian plum tree may be found in rural and sub-forest settings across India.

It is cultivated along the edges of paddy fields in various regions. Plum trees are used to raise kusmi and rangeeni lac crops in lac-growing regions. Farmers utilise the ground beneath the trees to cultivate crops such as potato and a variety of shade-loving plants such as colocasia as part of their routine agricultural operation. It is an important host plant for both the kusmi and rangeeni lac bug strains. There are around 93 million ber trees in the nation that might be employed for lac production to generate money. *Z. mauritiana* is a hardy tree that can withstand severe temperatures and thrives in hot, dry circumstances.

Kusum (*Schleichera oleosa*) is also the most common host tree for the growth of kusmi lac. Only kusum in India produces the highest quality resin generated by the kusmi strain. It belongs to the Sapindaceae family. In the nation, there are around 36.7 million kusum trees.

Palas (*Butea monosperma*) is India's most widely planted tree and a primary lac host tree for rangeeni lac farming and is a member of the Fabaceae family (Papilionaceae). Rangeeni strain contributes roughly 45 percent of all lac generated in the country, with palas contributing about 65 percent, ber 30 percent, and other minor lac host trees/plants contributing about 5 percent. There are around 491.7 million palas trees in the nation. Semialata (*Flemingia semialata*) is a bushy plant with the potential to boost lac output via systematic plantation development. This host plant is mostly utilised in the growth of kusmi lac.

8.1. Merits of *Ziziphus mauritiana* for Lac Production

- Suitable for both rangeeni and kusmi lac insect can cultivated
- Productivity of both types of lac is substantially high and coppice ability isalso high
- No adverse effect on health of host trees, if harvesting of crop and pruningis carried out simultaneously.

8.2. Limitations of *Ziziphus mauritiana* for Lac Production

- Large number of thorns
- Naked condition during peak summer (May-June) due to leaf fall
- Pruning time does not match with crop maturity time of rangeeni lac insect.

9. Investment Prerequisite and Income Generation

The convenience of climate, (temperature, tropical and subtropical), rainfall, soil requirements, irrigation infrastructure, equipment, the seed for nursery rearing, transplanting schedules, pruning, inoculation, phunki removal, pest control, harvesting, and so forth, were calculated for the various lac host trees/plants. A tabular analysis is used to display various inputs like labour, machinery, broodlac, pesticide doses, fertilisers and manures, production, use and disposal patterns, and so forth. In monetary terms, a comprehensive bundle of lac growing procedures for one hectare is also provided.

10. Conclusion

The wild Indian plum tree has a minimal economic benefit for farmers in natural vegetation, particularly forest areas. Because of its low market value, the fruit is either eaten locally or thrown away. If these trees are used for lac cultivation, enough jobs and revenue may be produced at the local level. Furthermore, there is no negative impact on the host tree; rather, harvesting the lac crop serves as a form of pruning and aids in the overall health of the tree in forest/rural environments. The cash generated by this enables disadvantaged individuals to feed their families for 3-4 months each year.

References

FAO, 2001. State of the Woerld's Forest. Food and Agriculture Organization of the United Nations, Rome, Italy. (3), 143-183.

FAO, 2008.Contribution of the forestry sector to national economies 1990-2006, by A. Lebedys. FAO Forest Finance Working Paper FSFM/ACC/08.Non Wood Forest Products, Rome, Italy.(7), 215-238.

FSI, 2013. Indian State of Forest Report. Forest Survey of India, Government of India, Dehradun.

Jaiswal, A.K. 2015. Pest management in lac cultivation. In book: Advances in Lac Production, Processing, Product Development and Value Addition, Edition: First Edition 2015, Chapter: 12, Publisher: ICAR-Indian Institute of Natural Resins and Gums, Kumar, A.; Jaiswal, A. K.; Singh, A. K. and Yogi Raj, K. (Eds.), pp. 89-95.

Jaiswal, A.K., Sharma, K.K. Ramani, R., 2011. Lac culture operations-When, Why & How, IINRG (ICAR), Ranchi.

Jaiswal, A.K., Singh, J.P., Yogi, R.K., 2012. Project profile of Khunti project Lac cultivation and processing unit establishment. IINRG (ICAR), Ranchi. (4), 94-106

Killman, W. 2003. Non-wood News, No.10, March 2003, (1), 28-36.

Yogi, R.K., Bhattacharya, A., Jaiswal, A.K. Kumar, A. 2015. Lac, Plant Resins and Gums Statistics 2014: At a Glance. ICAR-Indian Institute of Natural Resins and Gums, Ranchi (Jharkhand), India. Bulletin 07/2015, pp. 1-68.

Yogi, R.K., Sharma, K.K. Ramani, R. 2016. Model Bankable Projects: Lac Cultivation for Livelihood Security. ICAR-Indian Institute of Natural Resins and Gums, Ranchi, Jharkhand. Bulletin: 17/2016, pp. 1-34.

Yogi, R.K., Singh, A.K., Kumar, A. 2015a.Endo and Exogenous Environmental Aspects of Natural Resins Production and Distribution in India. In book: Advances in Lac Production, Processing, Product Development and Value Addition, Edition: First Edition 2015, Chapter: 24, Publisher: ICAR-Indian Institute of Natural Resins and Gums, Kumar, A.; Jaiswal, A. K.; Singh, A. K. and Yogi Raj, K. (Eds.), pp. 198-206.

17

Turfgrass Sod Production A Profitable Enterprise

Simrat Singh, Rishu Sharma and K.K. Dhatt

[1]*Department of Floriculture and Landscaping*
Punjab Agricultural University, Ludhiana, Punjab, India

1. Introduction

A lawn is an indispensable feature of a landscape with a sizeable proportion of space relative to other garden features. It is regarded as the heart of a garden that enhances the beauty of a landscape providing both functional and aesthetic utility. A well maintained lawn at a residential, public or institutional places can be used both for active and passive recreation. It enhance the beauty of surroundings and contribute to real estate value. The turf industry is being viewed as an integrated sector of ornamental horticulture in the developed countries like USA, Australia, China, Netherland, UK, Japan and Denmark. Globally, the turf grass industry has employed skilled workforce for maintaining more than 700,000 athletic fields and over 17,000 golf courses, with an annual worth of $ 20 million each year (Aldous and Chivers, 2002). Presently, United States alone has around 62 million acres of land under turfgrass covering majority of parks, residential lawns, golf courses and amenity places. Globally, residential segment is dominating the market with revenue of $5444.9 million and expected to grow at a CAGR of 6.1% during 2018 to 2023. In India, 50 international and around 520 domestic cricket grounds and athletic fields are being maintained with different turfgrass species adapted to the region.

Golf courses forms an integral part of our domestic turfgrass industry. There are about 255 golf courses in India, out of which 25 are in Delhi NCR, 9 in Gujarat, 6 in Chandigarh, 8 each in Bangalore and Pune, 7 in Mumbai, 3 in Srinagar and 30 boutique golf courses in North Eastern India (Masoodi, 2021). Apart from that several golf courses are maintained in defence cantonment areas throughout the country. Turfgrasss industry in India has been estimated with a land cover of more than 30,000 acres, out of which around 4,500 acres is under turfgrass cultivation and sod production for providing grass sprigs and sod for developing diverse landscapes. Ganu Farm Group, is the leading turf supplier in India covering 300 acres under high quality turfgrass production. Apart from that good quality turf or sod is being produced in small farms and nurseries throughout India which caters our domestic needs of turf as well some suppliers are exporting it to Sri Lanka and Middle East countries on small scale. Due to rapid urbanization, the need for open spaces with well-maintained lawns for the people to walk and engage in active and passive recreation has led to increase in demand for quality turfgrass mat, technically known as 'sod' for turf establishment in a short period of time. The varied agro-climatic conditions in India, with availability of abundant sunlight and water, agri-inputs, skilled and un-skilled manpower, the production of turfgrass sod can prove to be a boon for the farmers. The commercial production of turfgrass sod production has evolved during early 1900s. As per University of Tennessee Extension, major consumers of sod are landscapers and building contractors accounting for 50% of total sod sales and secondly, homeowners accounting for 15% of total sod sales in United States (Mathews and Pemrick, 2007).

2. Sod Culture and its Relevance

Sod or turf consists of intermingled grass segments with a part of the soil adhered to the roots. Commercially, turfgrass sod is a weed free, lush green grass cover produced in intensively managed cultural practices. This sod when harvested or removed mechanically with intact roots and soil, establishes overnight over a well prepared surface and instantly forms a green cover. The sod culture is done in specialized large farms, particularly on large scale in USA, Australia, Canada, New Zealand and countries of European Union. The sod farms produces different varieties of grass adapted to the edaphic and climate of the region and also provide services as per the consumer's preference and intended use of the area to be turfed. Commercial farms are usually established in the vicinity of target market in order to reduce production cost and damage to turf due to longer transit distance.

Sodding is a rapid method to establish a green cover during active period of growth. The green cover can be laid on sloppy terrains and is best for certain sites where grass seeds fail to establish or where there is need to cover the exposed patches in lawn. A well-established sod protects the surface from wind and erosion of soil. The areas covered with sod require less irrigation after planting than those areas that are seeded or dibbled. Further, there is limited application of herbicides for managing the weeds. Well established and maintained sod covers discourages the activity of rodents and other pests, and prevent these from dwelling and forming a habitat. Sod also enriches soil with plant nutrients with its extensive root system that decays to form a humus.

2.1. Species Selection

Most of the warm-season grasses can be grown for sod production on account of their spreading growth habit. Bermuda grass is widely used for production of sod, however other warm season grass species such as seashore *paspalum, zoysia grass, centipede grass, St. Augustine grass, buffalo grass, carpet grass* and *bahia grass* can also be raised through sod. Selection of suitable cultivar for sod production varies by agro-climate of a region and tolerance of the species to abiotic and biotic stresses. Some of the turfgrass species are slower to establish, but are in great demand due to their quality or better stress tolerance. Choosing the appropriate grass species (Table 1) for sod production makes a significant difference in the overall success of a new lawn.

Table 1: Different types of grasses with sod production ability

Grass species	Growth habit	Sod production ability	Remarks
Kentucky bluegrass	Rhizomatous	Excellent	Slow to establish from seed.
Creeping bentgrass	Dense stolons	Good	Difficult to manage, limited market. High maintenance costs
Perennial ryegrass	Tillers forming a cluster	Preferably mixed with Kentucky bluegrass	Does not form rhizomes and therefore not mixed more than 25% in mixture (on weight basis).
Fine fescue	Bunching with short rhizomes	Fair to moderate	Much slower
Tall fescue	Bunching with short rhizomes	Mixing 5 to 10% Kentucky bluegrass	Shorter rhizomes, weak sod strength, use of netting is advisable
Bermuda grass	Rhizomatous and stoloniferous	Good	Not suitable for shaded locations
Seashore paspalum	Rhizomatous and stoloniferous	Good	Spreads excessively by stolons, therefore weekly edging required.
St. Augustine Grass	Rhizomatous and stoloniferous	Good	Chemicals weedicides harm the grass. Grows well in shade
Centipede Grass	Stoloniferous	Good	Extremely low maintenance grass

2.2. Bermuda Grass – A Versatile Grass for Diverse Landscapes

Bermuda grass (*Cynodon spp.*) is the most widely planted as turfgrass with several desirable traits including drought resistance, wear tolerance, disease resistance and relatively better recuperative potential. It actually originates from different areas of Asia, Africa, Australia and Southern Europe (Zhang et al., 2020). Bermuda grass is commonly seen as a prominent grass species for lawns, amenity places, institutional playgrounds, golf courses etc. where a neat and dense turf is desirable. The species is tolerant to lower mowing height. Certain cultivars of Bermuda grass have been successfully used in golf course putting greens. *Cynodon dactylon* [L.] Pers. is the most widely used turfgrass species among the nine species of the genus

Cynodon. This species has been known by several other names such as couch grass, kweek, doub (or doob) and green couch (Taliaferro, 1995). Ecologically, the Bermuda grasses is well suited for soil stabilization at steep slopes, checking soil erosion, providing habitat for several soil borne insects and micro-organisms. This grass responds well to frequent mowing, moderate fertilization and exhibits better performance under low-maintenance conditions. Common Bermuda grass can be established from seed, sod, or sprigs.

3. Considerations to Start A Sod Production Venture

There are several key factors that will help the prospective grower to undertake a decision, to plan a sod production farm and study its feasibility as a viable business enterprise. Another very important aspect to consider is who will buy the sod turf. The marketing of sod is different than those of traditional horticultural crops, as the production of sod is a specialized enterprise and need a considerable stretch of landholding for sizeable production. It is worthwhile to identify the potential markets to determine the additional expenditure on logistics. For example, Dhaliwal turfs in Ludhiana, Punjab is a regional enterprise and they are prospering in this venture year by year by defining suitable market strategies. Setting up the commercial sod farm near big city was their key to success as it provided them suitable market linkages. Now, this farm is fulfilling the turf demands in Punjab and other neighboring states (Haryana, Rajasthan, Himachal Pradesh, New Delhi).

Following key factors will help in decision making process for a prospective sod producer to venture in sod production business.

- Analysis of demand-supply in particular area or neighboring states
- Survey of real estate establishments and community parks
- Any professional sports stadiums nearby that require sod for turf establishment

Available varieties that respond to sod production under natural conditions

3.1. Initial Investment and Start up Cost

Establishing a sod farm takes a considerable planning and capital investment. The highest cost is incurred in land, manpower and specialized equipments that are needed for quality production and harvesting of sod for timely delivery to landscape designers and other customers. The equipments required to operate a sod production farm include tractor, cultivator, sod cutter, lawnmower, roller, fertilizer spreader and vehicles for delivery of inputs and sod.

Manual kick txpe sod cutter

Electric sod cutter

Sod roller

Skind steer loader

Lawn roller

Electric lawn mower

Fig. 1: Machinery required for sod production business

4. Sod Establishment

4.1. Site Selection

The method for establishment for sod is generally same as for laying other turfgrasses, however, the sod producer needs to pay a special attention for the selection of site and other associated factors. The site for sod production should be free of rocks or stones underneath the soil surface to a depth of up to 2 feet and should be fertile enough. Although, the grasses are adaptable to grow over rocky soil, but stones and pebbles over the soil will likely make sod production impossible due to unevenness and abrasion that may degrade the sod quality. The site should be relatively leveled or preferably with a gentle slope (1-2%). Soil testing should be done to ensure the physicochemical properties be in ideal range for sod production. Drainage is also very important for the site to ensure adequate runoff of excess water. If the site is poorly drained, sod harvesting cannot be done efficiently and would take time of several days for the soil to reach at field capacity after rainfall. After selection of suitable site, it is critical to make a fine seed bed to ensure even sod establishment. The field is tilled to a 4- to 6 inches depth and lightly rolled with a planker to level the land. Care should be taken to avoid any unnecessary movement over the prepared field to avoid compaction.

4.2. Seed Quality and Seed Rate

Selection of quality seed is very crucial for obtaining uniform and true to type sod quality. Sod producers generally broadcast less seed than usually recommended for turf sod establishment. The reason being that the seeding turf at lower seed rates likely produces greater density of stolons/rhizomes relative to the turf seeding done at higher rates. Often a times, dense sod mats are required for high performance golf courses, therefore higher seed rates are generally preferred to ensure rapid establishment. It has been observed that the lower seeding rates for sod establishment can be as effective as higher rates provided soil is at proper tilth and adequate fertilization and irrigation is being undertaken. Following the lower seed rates will save the cost incurred on seed and will likely add to the profit in sod production.

Application of higher seed rates than recommended not only waste seed but also delay the maturity of sod due to competition among seedlings. For instance, the recommended seeding rates of Kentucky bluegrass established as a lawn is in the range of 50 to 100 kg/ha, whereas for sod production, usually the grass is seeded at 30-35 kg/ha. Commonly used Bermuda grass is seeded at the rate 8-10kg/ acre. Usually, the sandy soils require higher seeding rates compared to the soils higher in organic matter. It is generally with experience that the sod producers come up with the threshold level of seed rate for production of good quality sod.

4.3. Timing of Sod Establishment

Timing of sod establishment is also a key factor affecting the seeding rates for sod production. The favorable time to establish sod of cool-season grasses is during late summer to early falls, whereas for warm-season grasses, the ideal time for seeding is during spring. Spring seeding of cool-season grasses can result in application of higher seeding rates and may incur additional cost for application of herbicides to check the growth of annual weeds from competing with the emerging turfgrass seedlings.

The seeding equipment for uniform application of seed includes tractor-mounted drop spreaders that uniformly place the seed and cover it lightly with soil. In sod production, the aim is to uniformly spread the seed without overlapping to ensure lowest possible quantity of seed dropped per unit area. Application of seed in both the directions, extending across North-South and East-West orientation ensures uniform application of seed. For example, to broadcast 30 kg seed of Kentucky bluegrass for preparing a sod, generally 15 kg of seed is broadcasted in one direction and the remaining 15 kg seed is broadcasted perpendicular to the first direction.

4.4. Cultural Practices for Sod Establishment

4.4.1. Irrigation

Sod production can be undertaken under natural condition in humid regions, without irrigation. However, the time required for sod production can considerably be reduced to 6 to 12 months, with timely and assured irrigation. Actively growing sod is irrigated once or twice weekly or depending upon the prevailing weather conditions. Large sod farms use center-pivot irrigation systems to effectively cover a large area. Smaller sod farms often utilize portable irrigation systems that operate under pressure and make a jet spray of water in 360 degrees, which can be adjusted to avoid overlapping. These portable systems are removed before sod is harvested.

4.4.2. Fertilization

Quality sod production requires supply of adequate nutrient elements for maximizing the photosynthetic efficiency. Over fertilization, a very common practice among the sod producer is a waste of money with degradation of subsoil water due to leaching of soluble nutrients. Nitrogen (N) is the key element required in a timely and in adequate proportions for quality sod production. Higher application of N will turn the grass to dark green, but it will delay the maturity, that eventually lengthens the time of harvesting of sod. Excess N will favor shoot growth at the expense of root growth. However, in sod production, the growth and density of root and rhizome is critical, for adhering the roots with soil and maintaining the integrity of the sod following harvest. A sod producer should remember that the color of the sod is secondary concern, compared to the need for stronger development of root system that eventually will lead to healthy establishment at the consumer level. A moderate level of N fertilization will reduce the succulence and bring the sod to maturity in lesser time. Over fertilization of N will only enhance the excessive shoot growth which increases the need for mowing, therefore adding to the cost of production.

N fertilization is also effective in increasing wear tolerance and recuperative ability of turfgrasses. Although N fertilization varies according to species, it can be applied @ 100 to 300 kg/ha/year (Glab et al., 2020). For cool season grasses, the standard amount of N applied varies from 146 to 244 kg N/ha/year (Table 2) and for warm season grasses it varies from, 25 to 50 kg/ha per growing month. Nitrogen leaching is the major problem encountered in turf fertilization especially in sports turf as it requires regular irrigation and fertilization (Barton and Colmer, 2006; Groenveld et al., 2019). In order to minimize it, fertilizers should be applied during the active growth period of turfgrass. Apart from that over fertilization with N causes eutrophication, increase in greenhouse gases and ozone-depleting gases. Annual N leaching rates for turfgrass range from 0 to 160 kg N ha^{-1} that represent up to 30 % losses of the total N fertilization (Barton and Colmer, 2006). Fertigation in turfgrasses is a precise way of fertilization ensuring uniform application, reducing N losses and obtaining desired quality, growth rate and colour of turfgrass sod.

Table 2: Nitrogen fertilization schedule for sod production of cool season grasses

Stage or season of fertilization	Dose (Kg/ha)
Establishment stage (late summer)	Apply Nitrogen @ 22-34 kg /ha. Follow the same dose for each stage.
At one inches height	
Spring season	
Early summer	
Three weeks before harvesting	Apply Nitrogen @ 28-39 kg /ha.
Autumn season	Apply Nitrogen @ 39-50 kg /ha.

For quality sod production, development of a strong root system is an absolute necessity, which will promote the efficient lifting of the sod. Therefore, for development of strong and 'networked' root system, applications of phosphorus should be made based on a soil test report. Grasses assimilate P primarily in the orthophosphate ($H_2PO_4^-$) form available from soil minerals and soil organic matter. Phosphorus availability is also influenced by soil pH and found to be most available between pH 6.0 and 7.0. It can be applied through monoammonium phosphate and diammonium phosphate.

High potassium (K) in plant tissue is associated with improved tolerance to abiotic and biotic stresses such as cold tolerance, drought tolerance, wear tolerance and disease resistance. Potassium is assimilated by the turfgrass in cationic form (K^+). As the grass root uptakes K^+ from the soil solution, additional K^+ is adhered to the clay particles from the soil solution. Soil colloids thus serve as a reservoir for K^+ and reduce the leaching of K^+. After mowing of sod, removal of mowed grass clippings also severely depletes the K reserves. To maintain adequate levels of K^+ in the soil apply light and frequent applications of potash fertilizer. It is also advisable to retain the grass clippings after mowing and avoid overwatering to reduce the leaching losses. Soils with limited availability of P and K cause reduction in tensile strength of sod due to decreased production of stolons and rhizomes.

4.4.3. Mowing

Frequent mowing is essential for quality sod production. Periodic mowing improves the turf density, checks weed emergence and removes surplus turf growth. The mowing height if kept little higher (3-5 cm) increases photosynthetic efficiency due to more of the green tissue, which increases the translocation of carbohydrates towards roots and rhizomes.

4.4.4. Weed Management

Sod production of a particular turf species also encounters weeds, disease, and insect infestation. Insect and disease pests that are common on older turf areas are rarely seen over turf maintained for sod production provided these are harvested on a regular basis. Leaf spot and rust are most common diseases, which may affect sod.

Insect problems are rare and should be treated as and when they appear. Weed infestation varies seasonally therefore timing of grass seeding for sod establishment is generally a decisive factor to evade weed competition. For instance, seeding cool-season grasses during spring will result in coincide with emergence of annual weeds, compared to seeding done during late-summers or fall season. With periodic harvesting of sod, the weed density tend to reduce, since much of the weed seeds are found close to the surface, that is periodically removed during subsequent harvest. A few weeds in sod are tolerable if these are not interfering with the sod establishment or re-establishment after harvesting. Often, sod growers spray selective herbicides one month before harvest to ensure physical purity and better quality of sod.

5. Sod Harvesting

Sod is ready to harvest when the grass has fully covered the surface, rhizomes have form a netting and has developed sufficient tensile strength to permit harvesting without tearing. The time required between planting and harvesting of sod varies according to the species. When the field has been established for the first time for sod culture, it takes longer period to develop a harvestable sod, compared to the field that is undergoing re-establishment phase from rhizomes after first harvesting (McCarty et al., 1999). Bermuda grass can be harvested 3 to 4 months after the previous harvest, while other species may take up to 24 months to yield a harvestable sod.

The sod producers have turned to mechanized harvesting of mature sod using specialized sod-harvesting machines. These specialized equipments (figure 1) range from simple walk-behind units to large tractor-mounted sod cutters and harvesters, with an ability to harvest several acres of sod a day. The sod harvester consists of two blades. The first blade cuts the sod at an angle into the ground to lift the sod. The second blade cuts straight up and down at intervals, separating the grass slabs of desired dimensions. The cut sod slabs are moved over a conveyer belt where two operators standing on a platform stack the grass slabs forming a pallet. Once a pallet is fully loaded, it left over ground and a new one is carried over to stack subsequent sod pieces. The loaded pallets are fork lifted and sent to the consumer or intended place for turfing. The sod harvesters auto stack the harvested rolls (16 to 24 in. wide measuring 72 to 100 inch long) and slabs (48 x 24 inches sod pieces), operated by a single person. The thickness of the sod during harvest depends on the turfgrass species, soil type and growth of rhizomes. Usually, a sod with thin thickness is easier to handle and also will root and establish faster compared to thicker sod, which otherwise offers more resistance to drought than the thin sod. The presence of buds on the nodes of severed rhizomes, when pressed and come in contact with the soil, began to grow and send down the roots into the underlying soil.

6. Sod Quality and Shelf Life

A quality sod should possess the following characteristics:

- A sod should be uniform in thickness with no more than 1 inch soil attached to it.
- Sod soil should be consistent. It should not crumble or form any cracks while lifting.
- Sod grass should be dense and at least 2" long
- Harvested sod should be free from weeds and disease symptoms.
- Sod should have even and acceptable green colour with adequate strength.
- Roots in the sod should be tied up together like a thick mesh in order to withstand cutting and rolling.
- Grasses should have adequate sod tensile strength (STS) for sod production as tensile strength allows the sod to stay unbroken throughout the production and handling chain.

The shelf life of sod is the time period between the sod harvest and its final planting. The shelf life of sod is influenced by several factors. The best way to ensure the quality establishment of sod is to reduce the time between the harvest and installations much as possible. The longer the sod slabs are stacked, the lower will be its tensile strength that will likely make the sod to tear apart during handling and transit. In addition, the stacked sod slabs tend to heat up due to microbial decomposition, which in turn impairs its quality.

Certain management practices that can be followed to reduce the built up of heat in a harvested sod are:

- The harvested sod should be transported immediately after harvest to the location where it is to be laid, especially during hot weather. During cooler climatic conditions, the sod can be kept stacked for two to three days, or longer, depending on the air temperatures. During hot weather, it is advisable to harvest the sod during evening hours so that the sod can be delivered at the site during the night for installation on site next day in morning.
- Proper mowing also helps to reduce heat built up in harvested sod. The higher mowing height during harvest, will raise the respiration of the sod green surface, leading to poor sod quality.
- Excessive grass clippings should be collected before the sod is harvested as the microbes harbouring in mat (clippings + soil) aid in decomposition of sod.
- High N levels encourage excess growth and higher respiration rates. Therefore, it is beneficial to provide moderate N fertility during sod production cycle.

- The sod should be harvested at right stage of soil moisture, as the sod with turgid leaves will remain cooler and exhibit better quality. However, excess moisture is detrimental, that can increase microbial activity and practically enhance the sod heating. It is advisable to follow proper irrigation management for a quality sod production.
- Sod undergoing seed head formation is more susceptible to heating and therefore should not be hold unnecessarily and must be moved quickly to the site for installation.

7. Economics of Sod Production

First and foremost input to be considered is farm land. The value of agricultural land needs to be considered when selecting a site for sod production. It is always preferred to have sod production farm near market place, therefore the location of land should be selected with careful consideration with respect to the market and ease of transport. However in order to reduce on farmland cost one can start the business by taking land on lease to avoid a large financial expenditure incurred on land purchase. This saved up amount can be utilized on other inputs such as, machinery, fertilizers, structures etc.

Cost of planting material also varies according to species, variety, method of planting and company. On an average Bermuda grass (Selection No. 1) seeds can be purchased @ Rs. 1900- 2000/500g in India. Cost of organic manures, fertilizers, weedicides and other pesticides need to be calculated as per use, dose and standard price in the market. Sod fertilization alone comprises nearly 24 % of the total expenditure of sod production (Falconer and Niemeyer, 2006). Therefore fertilization approach should be very precise and effective with new alternatives to costly fertilizers that can make sod production more economic.

Another major component in economics is labor cost as sod production is highly labor intensive work. Labour cost per day for hired labour in India is Rs. 250-300 per man-day. Machines used in sod production viz., sod harvester, roller, loader etc., are very costly, therefore can be taken on rental basis. However their price per unit varies (Table 3) and mostly these machines are imported in India. The rates for harvested sod vary with species and region in which sod production is undertaken. Under Indian conditions, sod grasses are marketed at an average cost of 100-150 Rs. per m^2 hence for 1-hectare area, an average gross income of Rs. 10-15 lakhs can be obtained (Mathew et al., 2020).

Table 3: Approximate cost of different equipments used in sod production

Equipment	Price per unit (Rs.)
Tow behind grass seed spreader	11,000- 15,000
Power operated mower	10,000-15,000
Lawn roller	12,000-14,000
Sod harvester	13 -15 lacs
Pop up sprinklers	1500-2000
Fertilizer spreader	14,000-20,000

8. Challenges

Though, the sod production is a profitable venture, however, there is a potential loss of top soil from sod farms following years of continuous sod production. Studies reveal that less than 10 mm of soil is removed during each sod harvest (Jordan and Solntseva, 2014). In order to replenish, layer of sieved soil and organic matter (after each sod harvest) can be added which will form a mat with the clippings and roots left after harvest. However soil scientists concluded that commercial turfgrass sod production does not deplete the topsoil but actually improves the farmland by adding nutrients with the decomposition of organic matter comprising primarily of grass clippings and roots. In addition, the sod production reduces soil erosion and permits rain water percolation in soil profile. Among the challenges, establishing the enterprise of sod production is itself a costly venture that includes machinery worth of lacs, fertilizer and weedicides with other managerial costs. Transportation and marketing could also be a challenge in small cities and production farms located far away. Technical know-how, inputs information and expertise in this field is also scarce without which this venture is difficult to initiate and succeed.

Fig. 2: Pictorial representation of cultural operation for turfgrass sod production

9. Conclusion

A sod production businessis a very lucrative enterprise in a 'blooming landscape' in India. The concept is fast emerging and has attracted well established nurserymen to undertake sod production to meet the rising demand from landscape and tourism industry. However, sod production requires an intense management of inputs, capital and operational cost of machinery, good market linkages to earn decent profits. The prospective growers should be aware of several risks associated with sod production. However, for beginners, it would be wise to undertake sod production enterprise based on their past experience in turfgrass production using conventional strategy. It is imperative to have a preliminary assessment of market demand and existing supply chain. The initial investment in imported machinery is not advisable for beginners, instead the rental options should be preferred. With experience and profits reaped through the sod production startup, a sod producer can gradually proceed to mechanized sod culture.

References

Aldous, D.E., Chivers, I. H. 2002. Sports Turf and Amenity Grasses: A Manual for Use and Identification. Landlinks Press, Collingwood Victoria, Australia.

Barton, L., Colmer, T.D. 2006. Irrigation and fertiliser strategies for minimising nitrogen leaching from turfgrass. Agric Water Manag. 80, 160-175.

Falconer, L., Niemeyer, M. 2006.Economic analysis, impact and agronomic profile of sod production in Texas. http:// www.txsod.com/docs/ecoanaylasisrpt.pdf

Glab, T., Szewczyk, W., Gondek, K., Mierzwa-Hersztek, M., Palmowska, J., Nęcka, K. 2020. Optimization of turfgrass fertigation rate and frequency. Agric Water Manag.234, 106107.

Groenveld, T., Kohn, Y.Y., Gross, A., Lazarovitch, N. 2019. Optimization of nitrogen use efficiency by means of fertigation management in an integrated aquaculture-agriculture system. J. Clean. Prod. 212, 401-408.

Jordan,K. S., Solntseva, I. 2014. Determining soil loss fromsod production in Ontario. Ontario Turfgrass Research Foundation, Guelph Ontario.

Masoodi.N. 2021. Turf Grass: A Million Dollar Industry. Rising Kasmir, e-magazine.

Mathew, S., Seetharamu, G.K., Patil, B. C., Dileepkumar, M., Amruta, S.B., Satish, D., Mukund, S. 2020. Economic Analysis of Profitability in Turf Grasses for Different Methods of Planting. Int. J. Curr. Microbiol.App. Sci. 9, 2551-2554.

Mathews,W., Pemrick, W. 2007.History of turfgrass producers international. Turfgrass Producers International, East Dundee, IL.

McCarty, L. B., G. Landry, Jr., J. Higgins, L. Miller. 1999. Sod production in the southern United States. Clemson University Extension Pub. EC 702.

Taliaferro, C.M. 1995.Diversity and vulnerability of Bermuda turfgrass species.Crop Sci.35:327–332.

Zhang. J.X., Chen, M.H, Lu Gan, Zhang, C.J., Shen, Yu., Qian, J., Han, M.L., Xia, Y.G., Yan, X.B. 2020. Diversity Patterns of Bermuda Grass along Latitudinal Gradients at Di-erent Temperatures in Southeastern China. Plants.9, 1778.doi:10.3390/plants9121778.

18

Multidimensional Benefits of Garden Cafeteria

Anita Kumari[1], Gautam Kunal[2], Lajja Vati[3], Praveen Kumar Singh Gurjar[4], Ravi Y[5], Arvind Kumar Verma[5] and Abhijeet Ghatak[3]

[1]Agricultural Technology Management Agency, Vikas Bhawan Collectorate Campus, Gandhi Maidan, Patna, Bihar, India
[2] Plant Protection (Entomology), KVK Jale, Darbhanga, RPCAU, Pusa Samastipur, Bihar, India
[3]Bihar Agricultural University, Sabour, Bhagalpur, Bihar, India
[4]RVS Krishi Vishwa Vidyalaya, Gwalior, Madhya Pradesh, India
[5]ICAR-National Research Centre on Seed Spices, Ajmer, Rajasthan, India

1. Introduction

A combination of art and science had been used for a longer but due to the exponential increase in human population and reduction in operational land holdings its practices became narrow. Nowadays, gardening has gained much popularity among people, as the Covid-19 pandemic has forced them to lock themselves in their houses. These days Mass population is involved in gardening for growing their own food, resulting in a more personal approach to more economical, better-tasting, and healthier products (Pawelek et al., 2009). Gardens have grown increasingly significant for producing nutrient-dense food, protecting biodiversity, and conducting biological and ecological research and education.

Biological communities, which are interacting groups of numerous species in the same place, can also be found in gardens. Pollinators and other visitors to flowers eat nectar and pollen from flowers, and some may use floral volatile oils, lipids, or resins for their food, shelter, and mating locations (Simpson and Neff, 1983). As an incentive for pollination, several plants give breeding places (Ishida et al., 2009, Hembry and Althoff, 2016). As a result, a symbiotic relationship between plants and their floral visitors aids in the preservation of both plant and animal variety, particularly among insects, birds, and mammals (Ollerton, 2017). Both plants and pollinators are unable to exist without their obligate partners in this connection. Pollinators' need on plants ranges from entirely required, as some species have a brooding-site pollination system and some have food supplies, to facultative, as certain species have specific food sources from specific blooms (Ishida et al., 2009). Hummingbirds and butterflies, for example, are attracted to gardens that are designed specifically to attract beautiful pollinators (Jeroen et al., 2016). Pollinators commonly visit the plants to obtain nectar from the blossoms. Their tubular tongues (siphonate proboscis) and slender, curved bills are perfectly fitted for exploring flower tubes and collecting nectar. The pollinators assist in the cross-pollination of the flowers while doing so. In over half of all plant species, cross-pollination is an important aspect of seed development, and it leads to improved seed output and progeny accomplishment in many self-compatible plant species (Aizen and Harder, 2007). Inbreeding depression, which is widespread in self-fertilized plant progeny species, is reduced by cross-fertilization (Darwin, 1895; Husband and Schemske, 1996; Keller and Waller, 2002). Cross-fertilization also promotes genetic variety, which improves a plant's ability to adapt to new and changing conditions (Jump and Penuelas, 2005; Morran et al., 2009). Gardens can be used for a variety of purposes, including recreation, community action, and biodiversity protection (Martin and Mardsen, 1999; Lawson, 2005; Hunter and Hunter, 2008; Tallamy, 2009). In urban areas where green spaces are increasingly uncommon, gardens can provide a return to nature experience. They can also be used to study natural systems and ecological processes (Lawson, 2005; Flores, 2006).

2. Pollinator Attraction During Off Season

Flowers provide a food source for a variety of insect and vertebrate species (Ollerton, 2017). The pollinators assist in the cross-pollination during the process of their feeding. In over half of all plant species, cross-pollination is an essential part of embryogenesis, and it leads to improved seed output and progeny accomplishment in several self-compatible species of plants (Aizen and Harder, 2007). Many flies having shortened siphonate proboscis, beetles and opportunistic avian insectivores are facultative nectarivores, while almost all bees, hawkmoths, long-proboscides flies, and specialist avian nectarivores like hummingbirds and sunbirds are obligate nectarivores.

Only a small number of the animals that eat flowers are obligately reliant on a single plant species. The proportion of oligolectic bees in distinct groups is estimated at approximately around 15 to 60%. Oligolectic bees feed on a single plant taxon, and their specialisation usually comprises pollen gathering and univoltine reproduction timed to coincide with the host plant's blossoming (Minckley and Roulston, 2006). Other pollinators, such as the hover fly and other insects, are drawn to such gardens in the offseason, in addition to bees. The diversity of plants in the garden cafeteria also attracts a variety of pollinators. A glimpses of floral diversity is presented in Figure 1.

Fig. 1: Maintaining floral diversity at the road side to attract pollinators

3. Crop Diversity And Occurrence Of Diseases And Pests

"Genetic vulnerability" has been bandied around a lot lately to conjure up images of devastating epidemics and dangers to global food security, but it's not always clear what it means (NASEM, 1993). Genetic vulnerability is the situation that arises when a crop's genetic makeup makes it uniformly sensitive to a pest, infection, or environmental hazard, posing a risk of tragedy. When determining genetic vulnerability, three factors must be taken into account. First, through actual farm surveys, the area devoted to important cultivars must be monitored by assessing the volume and distribution of seed sales (NASEM, 1993). Second, pest and pathogen evolution patterns must be tracked to determine which pests and pathogens pose the most threat to common cultivars under certain conditions (Ghatak, 2017). Third, the genetic uniformity of important cultivars must be determined (NASEM, 1993).

Natural enemies can help lower pest populations in crops, which has environmental and economic benefits (Naylor and Ehrlich, 1997; Osterman et al., 2003). Pest reduction can be achieved through 'top-down' improvement of natural enemy populations and other 'bottom-up' influences that directly affect pests. A streamlined natural enemy community has the same or greater control of pest populations as a complex of natural enemies. Planting different types of the same crop in different combinations can help small holder farmers reduce pests and diseases in their crops. For the smallholder, who cannot afford chemical inputs, agricultural biodiversity is an economically viable and accessible option. Using diversity to manage pests and diseases allows farmers to keep local diversity on their farms, which is a valuable source of genetic material that could be utilised to create resistant varieties in the future.

In this context, the owner gets benefit from a garden cafeteria when (i) natural enemy populations are higher and more diverse, (ii) natural enemies significantly colonise in the arable fields, (iii) as a result, the natural enemies significantly reduce pest densities, (iv) automatically, the damage levels are reduced, and (v) benefits can be obtained from the outweigh costs. As a result, achieving economic advantage through natural pest control requires a series of key actions. In addition to natural enemies, non-crop habitats may function as reservoirs for pest species that invade crops (van Emden, 1965). The benefits of natural pest control are thus not self-evident, but rather reliant on a number of circumstances and readily disrupted. Alternative hosts and prey may boost parasitoid and predator populations, improving pest control (Bianchi and Van der Werf, 2004; Ostman, 2004). Pollen and nectar are vital requirements for many insects, and woody and herbaceous flora can provide them. When nectar supplies are present, parasitoids' longevity and fecundity are significantly increased (Costuming and Landis, 2004; Lee et al., 2004). The effectiveness of natural enemies to decrease insect pest populations may be influenced by the presence of nectar and pollen (Tylianakis et al., 2004). In simple (monoculture) systems, insects are attracted to and concentrated on their feeding plant resources, which are more easily found or visible, whereas in diversified systems, natural enemies are more effective and numerous.

The significance of garden composition on natural enemy-herbivore interactions as well as crop productivity is vividly pertinent. All of the following factors should be included and quantified in studies: (i) natural enemy population stimulation; (ii) natural enemy colonisation of arable fields; (iii) assessments in decrease of pest density; (iv) damage reduction; (v) obtaining higher yield; and (vi) improved cost–benefit ratio. When compared to monocultures, the greater the degree of variety in agroecosystems, the more stable the systems that govern insect populations become, and productivity is not affected (Lai et al., 2011). Pests are one of the most important biotic elements that impact continuously farmed plants over the course of each growing season. The formation of specialised phytophages occurs when just one plant species is grown (particularly in monocultures), resulting in a decline in plant productivity. The development of the pest population may be totally or partially inhibited depending on plant succession. The growing use of plant protection chemicals has resulted in a number of negative changes and issues, the most notable of which are increased insect resistance and harm to plant pollinators. The primary goal of using alternatives to chemical-based pest control approaches, such as a greater emphasis on natural enemies, is to boost the biocenosis' biodiversity.

4. Meagre use of Synthetic Chemicals

Any ecosystem's diversity should be managed holistically, and all trophic levels' roles must be understood and properly considered. Biological pest control is the safest and most environmentally friendly option among pesticide reduction options. Growing plants, vegetables, and fruits without pesticides or synthetic chemical fertilisers are what organic gardening is all about. Simply said, it is gardening done in the most natural way possible. Gardening is crucial since it does not require the use of any chemicals. It has no negative effects on the environment or the individuals who consume it. Organic kitchen gardening, according to the UNC Healthcare Organization, has various physiological and psychological benefits, including stress reduction and improved heart health. In addition to these advantages, they aid in the reduction of pollutants and soil erosion. Above all, the oxygen they emit and the carbon dioxide they consume have more worth.

5. Ensuring Family Nourishment

The continual rise in the price of vegetables is another reason to cultivate. Vegetable costs on the market can change from week to week. By the end of the year, onions are often at their most expensive. There have been multiple cases where vegetable prices have risen by 40% in a week, with onions being the most expensive. Furthermore, during that time, the price of the vegetable does not correspond to its quality. A garden and the culture of the surrounding town have a close relationship, and the two are completely intertwined. Small vegetable gardens in the backyards of marginalised and resource-poor populations demonstrate the path to a long-term solution. The world came to a halt as governments, medical professionals, health workers, and scientists waged a long war against the new coronavirus known as SARS-CoV-2. The globe has not been the same since most countries, including India, ordered a state-wide lockdown to stop the virus from spreading. We had to adjust to the 'new normal' at home and at work, while also taking steps to avoid infection. It was especially crucial for us Indians because we have the world's highest population of malnourished children. One out of every three children in our country is malnourished.

Gardens can furnish strengthen to community food security, nutritional security, and livelihoods by providing supplemental sources and key nutrients. It enhances communal dietary diversity by delivering necessary micronutrients through a year-round supply of products that are sufficient to meet one's needs (Susan, 2005). As a result, gardens have become an important part of community farming, providing food supply and diversity while also assisting communities in overcoming hunger. Micronutrients are abundant in fruits and vegetables grown in the garden cafeteria, especially for those who come from low-income homes and are malnourished. Rural communities offer plenty of open land and backyard farms where people can start a garden cafeteria quickly (Landauer and Brazil, 1985). The majority of the populations in India come under malnourished and substandard living conditions;

garden cafeteria can play a pivotal role. The oldest and most durable form of farming is food production on tiny plots next to human settlements.

6. Round-Year Supply of Organic Food Items

The garden cafeteria is a magic potion that promotes self-sufficiency, sovereignty, and dignity while also assisting with food security. Labour-power refers to a household's members' physical ability to generate money (Christopher, 2006). The food security and nutritional diversity of a household can be improved by employing household labour in the garden cafeteria. Small places around the house can make a difference in people's lives, even if land resources are limited. Households and small communities use vacant land to contribute to their own food needs as well as the needs of the city they reside in (Dresche, 2000). Vegetables are high in vitamins, minerals, and fibre, and their nutritional and medicinal benefits to humans are well proven. Because vegetables are typically short-duration crops, they are ideal for growing in the garden cafeteria.

The garden is usually close to the house and is used to cultivate seasonal vegetables, fruits, herbs, medicinal plants, and other plants required by the household (Figure 2). It not only saves us money and time, but it also provides a healthy, pollution-free atmosphere for the entire family. When a compost pit is dug in the side of the garden cafeteria, it aids in the recycling of kitchen trash. Gardening is beneficial to both individuals and communities, contributing to overall community health. The benefits of food production go beyond an individual's physical, mental, and emotional well-being to leave a lasting impact on others as well as the community's physical and social environment (Armstrong, 2000). On a low-cost level, the garden can be fertilised on a regular basis with a composite made from home biodegradable waste and vermicompost. Improved productivity could be achieved by combining the usage of household garbage as compost and manure.

Producing a diverse range of foods in the garden cafeteria is one of the simplest methods to ensure access to a balanced diet rich in macro and micronutrients (Figure 3 and 4). This is especially significant for those in rural and urban areas, where purchasing power is limited, markets are far away, and chemically treated vegetables are readily available. By providing access to products that can be collected quickly, processed, and served to family members, the garden cafeteria immediately provides food and nutritional security. Garden cafeteria can be cultivated in the backyard or on the roof of a house, as well as in containers. Such gardening has several societal benefits, including improved health and nutrition, greater income, employment, and food security within the household. Garden cafeteria is naturally an environmentally beneficial project due to its reduced reliance on chemical fertilisers and pesticides. Additionally, garden cafeteria furnishes environmental-friendly trash disposal options. Composting is a typical method of enriching the soil with domestic wastes such as kitchen garbage, paper, and even animal manure. Garden cafeteria is an environmentally friendly and sustainable agricultural strategy that helps rural households improve their food security and economic growth. Gardening is a good innovative solution for ensuring food security, providing work for young people, and providing an additional source of income for households in poor nations.

As all economic activity came to a halt in the midst of this, the lives of people from the most marginalised sections of our society were severely harmed. The importance of healthy eating habits and nutritious food intake has never been — or seen to be — so vital as it has been in the last two years. Culinary and botanic varieties have strong linkages, and a thorough grasp of both is essential for effective conservation management. Women's participation in conservation strategies is clear, and it must be given proper consideration during the planning and implementation stages. The garden's role in terms of security is another facet of the strong interaction between house, garden, and family. The garden is typically included in the homestead's protected area, which commonly includes a barrier to keep children and cattle in while keeping others out, preserving genetic variation. Garden output has expanded dramatically in the country, helping to reduce "hidden hunger" and disease caused by vitamin insufficiency.

The garden cafeteria is an on-site, in-situ fruit and vegetable production system that uses low-cost local materials and traditional expertise. This gradually promotes self-sufficiency and allows human capital to take the place of expensive external inputs. It is primarily designed to provide a consistent supply of fresh veggies for family consumption. A variety of fruits and vegetables are cultivated on tiny plots of land near homes. Fruit and vegetable gardening in the backyard contributes to a well-balanced diet by supplying energy-dense foods and essential protective nutrients. Fruits and vegetables not only beautify the table, but they also improve health by providing the most nutritious menu and increasing vitality and vigour. Furthermore, garden cafeteria could increase home food security while also providing an extra source of revenue by allowing the household to sell a portion of the garden's products and achieve household self-sufficiency (Rahaman and Islam, 2014). The scientific model for gardens would aid in increasing vegetable and fruit consumption, hence boosting farmers' livelihoods through improved nutrition (Gillespie et al., 2012).

In developing countries, gardening offers an innovative solution to ensure food security, young employment, and an additional source of cash (Leregger, 2018). Case studies from Bihar, Odisha's tribal and rural sector (Prashant et al., 2012), and Dharmapuri, Tamil Nadu's rainfed areas showed that they not only provided a source of income but also increased food intake.

Fig. 2: The diversity of a garden cafeteria established aside to the residential quarters. Images are showing plant diversity during winter (left) and summer to rainy (right) season. The area attracts pollinators, birds and other wild lives like squirrel, mongoose, chameleon etc.

Fig. 3: Variation in produce of obtained from a garden cafeteria. The vegetables and fruit (guava in blue tray) were harvested during Aril to September. Repeated harvesting was performed for the cucurbits, beans, okra and leafy vegetables.

Fig. 4: The glimpses of winter crops in a garden cafeteria before harvesting

7. Key Garden Cafeteria Plants and Their Functions

A garden cafeteria is an area surrounding the house where herbs, plants, and vegetables are produced for domestic use. There are many plants that we may cultivate in our garden, on our patio, or even on our windowsill that are herbs. Herbaceous plants in our garden cafeteria include parsley, mint, sage, basil, and many others. These botanicals add flavour to dishes and are perfect for soups, stews, and salads. Among shrubs plants like *Hibiscus syriacus*, *Thuja occidentali*, *Ilex aquifolium* and others are generally used to give boundary walls to serve as a protective fence. Apart from these, seasonal vegetables like cucumber, carrot, radish, tomato etc are grown. We must stay healthy in today's hectic world, and a balanced diet is essential for people to stay healthy. Vegetables are an essential part of a healthy diet since they include diverse nutrients that are necessary for many bodily functions. Vegetables are necessary for growth, energy, and disease protection. Vegetables are particularly beneficial for children, pregnant women, and nursing mothers.

8. Weeds an Inconceivable Source of Pollinators

The garden cafeteria is home to many weeds which are of no use but serve as an important source of pollen, nectar and propolis for the honey bees and pollinators. Weeds like *Aeschynomene aspera* L., *Leucaena leucocephala* (Lam.) de Wit., *Mimosa pudica* L., *Coccinia grandis* (L.) *Trichosanthes dioica* Roxb, *Polygonum hydropiper* L., *Trianthema portulacastrum* L., *Alstonia scholaris*, *Lantana camera* L., etc serve as an excellent source of food to the pollinators. In bee-keeping, hot-summer and rainy seasons are being considered as dearth periods in Bihar. It generally starts from the month of June and continues up to November. This is the period when neither nectariferous nor polliniferous crops are available barring in some limited patches of up-land paddy in a vegetable-based cropping system. In this period these weeds serve as life-saving sites for many pollinators, viz., *Apis mellifera*, *A. cerana indica*, *A. florea*, *A. dorsata*, *Tetragonula iridipennis* etc.

Garden cafeteria is so rich in floral diversity during the blooming of herbs, shrubs, vegetables and weeds it gives multicolour look to the garden. Flower colour attracts flower visitors. Bright colours, for example, red and blue, are attractive (Brody and Mitchell, 1997; Glover, 2008), but not for bees. Kiew and Muid (1991) who documented that 66% of flowers that were visited by bees (honey bees) in Malaysia are white, creamy white or yellow in colour. Also, stingless bees *Heterotrigona intima* showed higher efficiencies toward white colour (Ghazi, 2018).

9. Initiatives of Indian Government in Promoting Garden Cafeteria

The Indian government launched the National Nutrition Mission, known by the POSHAN Abhiyaan, with the goal of creating an undernourished-free India through a multi-ministerial convergence mission. Building synergy and merging the ideas of other different authorities through a convergence strategy is required to achieve a united goal. Garden cafeteria is being promoted by the National Rural Livelihood Mission (GoI) as part of the National Nutrition Mission's farm livelihood intervention approach to prevent malnutrition. To boost the nutritious content of meals supplied to children, Mizoram has begun building school areas for garden cafeteria named 'KanSikul, KanHuan' or My School, My Farm (Shashanka, 2000). Bijapur, the district in Chhattisgarh with the greatest prevalence of malnutrition, has built garden cafeteria (Rashmi) in anganwadi centres in rural areas. To supply for their mid-day meals, government schools in Chandigarh are building little farms. (Kamini) Under the mid-day meal scheme, the Ministry of Human Resource Development has produced guidelines for school nutrition gardens in government and aided schools. There are established projects in place around the country that can learn from one another and replicate excellent practises. The Jharkhand Poshan Vari initiative encourages women to grow grains, pulses, and vegetables in their backyard garden cafeteria in order to combat malnutrition and earn money. To create awareness about the value of fruits and vegetables, the Tamil Nadu Horticulture Department has teamed with the School Education Department to develop roof gardens in schools. The gardens' food is used in school kitchens that serve a nutritious in-school meal plan.

10. Some Vibrant Instances

Bihar's government is also pushing organic garden cafeteria as a source of midday meals for over 4,000 of the state's 70,000 government elementary and middle schools. The initiative, Bihar Rural Livelihood Promotion Mission (Jeevika), has been encouraging and giving required inputs to SHG HHs to take up similar gardening with the goal of promoting consumption of at least 7 food groups under the dietary diversity campaign to improve nutritional intake by SHG households. The Government of Bihar (GoB) promotes ApniKyariApniThali (AKAT) campaign, which endorse rural women's and children's nutritional security.

To promote the garden cafeteria in rural areas, government of Bihar in collaboration with Bihar Agricultural University, Sabour in Bhagalpur is conducting many extension programmes. For example, in Katihar district a programme named 'Nari' was organized by the KVK of Katihar with a view to provide emphasis on POSHAN in agriculture. Also, to aware the agricultural woman and rural youth, the concept of garden cafeteria has been disseminated and offer support to the interested individuals by helping them with technological intervention. In this programme, district agricultural coordinators and KVK scientists were actively participated and spread awareness about growing different vegetable and fruit crops in garden cafeteria. In a receiving positive note, so many such extension programmes are being conducted in different KVK's of Bihar. In another instance, Mr. Rajiv Kumar, a successful gardener efficiently works on the background of garden cafeteria from Bodhgaya of Gaya district (Bihar). He began garden cafeteria in March 2021, in the backyard of his house. Earlier, he has never been to a garden cafeteria. However, Mr. Kumar met with KVK scientists and showed interest in a garden cafeteria as well as expert technical help. After gathering knowledge, he began gardening, which has encouraged in family members too and by such additional earning he has successfully uplifted the socio-economic status of his family. He claims that his monthly savings, which he used to spend on daily vegetable purchases, have increased. His family now consumes a well-balanced diet that includes green vegetables such as bitter guard, okra, sponge gourd, ridge gourd, cucumber, pumpkin, bottle guard, brinjal, chilli, tomato, radish, carrot, and coriander, and others. Moreover, he also planted fruit plants such as mango, banana, guava, papaya, lemon in his premises. He has now become a source of inspiration for many of the local farmers. His eagerness to swiftly acquire and absorb new information, as well as his hard effort and positive attitude, appear to be the secrets to his success.

11. Conclusion

Non-availability of healthy food is the main issue for everyone and the garden cafeteria not only provides healthy food but also saves money. The garden cafeteria may provide the daily requirement of protein, fat, carbohydrate, vitamins and minerals to the whole family. Gardens can furnish strengthen community food security, nutritional security, and livelihoods by providing supplemental sources and key nutrients. It enhances communal dietary diversity by delivering necessary micronutrients through a year-round supply of products that are sufficient to meet one's needs. It helps to boost immunity, provide a balanced diet and protect humans from several diseases. If the garden cafeteria increased at the community level it can save biodiversity by providing shelters for different kinds of life.

References

Aizen, M.A., Harder. L.D., 2007. Expanding the limits of the pollen-limitation concept: Effects of pollen quantity and qulity. Ecology. 88, 271-281.

Armstrong, A.D., 2000. A survey of community gardens in upstate New York: Implications for health promotion and community development. Health and Place. 6.

Bianchi, F.J.J.A., van der Werf, W., 2004. Model evaluation of the function of prey in non-crop habitats for biological control by ladybeetles in agricultural landscapes. Ecol. Model. 171, 177–193.

Brody, A.K., Mitchell, R.J., 1997. Effects of experimental manipulation of inflorescence size on pollination and pre-dispersal seed predation in the humming bird-pollination plant Ipomopsis aggregate.Oecologia.110, 86–93.

Cardinale, B.J., Harvey, C.T., Gross, K., Ives, A.R., 2003. Biodiversity and biocontrol: emergent impacts of a multi- enemy assemblage on pest suppression and crop yield in an agroecosystem. Ecol. Lett. 6, 857–865.

Chikako, I., Masumi, K., Shoko Sakai, 2009. A new pollination system: brood-site pollination by flower bugs in Macaranga (Euphorbiaceae). Ann. Botany. 103, 39–44.

Christopher, B., 2006. Food Aid's Intended and Unintended. Rome: Agricultural and Development Economics Division (ESA).

Costamagna, A.C., Landis, D.A., 2004. Effect of food resources on adult Glyptapantelesmilitaris and Meteoruscommunis (Hymenoptera: Braconidae), parasitoids of Pseudaletiaunipuncta (Lepidoptera: Noctuidae). Environ. Entomol. 33, 128–137.

Darwin, C., 1895. The Effects of Cross and Self-fertilsation in the Vegetable Kingdom. D. Appleton and Company, New York.

Emden, H.F., 1965. The role of uncultivated land in the biology of crop pests and beneficial insects. Sci. Hort. 17, 121–136.

Flores, H.C., 2006. Food not Lawns: How to Turn Your Yard into a Garden and Your Neighborhood into a Community. Chelsea Green Publishing Company. White River Junction, VT. 334 pp.

Ghatak, A. 2017. Evolution and adaptation in phytopathosystems. In: The Phytopathogen: Evolution and Adaptation. Eds.: Ghatak, A., and Ansar, M. Apple Academic Press, USA. ISBN: 978-1-77188-406-8. Pp 3–20.

Ghazi, R., Zulqurnain, N.S., Azmi, W.A., 2018. Melittopalynological Studies of Stingless Bees from the East Coast of Peninsular Malaysia. In: Pot-pollen in stingless bee Melittology (Eds.: P. Vit., S.R.M. Pedro. D.W. Roubik). E-book, Springer- Nature, Switzerland, pp. 77-88.

Gillespie, S., Harris, J., Kadiyala, S., 2012. The agriculture-nutrition disconnect in India: What do we know.

Glover, B.J., 2008. Understanding flowers and flowering: An intergrated approach. Oxford University Press; Oxford, USA. pp. 227.

Hembry, D. H., Althoff, D.M., 2016. Diversification and coevolution in brood pollination mutualisms: Windows into the role of biotic interactions in generating biological diversity. Amer. J. Botany.103, 1783-1792.

Hunter, M.C., Hunter, M.D., 2008. Designing for conservation of insects in the built environment. Insect Conservation and Diversity. 1(4),189-196.

Husband, B.C., Schemske, D.W., 1996. Evolution of the magnitude and timing of inbreeding depression in plants. Evolution. 50, 54-70.

ICAR-Indian Institute of Horticulture Research, "Backyard Nutritional Kitchen Gardening-A success story," https://iihr.res.in/backyard-nutritional-kitchen-gardening-success-story-0

Jeroen, P., van der Sluijs., Nora S. V., 2016. Pollinators and Global Food Security: the Need for Holistic Global Stewardship. Food ethics. 1, 75–91.

Jump, A.S., Penuelas J., 2005. Running to stand still: adaptation and the response of plants to rapid climate change. Ecology Letters. 8, 1010-1020.

Keller, L.F., Waller, D.M., 2002. Inbreeding effects in wild populations. Trends in Eco.& Evolution. 17, 230-241.

Kiew, R., Muid, M., 1991. Beekeping in Malaysia: Pollen atlas. United Selangor press SdnBhd; Kuala Lumpur, Malaysia. 186 pp.

Lai, R.Q., You, M.S. Lotz, L.A.P.,Vasseur, L., 2011. Responses of green peach aphids and other arthropods to garlic intercropped with tobacco. Agron. J. 103, 856–863.

Lai, R.Q., Zeng, W.L., Jiang, G.H., Li, L.Y., Xie, X.H., 2011. Preliminary studyoncontrol effects of ethanol extracts from garlic plant against Ralstoniasolanacearum and TMV. J. Yunnan Agric. Univ. 26, 284–287.

Landauer, K., Brazil, M., 1985. Tropical home gardens. Selected papers from an international workshop at the Institute of Ecology, Padjadjaran University, Indonesia, December 1985, United Nations University Press, Japan.

Lawson, L.J., 2005. City Bountiful: A century of community gardening in America. University of California Press, Berkeley and Los Angeles, CA. pp. 363.

Lee, J.C., Heimpel, G.E., Leibee, G.L., 2004. Comparing floral nectar and aphid honeydew diets on the longevity and nutrient levels of a parasitoid wasp. Entomol. Exp. Appl. 111, 189–199.

Leregger, F., 2018. Micro gardening as a contribution to food security and nutrition in Africa. Institute for Environment Peace Development, Vienna, Austria.

Losey, J.E., Denno, R.F., 1998. Positive predator–predator interactions: enhanced predation rates and synergistic suppression of aphid populations. Ecology. 79, 2143–2152.

Martin, R., Mardsen, T., 1999. Food for Urban Spaces: The Development of Urban Food Production in England and Wales. Int. Planning Studies. 4(3), 389-412.

Minckley, R.L., Roulston, T.H., 2006. Incidental mutualisms and pollen specialization among bees. In: Waser, N.M. and Ollerton. Plant-Pollinator Interactions: From Specialization to Generalization. Chicago University Press, Chicago. Pp. 69-98.

Ministry of Rural Development, Government of India.

Morran, L.T., Parmenter,M.D., Phillips, P.C., 2009. Mutation load and rapid adaptation favour outcrossing over self-fertilization. Nature. 462, 350-352.

Myers, J.H., Higgins, C., Kovacs, E., 1989. How many insect species are necessary for the biological control of insects? Environ. Entomol. 18, 541–547.

NASEM (National Academies of Sciences, Engineering, and Medicine). 1993. Managing Global Genetic Resources: Agricultural Crop Issues and Policies. Washington, DC: The National Academies Press. https://doi.org/10.17226/2116.

Naylor, R.L., Ehrlich, P.R., 1997. Natural pest control services and agriculture. In Nature's services: societal dependence on natural ecosystems (ed. G. C. Daily), pp. 151–174. Washington, DC: Island Press.

Ollerton, J., 2017. Pollinator Diversity: Distribution, Ecological Function, and Conservation. In D. J. Futuyma. (edi). Annual Review of Ecology, Evolution, and Systematics. 48, pp. 353-376.

Ostman, O, Ekbom, B., Bengtsson, J., 2003. Yield increase attributable to aphid predation by ground- living polyphagous natural enemies in spring barley in Sweden. Ecol. Econ. 45, 149–158.

Pawelek, J., Frankie, G.W., Gobbin, W., Przybylski, M., 2009. Modification of a community garden to attract native bee pollinators in urban San Luis Obispo, California. Cities and the Environment. 2(1), pp. 21.

Prashant, K., Parida, S., Malarvannan, Ajit, K.S., Rajdeep, B., Jaganath N., Oliver, K., 2018. Nutrition Sensitive Annadata Kitchen Garden Model: Growth for health and wellbeing of tribal poor–A couple of case studies." Current Res. Nutrition and Food Sci. 6, (3), 882–890.

Rahman, K.M.M., Islam, M.A., 2014. Nutrition-sensitive agriculture in Bangladesh: a review. Food Security, 6(5), 671-683.

Rodrı´guez, M.A, Hawkins, B.A., 2000. Diversity, function and stability in parasitoid communities. Ecol. Lett. 3, 35–40.

Sakai, S., 2002. A review of brood-site pollination mutualism: plants providing breeding sites for their pollinators. J. Plant Res. 115:161-168.

Schmidt, M.H., Lauer, A., Purtauf, T., Thies, C., Schaefer, M. &Tscharntke, T., 2003. Relative importance of predators and parasitoids for cereal aphid control. Proc. R. Soc. B. 270, 1905–1909.

Shashanka, A., 2009. An IAS Officer who has sparked a Poshan revolution in a remote district of Mizoram with her My School, My Farm innovation. September. Outlookposhan 2.0.

Shoba, S., 2020. Nutrition Gardens: A Sustainable Model for Food Security and Diversity. Observer Research Foundation. June.369.

Simpson, B.B., Neff, J.L., 1983. Evolution and diversity of floral rewards. In; Jones, C.E., and R. J. Little (edi). Handbook of Experimental Pollination Biology. New York.142-159.

Snyder, W.E., Ives, A.R., 2003. Interactions between specialist and generalist natural enemies: parasitoids, predators, and pea aphid biocontrol. Ecology. 84, 91–107.

Stman, O., 2004.The relative effects of natural enemy abundance and alternative prey abundance on aphid predation rates. Biol. Control. 30, 281–287.

Susan, B., 2005. Women (re)construct the plot: The regen (d)eration of urban food growing," Area. 37 (2), 171–179.

Tallamy, D.W., 2009. Bringing Nature Home. Timber Press, Portland, OR. pp. 358.

Tylianakis, J.M., Didham, R.K., Wratten, S.D., 2004. Improved fitness of aphid parasitoids receiving resource subsidies. Ecology. 85, 658–666.

Willmer, P.G., 1991. Constraints on foraging by solitary bees.In: The Behaviour and Physiology of Bees (Eds L. J. Goodman andR. C. Fisher). CAB International, Wallingford, UK. pp. 131–148.

19

Genome Evolution and Domestication of Fenugreek Crop

Radheshyam Sharma[1], Stuti Sharma[1], Ashish Kumar[1], R. Shiv Ramakrishnan[1], Sharda Choudhary[2] and Madhu Choudhary[3]

[1]Jawaharlal Nehru Krishi Vishwa Vidhalya, Jabalpur, Madhya Pradesh, India
[2]ICAR-National Research Centre on Seed Spices, Ajmer, Rajasthan, India
[3]ICAR-Central Soil Salinity Research Institute, Karnal, Haryana, India

1. Introduction

Fenugreek (*Trigonella* sp.) is one among the world's oldest seed spice crops and has gained a supreme importance due to its extensive pharmacological properties (Basu and Prasad, 2011; Maloo et al., 2020). In ancient time's fenugreek plant was well known and used as a medicinal drug by Egyptians (Rouk and Mangesha 1963). The Ebers Papyrus (one of the oldest preserved pharmaceutical documents) detailed therapeutic advantages and descriptions in Egypt circa 1500 BC (Betty, 2008). It has gained great importance due to its multifaceted uses which are being exploited by farmer's and industrialist for different purposes. Every part of the plants are useful in term of medicinal as well as for consumption of human and animal feed, while the main useful part, seed is utilized for preparation of sauces, condiments, artificial flavoring and production of hormone (Kearns et al., 1998). Cultivation of fenugreek has been observed all over the world in a wide range of soil and climatic conditions, mainly in India and Mediterranean countries as cash and good renovator of soil crop. It is a multipurpose legume used as spice, forage, food and industrial purpose. Natural medicines comprehensive database 2010 described the 523 fenugreek based commercial products are available in market and 50 have been licensed under Canadian licensed products.

2. Origin of Fenugreek

Fenugreek is believed to be a native of southern Europe and the Mediterranean region, but it is also cultivated in central and southeastern Europe, India, northern Africa and western Asia. India has also been reported to be the native home of fenugreek and exists in wild form in Punjab, Kashmir and the upper Gangetics Plains (Shanmugavelu et al., 2002). The seed of fenugreek were also found in the tomb of Tutankhamum (Manniche, 1989). Fenugreek has been also cultivated around the Saharan oasis of north Africa since ancient times (Duke 1986). Charlemagne promoted fenugreek cultivation during ancient times (Fazli and Hardman 1968). It is probably one of the oldest fodder crop grown before the era of recorded history. As a fodder crop, it is said to be the Hedysarum of Theophrastus and Dioscorides (Leyel 1987). In, Central Europe it is introduced during the initiation of the ninth century (Schauenberg and Paris 1990). However, no literature is available until the sixteenth century, when it was recorded as cultivation started in England. In a study by Dangi et al. (2004), germplasm of *Trigonella foenum-graecum* and *T. caerulea* found in Turkey shown hugediversity and hold Vavilov's hypothesis. Accessions of *Trigonella foenum-graecum* from Pakistan and Afghanistan were clustered together in one cluster, while those from India and Nepal were placed together in another. However, accessions from Turkey for both species did not cluster together and were found in separate clusters. It was also revealed that large seeded cultivar of *T. foenum-graecum*, is mostly found around the Mediterranean region, whereas the miniature seeded cultivars are predominant eastward. Huge genetic variability was seen in *Trigonella caerulea* in comparisons to *Trigonella foenum-graecum. Trigonella foenum-graecum* is most popular species L of fenugreek across the globe among the 25 reported species of the fenugreek (Basu et al., 2008a; Zandi et al., 2011b). Thus, theorigin of domesticated fenugreek and their evolution under domestication as well as their relationship with wild relatives have been a central question for researchers and ethno-botanist.

3. Cultivation of Fenugreek World Wide

The cultivation of fenugreek has been reported in Portugal, United Kingdom, Spain, Austria, Germany, Greece, Switzerland, Turkey, Sudan, Egypt, Kenya, Ethiopia, Lebanon, Tanzania, Israel, Tunisia, Morocco, China, India, Japan, Argentine, Russia, Pakistan, United States of America (Rouk and Mangesha, 1963; Fazli and Hardman,1968; Basu et al., 2014). Diverse genotypes of fenugreek are present in the world wider adaptability, morphology and seed quality and seed production capability. Fenugreek is generally grown in autumn (October to December) while seeding can be done in in autumn and spring in southern Europe. On the other hand, Acharya et al. (2008) reported that when sown in the spring, the fenugreek crop performs well in western Canada (from late April to mid May). Since, fenugreek is a self-pollinated, small size and low duration crop therefore lot of work has been done for the improvement of crop. Some of the important fenugreek varieties published in theresearch papers and books all over the globe have been summarized in the following Table 1.

4. Fenugreek Species and Varieties

Petropoulos described 18 common species of the genus Trigonella which are mentioned in table 1 (Petropoulos, 2002).

Table 1: List of 18 well known fenugreek species as per Petropoulos

S.No	Species Name	S. No	Species Name	S. No	Species Name
1	*T. foenum-graecum*	7	*T. spinosa*	13	*T. striata*
2	*T. anguina*	8	*T. polycerata*	14	*T. torulosa*
3	*T. Arabica*	9	*T. radiata*	15	*T. arcuata*
4	*T. caerulea*	10	*T. platycarpos*	16	*T. suavissima*
5	*T. corniculata*	11	*T. hamosa*	17	*T. occulta*
6	*T. cariensis*	12	*T. cretica*	18	*T. rigida*

These 18 species of fenugreek have been used by many researchers and they have given many improved local varieties. Different species and cultivars of fenugreek with the somatic chromosomal number 2n=16 have been studied cytologically, including karyotype analysis (Najafi et al., 2013). Different varieties of fenugreek grown in India and other continents have been reported are listed in the Table 2.

Table 2: Name of fenugreek varieties released worldwide

Fenugreek Variety Name	Reference	Country Name
RajendraKanti, LamSel 1 Early, CO-1,RMt-1, RMt-143, Bunching Pusa	Edison, 1995	India
UM-70, UM-23, UM-32, UM-9, UM-18, UM-113, UM-17, UM-26, UM-25, UM-27, UM-58, UM-33, UM-50, UM-36, UM-52, UM-79, CVTUM-5, UM-67, CVTUM-17, UM-75, UM-77, UM-83, UM-84, CO-1, UM-105, CVTUM-35, UM-112, UM-114, CVTUM TC 2336, UM-115, CVTUM-32, CVTUM-34, CVT NLM, CVTTG 1084,CVTGF 1, CVTCC, NLM, CT Lam Sel 1, Local check	Kamal et al. 1987	India
TG-3, RG-07, TG-13, TG-24, TG-18, TG-34, UM-34, UM-5, UM-17, UM-6, UM-20, UM-35,UM-38, NI-01, MP-14, IC-99, LamSel 1, Pusa Earlier, Bangalore-Local, Local Bobes	Prasad and Hiremath, 1985	India
T-8	Parodaand Karwasra, 1975	India
AFg-1, AFg-2, AFg-3, GM-1, RMt-305, PRM-45, ML-150, Methi No-47, Hisar Sonali, Methi No-14, HM 219, Methi No-47, Hisar Mukta, EC-4911, HM-103, Hisar Suvarna, Pant Ragini, Hisar Madhavi (HM-350)		India
HM-46	Singh et al., 1994	India
IC-74	Singh and Singh,1974	
Gharbin-6	Bunting, 1972	Egypt
Ali Corte, Ali Lunghe	DelGaudio, 1953	Italy

Fenugreek Variety Name	Reference	Country Name
Ionia	Vaitsis, 1985	Greece
Gouta	Haefele et al., 1997	France
Fluorescent Ethiopian etropoulos	Hardman, 1980	England
Paul, Margaret, Barbara	Evans, 1991	England
Moroccan, Kenyan, Fluorescent, Ethiopian	Petropoulos, 2003	England

5. Blueprint of Domestication of Fenugreek

5.1. The Concept of Domestication

Agriculture has been and continues to be humanity's primary resource of food, fiber, and feed from ancient times. So far, it has been able to keep up with the massive population growth. The global human population was believed to be approximately 6 million individuals at the start of the Neolithic Revolution (Livi-Bacci, 2012), but it today stands at 7.9 billion (World Population Clock, US Census Bureau), with a practically likely growth of 2 billion by 2050 (World Population Clock, US Census Bureau) (United Nations, Department of Economic and Social Affairs). Other issues influencing food security are to blame for the present hunger episodes, not in sufficient production (FAO, The State of Food Insecurity). Nonetheless, agriculture's dominance as a human activity has come at a cost to the environment, including habitat destruction, a decline in biodiversity in both natural and agricultural ecosystems, widespread contamination with agricultural chemicals such as fertilizers and pesticides, and a significant contribution to greenhouse gas emissions. Domestication of crop plants (and farm animals) happened in this setting, not only as part of the adaptation process to which people subjected them through cultivation (or captive raising), but also as part of the Neolithic Revolution, which involved the move from HG to agriculture.

Domestication is fundamentally an evolutionary process of selection for adaptability to changing environmental conditions as plants are not only consumed but also actively developed. Cultivation is a vital step in the process of domestication. The selecting pressures that work on plant populations in their natural settings are vastly different from those that act on plant populations in later developed environments. Plants compete for light, water, and soil nutrients in natural habitats, and there is fierce intraspecific and interspecific competition for these resources. In contrast, even in well-known, diversified arrangements like agroforestry, agroecosystems and gardens are characterized by less competition (Perfecto and Vandermeer, 2010). Reduced competition leads to greater production of harvested organs (e.g., grain, leaves, roots, etc.) via altered photosynthate partitioning and a higher harvest index (ratio of harvested biomass to total biomass). The necessity to build a more efficient production system in terms of biomass per unit of surface and time is a second fundamental distinction between wild and developed ecosystems. The main objective has been to increase collected biomass while lowering the overall effort required to gather it.

5.2. Why we do Domestication

To know and understand why, we have to look at what it means for a plant to be domesticated. Plants selected for human use share a suite of features. They have reduced chemical or physical defenses (less bitter, less spiky, thinner rinds) and bigger edible parts. Many of them develop synchronous maturation, that is, the fruits all ripen together for easier picking. In grains and legumes the seeds are retained by the stalk or pod, unlike wild cousins, which use the wind to disperse seeds that break off easily.

Why, then, would fruits and vegetables be domesticated more quickly? The reason may well be that ancestral gardeners were pickier and more active in selecting the best produce than the early farmers who were out gathering grains and legumes in large numbers in the grasslands. Think about it: a gardener is visiting his or her plot every day or two to weed and manage the plants. S/he easily sees the best-producing individuals. On-site taste testing can quickly eliminate the bitter from the sweet. The thorny ones can be avoided or chopped down. Those plants that mature together are more likely to be gathered. In the fields, on the other hand, seeds from a target species are gathered with less discrimination. Seed gatherers only perform one selection event each year (i.e., harvest). They'll gather a wider variety of that species each year. They won't necessarily sort the best from the middling, so a greater diversity of genes gets passed on to the next year.

The development of increasingly sophisticated tools, the mastery of fire, the acquisition of writing, and the current digital revolution have all marked the evolutionary lineage leading to the modern human species (*Homo sapiens*) (Berna et al., 2012; Twomey, 2013). Agriculture, or the move from hunting-gathering (HG) to crop cultivation and animal care, is one of these inventions. This transition ushered in a shift that can't be overstated in all elements of life on the planet, including the biological and physical environment, as well as, of course, human culture. As a result, it's no surprise that this period is known as the Neolithic Revolution. There are various indicators of the significant changes brought about by the introduction of agriculture. Many renowned 'natural' landscapes on the planet, for example, are essentially agricultural landscapes, the rice terraces of Yunnan, China, cypress-studded fields of Tuscany, Italy, and the bigsky, brilliant plains of the Palouse, Washington etc. Agriculture today occupies 38% of total land area, making it the single most significant land use on the planet. Forests, for example, account for just 30% of the land area (the majority of which is subjected to some form of human exploitation) (FAO, Land use database). Sedentism (as opposed to nomadism) and the appearance of villages and, eventually, cities, the introduction of ceramics, the transformation of more egalitarian societies into more hierarchical societies, a stronger division of labour by the addition of nonfood production related occupations, and, later, the introduction of writing were among the changes that occurred more or less concurrently with the origination of agriculture.

5.3. Seed Dispersion and Domestication of Fenugreek

Trigonella foenum-graecum, or fenugreek, is an oldest crop in the World with trifoliate leaves and long tapering pods. Unlike lentil, chick pea and pea, seeds of fenugreek are more commonly used for flavouring and as a component in curry powder than for direct consumption. Seeds of fenugreek have also been used as a medicine, notably for their digestive stimulant properties. The plant is frequently used as a source of fodder, and the young tops are eaten as a vegetable. Botanically, fenugreek belongs to section *Foenum-graecum* of Trigonella which contains five more wild species and are endemic to the Middle East (Davis, 1970). The wild ancestor of fenugreek has not been known, however it is thought to be one of the wild species of the *Foenum-graecum* group. The wild species in the botanical area lack specialized dispersion mechanisms. Dehiscence does not occur in wild species of fenugreek because the dried ripe pods stay connected to the plant. The seeds are released only after the pod's disintegration in the next season, whereas in cultivation, seeds are disintegrated and released during the threshing and winnowing. Modification of seed dispersal pattern and domestication can not relate in fenugreek. Generally seed size and shape are the main factor for seed dispersal pattern and domestication but it is in appropriate in wild species of fenugreek.

The first sign of fenugreek was identified around 4000 B.C. in Tel Halaf, Mesopotamia. However, because Tel Halaf is inside the sharing area of *T. macrorrhyncha*, a member of the section *Foenum-graecum*, these remnants are not necessarily for the cultivated species. The domestication of fenugreek appears to have been less evident than pea, lentil and chickpea. The cultivated form differs from the wild type primarily by having a more upright growth habit, a higher number of pods per plant, pod length and size. Seed dispersion has been somewhat altered. Bitter vetch (*Vicia ervilia*) and common vetch (*Vicia sativa*) are fodder plants mostly used for hay production in Southern Europe, West Asia and North Africa. The entire plants are cut from the field during late maturity stage and dried in open condition under natural sunlight for hey preparation. This hey are used for fodder supplements in indigenous Mediterranean agriculture for animal during summer when there is scarcity of green fodder and natural pasture and new stands of grass are still too small. The common vetch is associated with a polytypic species generally known *Vicia sativa* aggregate (Ladizinsky and Tamkin, 1978). This aggregate comprises wild weeds and cultivated varieties, some of which have differing chromosome numbers and shapes but may still be crossed (Ladizinsky and Tamkin, 1978). The wild and weedy species, of wild pea and lentil, burst their seeds as they reach maturity. The dried mature pods of most current vetch cultivars remain intact, but varying degrees of breaking are prevalent, especially among land races and local variants. Bohrer (1972) hypothesized that production of fodder for animals and food for human consumption are the main reason for collection of this crop species. Bitter vetch and common vetch appear to have been used as fodder plants for centuries. The immature plants cut and naturally dried under sunlight and store to use for next season.The seeds are collected of these plants only for growing the crop in the subsequent year.

6. Genome Evolution of Fenugreek

6.1. Molecular Marker-Based Association Mapping of Seed Traits in Fenugreek

Fenugreek (*Trigonella foenum-graecum* L.) is a dicotyledonous annual legume with tiny seeds that belongs to the Leguminosae family (Fabaceae). It is a well-known aromatic spice derived from dried seeds and green leaves that has been used as a human food and herbal medicine from ancient times in Indian, Roman, Egyptian and Chinese history. It may be used as a supplement for maize and wheat flour in bread preparation as a source of flavour, colour, and to change the texture of food products in the modern food industry.

Despite advances in sequencing technology, very little effort has been undertaken in recent years to quantify genetic diversity across fenugreek genotypes. Several studies, however, have used standard molecular markers to explore fenugreek genetic diversity. For example, Restriction Fragment Length Polymorphisms (RFLPs), Randomly Amplified Polymorphic DNA (RAPD), Inter Simple Sequence Repeat (ISSR), Simple Sequence Repeat (SSR), Sequence-Related Amplified Polymorphism (SRAP), markers were used to analyze the genetic diversity among a set of accessions. However, due to the small set of genotypes and poor marker density, these experimental results lacked statistical power to provide accurate information for mapping complex quantitative traits and loci. The traditional approach to genetically dissect complex traits and identifying the underlying quantitative trait loci (QTL) or genes is to use the progeny of selected crosses such as near isogenic lines, recombinant inbred lines, backcrosses, or double haploid populations. However, this method is restricted by the lack parents of the many varieties and requires lot of time to create such populations. As a result, the alternative strategy known as association mapping (AM), which relies on linkage disequilibrium (LD) between polymorphic molecular markers and causal variations. AM has an advantage over traditional QTL mapping in that it uses already-existing naturally developed and adapted genotypes with more genetically diverse, avoiding the need to create new mapping populations. This criteria makes excellent use of historical recombination that has occurred over hundreds or thousands of generations across a wide range of genotypes. To discover such crossing-overs, genotypes must be densely genotyped in order to achieve high statistical mapping resolution and find single nucleotide polymorphisms (SNPs) linked to the characteristic under study. Several sequencing methods, like as restriction site-associated DNA sequencing (RAD-seq), which uses a single restriction enzyme to create genomic DNA fragments at random, have made genotyping more possible than before. Plants lacking reference genomes can be studied using RAD-seq. However, due to sequence read mistakes, it displays a large loss of data. The double digest RAD-seq (ddRAD-seq) technology, which utilizes two restriction enzymes to digest the genomic DNA, was designed to overcome these flaws. Adaptor ligations and precision size selection are performed on the fragments, and only a tiny percentage of the fragments are sequenced.The fenugreek genome has very little information and recently. Majority of research employed de novo transcriptome sequencing and next-generation sequencing (NGS) of representational difference analysis to discover genes involved in diosgenin production (RDA-NGS). Comparative transcriptome analysis and qRT-PCR were employed in two recent investigations. Lever ageing homology comparisons with known genomes could potentially fill some gaps in our knowledge of fenugreek genome structure and function, similar to recent studies in the Trifolieae tribe of the Fabaceae family between *Medicago truncatula* (barrel clover) and *Trifolium repens* L. (white clover) and *Trifolium pretence* and *Trifolium medium* (zigzag clover).

7. Conclusion

Fenugreek is an herb with plenty of health benefits and is widely cultivated in different parts of the globe. Plant is small in size and adapted to diverse soil and climatic conditions. It is believed to be a native of southern Europe and the Mediterranean region, from where it is spread throughout the world. Different fenugreek varieties grown in India and other continents have been reported and growing. Several studies have been carried out using cytological, biochemical and molecular markers to assess the diversity of fenugreek genome. Recently, genomics study will provide the correct blueprint of the fenugreek genome and be helpful in gene mining for desirable traits for the researchers.

References

Acharya, S.N., Thomas, J.E.,Basu, S.K. 2008. Fenugreek, an alternative crop for semiarid regions of North America. Crop Sci. 48, 841-853.

Francesco, B., Goldberg P., Horwitz K.L., Sharon J.B., Holt, B.M., Michael, C. 2012. Microstratigraphic evidence of in situ fire in the Acheulean strata of Wonderwerk Cave, Northern Cape Province, South Africa. Proceedings of the National Academy of Science of the USA. 109, 7593–7594.

Betty, R. 2008. The Many Healing Virtues of Fenugreek. Spice India, pp. 17–19.

Basu, S.K., Acharya, S.N., Thomas, J.E. 2008. Application of phosphate fertilizer and harvest management for improving fenugreek (*Trigonella foenum-graecum*L.) seed and forage yield in a dark brown soil zone of Canada. KMITL Sci. & Tech. J.8,1–7.

Basu, S.K., Prasad, R. 2011. Trends in new technological approaches for forage improvement. Aust J Agric Eng. 2, 176-185.

Basu, A., Basu, S.K., Kumar, A., Sharma, M., Chalghoumi, R., Hedi, A., Solorio-Sánchez, F., J., Ramírez-Aviles L.,Balogun, M.O., Hafez, E.E., Cetzal-Ix, W. 2014. Fenugreek (*Trigonella foenum-graecum*L.), a potential new crop for Latin America. AJSIH. 4: 148-162.

Bunting, E.S. 1972. Cultivation of fenugreek and some existing its varieties. Univ. of Feed. Lab., Oxford (personal communication).

Davis, P. H. 1970. Flora of Turkey and the East Aegean Islands. Edinburgh Univ. Press, Edinburgh. 3,65-76.

Del Gaudio, S. 1953. Sui, R., consumiidrici e indaginisull'autofertilitadelfieno Greco, Investigations on water consumption and surveys on the autochthonous hay greco. Annali Della SperimentazioneAgraria 7:1273-1287.

Dangi, R.S., Lagu M.D., Choudhary, B.L.,Ranjekar, P.K., Gupta, V.S. 2004. Assessment of genetic diversity in *Trigonella foenum-graecum* and *Trigonella caerulea* using ISSR and RAPD markers. BMC Plant Biology. 4,4-13.

Duke, A.J. 1986. Handbook of Legumes of World Economic Importance, Plemus Press, New York and London.

Edison, S. 1995. Spices research support to productivity. In: editor Ravi N. *The Hindu survey of Indian agriculture.* Madras: Kasturi & Sons Ltd., *National Press*

Evans, W.C. 1991. Trease and Evan's pharmacognosy, 13th edition. *BalliereTindall, London.*

Fazli, F.R.Y., Hardman R. 1968. The spice fenugreek (*Trigonella foenum-graecum* L). Its commercial varieties of seed as a source of diosgenin. Tropical Science 10, 66–78.

Haefele, C., Bonfils, C., Sauvaire, Y. 1997. Characterization of a dioxygenase from *Trigonella foenum-graecum* involved in 4-hydroxyisoleucine biosynthesis. Phytochemistry 44:563-6.

Hardman, R. 1980. Fenugreek -a multi-purpose annual legume for Europe and other countries. Cereal Unit Publication, *Royal Agricultural Show*, Stoneleigh, UK.

Kamal, R.,Yadav, R., Sharma, G.L. 1987. Diosgenin content in fenugreek collected from different geographical regions of south India. Indian Journal of Agricultural Sciences, 57, 674-676.

Gnanamurthi, K., Shanmugavelu, N. Kumar, K., Peter V. 2002. Production Technology of Spices and Plantation Crops. Laurier Books Limited, ISBN 81775-41617.

Kearns, C.A., Inouye D.W.,Waser M.N. 1998. Endangered mutualism: the conservation of plant-pollinator interactions. Annu. Rev. Ecol. Syst. 29, 83-112.

Leyel, C.F. 1987. Elixirs of life. London: Faber & Faber.

Ladizinsky, G.,Tamkin R.1978. The cytogenetic structure of *Vicia sativa* aggregate. *Theor. Appl. Genet*., 53, 33-42.

Maloo, S. R., Sharma, R., Jain, D., Chaudhary, S.,Soan,H. 2020. Assessment of genetic diversity in fenugreek (*Trigonella foenum-graecum*) genotypes using morphological and molecular markers. Indian Journal of Agricultural Sciences 90, 25–30.

Manniche, L., 1989. An Ancient Egyptian HerbalBritish Museum Press, London.

Najafi, S., Zarifi E.A., Birsin A.M. 2013. Karyotype characterization of reputed variety of Fenugreek (*Trigonella foenum-graecum*) in West Azerbaijan-Iran. Journal of Applied Biological Sciences, 7, 23-26.

Paroda, R.S., Karwasra R.R. 1975. Prediction through genotype environment interactions in fenugreek (*Trigonella foenum-graecum* L.). Forage Research, 1, 31-39.

Perfecto, V. 2010. The agroecologicalmatrix as alternative to the land-sparing/agriculture intensification model. Proceedings of the National Academy of Sciences. 107, 5786-91.

Petropoulos, G.A.2003. Fenugreek: the genus *Trigonella* CRC Press, Science.

Prasad, C.K.P.S., Hiremath, P.C. 1985. Varietal screening and chemical control foot-rot and damping-off caused by *Rhizoctonia solani*. Pesticides19, 34-6.

Rouk,H.F., Manresa H. 1963. Fenugreek (*Trigonella foenum-graecum* L.): its relationships, geography and economic importance. Dire Dawa, Ethiopia Imperial Ethiopian College of Agriculture and Mechanical Arts.

Singh, D., Singh, A. 1974. A green tilling mutant of *Trigonella foenum-graecum* L. (Methi). Crop Improv.2, 98-100.

Singh, J., Gupta, K., Arora S.K.1994. Changes in the anti-nutritional factors of developing seeds and pod walls of fenugreek (*Trigonella foenu-graecum* L.). Plant Foods for Human Nutrition 46, 77-84.

Scherbukhin, V.D.,Anulov, O.V. 1990. Legume seed galactomannans. Applied Biochemistry and Microbiology 35, 257–274.

Shanmugavelu, K.G., Kumar N., Peter, K.V. 2002. Production techniques of spices and plantation crops, Jodhpur, India: Agrobios, 109–130.

Twomey. 2013. The Cognitive Implications of Controlled Fire Use by Early Humans. Cambridge Archaeological Journal 23(01).

Vaitsis, T.1985. Creation of a new variety of fenugreek, named 'Ionia', resistant to *Sclerotinia sclerotiorum* (unpublished data). Fodder and Pastures Research Institute, Larissa, Greece.

Zandi, P., Basu, S.K., Khatibani B., Balogun, L., Aremu, M., Sharma, M.O., Kumar, M.A., Sengupta, R., Li, X., Li, Y., Tashi, S., Hedi, A., Cetzal, Ix.W.2015. Fenugreek (*Trigonella foenum-graecum*L.) seed: a review of physiological and biochemical properties and their genetic improvement. ActaPhysiologiaPlantarum 37, 1714.

Zandi, P., Shirani R., Daneshian, A.H.J., Bazrkar K.L. 2011. Agronomic and morphologic analysis of Fenugreek (*Trigonella foenum-graecum* L.) under nitrogen fertilizer and plant density via factor analysis. African Journal of Agricultural Research 6,1134–1140.

20

Integrated Nutrient Management and its Implications in Seed Spices

Chaman Kumari Jadon[1], Pratap Singh[1], Rajesh Kumar[1] and Ravindra Singh[2]

[1]Agriculture University, Kota, Rajasthan, India
[2]ICAR-National Research Centre on Seed Spices, Ajmer, Rajasthan, India

1. Introduction

India is known as seed spices hub because as far as International level of concern, our country is the largest producer, consumer and exporter of seed spices. Seed spices contribute about 51.79 % of total area and 19.06 % of production of total spices in the country. The area under seed spices is about 1.74 million hectare and production is about 1.45 million tons (Lal, 2018). In our national economy, seed spices are play an important role, the demand is increasing not only due to its large domestic consumption, but also for export. Since long, huge demand of seed spices and importing countries in global market look at India for quality produce of seed spices. The ISO has been listed a total of 109 spices, out of which 63 spices are being grown in India and 20 spices are being classified as seed spices. The major growing states of seed spices are Rajasthan and Gujarat besides this, other significant areas comes under Uttar Pradesh, Madhya Pradesh, Bihar, West Bengal, Orissa, Punjab, Karnataka and Tamil Nadu (Lal, 2018). Various kind of constraints are involved in cultivation of these crops namely seasonal crops under rainfed/irrigated conditions, grown by small and marginal farmers on less fertile and problematic soils without using balance nutritional supply, non-judicious use of pesticides, poor export quality product, post-harvest losses, marketing and policy support issues, etc.

Nowadays, as growing human population, it is quite difficult to maintain food security due to increasing stress on natural resources. A equilibrium between enhancing crop production while maintaining soil health and environmental sustainability is pre-requisite for long-term food security. Though, imbalanced and excessive nutrients usage directed to decreasing NUE (nutrient-use efficiency), producing fertilizer consumption uneconomical and causing adverse effects on the environment (Aulakh and Adhya, 2005) and groundwater quality may cause health hazards (Aulakh et al., 2009) and climate change. Availability of food for human under sustainable production systems having a remarkable challenge in the developing world and is extremely critical for alleviating poverty. To dodge this challenge, farmers tended to excess use of certain inputs such as agrochemicals, agricultural inputs, which in turn have already started deteriorating the food, soil,, human health and environmental. Though, it is hardly possible to supply all the nutrient need of crops through organic manures. The sole use of either inorganic fertilizers or organic manures cannot sustain productivity. Therefore, judicious uses of organic manures and inorganic fertilizers are crucial to protect soil health and boost productivity and input use efficiency. Therefore, to maintain soil fertility and deliver optimum level of nutrients from all feasible sources in an integrated way is necessary to sustain the desired productivity of crop (Chundawat, 2001). A promising strategy for soil health management and sustained productivity can be gain through integration of different sources of nutrients. Hence, considering the above facts, integrated nutrient management (INM) has been developed. Integrated nutrient management is the combined application of chemical fertilizers along with organic resource materials like organic manures, green manures, bio-fertilizers and other organic decomposable materials for crop production. This is ecologically, socially and economically viable and environment friendly. The cropping system and farming system are the focus of attention in this approach for developing INM practices for various categories. The conjunctive (integrated) application of organics with inorganic sources of nutrients reduces the dependency on chemical inputs and it not only performs as a source of nutrients but also delivers micro nutrients as well as modifies the soil physical behavior and increases the effectiveness of applied nutrients (Pandey, 2007; Parihar et al., 2010). A wide-ranging literature research work revealed that INM augments crop yields by 8-150% as compared with conventional practices, progresses water and nutrients use efficiency and the hike economic returns to farmers, whereas improving food quality and soil health and sustainability. Considering the present constraints in seed spices production, farmers can be practiced INM in seed spices to derive higher productivity with simultaneously maintaining soil fertility. Although merger of research work had been done on INM in seed spices, so this chapter may draw a keen attention in this direction and positive efforts for future research in context.

2. Concept of INM

The prime concept of INM is the maintenance of soil fertility, sustainable agricultural productivity and improving profitability through judicious and efficient use of fertilizers as mentioned. It improves various aspects of nutrient cycling, including macro and micronutrient inputs and outputs, with the objective of synchronizing nutrient demand by the crop and its supply into the environment. Under integrated nutrient management practices, the losses due to leaching, runoff, volatilization, emissions and immobilization are minimized, while high nutrient-use efficiency is obtained (Zhang et al., 2012).There is now a larger cognizance that INM can't only rise crop productivity, but also concurrently and almost gradually preserve soil resources. Its practices use farmyard manures, farm wastes, soil amendments, crop residues, natural and chemical fertilizers, green manures, cover crops, intercropping, crop rotations, fallows, conservation tillage, irrigation and drainage to conserve available water and to boost plant nutrients (Wu and Ma, 2015). This policy also includes new techniques, such as deep placement of fertilizers and the use of inhibitors or urea coatings that have been established to reduce nutrient losses and improves plant uptake. These practices motivate farmers to focus on long-term planning and have greater deliberation of environmental influences, rather than only paying attention on yield-scaled profit.

3. Principles of INM System

Basic principles of sustainable INM system laid out by Dennis Greenland

- **Maintain the nutrient cycle:** Nutrients removed by crops must be given back to the soil. Its means spatial and temporal crop demand should be matched with soil nutrient supply system.
- **Restore the soil quality & properties:** Physical conditions of soil should be upheld and upgraded. Level of OC (organic carbon) in soil should be improved and maintained. Providing all possible sources of nutrients to optimize their inputs. Soil quality with respect to soil acidity, salinity and sodicity or toxic elements build-up must be minimized.
- **Minimize the abiotic stress**: Build-up of abiotic stress should be minimal.
- **Reduction in N losses, while improving the crop yield**: The major principle of INM is to control the N losses and their detrimental effects on environment as well as obtaining high crop productivity. The fate of N in field is an integrated consequence of crop N uptake, immobilization and residues in the soil, and N losses to the environment, such as ammonia volatilization, NO_X emissions, denitrification, N leaching and runoff (Li et al., 2015).
- **Control of soil erosion:** Degradation of land due to soil erosion must be controlled.

4. Goals of INM

The overall objective of INM is to maximize the use of soil nutrients to improve crop productivity and resource-use efficiency.

Besides this, following are the main goals of INM

- To ensure productive as well as sustainable agriculture.
- To decrease expenditure on costs of purchased inputs by using farm manure and crop residue, etc.
- To utilize the potential benefits of green manures, leguminous crops and bio-fertilizers.
- To check degradation of the environment.
- To fulfil the social and economic aspiration of the farmers without harming the natural resource base of agricultural production.
- To optimize the soil conditions by improving its physical, chemical, biological and hydrological properties to increase farm productivity and minimize land degradation.

5. Components of INM

Key components of integrated nutrient management are (i) balanced use of inorganic fertilizers, (ii) organic manures like FYM, vermicompost, compost, poultry manure, biogas, slurry, press mud, cakes, manure of bio gas plant, bone meal, fish manure Phospho-compost, etc (iii) soil fertility restoring crops such as green manures, legumes, etc.; (iv) crop residues incorporation and (v) biological agents like bio-fertilizers, etc.

5.1 Fertilizers

Fertilizers, being most important ingredient of INM. Dependability on fertilizers has been increasing continuously because of the need to supply huge amounts of nutrients in intensive cropping with high productivity. Domestic fertilizer production is insufficient to meet out the demand and the situation is not likely to recover in the near future. Besides restrictions like global price hike of fertilizers and raw materials would not permit fertilizer import in big quantities leading to a large gap between supply and consumption. While organics and bio-fertilizers are likely to bridge a part of this gap, effective use of fertilizers in narrowing the nutrient supply gap also needs greater importance. Use of fertilizer nutrients by the crops vary from 30-50% in case of N, 15- 20% in case of P and less than 5% in case of micronutrients. Thus, a considerable amount of applied nutrients is lost through various paths. Enhancing NUE therefore, be a prioritized area of research for restoration and improvement of soil health and minimizing the cost of crop production. Due to chemical fertilization, the crop productivity has improved undoubtedly, but to greater extent they have contributed to soil deterioration. Intensive agriculture, while enhancing food production, has resultant second generation problems in respect of nutrient imbalance.

India is the 3rd largest manufacturer and consumer of fertilizers in the world, after China and USA. However, India's yearly consumption of chemical fertilizers in nutrient terms (NPK), has augmented from 0.7 lakh MT in 1951-52 to 289.69 lakh MT 2019-20.

5.2. Organic Manures

Organic manures namely FYM, urban compost, human excreta, rural compost, city refuse, sewage-sludge, vermicompost, biogas slurry, press mud phospho-compost, bio-compost, press mud, oil cakes, and other agro-industrial wastes etc. have large nutrient potential. Compost and Farm Yard Manure have traditionally been considered as the most important manures for preserving soil fertility and ensuring yield stability. There are also other possible organic sources of nutrients such as non-edible oil cakes and wastes from food processing industry. In addition to this, there are many industrial by- products and municipal wastes with fair nutrient potential. Organic manures in addition to fertilizers sustain high crop yields over long periods as compared to application of only fertilizers as observed in many long term studies (Prasad, 2008). Organic manures are a dynamic source for supplying plant nutrients and for replacing organic matter content of most agricultural soils. This would more emphasize the necessity to use organic manures alone or in conjunction with chemical fertilization to maintain soil fertility for the sustainable crop production along with more direct environmental benefits. If other management practices, like incorporation of crop residues and conservation tillage are used with organic manure, it may also decrease GHG emissions, enhance the soil quality and C-sequestration, associated high crop yield. There is an urgent need to integrate these sources depending on their availability in different crops and cropping systems.

5.3. Green Manure and Legumes

Green manuring can be described as unrecompensed green plant tissues for improving physical structure as well as soil fertility, its principal means adding organic matter to the soil. The green-manure crop supplies organic matter as well as extra nitrogen for crops, extremely if it is a legume crop, due to its ability to fix nitrogen from the air with the help of its root nodule bacteria to the extent of 50-500 kg N ha^{-1} before the plant starts flowering (about 40-60 days growth). The legumes residues after harvest of grain contain N which is released at a steady rate when incorporated because of optimum lignin content. Besides nitrogen, the GM crop can enhance mobilization of less available soil phosphorus and potassium which can be recycled into the system. In general, a sixty days old GM crop when incorporated released 20-30% of biomass N after 15 days and 26-30% in 30 days. The green-manure plants also help to protect against leaching and soil erosion. Before the flowering stage, green manure to be mixed in the soil, which gives more nutrients and helps the soil protection. Green manure crops should be short duration, fast growing, and pest & disease resistant, efficient in water use, easy decomposition into the soil quickly, but gradually, and it gives higher nutrients availability to the soil for the next crop.

It contains in two ways: by growing green manure crops in agricultural land or by collecting green leaf from plants grown in forest wastelands, and field bunds. The crops cowpea, sunhemp, dhaincha, pillipesara, cluster beans, and *Sesbania rostrata* are the most significant green manure crops. Legumes could provide a significant ingredient of INM when grown for grain or fodder in a cropping system, or when used for green manuring. Legumes grown as green manure, forage or grain crops enhanced the productivity of rice-wheat cropping system (RWCS) and revived soil fertility (Yadav et al., 2000). Plants that grow in wastelands, field bunds, etc., are the other source of green leaf manure. The subabul, neem, mahua, glyricidia, wild indigo, karanj, calotropis and other shrubs are the very important plant trees useful for green leaf manure.

5.4. Crop Residues

Crop residues have many competitive uses and may not be always accessible as an ingredient of INM, in North-West India where mechanical harvesting is practiced, a considerable quantity of residues is left in the field, which can form a part of nutrient supply. Large amounts of residues of potato, sugarcane, vegetables, etc., are practically wasted in most of the cases. While cereal crop residues are valuable cattle feed, these could be used to supplement the fertilizers wherever available in surplus of local requirements. In case of different crops, stubbles left in the field in traditional harvesting methods range from 0.45 to 1.5t/ha. Under the mechanical harvesting, this amount is much larger. Coarse cereals stubbles such as sorghum, maize, pearl millet etc. which are difficult are normally collected and burnt during land preparation causing significant loss of plant nutrients. Recently, burning of crop residues is also a burning issue as environmental point of view.

5.5. Biofertilizers

During the recent years, uses of biofertilizers are increased as a supplement to nutrient supply in agricultural system due to increasing cost of chemical fertilizers and global concern for environmental pollution. Biofertilizer is a substance which comprises living microorganism which, when applied to plant surfaces, seed, or soil, colonizes the rhizosphere or the interior of the plant and promotes growth by increasing the supply or availability of primary nutrients to the host plant. In nutrient management systems, biofertilizers are an alternative low cost and eco-friendly input which can reduce the chemical fertilizer dose by 25–50% (Pattanayak et al., 2007). These are also renewable source of plant nutrients to supplement the chemical fertilizers for sustainable agriculture. Microorganisms are frequently not as effective in natural surroundings as one would expect them to be and therefore artificially multiplied cultures of efficient selected microorganisms play an vital role in accelerating the microbial processes in soil. A number of microorganisms are considered as beneficial for agriculture and used as biofertilizers viz. *Rhizobium, Azotobacter, Azospirillum, Cyanobacteria, Azolla*, Phosphate and potassium solubilizing microorganisms. Silicate solubilizing bacteria, plant growth promoting rhizobacteria and these are also available as liquid biofertilizers. Number and efficiency of these micro- organisms may affect its benefit, however, it is governed by a large number of soil and environmental factors. Biofertilizers have an ability to change nutritionally important elements from unavailable to available form through biological processes and they are able to exert beneficial effects on plant growth and also increase plant resistance to adverse environmental stresses, such as water and nutrient deficiency and heavy metal contamination.

6. Effect of INM on Soil Physical Properties and Soil Fertility

Soil physical properties are associated with SOC and OM, thus, any soil management practice that increases soil organic matter has direct bearing on soil physical properties and microbial biomass, in such way, combined use of organic and inorganic nutrient sources might be the right plan for these soils, primarily for the improvement of soil health. Many researchers were observed noteworthy improvement in the soil physical conditions, the addition of NPK fertilizers along with organic manure, lime, green manure and bio-fertilizers augmented SOC, WSA, moisture holding capacity, and infiltration rate in soil while dropping bulk density. It also has a significant effect on soil aggregation, soil structure and fertility status.

Owing to the imbalanced use of mineral fertilization, most of the agricultural soil fertility deteriorated day by day. A significant improvement in soil fertility status was reported by many of the researchers. Soil Organic Carbon of the soil improved by the use of organic manure with RDF, this combination significantly influenced crop growth, development and productivity. Most of the research results visibly confirmed that INM increases the yield potential of crops over and above achievable yield with recommended fertilizers resulted well synchrony of crop N needs due to slower mineralization of organics; reduced N losses via de-nitrification and nitrate leaching; enhanced NUE and recovery by crops, and improvements in soil health and productivity, and henceforth could sustain high crop yields in several cropping systems ensuring long term sustainability of the system. Incorporation of FYM, GM and BGA, through an inorganic source in the treatment increased organic carbon, mineralizable N, NH_4^+-N and reduced the bulk density (Rahman et al., 2013).

7. Effect of INM in Seed Spices Crops

This section of chapter is the highlights the research works on INM practices, which had been done by many researchers along with their findings.

- Application of 10t vermicompost in coriander + fennel mix grown on medium soil fertility, plant height, primary, secondary branches and yield parameters were higher. Seed yield enhancement was 48.7 per cent. Availability of N, K, Cu, Fe and Mn, soil organic carbon (SOC) in soil enhanced (Aishwath et al., 2019).

- The combination as soil application of inorganic N (100%) of RDF + Azospirrilum @ 15kg ha^{-1} + FYM @ 5 t ha^{-1} gave the maximum plant height, number of branches per plant, number of umbels per plant, number of umbellates per umbel, number of grains per umbel and yield per plot and increased the yield 71.65 per cent with the maximum net profit and benefit: cost over control (Singh, 2014).
- 100 per cent of RDN through FYM with biofertilizers (*Azatobacter* and PSB) can fulfil the nutrient need of cumin under organic farming condition. Organic manure with bio-fertilizers was significantly superior over organic manure alone. Both the organic sources (FYM and vermicompost) were statistically at par, but FYM was slightly more effective than vermicompost (Patel et al., 2013).
- Sunanda et al. (2014), was reported that INM significantly helped to improve growth, yield and quality of kasuri methi. Maximum plant height, number of leaves, branches per plant, leaf area, leaf area index, chlorophyll content, Cumulative Growth Rate, dry herb yield, fresh herb yield (8.02 t/ha), number of pods per plant and attributes on quality like crude protein in herb (13.31 %) and in seed (21.33 %) were recorded with 75% Nitrogen+ Recommended Dose of Phosphorus and Potassium + Farm Yard Manure (7.5 t ha^{-1}) + Rhizobium (1.5t ha^{-1})+ Azospirillum (5 kg ha^{-1})+ Phosphate Solubilizing Bacteria (5 kg ha^{-1}).
- Use of *Azospirillum* + 50% nitrogen+ 5t farm yard manure ha^{-1} produced higher seed yield of coriander (Malhotra et al., 2006).
- Fenugreek is also a good soil renovator and reported to fix 48% of its total N2 during the growing season (Pawar et al., 2012).
- Use of *Azotobacter*, PSB and their combinations of both was increased seed yield by 8.8, 11.6 and 18.5per cent, respectively, as compared to control. *Azotobacter* and PSB combination enhanced plant height at different stages and number of primary and secondary branches of coriander. Numbers of umbellates/umbel, seed, straw and biological yields were highest with the combine use of biofertilizers (Aishwath et al., 2012).
- Using of BD 500 and BD 501 along with either FYM @ 6t/ha or vermicompost @ 2t/ha recorded a significantly increase in seed yield of cumin by 20.56 and 12.85 per cent over the application of FYM @ 6t/ha and Vermicompost @ 2t/ha alone, respectively. Integrated use of organic manure + biodynamic preparations (BD 500 and BD 501) resulted into the maximum increase in yield attributes and yield of cumin under organic production system (Sharma et al., 2012).
- Desai et al. (2020) reported that use of 50 % RDN through vermicompost + 50 % RDN through fertilizer + *Azotobacter*+ PSB gave significantly higher plant height at harvest, number of umbels per plant, number of branches per

plant, number of umbellates per umbel, number of seeds per umbellate, seed yield, straw yield, gross and net income.

- The highest seed yield of 19.16 q ha^{-1} was recorded in coriander variety JD – 1 in treatment combination Poultry manure @ 5 t ha $^{-1}$ + 100 % RDF along with cost benefit ratio 1:2.98 (Dadiga and Jain, 2017).
- Recommended dose of nutrients (100%) applied in coriander through fertilizers exhibited highest vegetative growth and yield attributes with maximum yield, net returns and benefit cost ratio closely followed by 50 % RDF through fertilizers + 50 % RDF through vermicompost (Godara et al., 2014).
- The combination of Azospirillum + PSB + KS with 75% RDF was best for economics (maximum net return and B: C ratio) and optimum production of black cumin without loss in yield and reduction of 25% of inorganic fertilizers through application of bio-fertilizers and environmental pollution to some extent (Jamir et al., 2021).
- Use of biofertilizer alone as well as in combination with sheep manure, vermicompost and RDF (recommended doses of fertilizer) resulting superiority in terms of growth, yield attributes and yield over absolute control. Application of recommended doses of fertilizer with seed inoculation by biofertilizer proved to be superior for realizing higher net return and benefit:cost ratio (Mehta et al., 2012). Similarly, Mehta et. al., 2007a, 2007b & Mehta et. al., 2011 worked out influence of sheep manure, vermicompost and biofertilizers on productivity of ajwain, dill and coriander, respectively.
- Lal, 2018 reported that application of 75% N & P through vermin-compost and enriched vermi compost with seed inoculation by *Azotobactor and PSB* proved economical organic nutrient management strategy in coriander and cumin.
- Prabhu et al. (2000) recorded significantly higher yield of coriander with 25% recommended dose of fertilizer + FYM @ 10 t ha^{-1}+ Azospirillum sp. + VAM over other combination of nutrient sources.
- Inoculation with *A. chroococcum* or *A. brasilense*, combined with *G. mossese* (VAM), gave significant increases in vegetative growth, total carbohydrates, photosynthetic characteristics, essential oils, seed parameters, N, P and K contents, and linalool (Abou and Gomma, 2002).
- In red, sandy loam soils, Manure et al. (2000) reported that yield attributes, seed and oil yield, and oil content in coriander were all enhanced by the application of N, S, and Zn fertilisers.
- Application of 15 to 20 t of FYM is sufficient for the nutrition of rain-fed cumin, but an additional 30 kg of N fertilizer/ha is recommended for irrigated conditions (Peter et al., 2000).

- In loamy sand soils condition, application of 20 kg/ha of ZnSO4 along with 30 kg N/ha significantly increased the seed yield of cumin. A combined application of 50% inorganic N (90 kg/ha) + *Azospiriullum* + 5 t FYM/ha produced yields of fennel on par with 75 and 100% N through inorganic source + *Azospirillum* + 5 t FYM/ha as well as with 100% inorganic N alone, but proved superior to applied 5 t FYM+ *Azospirillum*, 5 and 10 t FYM alone, *Azospirillum* alone and control. Highest fenugreek seed yield, net returns and B:C ratio (6:1) recorded with 100% inorganic N alone closely followed by 100% inorganic N + Rhizobium applied at 1.5 kg/ha + 5 t FYM/ha (Aishwath et al., 2011).

8. Advantages of Integrated Nutrient Management

- The INM supports to restore and sustain soil fertility and crop productivity in farming systems.
- It enables combined and harmonious use of organic and inorganic nutrient resources to serve the simultaneous needs of food production and economic, environmental and social viability. It brings economy and efficiency in fertilizer use and positively affects the physical, chemical and biological environment of soil.
- It provides sustainable production system, increasing nutritional quality of products, improving soil health, and reduction in environmental pollution.
- It helps to check the emerging deficiency of nutrients other than NPK.
- It empowers farmers by improving their skill, technical and decision-making expertise.
- It promotes changes in land use and synergistic interactions between forestry, livestock and cropping systems as part of agricultural intensification and diversification.
- INM is also profitable for marginal farmers who cannot afford to supply all crop nutrients through expensive chemical fertilization.

9. Major Constraints for Adoption of INM by Farming Communities

- Non-availability of FYM because use cow dung as a source of fuel as well as mishandling of bulk organic manures.
- Difficulties in growing green manure crops.
- Non-availability of bio-fertilizers as well as low-quality bio-fertilizers that reach to farmers are ineffective.
- Non-availability of soil testing facilities at local level.
- Costly chemical fertilizers.
- Unavailability of improved seeds at the time of sowing.

- Increasing competitive value of crop residue as animal feed therefore affects the recycling of agricultural wastes
- The measure issues in the adoption of INM system are like large amounts of insoluble constituents of organic matter like cellulose, hemicellulose and lignin which delay the process of composting due to high C/N ratio, and development of other worms due to lack of proper air circulation in the container during vermicomposting.
- Complications in timely preparation of fields due to incorporation and decomposition of agricultural wastes and green manures which require irrigation.
- Lack of knowledge and poor advisory services.

10. Conclusion

It is well known that the nutrients play a vital role in the crop production but under intensive cultivation non judicious utilization of chemical fertilizers alone for long period could result in deterioration of soil fertility and quality of produce. For ensuring sustainability of soil-plant-microbes-environmental, INM is a basic prerequisite or solution which is not only increases in crop productivity and resource-use efficiency but substantially also helps in balancing soil fertility, environment and reduce the cost of inputs. Balanced use of organic manures, bio-fertilizer will be of fundamental importance for crop productivity and environmental concerns, which should be a priority for INM practices, provides a "win–win "opportunity to simultaneously enhance seed spices crop productivity and improving quality and overall achieving agricultural sustainability.

References

Abou A.H.E. and Gomaa, A.O. 2002. Influence of combined inoculation with diazotrophs and phosphate solubilizers on growth, yield and volatile oil content of coriander plants (*Coriandrum sativum* L.) Bulletin of Faculty of Agriculture Cairo University, 53(1), 93-113.

Aulakh, M.S., Adhya, T.K. 2005. Impact of agricultural activities on emission of green house gases - Indian perspective. In International Conference on Soil, Water and Environmental Quality - Issues and Strategies', 2005, 319-335.

Aulakh, M.S., Khurana, M.P.S., Singh, D. 2009. Water pollution related to agricultural, industrial and urban activities, and its effects on food chain: Case studies from Punjab. J. New Seeds.10, 112-137.

Aishwath, O.P., Khurana, H.S., Anwer, M.M. 2011. A Review on the Effect of Integrated Nutrient Management on Yield and Quality of Major Seed Spice Crops in India. Better crops-South Asia: 4-6.

Aishwath, O.P., Lal, G., Kant, K., Sharma, Y.K., Ali, S.F., Naimuddin. 2012. Influence of bio-fertilizers on growth and yield of coriander (*Coriandrum sativum* L.) under Typic Haplustepts. International J. Seed Spices. 2(2), 9-14

Aishwath, O.P., Mehta, R.S., Lal, G. 2019. Effects of on-farm composted seed spices residues on coriander, nutritional parameters and seasonal carbon offset by the crop and soil. International J. Seed Spices. 9(1), 91-98.

Chundawat, B.S. 2001. Integrated nutrient management in tropical and subtropical Fruits. Indian Journal of Horticulture. 58, 59-69.

Dadiga, Ashwini and Jain, P.K. 2017. Interaction effect of organic manures and fertilizers levels on growth and yield of coriander (*Coriander sativum* L.). Agric. Update, 12 (TECHSEAR-8), 2194-2201.

Desai, S.S., Chaudhary, M.G., Patel, R.S., Joshi, J.R. 2020. Effect of integrated nutrient management on growth, yield and economics of cumin (*Cuminum cyminum* L.). Int. J. Curr. Microbiol. App. Sci. *9* (11), 3712-3719.

Godara, A.S., Gupta, U.S., Lal G., Singh, R. *2014.* Influence of organic and inorganic source of fertilizers on growth, yield and economics of coriander (*Coriandrum sativum* L.).International J. Seed Spices. 4(2), 77-80.

Jamir, A.R., Chattopadhyay, N., Momin, M.C. 2021. Effect of biofertilizers with various levels of inorganic nutrients on growth and yield of black cumin (*Nigella sativa* L.) var. Azad Kalonji. Environment Conservation Journal. 22 (1&2), 73-78

Lal, G. 2018. Scenario, Importance and Prospects of Seed Spices: A Review. Current Investigations in Agriculture and Current Research. 4(2), 491 -498.

Li, Y., Simunek, J., Zhang, Z., Jing, L. and Ni, L.2015. Evaluation of nitrogen balance in a direct-seeded rice field experiment using Hydrus-1D.*Agril.* Water Mgt. 2015; 148: 213–222.

Malhotra, S. K., Vashishtha, B.B. and Apparao, V.V. 2006. Influence of nitrogen, *Azospirillum*sp. and farmyard manure on growth, yield and incidence of stem gall disease in coriander (*Coriandrum sativum* L.). Journal of Spices and Aromatic Crops. 15 (2), 115-117.

Mehta, R.S., Anwer, M.M., Malhotra, S.K. 2012. Influence of sheep manure, vermicompost and biofertilizer on growth, yield and profitability of cumin (*Cuminum cyminum* L.) production. Journal of Spices and Aromatic Crops. 21 (1), 16–19

. Mehta, R.S., Malhotra, S.K., Lal, G., Aiswath, O.P., Meena, S.S., Kant, K., and Khan, M.A. 2011. Growth & yield of coriander as affected by sheep manure, vermicompost, biofertilizer. International Journal of Seed Spices. 1(1), 22-28.

Mehta, R.S., Malhotra, S.K., Vashishtha, B.B., Meena, S.S. 2007a. Effect of vermi-compost, sheep manure and biofertilizer on yield attributes and yield of Ajowan (*Trachyspermum ammi* Sprague) cv. NRCSS-AA-1. In: Malhotra S K and Vashishtha B B (Eds) Production, Development, Quality and Export of Seed Spices. (pp. 270–273). National Research Centre on Seed Spices, Ajmer.

Mehta, R.S., Malhotra, S.K., Vashishtha, B.B., Meena, S S. 2007b. Influence of sheep manure vermicompost and biofertilizer on growth and yield of dill (*Anethum graveolense* kurz) cv–NRCSS-AD-1. In: Malhotra S K &Vashishtha B B (Eds). Production, Development, Quality and Export of Seed Spices. (pp. 274–277). National Research Centre on Seed Spices, Ajmer.

Manure, G.R., Shivaraj, B., Farooqii, A.A., Surendra, H.S. 2000. Yield attributes, seed, essential oil yield and oil content of coriander (*Coriandrum sativum* L.) as influenced by the graded levels of nitrogen, sulphur and zinc nutrition in red sandy loam soils. *In* Proc. Centennial conference on spices and aromatic plants, Calicut, Kerala, India, 20-23 September 2000 pp 139-144.

Pandey, N., Verma, A.K., Anurag and Tripathi, R.S. 2007. Integrated nutrient management in transplanted hybrid rice (Oryza sativa). Indian J. Agron. 52(1), 40-42.

Panwar, A., Sharda, C., Manoj, S., Shrama, Y.K., Meena, R.S., Malhotra S.K., Mehta, R.S. and Aishwath, O.P. 2012. Morphological and biochemical characterization of *Rhizobium* isolates obtained from fenugreek (*Trigonellafoenum*). Seed Research 40 (2), 196-200.

Parihar, C.M., Rana, K.S. Kantwa, S.R. 2010.Nutrient management in pearlmillet (*Pennisetum glaucum*) -mustard (*Brassica juncea*) cropping system as affected by land configuration underlimited irrigation. Indian J. Agron. 55(3), 191-196.

Patel, S.G., Amin, A.U., Patel, S.P., Agalodiya A.V., Patel, S.M. 2013. Effect of different sources of organic manures with and without bio fertilizers in Cumin (*CuminumcyminumL.*). International J. Seed Spices. 3(2), 54-58

Pattanayak, S.K., Rao, D.L.N., Mishra, K.N. 2007. Effect of biofertilizers on yield, nutrient uptake and nitrogen economy of rice–peanut cropping sequence. J. Indian Soc. Soil Sci. 55, 184–9.

Peter, K.V., Srinivasan, V., Hamza, S. 2000. Nutrient management in spices. Fert. News. 45,13-28.

Prabhu, T, Narwadakar, P.R., Sajindranth, Rathore N.G. 2000. Effect of integrated nutrient management on growth and yield of coriander (*Coriandrum sativa*).South Indian Horti. 50, 680–684

Prasad, R. 2008. Integrated plant nutrient supply system (IPNS) for sustainable agriculture. Indian J. Ferti. 4 (12), 71-90.

Rahman, M.H., Islam, M.R., Jahiruddin, M., Rafii, M.Y., Hanafi, M.M., Malek, M.A. 2013. Integrated nutrient management in maize-legume-rice cropping pattern and its impact on soil fertility. J. Food Agric. Environ. 11, 648-652.

Sharma, S.K., Laddha, K.C., Sharma, R.K., Gupta, P.K., Chatta L.K and Pareeek, Priyanka.*2012.* Application of biodynamic preparations and organic manures for organic production of cumin *(Cuminum cyminumL.).* International J. Seed Spices 2(1), 7-11

Singh, S.P. 2014. Effect of bio-fertilizer azospirillum on growth and yield parameters of coriander (*Coriandrum sativum* L.) cv. Pant Haritima. International J. Seed Spices 4(2), 73-76

Sunanda, B.B., Shetty, G.R., Venkatesh, J. 2014. Influence of integrated nutrient management on growth, yield and quality of Kasuri Methi (*Trigonella corniculata* L.) under hill zone of Karnataka. International J. Seed Spices. 4(2), 62-67.

Wu, W., Ma, B. 2015. Integrated nutrient management (INM) for sustaining crop productivity and reducing environmental impact: A review. Sci. Total Environ.512, 415–427.

Yadav, R.L., Singh, S.R., Prasad, K., Dwivedi, B.S., Batta, R.K., Singh, A.K., Patil, N.G. and Chaudary, S. 2000. In Yadav JSP and Singh GB (eds), Natural Resource Management for Agricultural Production in India. pp. 775-880. ICMNR, New Delhi.

Zhang, F., Cui, Z., Chen, X., Ju, X., Shen, J., Chen, Q., Liu, X., Zhang, W., Mi, G., Fan, M. and Jiang, R. 2012. Integrated nutrient management for food security and environmental quality in China. *Adv. Agron.*116, 1-40.

21

Biostimulants Application and Its Impact on Horticultural Crops

Pooja S. and Vijaykumar B. Narayanapur

Department of Plantation, Spices, Medicinal and Aromatic Crops College of Horticulture, UHS, Bagalkot, Karnataka, India

1. Introduction

A primary issue for modern horticulture is facing two contradictory objectives, such as the need to produce food for the increasing world population and to minimize damage to the environment, which can in turn negatively impact horticulture. Fertilizer and chemical use and manufacture are hazardous to the environment and humans. As a result, new and effective environmentally friendly compounds are required. Plant biostimulants, which are harmless for humans and the environment, are one answer. They are particularly useful for lowering pesticides in agriculture. Plant biostimulants have become popular in farming and cultivation in recent years. Many investigations have proven the positive effects of its application, which have produced amazing outcomes (Smolen and Sady, 2010).

Biostimulants are chemicals that increase plant growth, water resistance, and abiotic stress tolerance while reducing fertiliser use. These chemicals are effective in little amounts, allowing for high yields and high-quality products by encouraging the smooth operation of the plant's key functions. Furthermore, biostimulants applied to plants improve nutrition efficiency, abiotic stress tolerance, and/or plant quality features regardless of nutrient availability. Several research have been carried out to determine the usefulness of biostimulants in boosting plant development under stress, in a saline environment, and in seedling development, among other things. Various raw ingredients, such as humic acids, hormones, algal extracts, and plant growth-promoting microorganisms, have also been employed in biostimulant compositions. In this regard, the purpose of this study is to examine the use of biostimulants in plant growth in terms of the raw materials utilised in their formulations and their application effects in the horticultural crops with respect to yield attributes.

2. Defining Plant Biostimulants

"A plant biostimulant is any substance or microorganism applied to plants with the aim to enhance nutrition efficiency, abiotic stress tolerance and crop quality traits, regardless of its nutrients content" (Patrick, 2015).

3. Common Features of Biostimulants

3.1. Diverse Nature

Biostimulants involve substances and microorganisms. Single compounds (for example, glycine betaine) or groupings of compounds of single natural origin (for example, seaweed extracts) whose composition and bioactive components are unknown. Single strains (e.g. Bacillus subtilis) or mixes of microorganisms with additive or synergistic effects can be found in microbial inoculants (e.g. several products on the market).

3.2. Diverse Physiological Functions

Any influence on plant processes is referred to as physiological function. The preservation of photosynthetic apparatus from light damage and the start of lateral roots are examples of physiological activities. Cellular processes like antioxidants scavenging reactive oxygen or enhanced auxin transporter production support functions.

3.3. Agricultural/Horticultural Function

They improve nutrient efficiency, resistance to abiotic stress, and/or crop quality attributes. Nutritional value, grain protein content, shelf life, and other quality attributes are examples. Any definition of biostimulants should be based on these convergent activities.

4. Categories of Plant Biostimulants

1. Humic substances (HS)
2. Protein hydrolysates (PH)
3. Seaweed extracts
4. Chitosan
5. Beneficial fungi
6. Beneficial bacteria

4.1. Humic Substances (HS)

Humic substances (HS) are organic matter elements that occur not only from the decomposition of plant, animal, and microbial leftovers, but also from the metabolic activity of soil bacteria that use these substrates. HS are groups of heterogeneous substances that were previously classified into humins, humic acids, and fulvic acids based on their molecular weights and solubility. HS is derived from naturally humified organic matter (e.g., peat or volcanic soils), composts and vermicomposts, or mineral deposits (leonardite, an oxidation form of lignite) (Rose *et al.,* 2014).

4.1.1. Humin

It is the insoluble component of soil organic matter (SOM) that remains after extraction of the other components of SOM that are soluble in aqueous base. It is insoluble in both acid and alkali.

4.1.2. Humic Acids

Humic acid is an organic chemical produced by decaying plants and animals residues by a process called "Humification". Humic acid is one of the major components of humic substances which is dark brown in colour and the major constituents of soil organic matter. Humic acid is alkaline soluble and acidinsoluble. It is also a natural biostimulant that is derived from leonardite (These layers were originally deep in the earth's crust, but over many years have been exhumed to near-surface location).

4.1.3. Fulvic Acid

Fulvic acid is a naturally occurring organic product derived from humus. They are similar to humic acids, with differences being the carbon and oxygen contents, acidity, degree of polymerization, molecular weight, and color. Fulvic acid is alkaline and acid soluble.

4.1.4. Impacts

Humic substances have been recognized as essential contributors to soil fertility, acting on physical, physicochemical, chemical and biological properties of the soil. Humic and fulvic acids are both essential for healthy plants. For instance, humic acids increase the permeability of cell walls, making it easier for fulvic acids to carry nutrients into the plant. While fulvic acids are the carrier for nutrients, humic acids make those nutrients more readily available in the soil.

The majority of HS's biostimulant effects are related to improving root nourishment through several routes. One of them is improved macro- and micronutrient absorption due to increased cation exchange capacity of soil containing polyanionic HS and increased phosphorus availability due to HS interfering with calcium phosphate precipitation. The stimulation of plasma membrane H+-ATPases, which transform the free energy supplied by ATP hydrolysis into a transmembrane electrochemical potential used for the import of nitrate and other nutrients, is another essential contribution of HS to root nutrition. Proton pumping via plasma membrane ATPases contributes to cell wall loosening, cell expansion, and organ growth in addition to nutrition intake (Jindo *et al.,* 2012).

Table 1: Impacts on horticultural crops

Crops	Simplified assay description	Effects	References
Apricot	Soil and foliar spray of commercial humic acids at different doses	Increased yield which ranged from 16 to 33%	Fathy et al. (2010)
Grape	Foliar spray of HA at different plant stages	Increased berry size and improved fruit quality	Ferrara and Brunetti (2010)
Eggplant	Foliar spray of commercial HA and different N doses	Enhanced fruit yield from 23 to 63% and N use efficiency	Azarpour et al. (2012)
Gladiolus	HA from composts at different doses; corms soaked for 24 h in treatment solutions	Increased growth and promoted early flowering	Baldotto and Baldotto (2013)
Pepper	Foliar and soil application of HA at different doses; greenhouse experiment	Higher mean fruit weight, and early and total yield than control	Karakurt et al. (2009)

4.2. Protein Hydrolysates

Chemical, enzymatic, and thermal hydrolysis can produce a combination of amino acids, peptides, polypeptides, and denatured proteins from proteins from both plant and animal sources (Nardi *et al.,* 2016).

4.2.1. Classification and Chemical Characteristics

PHs can be classified on the basis of protein source, and method of protein hydrolysis. Chemical hydrolysis of proteins under acid or alkaline conditions is usually preferred for producing animal-based PHs. Acid hydrolysis is a very aggressive process carried out at high temperature (>121°C) and pressure (>220.6 kPa). In acid hydrolysis, hydrochloric and sulphuric acid are mainly used to hydrolyse proteins, the most common being hydrochloric acid. Alkaline hydrolysis is a fairly simple and straightforward process where proteins are solubilised by heating followed by the addition of alkaline agents, such as Ca, Na or potassium hydroxide, and maintaining the temperature to a desired set point.

Enzymatic hydrolysis is usually adopted for producing plant-based PHs. The protein hydrolysis is carried out by proteolytic enzymes originating from animal organs (e.g., pancreatin, pepsin), plants (e.g., papain, ficin, bromelain) or microorganisms (e.g., alcalase, flavourzyme) (Pasupuleti and Braun, 2010).

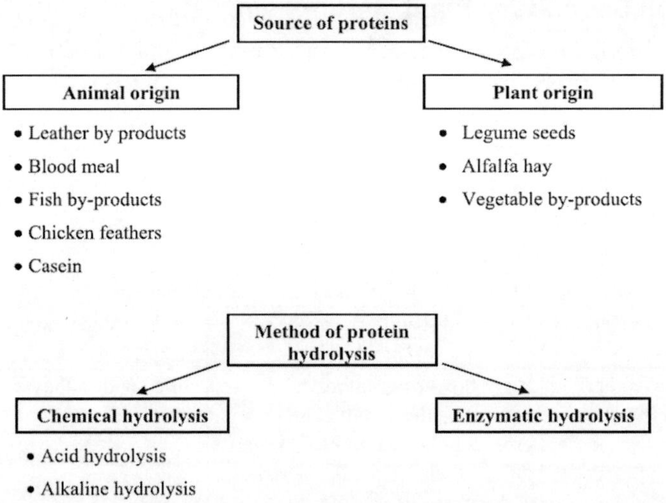

Fig. 1: Effects of protein hydrolysates on plant metabolism and physiology

4.2.1.1. Plant Uptake and Transport of Amino Acids and Peptides

Amino acids and small peptides are absorbed by both roots and leaves and then translocated into the plant. Foliar application of PHs can increase amino acid and peptide availability for plant uptake by reducing the competition with microorganisms. In a study, peach leaf uptakes of amino acids after foliar application were 14, 10, 25 and 26% for alanine, glutamic acid, glycine and lysine, respectively.

Amino acids and peptides can play an important role as signaling compounds. Microorganisms living in the phyllosphere and rhizosphere can also affect crop response to PHs application not only because they are competing with plants for amino acids and peptides but also because they secret enzymes that can hydrolyse peptides into small fragments that may act as signaling compounds, modulating crop response. The above findings suggest that PHs may also act as plant biostimulants through a microorganism-mediated enhancement of plant growth (Furuya and Umemiya, 2002).

4.2.1.2. Effects on Primary Plant Metabolism

In plants, all inorganic N is first reduced to ammonia before it is incorporated into organic form. The N assimilation is a key process controlling plant growth and development. Inorganic N derived from fertilizers is assimilated into amino acids glutamine, glutamate, asparagine and aspartate, which serve as important N carriers in plants. Enzymatic and molecular approaches have recently demonstrated that PHs supplied to maize plants influence different metabolic pathways, such as glycolysis and the Krebs cycle (Ertani *et al.,* 2013).

4.2.1.3. Effects on Secondary Plant Metabolism

Ertani *et al.* (2013) reported the positive effects exerted by PHs and amino acids, including proline and betaine, in inducing secondary plant metabolism and increasing plant defence responses and tolerance to stresses, including salinity, drought and temperature and oxidative conditions. According to this hypothesis, the increase of proline under stress would be partially responsible for the triggering of the phenolic compound build-up.

Table 2: Effects on PHs on horticultural crops

Crops	Type of PHs	Effects	References
Passion fruit	Animal derived PH	Increased seedling growth	Morales-Pajan and Stall (2004)
Banana	Chicken feather derived PH	Early flowering; increased nutrient, chlorophyll content, and proline in leaves; reduced sugars, proteins, amino acids, phenolics and flavonoids in fruits	Gurav and Jadhav (2013)
Lettuce	Plant derived PH (Trainer)	Pot culture under greenhouse using saline and non-saline solution	Lucini et al. (2015)

4.3. Seaweed Extract

Seaweeds are a diverse category of multicellular, macroscopic sea algae that can be brown, red, or green in colour. They provide valuable organic matter and fertiliser nutrients. In agriculture, seaweed extracts have been employed as soil conditioners and plant stimulators however the majority are benthic, attached to hard substrata by holdfasts - the fucoids, and 'kelps' are perhaps the best known and most easily observed by the casual observer on the shoreline. Seaweeds are quintessential members of inshore, marine ecosystems as they provide shelter and food to numerous marine biota and can even contribute to the modification of physicochemical properties of seawater. A relatively small proportion of the total number of seaweed species are of significant importance as animal and human food/supplements and also in agriculture as mulches/manure and modified extracts (Craigie, 2011).

4.3.1. Origin, Production Process and Chemical Compositions of Seaweed Extracts

With over 10,000 species of red, brown, and green seaweeds known, seaweeds are a diverse collection (Khan et al., 2009). Brown seaweeds (Phaeophyta) are among the most often employed for the commercial fabrication of extracts for agricultural and horticultural uses due to their quantity and spread. The extract industries most regularly use Ascophyllum nodosum, Ecklonia maxima, Macrocystis pyrifera, and Durvillea potatorum among the brown seaweeds (Khan et al., 2009).

4.3.2. Why Brown Algae?

Brown algae also contain bioactive secondary metabolites, vitamins and vitamin precursors. Polysaccharides found in brown seaweed extracts include alginates, fucoidans, laminarans, lichenan like glucans and fucose containing glucans. Phenolics are secondary metabolites that are synthesized under stress which protect cells and cellular components. Important roles of phenolic compounds include antioxidant activity, scavenging radicals such as, single oxygen, superoxide, hydroxyl, alkoxyl and peroxy radicals. Brown seaweeds have high concentrations of total phenolics. Phloro tannins, complex polymers of phlororglucinol, such as eckol and dieckol, are phenolic compounds found in brown seaweeds (Balboa et al., 2013).

4.3.3. Modes of Application of Seaweeds and Their Extracts in Horticulture

4.3.3.1. Seaweed Biomass and Seaweed Meal

A number of application methods have been used to deliver sea weed products to horticultural crops. The type of application employed depends on the nature of the seaweed product (e.g. meal/powder or extract). Application of whole seaweed biomass or meal is most common near coastal areas where the seaweeds are abundant. Whole seaweeds or seaweed meal are spread on the ground and are usually worked into the soil to facilitate microbial decomposition of the seaweed. Seaweeds are incorporated into the soil for considerable time in advance of planting, because during the decomposition phase, soil bacteria deplete nitrogen in the soil resulting in a temporary nutrient deficiency that will negatively impact plant growth. The decomposed seaweed, as an organic matter, generally improved soil physicochemical properties, water holding capacity, microbial activity and also protected plant against unfavorable environmental conditions such as extreme temperatures, water stress or excess of nutrients (Anderson, 2009).

4.3.3.2. Seaweed Extracts

Seaweed extracts are the most widely used seaweed production horticultural crops, available as liquid extracts or in a soluble powder form. Liquid extracts may be applied near the root of the plant, this can be achieved by mixing the extracts with irrigation water and applied as drip irrigation to crops. Seaweed extracts are also used as foliar sprays on a variety of flower, vegetable and tree crops including potato, tomato, plum, cherry, almonds and mango. Foliar application of seaweed extracts appeared to be most effective if applied in the morning when the leaf stomata are open. The efficacy of seaweed extracts also depend on the growth stage of the plant (Haider et al., 2012).

4.3.4. Production Process

Various solvents for sea weed extraction have been found to work well, including cold and hot water, alcohols, alkali, and acids. Sea weeds with a weight to volume ratio of 3 to 10% seaweed were treated with 0-25 percent H2SO4 at higher temperatures (100-120oC) for 30 minutes in an autoclave. The pH was adjusted by the addition of NaOH. Seaweed extraction has also been attempted using a mixture of solvents such as hexane and acetone.

Alkaline extraction is commonly used to extract bio stimulants high in auxins, and it is done under low pressure with the help of a microwave for 30 minutes. Microwave-assisted hot water extraction is commonly used to extract bio stimulants rich in fucoidan at greater pressures. High temperatures are not required, and only mild solvents are employed in the process, resulting in cost savings and an environmentally benign procedure (Rodriguez et al., 2011).

Table 3: Effect of seaweed extracts on horticultural crops

Crops	Seaweed extract	Effects	Reference
Potato	*Ecklonia maxima*	Improved seedling establishment after transplanting	Kowalski et al. (1999)
Carrot	*Ascophyllum nodosum*	Increased peroxidase, phenylalanine ammonia lyase, chitinase, PR1 and PR5; reduced growth of *Alternaria radicina and, Botrytis cinerea*	Jayaraj et al. (2008)
Strawberry	Commercial extract of *Ascophyllum nodosum*	Increased in fruit yield, size and, total anthocyanin	Spinelli et al. (2010)
Marigold	*Ecklonia maxima*	Increased in vegetative and reproductive growth	Van Staden et al. (1994)

4.4. Chitosan

Chitosan is formed from chitin, a co-polymer of N-acetyl-d-glucosamine and d-glucosamine, when over 80% of the acetyl groups of the N-acetyl-d-glucosamine residues are removed. Chitosan is the deacetylated form of chitin, a biopolymer that occurs naturally as a component of fungal cell walls, insect exoskeletons and crustacean shells. It is produced naturally and industrially. Polymers and oligomers of variable, controlled sizes are used in the food, cosmetic, medical and agricultural sectors (Hadwiger, 2013).

4.4.1. Extraction of Chitosan

4.4.2. Isolation of Chitin

To extract chitin, scales were gathered from various locally available sources. Scales are collected, cleaned, and dried in the sun for four days before being demineralized and deproteinized.

4.4.3. Demineralization

In a dilute HCl solution, demineralization was done. Because the minerals contained in fisheries scales are not always the same for each species, several treatment procedures for the chitin resources collected were used. For demineralization, fish scales were treated with 1.0 M HCl. The solid that resulted was rinsed with distilled water until it was acid-free. The dehydrated samples were then dried and weighed. Depending on the species, the number of washings and their duration (15–180 minutes) varied. The amount of mineral matter present and the degree of contact between the acid and the mineral matter, which varies depending on the species utilised, are indicated by gas emissions, mainly CO2 (Kumari et al., 2015).

4.4.4. Deproteinization

Chitin was deproteinized for 18 hours at room temperature using 0.5 N NaOH. The absence of any colour in the solution after repeated treatments of the sample indicated full deproteinization, even after leaving it undisturbed overnight. The sample was washed multiple times with distilled water and made neutral after the chitin was removed from the liquid medium. The chitin content was then determined after pure chitin was dried in an oven at 110 °C. A conventional technique was used to determine the ash content of dried chitin (Kumari et al., 2015).

4.4.5. Deacetylation of Chitin to Chitosan

Chitin was refluxed in a strong NaOH solution in a normal environment for preliminary studies. The studies took over 20 hours and resulted in low deacetylation levels, as well as significant deterioration of the final chitosan. The refluxing was done in alkaline solution to prevent long heating times. Even after heating for several hours (5–6 h), the chitosan was only partly soluble in acetic acid, indicating minimal deacetylation (Kumari et al., 2015.).

4.4.6. Effect of Chitosan on Horticultural Crops

- The addition of 1.0% high MW (600–900 kDa) chitosan to the soil enhanced plant height, canopy diameter and leaf area of chilli pepper (*Capsicum annuum* L.) (Chookhongkha et al., 2012)
- Chitosan @ 75mg L^{-1} when sprayed at 15 and 25 DAS on Indian spinach (*Basella alba* L.) resulted in an increase in plant height, leaf number, branch number, leaf area, and fresh and dry weight (Mondal et al., 2011)
- Foliar application of 100–125 mg L^{-1} chitosan at every 15 days increased okra fruit production, as well as plant height, leaf number, relative growth rate, photosynthesis rate and nitrate reductase activity (Mondal et al., 2012)
- Crab shell chitosan at 0.2% and 0.4% (w/v) in 0.5% (v/v) glacial acetic acid which is used to coat the fruits during post-harvest helped in extending the shelf-life of mango, maintained its eating quality up to four weeks of storage and decreased disease incidence (Abbasi et al., 2009)
- Sweet basil chitosan (30 cP) at 0.1% (w/v) for seed soaking and root dipping before transplanting, increased phenolic and terpenic compounds, especially rosmarinic acid and eugenol which increased the antioxidant activity and enhanced the growth (Kim et al., 05).

4.5. Beneficial Fungi

Fungi interact with plant roots in a variety of ways, including mutualistic symbiosis (where both organisms live in close proximity and form mutually beneficial connections) and parasitism. Mycorrhizal fungi are a diverse collection of organisms that form symbiotic relationships with about 90% of all plant species. Arbuscule Forming Mycorrhiza (AMF) are a common type of endomycorrhiza linked with crop and horticultural plants, in which fungal hyphae of Glomeromycota species penetrate root cortical cells and create branched structures known as arbuscules (Behie and Bidochka, 2014).

Among beneficial microorganisms, arbuscular mycorrhizal fungi (AMF) play a key role in plant performance and nutrition due to their capacity to improve plant mineral uptake (Smith and Read, 2008). AMF can only be grown in the presence of host plants (*i.e.,* obligate symbionts; Owen et al., 2015). AMF symbiosis is particularly important for enhancing the uptake of the relatively immobile and insoluble phosphate ions in soil, due to interactions with soil bi and trivalent cations, principally Ca^{2+}, Fe^{3+}, and Al^{3+}(Fitter et al., 2011).The basis of this symbiosis is the capacity of AMF to develop a network of external hyphae capable of extending the surface area (upto 40 times) and also the explorable soil volume for nutrient uptake throughout the production of enzymes and/or excretions of organic substances (Smith et al., 2011; Smith and Read, 2008).

Piriformospora indica, a mycorrhizal-like fungus able to establish associations with roots of a wide range of horticultural plants, supporting plant nutrition and increasing plant resistance and tolerance to stress (Vassilev et al., 2017). Phytostimulation due to release of auxins, small peptides, volatiles and other active metabolites into rhizosphere, which promote root branching and nutrient uptake capacity, thereby boosting plant growth and yield (Lopez-Bucio et al., 2015).

Table 4: Effects AMF on horticultural crops under drought conditions

Crops	Mycorrhizal species	Crop performance	Reference
Tomato	*Funneliformis mosseae Glomus versiforme*	Increased growth responses and yield by 19–32% compared to non-inoculated plants under various water stress conditions	Wang et al. (2014)
Capsicum annuum	Glomus mix (*G. albidium, G. claroides* and *G. diaphanum*)	as indicated by higher leaf water potential and higher root-to-shoot ratio in comparison to non-inoculated plants	Davies et al. (2002)
Cucumis melo	*F. mosseae, G. versiforme, R. intraradices*	enhanced growth parameters, antioxidant activities, soluble sugars contents, net phostosynthetic rate and photosynthetic water use efficiency	Huang et al. (2011)

Table 5: Effects AMF on horticultural crops under saline conditions

Crops	Mycorrhizal species	Crop performance	Reference
Tangerine orange	*F. mosseae, Paraglomus occultum*	Better plant growth, root morphology, photosynthesis and nutritional status (higher leaf K, Mg and K/Na ratio and lower Na)	Wu et al. (2010)
Olive	*F. mosseae, R. intraradices, Claroideoglomus claroideum*	Mycorrhizal plants showed the lowest biomass production reduction (−34%) under salinity in comparison to control plants (−78%), with *F. mosseae* being the most efficient.	Porras-Soriano et al. (2009)
Pumpkin	*R. intraradices*	improved nutritional (higher K and lower Na in leaf tissue) and leaf water status	Colla et al. (2008)

Table 6: Effects AMF on horticultural crops under nutrient deficiency conditions

Crops	Mycorrhizal species	Crop performance	Reference
Asparagus officinalis	*F. mosseae*	The soil P concentration required for maximum yield growth of asparagus seedlings could be lowered by inoculation with *F. mosseae*, associated with increased phosphorus utilization efficiency	Xu et al. (2014)
Citrus sinensis and *C. reticulate*	*G. versiforme*	Increase plant growth parameters, photosynthesis and Mg concentration in plant tissues under low magnesium conditions	Xiao et al. (2014)
Solanum lycopersicum	*F. mosseae, R. intraradices*	Increased the marketable fresh yield of tomato in particular at low fertilization regimes	Nedorost and Pokluda (2012)

Table 7: Effects AMF on horticultural crops under adverse soil pH conditions

Crops	Mycorrhizal species	Crop performance	Reference
Musa acuminata	*R. intraradices*	The higher crop performance of inoculated plants under Al stress was attributed to the reduced Al concentration in shoots and roots	Rufyikiri et al. (2000)
Cucumis sativus	*R. intraradices*	Inoculating plants were able to maintain growth and yield under alkalinity conditions. The AMF improved the photosynthesis and the nutritional status (high P, K, Mg, Fe, Zn and Mn, and low Na) in response to bicarbonate	Rouphael et al. (2010)
Catharantus roseus	ZAC-19*: G. albidum, C. claroideum* and *G. diaphanum*	Effectiveness of inoculated vinca plants to high alkalinity was associated to an increase in P uptake and to maintain the detoxifying activity through increased antioxidant activity	Cartmill et al. (2008)

Crops	Mycorrhizal species	Crop performance	Reference
Symplocoscoc-hinchinensis	AMF	Better establishment in nursery	Ajil (2018)
Black pepper	*P. indica* *Trichoderma harzhianum*	Plant growth promotion	Anith et al. (2011)
Centella asiatica	*P. indica*	Enhanced root rot and shot biomass And asiaticoside	Jisha et al. (2011)

4.6. Beneficial Bacteria

In soil microorganisms increase the nutrient and water use efficiency and uptake capacity. Among these potential soil microorganisms, plant growth promoting rhizobacteria (PGPR) are the most promising. Agricultural uses of biostimulants, two main types should be considered within this taxonomic, functional and ecological diversity: (i) mutualistic endosymbionts of the type *Rhizobium* and (ii) mutualistic, rhizospheric PGPRs ('plant growth-promoting rhizobacteria). *Rhizobium* and related taxa are commercialised as biofertilisers. PGPR mechanisms include hormone release or hormonal changes within plants, the production of volatile organic compounds, the improvement in the availability of nutrients and the enhancement of tolerance to abiotic stresses (Ahmad et al., 2008).

The term PGPR includes three types of soil bacteria, depending on their lifestyle: free-living bacteria inhabiting the zone around the root (rhizosphere), those that colonize the root surface (rhizoplane), and endophytic bacteria that live within roots (Alavi et al., 2013). Rhizobacteria with PGP-activity occur in a number of bacterial phyla (Actinobacteria, Proteobacteria and Firmicutes), includingstrains belonging to genera *Bacillus, Pseudomonas, Azospirillum, Azotobacter, Alcaligens, Arthobacter, Agrobacterium, Burkholderia, Comamonas, Pantoea, Rhizobium, Serratia*, and *Variovorax* (Kloepper et al., 1989).

4.6.1. Mode of Action

4.6.1.1. Plant Hormonal Changes

It is well known that plant growth is regulated by hormonal changes, therefore, one of the possible mechanisms used by PGPR to promote plant growth might be changes in the hormonal content of the host plant. In 1994, Glick et al. (1994) found that a *Pseudomonas putida* mutant strain lacking the enzyme 1-aminocyclopropane-1-carboxylic acid (ACC) deaminase was unable to promote the root growth of canola seedlings (Glick et al., 1994). This provided the first evidence for a role of a reduction in ethylene content by PGPR in increasing plant growth. ACC is the immediate precursor of ethylene, therefore, decreasing the levels of ACC will decrease the levels of ethylene and inhibit the growth reduction effect of this plant hormone. And also, *Phyllobacterium brassicacearum* cause increased levels of auxins in the host plant (Contesto et al., 2010).

4.6.1.2. Volatile Organic Compounds

Several PGPR emit volatile organic compounds (VOCs), which promote plant growth. These VOCs have been largely studied in relation to their function as inducers of resistance against plant pathogens. However, they also promote plant growth in the absence of pathogens, and can confer tolerance against abiotic stresses (Bhattacharyya et al., 2015). The VOCs for which a plant growth promotion effect has been confirmed by using bacterial mutants unable to synthesize these particular VOCsor by application of the pure compounds.

4.6.1.3. Nutrient Availability and Uptake by Plants

Growth promotion by PGPRs is actually enhanced under non-fertilized conditions. It has also been shown that exudates from nutrient-deficient plants (especially N and P deficientplants) modify the transcriptome of the PGPR *Bacillus amyloliquefaciens* (Talboys et al., 2009).

4.6.1.4. Abiotic Stress Tolerance in Plants

The mechanisms behind the higher tolerance of plants to several abiotic stresses following inoculation by some PGPR are extremely broad. It consists of effects on the antioxidant system and the alleviation of stress caused by leaf dehydration. Other mechanisms such as the regulation of hormonal changes or improvements in nutrient acquisition are also present. It should be borne in mind that all the mechanisms are interconnected, and all favour to some extent the promotion of growth and tolerance to abiotic stress conferred by PGPR (Zhu andGong, 2014).

Table 8: Effects of plant growth-promoting rhizobacteria (PGPR) application on vegetable crops

Crop	PGPR (species/strain)	Application mode	Effects	References
Cabbage	*Pantoeaagglomerans* strain RK-92	Seed-dipping (108 CFU mL−1) before planting	Enhancement of growth, nutrient, and hormone content	Turan et al. (2014)
Broccoli	*Brevibacillusreuszeria* and *Rhizobium rubi*	Root-dipping of seedlings for 60 min (108 CFU ml−1)	Increased yield, plant weight, head diameter, chlorophyll content, macronutrient and micronutrient uptake	Yildirim et al. (2011)

Table 9: Effects of plant growth-promoting rhizobacteria (PGPR) application on fruit crops

Crop	PGPR (species/ strain)	Application mode	Effects	References
Hazelnut	N_2-fixing and P-solubilizing bacteria	Seed-dipping (109 CFU mL−1), on one-year old seedlings	Increased seedling and total branch length, branch number, trunk diameter, and nutrient uptake	Erturk et al. (2011)
Cherry	*Pseudomonas* sp. strain BA-8 and *Bacillus* sp. strain OSU-142 (alone or combinations)	Foliar application (spray; 109 CFU mL−1)	Stimulated plant growth, increased yield per trunk, fruit weight and shoot length and resulted in significant yield increase	Esitken et al. (2006)

Table 10: Effects of plant growth-promoting rhizobacteria (PGPR) application on ornamental plants

Crop	PGPR (species/strain)	Application mode	Effects	References
Chrysanthemum	*Pseudomons fluorescens* strain 51	Soil drench (1011 CFU per plant)	Increased plant height, leaf surface area and number of flowers per plant	Gore and Altin (2006)
Petunia	*Azospirillumlipoferum* and *Bacillus polymxaa*	Foliar spray	Increased plant height, leaf area, shoot and root dry weight, flowering date, number of flower/branches, and flowering period	El-Mokadem and Mona (2014)
Black pepper	*Azospirillum lipoferum*	Priming of planting material	Better rooting and growth of orthotropic shoot	Mathew (2014)
Curcuma aromatica	*Azospirillum brasilenses, Pseudomonas flurescences*	Soil application	Better growth, yield and quality	Nirmalatha (2009)
Tomato	*Bacillus pumulus p. flurescence along with p. indica*	Soil	Surpress root knot nematode infection	Nirmalatha (2009)

5. Application Method of Biostimulants

Soil application, foliar application, seed soaking, fruit coating. Biostimulants may be used in the form of soil preparations (powders, granules, or solutions added to the soil) or as liquid foliar application products. Biostimulants containing humic substances and nitrogen compounds are often applied directly onto the soil, whereas various types of extracts from plants and seaweed are used in the form of foliar applications. Biostimulants can be introduced into the irrigation system and taken up by plants along with water. One example is the Kelpak SL (*Ecklonia maxima* extract) biostimulant which was sprayed in an aqueous solution of *Phaseolus vulgaris* L. (Kocira et al., 2018). Biostimulants can also be used in the form of biomass or meal from seaweed, however, this method has some limitations. Biomass and meal may be used in areas located close to the source of seaweed acquisition due to transport problems. Biomass or meal is applied directly to the soil long before planting in order to enrich the substrate with nutrients. Agro-technical measures such as ploughing are used to mix biomass or meal with the topsoil (Battacharyya et al., 2015).

6. Common Benefits of Biostimulants

Stimulate plant responses and work in all weather conditions, stimulate plants' immune system, increase profits, cut operating costs, lead to 50% reduction in fertilizer, produce better colour, increase natural plant toxins, repelling pests, result in better performance, increase microbial root protection from soil pathogens, produce deeper root, increase soil nutrient reserve up to 3000%, improve stress tolerance, improve root development, accelerate establishment, build yields, increases cation exchange, improve taste and shelf life, enhances fertilization and reduces leaching, improve drought tolerance, detoxify chemical residues and heavy metals, increase nutrient uptake, make urea a long-life nitrogen, improve seed germination rate, increase stomata opening and plant transpiration (Rafiee et al., 2016).

7. Challenges

The exact mechanism of action of biostimulants is yet to be clearly unravelled Formulation and blending of biostimulants with other fertilizer materials and/or plant protection products. Limited awareness regarding the potential benefinits among farmers

8. Conclusion

Biostimulants would be an effective and sustainable tool to enhance plant growth and productivity. It, so augments the efficiency and quality of horticultural crop production that the farmers obtain better return on their investment in agrochemicals. In the current scenario of declining farm land and requirement of sustainable solution for improving productivity, horticutural/ agricultural systems would certainly witness the widespread application of biostimulants.

References

Abbasi, N.A., Iqbal, Z., Maqbool, M., Hafiz, I.A. 2009. Postharvest quality of mango (*Mangifera indica* L.) fruit as affected by chitosan coating. Pakist. J. Bot. 41, 343–357.

Ahmad, I., Pichtel, J., Hayat, S. 2008. *Plant-Bacteria Interactions, Strategies and Techniques to Promote Plant Growth*. Wiley Publishers, New Jersey, pp. 84-89.

Alavi, P., Starcher, M.R., Zachow, C., Muller, H., Berg, G. 2013. Root-microbe systems: the effect and mode of interaction of stress protecting agent (SPA) *Stenotrophomonas rhizophila*DSM 14405(T). Front. Plant Sci. 4, 141.

Anderson, G. 2009. Seaweed extract shows improved fruit quality at McLaren Vale vineyard trial. *Aust. N. Z. Grape grower.* Winemaker 548, 17–22.

Azarpour, E., Motamed, M.K., Moraditochaee, M., Bozorgi, H.R. 2012. Effects of biomineral nitrogen fertilizer management, under humic acid foliar spraying on fruit yield and several traits of eggplant (*Solanum melongena* L.). Afr. J. Agric. Res. 7, 1104–1109.

Balboa, E.M., Enma, C., Moure, A., Falqué, E., Domínguez, H. 2013. Invitro antioxidant properties of crude extracts and compounds from brown algae. Food Chem. 138, 1764–1785.

Baldotto, M.A., Baldotto, L.E.B. 2013. Gladiolus development in response to bulb treatment with different concentrations of humic acids. Rev. Ceres. 60, 138-142.

Battacharyya, D., Babgohari, M.Z., Rathor, P., Prithiviraj, B. 2015. Seaweed extracts as biostimulants in horticulture. *Sci. Hortic.* 196, 39-48.

Behie, S.W., Bidochka, M.J. 2014. Nutrient transfer in plant-fungal symbioses. Trends Plant Sci.19, 734–740.

Bhattacharyya, D., Yu, S.M., Lee, Y.H. 2015. Volatile compounds from *Alcaligenes faecalis* JBCS1294 confer salt tolerance in *Arabidopsis thaliana* through the auxin and gibberellin pathways and differential modulation of gene expression in root and shoot tissues. Plant Growth Reg. 75, 297–306.

Blunden, G., Currie, M., Mathe, J., Hohmann, J., Critchley, A.T. 2009. Variations inbetaine yields from marine algal species commonly used in the preparation ofseaweed extracts used in agriculture. Phycologist 76, 14.

Cartmill, A.D.,Valdez-Aguilar, L.A., Bryan, D.L., Alarcón, A. 2008. *Arbuscular mycorrhizal* fungi enhance tolerance of vinca to high alkalinity in irrigation water. Sci. Hortic. 115, 275-284.

Chookhongkha, N., Miyagawa, S., Jirakiattikul, Y., Photchanachai, S. 2012. Chilli growth and seed productivity as affected by chitosan. In: Proceedings of International Conferenceon Agriculture Technology and Food Sciences (ICATFS), November 2012, Manila. Planetary Scientific Research Centre, Manila, Philippines, pp. 17–18.

Colla, G., Rouphael, Y., Cardarelli, M., Tullio, M., Rivera, C.M., Rea, E. 2008. Alleviation of salt stress by arbuscular mycorrhizal in zucchini plants grown at low and high phosphorus concentration. Biol. Fertil. Soils 44, 501-509.

Contesto, C., Milesi, S., Mantelin, S., Zancarini, A., Desbrosses, G., Varoquaux, F., Bellini, C., Kowalczyk, M., Touraine, B. 2010. The auxin-1 signaling pathway is required for the lateral root response of Arabidopsis to the *rhizobacterium Phyllobacteriumbrassicacearum.* Planta. 232, 1455-1470.

Craigie:J.S. 2011. Seaweed extract stimuli in plant science and agriculture. J. Appl. Phycol. 23: 371-393.

Davies, Jr. F.T., Portugal-Olalde, V., Aguilera-Gomez, L., Alvarado, M.J., Ferrera-Cerrato, R.C., Bouton, T.W. 2002. Alleviation of drought stress of Chileancho pepper (*Capsicum annuum* cv. San Luis) with arbuscular mycorrhiza indigenous to Mexico. Sci. Hortic. 92: 347-359.

Dogra, B.S., Mandradia, R.K. 2014. Effect of seaweed extract on growth and yield ofonion. Int. J. Farm Sci. 2, 59–64.

El-Mokadem, E., Mona, S. 2014. Effect of bio and chemical fertilizers on growth and flowering of *Petunia hybrid* plants. Am. J. Plant Physiol. 9, 68–77.

Ertani, A., Schiavon, M., Muscolo, A., Nardi, S. 2013. Alfalfa plant-derived biostimulant stimulate short-term growth of salt stressed *Zea mays* L. plants.Plant Soil. 364, 145-158.

Erturk, Y., Cakmakci, R., Duyar, O., Turan, M. 2011. The effects of plant growth promotion rhizobacteria on vegetative growth and leaf nutrient contents of hazelnut seedlings (*Turkish hazelnut* cv. Tombul and Sivri). Int. J. Soil Sci. 6, 188-198.

Esitken, A., Pirlak, L., Turan, M., Sahin, F. 2006. Effects of floral and foliar application of plant growth promoting rhizobacteria (PGPR) on yield, growth and nutrition of sweet cherry. Sci. Hortic. 110, 324-327.

Fathy, M.A., Gabr, M.A., El Shall, S.A. 2010. Effect of humic acid treatments on'canino' apricot growth, yield and fruit quality. New York Sci. J. 3(12), 109-115.

Fernando, D., Milagrosa, S., Francisco, C., Francisco, M. 2018. Biostimulant activity of *Trichoderma saturnisporum* in melon (*Cucumis melo*). Hort. Science *53*(6), 810-815.

Ferrara, G., Brunetti, G. 2010. Effects of the times of application of a soil humic acid on berry quality of table grape (*Vitis vinifera L.*) cv. Italia. *Span*. J. Agric. Res. 8(3), 817–822.

Fitter, A.H., Helgason, T., Hodge, A. 2011. Nutritional exchanges in the arbuscular mycorrhizal symbiosis: implications for sustainable agriculture. Fungal Biol.Rev. 25(1), 68-72.

Furuya, S., Umemiya, Y. 2002. The influence of chemical forms on foliar-appliednitrogen absorption for peach trees. Acta Horticulturae 594, 97-103.

Glick, B.R., Jacobson, C.B., Schwarze, M.M.K., Pasternak, J.J. 1994. 1-aminocyclopropane-1-carboxylic acid deaminase mutants of the plant-growth-promoting rhizobacterium *Pseudomonas putida* GR12-2 do not stimulate canola root elongation. Can. J. Microbiol 91, 1-915.

Gore, E.M., Altin, N. 2006. Growth promoting of some ornamental plants by root treatment with specific *fluorescent pseudomonads*. J. Biol. Sci. 6(3), 610-615.

Gurav, R.G., Jadhav, J.P. 2013. A novel source of biofertilizer from feather biomassfor banana cultivation. Environ. Sci. Pollut. Res. 20(7), 4532-4539.

Hadwiger, L.A. 2013. Multiple effects of chitosan on plant systems: Solid science orhype. Plant Sci. 208, 42-49.

Haider, M.W., Ayyub, C.M., Pervez, M.A., Asad, H.U., Manan, A., Raza, S.A., Ashraf, I. 2012. Impact of foliar application of seaweed extract on growth, yield andquality of potato (*Solanum tuberosum* L.). Soil Environ. 31(2), 157-162.

Halpern, M., Bar-Tal, A., Ofek, M., Minz, D., Muller, T., Yermiyahu, U. 2015. The use of biostimulants for enhancing nutrient uptake. Adv. Agron. 130, 141-174.

Huang, Z., Zou, Z.R., He, C.X., He, Z.Q., Zhang, Z.B.and Li, J.M. 2011. Physiological and photosynthetic responses of melon (*Cucumis melo* L.) seedlings to three glomus species under water deficit. Plant Soil 339(1), 391-399.

Jayaraj, J., Wan, A., Rahman, M., Punja, Z.K. 2008. Seaweed extract reduces foliar fungal diseases on carrot. Crop Prot. 27(10), 1360-1366.

Jindo, K., Martim, S.A., Navarro, E.C., Aguiar, N.O., Canellas, L.P. 2012. Rootgrowthpromotion by humic acids from composted and non-composted urbanorganic wastes. Plant Soil 353, 209-220.

Karakurt, Y., Unlu, H., Unlu, H.,Padem, H. 2009. The influence of foliar and soilfertilization of humic acid on yield and quality of pepper. Acta Agric. Scand. Section B Soil Plant Sci. 59(3), 233-237.

Khan, W., Rayirath, U.P., Subramanian, S., Jithesh, M.N., Rayorath, P., Hodges, D.M., Critchley, A.T., Craigie, J.S., Norrie, J., Prithiviraj, B. 2009. Seaweed extracts as biostimulants of plant growth and development. J. Plant Growth Reg. 28, 386-399.

Kim, H., Chen, F., Wang, X., Rajapakse, N.C. 2005. Effect of chitosan on the biological properties of sweet basil (*Ocimum basilicum* L.). J. Agric. Food Chem. 53(9), 3696-3701.

Kloepper, J.W., Lifshitz, R., Zablotowich, R.K. 1989. Free living bacterial inocula for enhancing crop productivity. Trends Biotechnol. 7, 39-43.

Kocira, A., Swieca, M., Kocira, S., Złotek, U., Jakubczyk, A. 2018. Enhancement of yield, nutritional and nutraceutical properties of two common bean cultivars following the application of seaweed extract (*Ecklonia maxima*). Saudi J. Biol. Sci. 25(3), 563-571.

Kowalski, B., Jager, A.K., Van Staden, J. 1999. The effect of a seaweed concentrateon the *in vitro* growth and acclimatization of potato plantlets. Potato Res. 42(1), 131–139.

Kumari, S., Rath, P., Kumar, A.S.H., Tiwari, T.N. 2015. Extraction and characterization of chitin and chitosan from fishery waste by chemical method. Environ. Technol. Innovation 3, 77-85.

Lopez-Bucio,J., Pelagio-Flores, R., Herrera-Estrella, A. 2015. *Trichoderma* as biostimulant: exploiting the multilevel properties of a plant beneficial fungus. Sci. Hortic. 196, 109-123.

Lucini, L., Rouphael, Y., Cardarelli, M., Canguier, R., Kumar, P., Colla, G. 2015. The effect of a plant-derived biostimulant on metabolic profiling and cropperformance of lettuce grown under saline conditions. Sci. Hortic. 182, 124-133.

Mondal, M.M.A., Malek, M.A., Puteh, A.B., Ismail, M.R., Ashrafuzzaman, M., Naher, L. 2012. Effect of foliar application of chitosan on growth and yield in okra. Aust. J. Crop Sci. 6, 918-921.

Mondal, M.M.A., Rana, M.I.K., Dafader, N.C., Haque, M.E. 2011. Effect of foliarapplication of chitosan on growth and yield in Indian spinach. J. Agrofor. Environ. 5, 99-102.

Morales-Pajan, J.P., Stall, W. 2004. passion fruit (*Passiflora edulis*) transplant production is affected by selected biostimulants. Proc. Fla. State Hortic. Soc. 117, 224-227.

Nardi, S., Pizzeghello, D., Schiavon, M., Ertani, A. 2016. Plant biostimulants: Physiological responses induced by protein hydrolyzed-based products and humic substance in plant metabolism. Sci. Agric. 73, 18-23.

Nedorost, L., Pokluda, R. 2012. Effect of arbuscular mycorrhizal fungi on tomato yield and nutrient uptake under different fertilization levels. Acta Univ. Agric.Silvic. Mendel. Brun. 60, 181-186.

Owen, D., Williams, A.P., Griffith, G.W., Withers, P.J.A., 2015. Use of commercial bio-inoculants to increase agricultural production through improved phosphorus acquisition. Appl. Soil Ecol. 86, 41-54.

Pasupuleti, V.K., Braun, S. 2010. State of the art manufacturing of proteinhydrolysates. In: Pasupuleti, V.K., Demain, A.L. (eds.), *Protein Hydrolysates Biotechnology*. Springer, Dordrecht, New York, USA, pp. 11-32.

Patrick, D.J. 2015. Plant biostimulants: definition, concept, main categories and regulation. Sci. Hortic. 196, 3-14.

Porras-Soriano, A., Sorano-Marintin, M.L., Porras-Piedra, A., Azcon, P. 2009. Arbuscular mycorrhizal fungi increased growth, nutrient uptake and toleranceto salinity in olive trees under nursery conditions. J. Plant Physiol. 166, 1350-1359.

Rafiee, H., Naghdi Badi, H., Mehrafarin, A., Qaderi, A., Zarinpanjeh, N., Sekara, A., Zand, E. 2016. Application of plant biostimulants as new approach to improve the biological responses of medicinal plants-A critical review. J. Med.Plant *3*(59), 6-39.

Rodriguez Jasso, R.M., Mussatto, S.I., Pastrana, L., Aguilar, C.N., Teixeira, J.A. 2011. Microwave assisted extraction of sulfated polysaccharides (fucoidan) from brown seaweed, Carbohydrate Polymers 86, 1137-1144.

Rose, M.T., Patti, A.F., Little, K.R., Brown, A.L., Jackson, W.R.,Cavagnaro, T.R. 2014. A meta-analysis and review of plant-growth response to humic substances: practical implications for agriculture. Adv. Agron. 124, 37-9.

Rouphael, Y., Schwarz, D., Krumbein, A., Colla, G. 2010. Impact of grafting onproduct quality of fruit vegetables. Sci. Hortic. 127, 172-179.

Rufyikiri, G., Declerck, S., Dufey, J.E., Delvaux, B. 2000. Arbuscularmycorrhizalfungi might alleviate aluminium toxicity in banana plants. New Phytol. 148, 343–352.

Smith, E., Read, D.J. 2008. *Mycorrhizal Symbiosis* (3rd Ed). Academic Press, London.

Smith, S.E., Jakobsen, I., Gronlund, M., Smith, F.A. 2011. Roles of arbuscular mycorrhiza in plant phosphorus nutrition: interactions between pathways ofphosphorus uptake in arbuscularmycorrhizal roots have importantimplications for understanding and manipulating plant phosphorusacquisition. Plant Physiol. 156, 1050-1057.

Smolen, S., Sady, W. 2010. Effect of plantbiostimulation with Pentakeep V fertilizer and nitrogen fertilization on the content of macroand micronutrients in spinach. J. Elementol. 15(2): 343 - 53.

Spinelli, F., Fiori, G., Noferini, M., Sprocatti, M., Costa, G. 2010. A novel type ofseaweed extract as a natural alternative to the use of iron chelates instrawberry production. Sci. Hortic. 125, 263-269.

Talboys, P.J., Owen, D.W., Healey, J.R., Withers, P.J.A., Jones, D.L. 2009. Auxin secretion by *Bacillus amyloliquefaciens* FZB42 both stimulates root exudation and limits phosphorous uptake in *Triticumaestivum*BMC. Plant Biol. 14, 51.

Turan, M., Ekinci, M., Yildirim, E., Günes, A., Karagoz, K., Kotan, R., Dursun, A. 2014. Plant growth-promoting rhizobacteria improved growth, nutrient, and hormone content of cabbage (*Brassica oleracea*) seedlings. Turkish J. Agric. For. 38, 327-333.

Vassilev, N., Eichler-Lobermann, B., Flor-Peregrin, E., Martos, V., Reyes, A., Vassileva, M. 2017. Production of a potential liquid plant bio-stimulant by immobilized *Piriformosporaindica* in repeated-batch fermentation process. AMB Express. 7(1), 106.

Wang, B., Yao, Z., Zhao, S., Guo, K., Sun, J., Zhang, H. 2014. *Arbuscular mycorrhizal* pplication to improve growth and tolerance of processing tomato (*Lycopersicum esculentum* Miller) under drought stress. J. Food Agric. Environ. 12, 452-457.

Wu, Q.S., Zou, Y.N., He, X.H. 2010. Contributions of *Arbuscular mycorrhizal* fungi togrowth, photosynthesis, root morphology and ionic balance of citrus seedlings under salt stress. Acta Physiol. Plant. 32, 297-304.

Xiao, J.X., Hu, C.Y., Chen, Y.Y., Yang, B., Hua, J. 2014. Effects of low mangnesium and *Arbuscular mycorrhizal* fungus on the growth, magnesium distribution and photosynthesis of two citrus cultivars. Sci. Hortic. 177, 14-20.

Xu, P., Liang, L.Z., Dong, X.X., Xu, J., Jiang, P.K., Shen, R.F. 2014. Response of soil phosphorus required for maximum growth of *Asparagus officinalis* L. to *Arbuscular mycorrhizal* fungi. Pedosphere 24, 776-782.

Yildirim, E., Karlidag, H., Turan, M., Dursun, A., Goktepe, F. 2011. Growth, nutrient uptake and yield promotion of broccoli by plant growth promoting rhizobacteria with manure. Hort. Sci. 46, 932-936.

Zhu, Y.X., Gong, H.J. 2014. Beneficial effects of silicon on salt and drought tolerance in plants. Agron. Sustain. Dev. 34, 455-472.

22

Role of Nanofertilizer for Sustainable Crop Production

Rajesh Kumar[1], Ravindra Singh[2], Chaman Jadon[1], Pratap Singh[1] and Sharda Choudhary[2]

[1]Agriculture University, Kota, Rajasthan, India
[2]ICAR-National Research Centre on Seed Spices, Ajmer, Rajasthan, India

1. Introduction

Agriculture in India is livelihood for a majority of the population and can never be underestimated. Increasing population, increasing average income and globalisation effects in India will increase demand for quantity, quality and nutritious food, and variety of food. Therefore, pressure on decreasing available cultivable land to produce more quantity, variety and quality of food will keep on increasing. India's food security depends on producing cereal crops, as well as increasing its production of fruits, vegetables and milk to meet the demands of a growing population with rising incomes. To reach world-wide food security, improved food production and food productivity, we need to approve innovative and futuristic advances and technologies to improve crop cultivation, yield and productivity.

To feed the growing world population, we need to tackle the problems of imbalanced fertilization, malnutrition, inefficient fertilizer use and depletion of soil organic matter. To do this, it is necessary to develop nano-based fertilizer formulations with multiple functions. Chemical fertilizers are needed to replenish the world's soil fertility, but intensive agricultural practices reduce it significantly to about 40% (Dijk and Meijerinck, 2014). In addition to the efficiency of other agricultural inputs, fertilizers contribute one-third to agricultural and horticultural productivity. However, the performance of conventional fertilizers rarely exceeds 30-40%. The performance of conventional fertilizers has been constant over the past few decades, with nitrogen (N) 30-35%, phosphorus (P) 18-20% and potassium (K) 35-40% (Subramanian et al., 2015). Traditional fertilizers have many drawbacks, and the aquifer is contaminated with nutrients lost by leaching.

In other words, fertilizers have environmental implications such as greenhouse gas emissions and hypoxia, and these issues need to be addressed urgently. Therefore, searching for alternatives such as nanofertilizers (Suppan 2013).

Materials with a particle size of up to 110 nm in at least one dimension are generally classified as nanomaterials and form the basis of nanotechnology. There are various types of nanomaterials such as single or multiwalled nanotubes, magnetized iron nanoparticles, copper (Cu), aluminum (Al), silver (Ag), gold (Au), zinc (Zn) and zinc oxide (ZnO), silica (Si), cerium oxide (Ce2O3), and titanium dioxide (TiO2), among others. Given the unique properties of nano materials such as high surface-to-volume ratio, controlled-release kinetics to targeted sites and sorption capacity, nanotechnology has a high relevance for the design and use of new fertilizers. Nanotechnology is highly relevant to the development and use of new fertilizers. The component of nanofertilizer includes zinc oxide nanoparticles (ZnONP), silica, iron, and titanium dioxide, ZnS / ZnCdSe core-shell quantum dots (QD), InP / ZnS coreshell QD, Mn / ZnSe QDs include gold nanorods, Al_2O_3, TiO_2, CeO_2, and FeO (Prasad et al., 2017).

2. Nanofertilizers

Nanofertilizers are nutrients encapsulated/coated with nanomaterial for the control and slow delivery of one or more nutrients in order to satisfy the imperative nutrient requirements of plants. Nano fertilizers are synthesized by fortifying nutrients singly or combination on to the adsorbents with nano-dimension. Nanofertilizers are products that supply nutrients to plants in one of three ways. Nutrients are encapsulated in nanomaterials such as nanotubes and nanoporous materials, coated with a thin protective polymer film, and delivered as particles or emulsions of nanoscale dimensions. Nanofertilizers are defined as nanometer-scale materials in the form of nanoparticles containing macronutrients and micronutrients released into plants in a controlled manner (Shang et al., 2019). Nanofertilizers are nutrient carriers designed using a 1-100 nm nanoscale substrate that can be used alone or in combination to improve plant growth, performance and yield. They do not provide nutrients directly to the plants, but they have better performance compared to traditional fertilizers. Nanofertilizers are products synthesized using nanoparticles or nanotechnology by adding nutrients to adsorbents to enhance nutrient performance and improve plant nutrients compared to traditional fertilizers. is. The large surface area of nanoparticles allows them to retain nutrients abundantly and deliberately release nutrients to meet the needs of plants without adverse effects. Nanofertilizer and nanobiofertilizers contain synthetic and natural substances, respectively, and enhance the soil fertility as well as bioavailability more efficiently than the conventional fertilizers (Chhipa, 2017). However, to be declared as nanofertilizer, the individual size of the molecules, in dispersed and/or aggregated state, should be $\leq$100 nm, and the bulk size should be closer to 100 nm, and the nanoproduct should be stable, i.e. the nanofertilizer should retain its nanoscale dimension and intactness during pre- and post-interaction with the soil or the crop. However, the rate of dissolution and bioactivity are inevitably influenced by the shape. The design of the nanofertilizer where the products are functionalized with certain target agents and/or for control release agents to develop a hybrid product (Dimkpa and Bindraban, 2017). Among the abovementioned features of nanofertilizer, the most spectacularapplication is the optimization of nutrient use efficiency (NUE), control release ofthe active substances and no effect on soil biodiversity (Sempeho et al., 2014). Thesenanofertilizers, based on their biodistribution and nutrient composition, by the plantmay be subcategorized into macronutrient nanofertilizers, micronutrient nanofertilizersand nanoparticulate fertilizers (Chhipa, 2017).

Nanofertilizers can be divided into three categories based on the nutritional requirements of plants: (a) macro-nano fertilizers, (b) micro-nano fertilizers, and (c) nano-particle fertilizers.

2.1. Macronutrient Nanofertilizers

Macronutrient nanofertilizers contains nitrogen, phosphorus, potassium, calcium, magnesium and sulfur. Plants require relatively large amounts of these nutrients. Due to the increased surface area that correlates with conventional fertilizers, nanofertilizers are more efficient at using nutrients. Kottegoda et al. (2011) and Kottegoda et al. (2017) discovered that nanourea-coated zeolite chips and urea-modified hydroxyapatite can be used to control the release of major nutrients in nitrogen. The effectiveness of nanourea-modified hydroxyapatite encapsulated under pressure in Gliricidia sepium was reported by Kottegoda et al. (2011). The nitrogen was slowly released over 60 days, demonstrating that slow release effectively increased plant uptake of nitrogen. This significantly increased the yield of plants compared to conventional fertilizers (Kottegoda et al., 2011). This study shows that the formulation of key nutrients in nanofertilizers is promising for increasing crop yields. Development or production of highly efficient nano-major nutrients, *i.e.* Low leaching rate, low soil immobilization rate, high plant uptake rate, low environmental risk *i.e.* requires low eutrophication potential and low nitrogen leaching rate.

2.2. Micronutrient Nanofertilizers

Micronutrients nanofertilizers contains iron, manganese, zinc, copper, molybdenum and other nutrients. These nutrients are required by plants in relatively small amounts. Only a small amount is needed, but it is necessary for proper plant growth. Dimkpa and Bindraban (2017) state that nanoscale forms of micronutrients can be used to enhance plant growth, development, proper metabolism and increase plant nutrient levels. Micronutrients as soluble salts are often added in small amounts to major nutrient fertilizers such as nitrogen, phosphorus and potassium. i.e., 5 mg / L for ingestion by plants. Excessive use of fertilizers reduces the amount of these micronutrients present in the soil. As a treatment, nanoforms of micronutrients can be applied to the soil or sprayed onto plants to improve the health and vitality of the soil. Zinc deficiency, a micronutrient essential for plant growth, is commonly found in soil. Zinc oxide nanoparticles can be combined with complex fertilizers to increase the efficiency of recommended fertilizer doses (Kale et al., 2016). It should be formulated with nanomicronutrients that control the problems of molybdenum field leaching and soil fixation of Fe, Zn, or Cu.

Since the nutrient release rate of nano-fertilizer can be controlled according to environmental conditions such as soil moisture content, temperature change, and soil acidity, it effectively enhances nutrient efficiency for plant growth compared to conventional fertilizers. can do. Nanofertilizers are formulated to release and emit nutrients intentionally and regularly for over 35 days. This helps reduce adverse effects on soil, crops and the environment, improve the efficiency of applied nutrients and reduce the loss of nutrients due to leaching (Siddiqui et al., 2015).

Nanofertilizers provide the proper nutrients to promote plant growth through application to leaves and soil. They are a cheap and sustainable source of phytonutrients. They are easy to use. The efficiency of fertilizer use is high. They prevent environmental pollution. IFFCO recently developed the world's first nanourea. Nanourea can only be applied to leaves. It cannot be used on the ground. The recommended dose of foliar application of nanourea is 2-4 ml/lt.

3. Preparation of Nano Fertilizers

Nanoparticles are mainly synthesized by two approaches: (a) top-down approach and (b) bottom-up approach. In the first approach, nano objects and materials are generated from larger entities without repelling atomic reactions. The top-down approach is generally less common than the bottom-up approach. Top-down approach processes include grinding or abrasion, chemical processes, and the volatilization of solids followed by the condensation of volatile components. (Ghorbani, 2014). The second bottom-up approach builds a variety of materials and devices from unique molecular components. Methods include sol-gel treatment, chemical vapor deposition, plasma or frame spray synthesis, laser pyrolysis, and atomic or molecular condensation.

Table 1: Comparison of applications between nano-fertilizer and conventional fertilizer

Properties	Nano fertilizers	Conventional fertilizers
Solubility and dispersion of micronutrients	Improves the solubility and distribution of insoluble nutrients in soil, reduces soil uptake and fixation, and enhances bioavailability	Due to the large particle size and low solubility, plants have low bioavailability.
Nutrient use efficiency	Nano fertilizer increases fertilizer utilization efficiency and soil nutrient absorption rate in crop production and saves fertilizer resources	Mass composites cannot be used by roots and are less efficient
Controlled-release modes	Nutrient release rates and patterns of water-soluble fertilizers can be precisely controlled by encapsulation in envelope forms	Careless or excessive release of fertilizer can create toxicity and disrupt the ecological balance of the soil.
Effective duration of nutrient release	Duration of fertilizer nutrient supply duration into soil may be increased.	Consumed by plants at the time of delivery, the rest is converted to insoluble salt in the soil
Fertilizer Nutrient loss rate	Losses of fertilizer nutrient are less into the soil by leaching and/or leaking.	Loss of fertilizer nutrients into the soil due to leaching and/or seepage are more.

Source: Cui et al., 2010

The application of nano-fertilizer poses no health risk and is suitable for all kinds of crops including food grains, vegetables and horticulture.

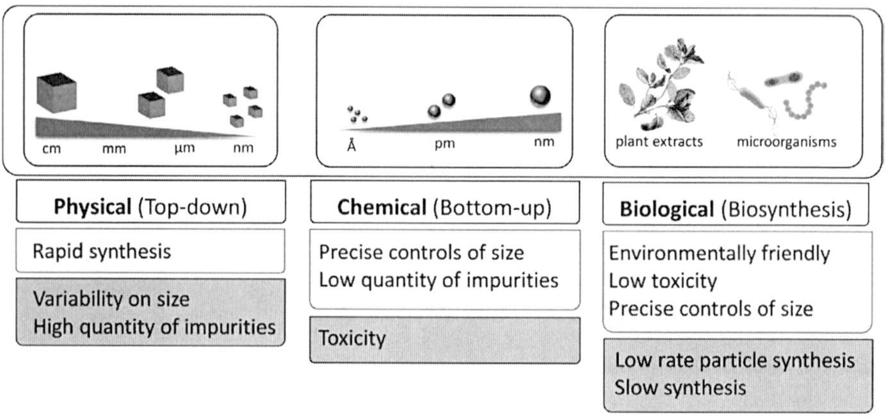

Fig. 1: Methods for the synthesis of nanofertilizers. Physical (top-down), chemical (bottom-up) and biological approached for the production of nanofertilizers (Source: Zulfiqar et al., 2019)

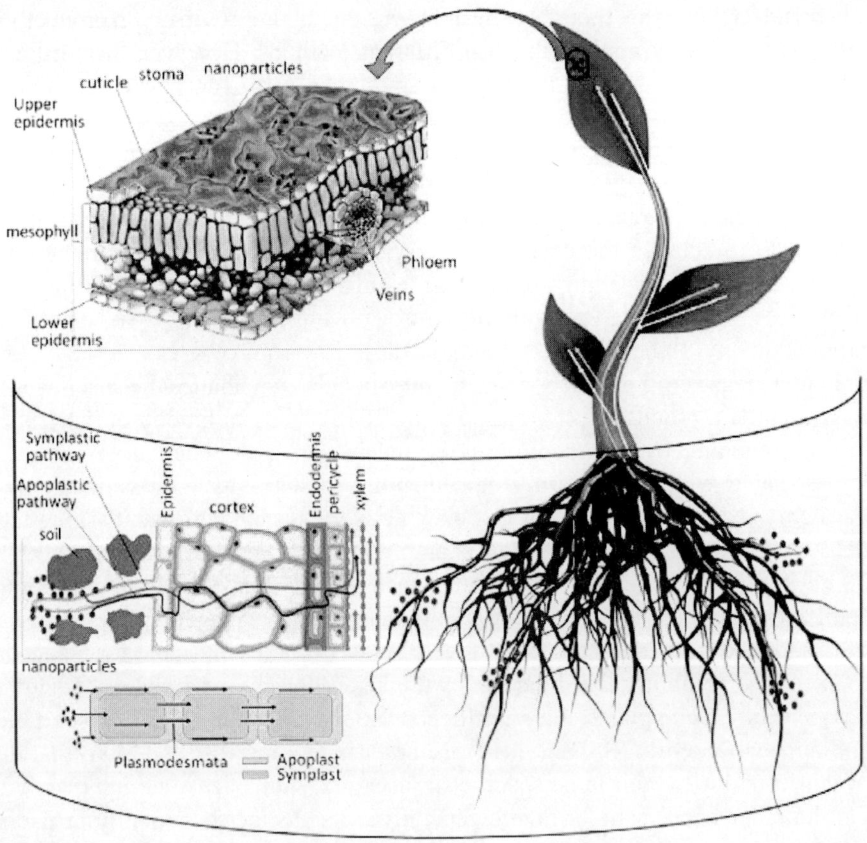

Fig. 2: Mechanism of action of nanofertilizer

3.1. Advantages of Nanofertilizers

Nano fertilizers are more beneficial than traditional fertilizers because they triple nutritional efficiency, reduce the need for chemical fertilizers, increase resilience to drought and disease, and reduce environmental damage. Due to its high surface area to volume ratio, nanofertilizers are more easily absorbed by plants. However, the size and shape of the nanoparticles are powerful factors in determining the level of bioavailability of plants in the soil. Nanoparticles cannot be activated immediately due to uptake by plants. Instead, a series of reactions take place, causing oxidation and recombination to provide the plant with the micronutrients it needs. Plants not only grow, but also store nutrients to make up for undernourishment. Nanofertilizers can also be designed to supplement the deficiencies of certain nutrients in plants. This is possible because the atoms on the surface of the nanomaterial can be patterned to create other characteristic properties.

Chemical fertilizers are thought to be indispensable for improving crop productivity and are extensively applied through different methods. However, crop usage is generally less than half of the applied amount of fertilizer, and the remaining amount of minerals intended to reach the targeted site may leach down, so that they become fixed in soil or contribute to water pollution. Nutrients such as N, P and K applied to the soil lose 40-70%, 80-90% and 50-90% respectively, resulting in significant resource loss. In addition, growers tend to use these fertilizers repeatedly to achieve the desired higher yield. This can conversely reduce soil fertility and increase salinity, which can lead to future crop losses. Furthermore, uneven use of fertilization without control on nutrient release can deteriorate product quality. Hence, it is crucial to develop slow/control release fertilizers not only to increase crop production and quality, but also to enhance the sustainability in horticultural production (Feregrino et al., 2018). The availability of nutrients in crops is regulated by slow/control release mechanisms. Slow delivery of nutrients is associated with the covering or cementing of nutrients with nanomaterials. Farmers may take advantage of this slow delivery system consistently long-term delivery of nutrients to plants. For example, nutrients can be released over 40–50 days in a slow release fashion rather than the 4–10 days by the conventional fertilizers (X. Chen, Wei, 2018). In traditional nutrient management systems, half of the applied fertilizer is lost due to leaching or becomes unusable for plants due to excessive solubility that interferes with absorption by the roots or sometimes has toxic effects on plants. Nanofertilizers also reduce the need for transportation and application costs (S. Fan, 2014). Nanofertilizers increase the bioavailability of nutrients due to their large specific surface area, small size and high reactivity. On the other hand, nanofertilizers generally offer distinct benefits by providing a balanced diet, allowing plants to deal with a variety of biological and abiotic stresses.

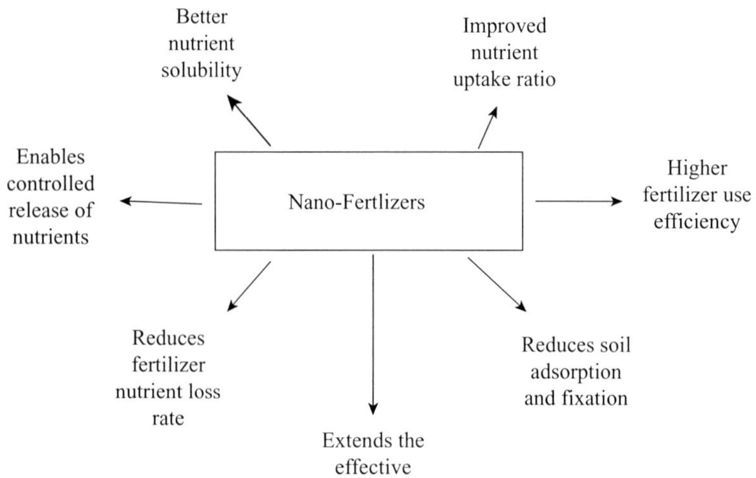

Fig. 3: Advantages of Nanofertilizers

Table 2: Some approved nanofertilizers and their formulations used in the world today

Nanofertilizers	Constituents	Name of manufacturer
Nano Ultra-Fertilizer (500) g	Organic matter 5.5%, Nitrogen 10%, P_2O_5 9%, K_2O 14%, P_2O_5 8%, K_2O 14% and MgO 3%	SMTET Eco-technologies Co., Ltd., Taiwan
Nano Calcium (Magic Green) (1) kg	$CaCO_3$ 77.9%, $MgCO_3$ 7.4%, SiO_2 7.47%, K 0.2%, Na 0.03%, P 0.02%, Fe-7.4 ppm, Al2O3 6.3 ppm, Sr 804 ppm, sulfate 278 ppm, Ba 174 ppm, Mn 172 ppm and Zn 10 ppm	AC International Network Co., Ltd., Germany
Nano Capsule	N 0.5%, P2O5 0.7%, K_2O 3.9%, Ca 2.0%, Mg 0.2%, S 0.8%, Fe 2.0%, Mn 0.004%, Cu 0.007% and Zn 0.004%	The Best International Network Co., Ltd., Thailand
Nano Micro Nutrient (EcoStar) (500) g	Zn 6%, B 2%, Cu 1%, Fe 6%, EDTA Mo 0.05%, Mn, 5% and Aminos 5%	Shan Maw Myae Trading Co., Ltd., India
Nano Max NPK Fertilizer	Multiple organic acids chelated with essential nutrients, amino acids, organic carbon, organic micronutrients/trace elements, vitamins and probiotics.	JU Agri Sciences Pvt. Ltd., Janakpuri, New Delhi, India
TAG NANO (NPK, PhoS, Zinc, Cal, etc.) fertilizers	Protein lactogluconate chelated with trace elements, vitamins, probiotics, seaweed extract and humic acid	Tropical Agrosystem India (P) Ltd., India
Nano Green	Corn, Grain, Soybean, Potato, Coconut and Palm Extracts	Nano Green Sciences, Inc., India
Biozar Nano-Fertilizer	Combination of organic substances, trace elements and macromolecules	Fanavar Nano-Pazhoohesh Markazi Company, Iran

Source: Prasad et al., 2017; Azam, 2002

3.2. Nanoporous Zeolites in Farming

Zeolites are crystalline alumina silicates among the most common minerals in sedimentary rocks.

Other useful chemical and physical properties include

- High void volume (50%),
- Low density (2.1-2.2 g/cm^3),
- Excellent molecular sieve properties and
- High cation exchange capacity(CEC) of 150-250 cmol/kg.

3.3. Applications of Zeolites in Agriculture

3.3.1. Input Use Efficiency

i) As a carrier of nutrients

- Zeolites enhances the action of N, P& K as slow release fertilizers.
- Reduces the losses of N to environment.

3.3.2. Improving Organic Manure Efficiency

- Zeolites mixed with manure increases the effectiveness of organic fertilizers
- Increases the solubilisation of phosphate minerals.

3.3.3. Improving Nitrogen Use Efficiency

Urea impregnated zeolite chips can be used as sustained release of nitrogen fertilizers.

3.3.4. Improving Phosphorus Use Efficiency

Increase the solubilization of phosphate minerals, leading to improved phosphorus uptake

3.4. Zeolites As An Input In Farming

3.4.1. As A Fertilizer

Zeolites in soils exchange sodium and potassium cations for NH4.

- High potential of zeolite minerals as nitrogen fertilizers

3.4.2. Zeolites For Soil Amelioration

Improving soil physical properties- Water holding capacity

- Remediation of contaminated soil- heavy metals
- As soil amendment
- Buffering soil pH levels

3.4. Slow Release Nano Fertilizers

Nan membranes (nano-encapsulation) assist in slow and steady release of nutrients

- This is a great alternative to soluble fertilizers.
- They release nutrients more slowly throughout the season. Plants can absorb most of their nutrients through leaching without wasting.
- Slow release fertilizers are more convenient because they require less frequent application.
- Sustained release fertilizers are more expensive than soluble fertilizers, but the advantages outweigh the disadvantages.

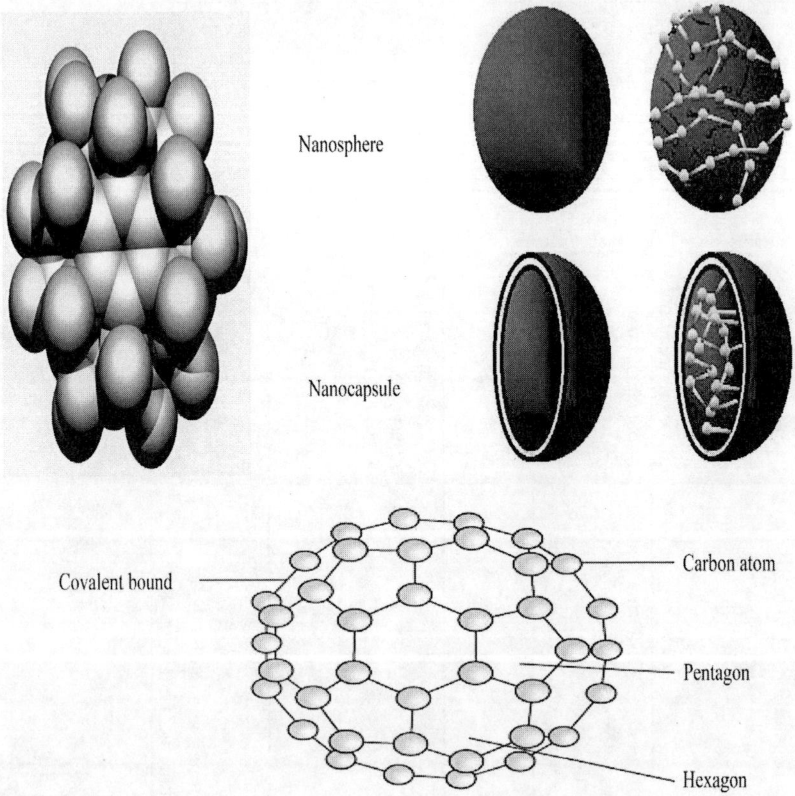

Fig. 4: Bucky Ball carries 60 carbon atoms (20 hexagons + 12 Pentagons)

Table 3: Effects of some engineered nanomaterials on crops

Engineered Nanomaterials	Positive impact on crop production	Negative impacts on Plants	Dosage Required for Plants	References
SWCNT	Crop germination improvement	Higher concentrations (100 mg/l) may cause toxic effects such as necrosis and cell death	10 mg/L for pepper (C. annuum) and 30 mg/L for salvia (S. macrosiphon) and tall fescue (F. arundinacea)	Zaytseva and Neumann (2016)
MWCNT	Nitrogen and phosphorus absorption in waste water to deliver to crops	Increased possible ROS formation, decreased chlorophyll content, and cell viability. DNA damage was recorded in onion roots.	In tobacco, 100 mg/ L	Zaytseva and Neumann (2016)
Nanozeolite (building blocks of SiO4 and AlO4)	Improving soil quality			Rastogi et al. (2017)
FeO NPs, nano-zero-valent iron (nZVI)	Elongation of roots, chemical modification and Neutralization in the soil		Decrease in germination was observed at at 250 mg/l (seeds of *Hordeum vulgare* and *Linum usitatissimum*).	Denget al. (2014)
ZnO NPs	This can result in higher yields of plants.	It can reduce the number of roots, the length of rice seedlings, and inhibit the photosynthesis of chlorophyll	$\leq$ 200 mg/kg Reduced germination in Zea mays at 2000 mg/l	Raddy (2014)

3.5. Limitation of Nano Fertilizer

Though Nanofetilizers have potential to revolutionize the agriculture sector and its productivity, some of these advantages may come at a high cost. There are several safety issues pertaining to the application of Nanofertilizers in agriculture systems, as exposure to Nanofertilizers could result in health risks. In this situation, unintentional health safety and environmental issues can limit the application of Nanofertilizer in agricultural crop production (Diatta et al., 2020). NFs can enter the food chain, thereby increasing their dissemination in non-targeted living organisms. NPs affect living organisms in different ways, e.g., carbon-based NPs modify DNA structure and the expression levels of genes in plant tissue (Lahiani et al., 2015)

4. Conclusion

From the sustainable agriculture perspective, Nanofertilizers have potential as part of smart crop productionsystems under the framework of sustainable agriculture because they have largesurface areas and a characteristic slow and steady release of nutrients. These promisingcharacteristics make them highly suitable for use in modern agriculture.The use of Nanofertilizers can increase agricultural productivity and resistance against bioticand abiotic stresses. Therefore, the use of NFs in the agriculture sector should be increased. The application of NFs may help to decrease the amount of fertilizers via the smart deliveryof active ingredients, to increase nutrient uptake and NUE values, and to decrease fertilizerlosses from volatilization, leaching, runoff, and consumed energy during production. The foliar application of NFs is much better as compared to the soil application of NFs due to its significant enhancements in the growth, physiological and biochemical traits,yield, and quality of crops—particularly in smart agriculture.

References

Azam, F. 2002. Added nitrogen interaction in the soil-plant system—A review. Pakistan J. Agron. 1,54–59.

Chhipa, H. 2017. Nanofertilizers and nanopesticides for agriculture. Environ. Chem. Lett. 15(1), 15-22.

Cui, H.X., Sun, C.J., Liu, Q., Jiang, J., Gu, W. 2010. Applications of nanotechnology in agrochemical formulation, perspectives, challenges and strategies. In: International conference on Nanoagri, Sao Pedro, Brazil, pp. 28–33.

Deng, Y., White, J.C., Xing, B. 2014. Interactions between engineered nanomaterials and agricultural crops: Implications for food safety. J. Zhejiang Univ. Sci. A, 15,552-572.

Dijk Van, M.,Meijerink, G.W. 2014. A review of food security scenario studies: Gaps and ways forward. In: The Food Puzzle:Pathways to Securing Food for All. T.J. Achterbosch, M. Dorp, W.F. van Driel, J.J. van Groot, J. Lee, A. van der Verhagen, I. Bezlepkina, (eds.). Wageningen UR: Wageningen, The Netherlands. p. 30-32.

Dimkpa, C.O., Bindraban, P.S. 2017. Nanofertilizers: new products for the industry? J Agric Food Chem. 66(26), 6462–6473

Zulfiqar,F.l, Navarro, M.,Ashraf, M., Akram, N.A.,Munné-Bosch,S. 2019. Nanofertilizer use for sustainable agriculture: Advantages and limitations. Plant Science 289 (2019) 110270.

Kale, A.P.S.N. Gawade, 2016. Studies on nanoparticle induced nutrient use e□ciency of fertilizer and crop productivity. Green Chem. Tech. Lett. 2, 88-92.

Kottegoda, N., Sandaruwan, C. Priyadarshana, G.,Siriwardhana, A.,Rathnayake, U.A.,Berugoda Arachchige,D.M.,Amaratunga, G.A. 2017. Urea-hydroxyapatite nanohybrids for slow release of nitrogen, ACS Nano. 11(2), 1214-1221.

Kottegoda, N., Munaweera,I.,Madusanka,N.,Karunaratne,V. 2011.A green slow-release fertilizer composition based on urea-modified hydroxyapatite nanoparticles encapsulated wood. Curr. Sci. 101(1), 73-78.

Lahiani, M.H., Chen, J., Irin, F., Puretzky, A.A., Green, M.J., Khodakovskaya, M.V. 2015. Interaction of carbon nanohorns with plants: Uptake and biological effects. Carbon. 81, 607–619.

Prasad, R. Bhattacharyya, A. Nguyen, Q.D. 2017. Nanotechnology in sustainable agriculture: Recent developments, challenges, and perspectives. Front Microbiol. 8, 1-13.

Raddy, R. 2014. Efficacy of nano zinc particle on growth. Master's Thesis, University of Agricultural Sciences, Bangalore, India.

Rastogi, A., Zivcak, M., Sytar, O., Kalaji, H.M. He, X., Mbarki, S. Brestic, M. 2017. Impact of Metal and Metal Oxide Nanoparticles on Plant: A Critical Review. Front Chem. 5, 1–16.

Shang, Y., Hasan, M.K., Ahammed, G.J., Li, M., Yin, H., Zhou, J. 2019. Applications of nanotechnology in plant growth and crop protection: a review. Molecules 24: 2558–2580.

Siddiqui, M.H., Al-Whaibi, M.H.,Firoz,M.,Al-Khaishany, M.Y. 2015. Role of nanoparticles in plants. In: Nanotechnology and Plant Science. Springer International Publishing. P.19-35.

Subramanian, K.S., Manikandan, A.,Thirunavukkarasu,M.,Rahale, C.S. 2015. Nano-fertilizers for balanced crop nutrition. In: Nanotechnologies in food and agriculture. M. Rai et al.(eds). Springer, Cham. p. 69-80.

Zaytseva, O.; Neumann, G. 2016. Carbon nanomaterials: Production, impact on plant development, agricultural and environmental applications. Chem. Biol. Technol. Agric.3,17. https://doi.org/10.1186/s40538-016-0070-8

23

Post Harvest Diseases of Horticulture Crops and Their Management Strategies

***Pranjali Sinha*[1] *and Arunima Tripathi*[2]**

[1]*Department of Plant Pathology, Indira Gandhi Krishi Vishwavidyalaya Raipur, Chhattisgarh, India*
[2]*RMD, CARS, Indira Gandhi Krishi Vishwavidyalaya, Ambikapur Chhattisgarh, India*

1. Introduction

Post harvest storage is a powerful challenge in agriculture, about one third of the fresh fruits and vegetables are discarded because of the deterioration of quality of produce beyond a reasonable standard. (FAO, 2011; OCD, 2014) Any change or damage in the quality of product from harvest till consumption is considered as loss. Post harvest handling practices do not increase the quality of product, rather just slows down the process of deterioration. The quality and shelf life of produce is determined prior to the time of harvest. Various factors such as climate, soil preparation and cultivation, fertigation, pest and disease incidence, crop loads etc influence the quality and flavour aspects of fresh food. Temperature treatment, regulated and altered environment, surface coating, physical therapy, biocontrol and other procedures can all influence the marketing and shelf life of the product. However, mechanical damage, shipping, and physiological abnormalities such as maturation, water loss, and texture loss account for a significant portion of fruit loss.

In India post harvest spoilage of fresh produce share about 4.6 -15.9% (Jha et al., 2015). The primary losses of produce is caused by fungal pathogens such as *Alternaria, Aspergillus, Fusarium, Penicillium, Mucor, Colletotrichum* etc that also produce mycotoxins and Citrin, Patulin, Chaetoglobosins, Aflatoxins, Orchratoxins, and Fumonisin are all possible carcinogenic metabolites, that contaminate the food and can cause health risk.(Dukare et al., 2019) many bacteria like *Erwinia* and *Pseudomona* although weak pathogen, invade the wounded products.

2. Major Post Harvest Diseases

The major post-harvest fungi diseases associated with different fruit and vegetable crops include-

Disease	Causal organism	Affected Fruit	Symptom	Reference
Bitter rot	*Colletotrichum gleosporiodes, Pestalotiopsis mangifera*	Pome ,stone fruits, mango, sapota	Brown superficial discoloration	Masoud et al. (2013)
Black mould	*Aspergillus niger, A. alliaceous*	Garlic, tomato, ginger, onion	Black powdery mass ,black, pink or white coloured rotting	Prajapati et al. (2015)
Black lesion, dark spots	*Stemphyium botryosum*	Pome, grape	Light oval blotches that turn brown on sporulation	Toselli et al. (2012)
Black rot	*Botryosphaeria obtuse*	Apple	Circular brown to black decay spot in concentric ring	Gaikwad et al. (2021)
Blue mold	*Penicillium expansum, Penicillium waksmanii*	Mainly pome ,Orange, stone fruits	Blue green spore masses	Yu. (2012)
Brown rot	*Monilinia fruicticola*	Stone fruits	Ash grey to brown tuft mold	Zhu, (2016); Li (2021)
Black spot rot,sooty mold	*Alternaria alternata*	Apple, plum ,peach pear, cherry	Blackish brown spot of rotting	Huaiyu et al. (2020)
Lenticel rot	*Gleosporium album*	Apple ,pear, persimmon, carrot	Soft watery colourless decay, white spores covereing decayed area	Edney et al. (1967)
Olive green, sooty mold	*Cladosporium herbarum*	Pear, Apple ,cherry, peach, plum, stone fruit		Latorre et al. (2011)
Pink mold	*Trichoderma roseum*	Pome and stone fruit	Pinkish to white powdery mycelium, water soaked lesions	Wang et al. (2008)
Watery white rot	*Rhizopus stolonifera*	Apple, Plum, Pear, cherry, Tomato, Peach,	Brown with soft and watery flesh	Zhang et al. (2009)
Mouldy nose	*Sclerotium cepivorum*	Onion, Carrot	Small black sclerotia appear ,fluffy white mycelium covering the fruit	Nabi et al. (2017)

Disease	Causal organism	Affected Fruit	Symptom	Reference
Stem rot of pawpaw	*Phomopsis caricae-papayae*	Papaya	Water soaked rotten are black to brown and later get depressed and cracked	Nabi et al. (2017)
Grey mold	*Botrytis cinerea*	Grape	Grey mycelium and greyish spore mass	Liu (2014)
Crown Rot,Basal rot	*Fusarium roseum, Lasidiplodia theobrome, Deightonialla torula*	Banana, Garlic	Darkening of the peduncle and appearance of black tip	Nabi et al. (2017)
Blister spot	*Pseudomonas syringae pv papulans*	Apple	Raised purplish black blister on fruit lenticels	Nabi et al. (2017)
Bacterial Soft rot	*Erwinia carotovora*	Carrot, cucurbit	Water soaked ,rotten tissue, foul odour	Nabi (2017)

3. Potential Strategies of Management

The major goal of post-harvest fruit disease control is to preserve the product free of disease or symptoms until it is sold for consumption. As a result, management techniques should focus on disease eradication, prevention and postponement of symptoms during storage, processing and distribution of fruits and vegetables. The remedies for postharvest ailments in temperate fruits are categorized as follows: cultural, physical, chemical, biological, and resistance measures. The success of the treatment is determined by the agent's ability to reach the target pathogen, as well as the sensitivity and degree of the microbial infection and the host's vulnerability. Various methods for the control of post harvest spoilage is briefly described as follows-

3.1 Cultural Methods

Vegetable and Fruits that have been wounded during picking, grading, packaging, or shipping, but have avoided infection by wound pathogens are nonetheless susceptible to pathogen contact during loading and warehousing. Since disease development needs the presence of a specific pathogen as well as an open wound for entry, a reduction in any of these variables will result in disease suppression. When it comes to avoiding wounds, it should be kept in mind that physiological injuries can be produced by cold, heat, oxygen deficit, and other external conditions, all of which make host more vulnerable to pathogen attack. A decrease in wounds minimises the risk of pathogens infecting vegetables and fruits during warehousing and transmission; such aspects should be considered while packaging or storing fruits and vegetables. The pathogen source should be eliminated as soon as possible, whether it be by discarding rotting fruits or soaking them in a disinfectant solution, such as formaldehyde, sodium or calcium hypochlorite, ammonium compounds, isopropyl, alcohol, in a separate container (Nabi et al., 2017)

3.2. Physical Method

Physical treatments such as high temperature treatments, low temperature storage, radiation, magnetic fields, can be used to control postharvest spoilage. Other methods ultrasound sound infrared microwave, X rays ,visible light, ultraviolet, gamma rays, and cathode ray spectrum are among the different radiations could be chosen to serve the purpose.

3.3 Hot Water Treatment

Fruits are given heat treatment in various forms as vapour, water dip or hot air. But its treatment depends on various factors as thermal sensitivity of the microbe, site of presence of fungus and fruit sensitivity to heat. This is an eco-friendly, widely adopted method for the management of post harvest spoilage. The surface pathogens are controlled effectively and various pathogens that have penetrated to the surface to some extend in cell surface. The range of temperature for the effective management of damage involves 50 to 55 for water rinsing and brushing of fruits given for an duration of 10-50 seconds according to the fruit sensitivity whereas, water dip treatment involves 40-50 for 10-30 minutes which reduces storage pathogen slightl. (Nabi et al., 2017)

3.4. Irradiation

Irradiation, which includes exposing foodstuffs to X-rays, electron beams, gamma rays or UV radiation in bulk, has proved to minimise postharvest losses and is widely used in a variety of agricultural applications as demonstrated for the control of *Pencillium* spp. and *Aspergillus* spp.It is demonstrated thatthe combination of this technology with other post-harvest diseases management strategies will enhance the synergistic effect for fungal disease control in fresh fruit. Fruit exposed to 200Gy of gamma irradiation developed resistance to *P. expansum* and increased the expression of defense-related enzymes and pathogenesis-related (PR) genes. A synergistic antifungal activity of gamma irradiation and sodium dichloro isocyanurate at 2kGy and 70ppm completely reduced *P. expansum* spore germination, germ tube elongation, and mycelial growth.The gamma rays and electromagnetic wave have a benefit over disinfectant therapy because of their strong penetrating strength and ability to flow materials without leaving any trace (Oyom et al., 2022)

3.5. Pulse Electric Field

This is comparatively new storage method that is commonly employed in inactivation of microorganisms and preservation of horticulture crops. It substantially controls senescence and microbiological deterioration while maintaining, considerable colour and nutritional values in mango and tomato fruits (Oyom et al., 2022) Microorganisms are rendered inactive by invading the cell membrane, prompting the release of intracellular chemicals and cell death. However, this has limited application as its mechanism is not yet well.

3.6. Essential Oils and Application of Edible Coating

Essential oils are plant extracts with high antifungal and antioxidant properties that are released as secondary metabolites by aromatic plants. Cinnamaldehyde, citral, eugenol, limonene, and thymol are volatile chemicals found in several plant extracts that can damage the cellular membranes and cell organelles of pathogenic fungus, impairing cell viability. The antifungal properties of neem seed kernel extracts (NSKE) on postharvest diseases, effectively inhibited the growth of *Monilinia fructicola, Penicillium expansum, Trichothecium roseum,* and *Alternaria alternate* by inducing defense-related enzymes, thus delaying spore development (Oyom et al., 2022)

Use of edible coatings made from natural biopolymers has recently demonstrated a high level of resistance to plant pathogens in fruits and vegetables. After being treated with a variety of fruits, edible coatings made from protein, lipid components, and polysaccharides such as soy, chitosan, starch, gum, and others have shown promising effects. Coating the fruit with 0.2–0.4 percent cumin essential oil considerably decreased Alternaria rot growth and further slowed physiological changes over storage at 23-25 °C for 28 -30days.During optimal storage conditions, composite edible coating comprising olive oil, cellulose, and soy protein isolate enhanced the durability of fruits against microbial spoilage. These coatings created a stronger barrier for exchange of gas and moisture, limiting microbial attack on the cell surface of fruits (Oyom et al., 2022)

3.7. Chemical Treatment

Post harvest management should begin in the field before crop harvest. Because most fungicides are unable to penetrate deeply into tissues, and effective concentrations of the fungicide would not reach deep-seated mycelium, this treatment is unlikely to cure firmly penetrated diseases. The following are some of the chemicals/fungicides in combination with hot water treatment that are used to prevent postharvest deterioration in mature fruits.

Table 1: Chemical management of diseases

Crop	Fungicide	Treatment	Pathogen	Reference
Apple	Benomyl (500mg/L), TBZ	(54.4°C/4 min dip	*Penicillium expansum*	Spalkding et al. (1969)
Pear, Citrus Fruit	TBZ 1000 mg/L	55°C/5 min dip	Most fungi	Sharma, et al. (1990)
Oranges	SOPP 500-1000 mg/L, IMZ 400 mg/L	25-45°C/3 min dip) 55 °C/20s Drench	*Penicillium italicum*, blue mold, green mold	Schirra et al. (1996)
Grapefruit	TBZ or IMB ,1000 mg/L	53 °C/2 min	Green mold ,blue mold, Stem end rot	Barkai-Golan et al. (1991)
Lemons	IMZ 1000 mg/L	(52°C /3 min dip)	Blue mold, *Botrytis cinerea*	Schirra et al. (1995)
Guava	Benomyl, 500-2000 mg/l, Guazatine, 500 mg/L	48-50°C 2 mi 50°C/5 minn dips	Green mold, *Alternaria alternata,* fruit canker, *Pestalotia psidii*	Ben-yehoushua et al. (2000)
Litchi	Benomyl ,500 mg/L,+ packing	50°C/2 min	Most fungi	Ansari et al. (2008)

Crop	Fungicide	Treatment	Pathogen	Reference
Mango	TBZ or benomyl,1000 mg/L, IMZ,1000mg/ (52 □/1-3min dip) Prochlor,225 mg/L (48-64 □/15-20 s brushing)	54.5□ /5 min Hot air (core temp 46.5 □/10 min, relative humidity >95%), Carbendazin 1000 mg/L (55□ 5 min dip)	*Colletotrichum gleosporoides*/ stem end rot, *Lasiodiploidia theobrone, Alternaria alternata*	McDonald et al. (1991)
Muskmelon	SDDC,400 MG/l; IMZ ,1000 mg/L +polyethylene wrapping	60□/3 min dip	Fusarium rot/ phomopsis rot / Diaporthe melonis	Schirra et al. (1995)
Stone fruits (Peaches, Nectraines)	DCNZ or benomyl,100 mg/L	45 or 52□/4 min dip	Brown rot and Rhizopus rot	Smith et al. (1971)
Taro, Red tamarillo	Hot water followed by 250 mb/L unheated IMZ /1 min dip	50□ /8 min dip	*Colletotrichum gleosporoidies, Phomopsis spp* and *Phoma exigua*	Yearsley (1988)

SDDC, sodiun dimethyldithiocarbamate; SOPP, sodium orthophenylphenate; IMZ, imazalil; TBZ, thiabendazole; DCNA, 2,6-dichloro-4-nitroaniline

3.8. Biological Control

A large variety of microbial antagonists with antimicrobial action against pre-harvest and post-harvest diseases have been discovered. Pathogen-specific antifungal compounds/metabolites are produced by these bacteria, which limit pathogen development and metabolism. These microorganisms restrict, inhibit, or destroy disease propagules developing on fruit, reducing the risk of additional fruit deterioration during storage. In the biological suppression of a fungal infection, antifungal microbial agents use both direct and indirect inhibitory mechanisms.

Table 2: Antimicrobial antagonists used to treat postharvest infections

Antimicrobial Agents	Pathogen of interest	Host crop	Mechanism of action	References
Bacillus subtilis M4	*Botrytis cinera*	Apple	Antibiotic synthesis	Ongena et al. (2005_
B. subtilis	*P. digitatum*	Citrus		Waewthongrak et al. 2015
Bacillus amyloliquefaciens PPCB004	*Alternaria citri, Colletotrichum gloeosporioides, Penicillium crustosum*			Arrebola et al. (2010)
Bacillus atrophaeus CAB-1	*B. cinera*			Zhang et al. (2013)

Antimicrobial Agents	Pathogen of interest	Host crop	Mechanism of action	References
Cryptococcus laurentii	*Penicillium expansum*	Pear	Lytic enzyme synthesis	Yu et al. (2008)
Bacillus amyloliquefaciens	*B. cinera*			Qu et al. (2016)
Pichia guilliermondii	*Colletotrichum capsici*	Chilli		Chanchaichaovivat et al. (2008)
Halophilic bacteria	*B. cinerea*	Strawberry		Essghaier et al. (2009)
Rahnella aquatilis BNM	*B. cinerea*			Sansone et al. (2011)
P. guilliermondii	*B. cinerea*	Apple		Zhang et al. (2011)
Aureobasidium pullulans	*P. expansum, B. cinerea, Monilinia fructicola,*			Zhang et al. (2012)
Paenibacillus polymyxa Bacillus subtilis	*Colletotrichum gloeosporioides*			Kim et al. (2016)
C. oleophila	*P. expansum*			Urbina et al. (2016)
Gluconobacter cerinus	*B. cinerea*	Grapes		Guzzon et al. (2014)
Metschnikowia fructicola	*Monilinia laxa, M. fructicola*	Peach		Banani et al. (2015)
Rhodotorul aminuta, Candida azyma,	*Geotrichumcitri aurantii.*	Citrus		Ferraz et al. (2016)
Pichia membranaefaciens	*Penicillium expansum*	Peach	Host defense induction	Chan et al. (2007)
Pichia caribbica	*Rhizopus stolonifer*			Xu et al. (2013)
Bacillus thuringiensis	*Guignardia citricarpa*	Citrus		Lucon et al. (2010)
Pseudomonas fluorescens and Bacillus subtilis	*Lasiodiplodia theobromae*	Mango		Seethapathya et al. (2016)
Muscodralbus	*Penicillium italicum Geotrichium candidum*	Lemon		Mercier and Smilanick (2005)
Streptomyces globisporus	*Penicillium italicum*	Citrus		Li et al. (2010)
Candida intermedia	*Botrytis cinerea*	Strawberry		Huang et al. (2011)
Wickerhamomyces anomalus	*B. cinerea*	Grapes		Parafati et al. (2015)

Antimicrobial Agents	Pathogen of interest	Host crop	Mechanism of action	References
Pichia membranaefaciens, Cryptococcus laurentii, Candida guilliermondii, Rhodotorula glutinis	*Monilinia fructicola*	Peach	Alleviation of oxidative damage of fruit host	Xu et al. (2008)

4. Conclusion

The success of the remedial action depends on the close monitoring of crop before harvest and spray of fungicide only when in need. Continuous inspection, good technical assistance followed with accurate identification of the pathogens is the pre-requisite. Combination of remedies such as physical treatment with fungicidal treatment, Biocontrol with physical actions or fumigation followed with physical treatment in an integrated manner may prove to significantly decrease the spoilage of foods. However most fundamental measure is to avoid injuries of product during harvesting, transportation, packing or grading. Thus, need of combining multiple ways to improve disease control and achieve safe, and residue free food items is the urgent requirement.

References

Ansari, N.A.; Feridoon, H. 2008. Postharvest application of hot water, fungicide and waxing on the shelf life of Valencia and the local orange cv. Siavarz. Acta Hortic. 768, 271-277

Arrebola, E., R. Jacobs, Korsten, L. 2010. Iturin A is the principal inhibitor in the biocontrol activity of Bacillus amyloliquefaciens PPCB004 against postharvest fungal pathogens. J. Appl. Microbiol. 108, 386-395.

Banani, H., Spadaro, D., Zhang, D., Matic, S., Garibaldi, A., Gullino, M.L. 2015. Postharvest application of a novel chitinase cloned from Metschnikowia fructicola and over expressed in Pichia pastoris to control brown rot of peaches. Int. J. Food Microbiol. 199:

Barkai-Golan, R., Appelbaum, A. 1991. Synergistic effects of heat and sodium o-phenylphenate treatments to inactivate Penicillium spores and suppress decay in citrus fruits. Trop. Sci. 31, 229-233.

Ben-Yehoshua, S., Peretz, J., Rodov, V., Nafussi, B., Yekutieli, O., Wiseblum, A., Regev, R. 2000. Postharvest application of hot water treatment in citrus fruit: the road from the laboratory to the packinghouse. Acta Hortic. 518, 19-28.

Chan, Z., Qin, G., Xu, X., Li, B., Tian, S. 2007. Proteome approach to characterize proteins induced by antagonist yeast and salicylic acid in peach fruit. J. Proteome Res. 6, 1677–1688.

Chanchaichaovivat, A., Panijpan, B., Ruenwongsa, P. 2008. Putative mode of action of *Pichia guilliermondii* strain R13 in controlling chilli anthracnose after harvest. Biol. Control. 47, 207-215.

Dukare, A.S., Paul, S., Nambi, V.E., Gupta, R.K., Singh, R., Sharma, K., Vishwakarma, R.K. 2019. Exploitation of microbial antagonists for the control of postharvest diseases of fruits: a review. Critical Reviews in Food Science and Nutrition. *59*(9), 1498-1513.

Edney, K.L., Tan, A.M. and Burchill, R.T. 1967. Susceptibility of apples to infection by Gloeosporium album. Ann. Appl. Biol. 86(1), 129-132

Essghaier, B., M.L. Fardeau, J. L. Cayol, M. R. Hajlaoui, A. Boudabous, Jijakli, H. Sadfi-Zouaoui, N. 2009. Biological control of grey mould in strawberry fruits by halophilic bacteria. J. Appl. Microbiol. 106, 833-846

FAO. 2011. Global food losses and food waste-extent, causes and prevention. (http://www.fao.org/docrep/014/mb060e/mb060e00.pdf). Rome, Italy: Food and Agriculture Organization

Ferraz, L.P., Cunha, T.D., da Silva, A.C., Kupper, K.C. 2016. Biocontrol ability and putative mode of action of yeasts against Geotrichum citri-aurantii in citrus fruit. Microbiol. Res. 188, 72.79.

Gaikwad, S.S., 2021. Fungi Classification using Convolution Neural Network. Turkish Journal of Computer and Mathematics Education. 12(10), 4563-4569.

Guzzon, R., Franciosi, E., Larcher, R. 2014. A new resource from traditional wines: characterization of the microbiota of "Vino Santo" grapes as a biocontrol agent against Botrytis cinerea. Eur. Food. Res. Technol. 239, 117-126.

https://agritech.tnau.ac.in/crop_protection

https://www.apsnet.org/

Huang, R., Li, G.Q., Zhang, J., Yang, L., Che, H.J., Jiang, D.H., Huang, H.C. 2011. Control of postharvest Botrytis fruit rot of strawberry by volatile organic compounds of Candida Intermedia. Phytopathol. 101, 859–869

Jha, S.N., Vishwakarma, R.K., Ahmad, T., Rai, A., Dixit, A.K. 2015. Technical report on assessment of quantitative harvest and post-harvest losses of major crops and commodities in India. Ludhiana, India: ICAR-All India Coordinated Research Project on Post-Harvest Technology, ICAR-CIPHET.

Kim, Y.S., Balaraju, K., Jeon, Y. 2016. Effects of rhizobacteria Paenibacillus polymyxa APEC136 and Bacillus subtilis APEC170 on biocontrol of postharvest pathogens of apple fruits. Zhejiang Univ. Sci. B (Biomed and Biotechnology) 17(12), 931–940.

Latorre, B.A., Briceño, E.X., Torres, R. 2011. Increase in Cladosporium spp. populations and rot of wine grapes associated with leaf removal. Crop Protection, 30(1), 52-56.

Li, S., Wang, Y., Wu, F., Xiao, L., Peng, W., Xiang, M., Chen, J., Chen, M. 2021. First report of pyrus pyrifolia 'cuiguan' fruit rot caused by monilinia fructicola in Southern China. Plant Disease https://doi.org/10.1094/PDIS-05-21-1076-PDN

Liu, T., Li, L., Zhan, G., Li, T., Zhang, F., Wang, Y. 2014. Postharvest propionic acid and hot water treatment for the control of black spot disease in chinese ya pears. Advance Journal of Food Science and Technology 6(2), 265-270.

Macarisin, D., Droby, S., Bauchan, G., Wisniewski, M. 2010. Superoxide anion and hydrogen peroxide in the yeast antagonist–fruit interaction: a new role for reactive oxygenspecies in postharvest biocontrol? Postharvest Biol. Technol. 58, 194-202.

Masoud, A., Ibrahim, T. and Hilde, N., 2013. Impact of harvesting time and fruit firmness on the tolerance to fungal storage diseases in an apple germplasm collection. Postharvest Biol. Technol., 82, 51-58

McDonald, R.E., Miller, W.R., McCollum, T.G., Brown, G.E. 1991. Thiabendazole and imazalil applied at 53°C reduce chilling injury and decay of grapefruit. HortScience. 26, 397-399

Nabi, S.U., Raja, W.H., Kumawat, K.L., Mir, J.I., Sharma, O.C., Singh, D.B., Sheikh, M.A. 2017. Post harvest diseases of temperate fruits and their management strategies-a review.

OECD. 2014. Market and trade impacts of food loss and waste reduction. Paris, France: Organization for Economic Cooperation and Development

Ongena, M., Jacques, P., Toure, J., Destain, A., Jabrane, P., Thonart. 2005. Involvement of fengycin-type lipopeptides in the multifaceted biocontrol potential of Bacillus subtilis. Appl. Microbiol. Biotechnol. 69, 29-38

Oyom, W., Li, Y.C., Prusky, D., Zhang, Z., Bi, Y. and Tahergorabi, R., 2022. Recent advances in postharvest technology of Asia pears fungi disease control: A review. Physiological and Molecular Plant Pathology. 117, 170-771.

Parafati, L., Vitale, A., Restuccia, C., Cirvilleri, G. 2015. Biocontrol ability and action mechanism of food-isolated yeast strains against Botrytis cinerea causing postharvest bunch rot of table grape. Food Microbiol. 47, 85-92

Platania, C., Restuccia, C., Muccilli, S., Cirvilleri, G. 2012. Efficacy of killer yeasts in the biological control of Penicillium digitatum on Tarocco orange fruits (Citrus sinensis). Food Microbiol. 30, 219-225

Prajapati, B.K., Patil, R.K. 2015. Black mould rot: an important post harvest disease of onion and its management. Pop. Kheti, 3, 77-80.

Qu, H., Longyu Zhao, L., Zhao, F., Yufang Liu, Y., Yang, Z. 2016. Biocontrol of Gray Mold Decay in Pear by Bacillus amyloliquefaciens Strain BA3 and its Effect on Postharvest Quality Parameters. Pol. J. Microbiol. 65(2), 171-176.

Sansone, G., Rezza, I., Calvente, V., Benuzzi, D., SanzFerramola, M.I. 2011. Inhibitors of polygalacturonase and laccase of Botrytis cinerea and their application to the control of the fungus. Int. Biodeter. Biodegr. 65, 243-247.

Schirra, M., Barbera, G., D'Aquino, S., La Mantia, T., McDonald, R. E. 1996. Hot dips and high temperature conditioning to improve shelf quality of late-crop cactus pear fruit. Trop. Sci. 36, 159-165.

Schirra, M., D'Aquino, S., Cabras, P., Angioni, A. 2011. Control of postharvest diseases of fruit by heat and fungicides: efficacy, residue levels, and residue persistence. A review. Journal of agricultural and food chemistry. *59*(16), 8531-8542.

Schirra, M., Mulas, M. 1993. Keeping quality of Oroblanco grapefruit type as affected by hot-dip treatments. Adv. Hortic. Sci. 7, 73-76

Schirra, M., Mulas, M. 1995. Improving storability of 'Tarocco' oranges by postharvest hot-dip fungicide treatments. Postharvest Biol. Technol. 6, 129-138

Schirra, M., Mulas, M. 1995. Influence of postharvest hot water dip and imazalil-fungicide treatments on cold-stored 'Di Massa' lemons. Adv. Hortic. Sci. 9, 43-46.

Schirra, M., Mulas, M., Baghino, L. 1995. Influence of postharvest hot-dip fungicide treatments on Red blush grapefruit quality during long term storage. Food Sci. Technol. Int. 1, 35-40.

Seethapathya, P., Gurudevana, T., Subramanian, K.S., Kuppusamy, P. 2016. Bacterial antagonists and hexanal-induced systemic resistance of mango fruits against Lasiodiplodia theobromae causing stem-end rot. J. Plant Interact. 11(1),158-166.

Sharma, R.L., Kaul, J.L. 1990. Efficacy of hot water and carbendazim treatments in controlling brown rot of apple. Indian J. Mycol. Plant Pathol. 20, 47-48.

Spalding, D.H., Vaught, H.C., Day, R.H. and Brown, G.A. 1969. Control of blue mold rot development in apples treated with heated and unheated fungicides. Plant Disease Reporter. 738-742.

Toselli, M., Sorrenti, G., Quartieri, M., Baldi, E., Marcolini, G., Solieri, D., Marangoni, B., Collina, M. 2012. Use of soil-and foliar-applied calcium chloride to reduce pear susceptibility to brown spot (Stemphylium vesicarium). J. Plant Nutr. 35(12), 1819-1829

Urbina, C. T., Prieto, V. G., Lopez, C. G., Albores, F. V., Reyes, D. B., Muniz, C. A., Barrios, D. O. 2016. Purification and Characterization of b-1,3-glucanase from Candida Oleophila for the Biocontrol of Penicillium expansum. Res. Rev. J. Bot. Sci. 5(1), 38-45.

Waewthongrak, W., Pisuchpen, S., Leelasuphakul, W. 2015. Effect of Bacillus subtilis and chitosan applications on green mold (Penicillium digitatum Sacc.) decay in citrus fruit. Postharvest Biol. Technol. 99, 44-49.

Wang, Y., Li, X., Bi, Y., Ge, Y., Li, Y., Xie, F. 2008. Postharvest ASM or Harpin Treatment Induce Resistance of Muskmelons against *Trichothecium roseum*. Agric. Sci. China. 7(2), 217-223

Yearsley, C.W.; Huang, B.I.; McGrat, H.J. W, Fry, J., Stec, M.G., Dale, J.R. 1988. Red tamarillos (Cyphomandra betacea): comparison of two postharvest dipping strategies for the control of fungal storage disorders. N. Z. J. Exp. Agric. 16, 359-366.

Yu, S. M., Lee, Y. H. 2015. Genes involved in nutrient competition by Pseudomonas putida JBC17 to suppress green mold in postharvest satsuma mandarin. J. Basic Microbiol. 55(7), 898-906.

Yu, T., Yu, C., Chen, F., Sheng, K., Zhou, T., Zunun, M., Abudu, O., Yang, S., Zheng, X. 2012. Integrated control of blue mold in pear fruit by combined application of chitosan, a biocontrol yeast and calcium chloride. Postharvest Biology and Technology, 69, 49-53.

Zhang, D., Spadaro, D., Valente, S., Garibaldi, A., Gullino, M.L. 2012. Cloning, characterization, expression and antifungal activity of an alkaline serine protease of Aureobasidium pullulans PL5 involved in the biological control of postharvest pathogens. Int. J. Food Microbiol. 153, 453-464

Zhang, H., Wang, L., Dong, Y., Jiang, S., Zhang, H., Zheng, X. 2009. Biocontrol of major postharvest pathogens on apple using Rhodotorula glutinis and its effects on postharvest quality parameters. Biol. Control, 48(1), 79-83

Zhang, H., Zheng, H.D., Yu, T. 2007. Biological control of postharvest diseases of peach with Cryptococcus laurentii. Food Control. 18(4), 287-291.

Zhao, L., Zhang, H., Li, J., Cui, J., Zhang, X., Ren, X. 2012. Enhancement of biocontrol efficacy of Pichia carribbica to postharvest diseases of strawberries by addition of trehalose to the growth medium. Int. J. Mol. Sci. 13, 3916-3932.

Zheng, M., Shi, J., Shi, J., Wang, Q., Li, Y. 2013. Antimicrobial effects of volatiles produced by two antagonistic Bacillus strains on the anthracnose pathogen in postharvest mangos. Biol. Control. 65.

24

Trichoderma spp. Opportunities for Disease Management in Horticultural Crops

D.L. Yadav and Pratik Jaisani

Agricultural Research Station, Agriculture University, Kota, Rajasthan, India

1. Introduction

Agriculture is prime source of living for rural people world wide effective in raising incomes among the poorest compared to other sectors. In Indian economy horticulture has been recognized as one of the most potential agriculture endeavours in terms of exponential growth. The Horticultural crops viz., fruits counting nuts, vegetables with potato, ornamental plants plus cut flowers, spices, plantation crops and medicinal and fragrant plants has developed a prime promoter for monetary growth in numerous of the states in the nation and it pays 30.4% to GDP of agriculture (ICAR website). Its role in the nutritious security, famine mitigation and employment gerenrate are becoming progressively significant. It offers a wide range of options for crop diversification and also provides huge scope for satisfying huge number of agro based industries which generate employments. Roughly estimates have suggested in India, on an average loss of about 25% in crop yield takes place due to plant pathogens like fungi, bacteria, viruses, viroids, virusoids, mollicutes and protozoa and opposed conservational circumstances such as lack or excess nutrients, moisture, light etc. Losses caused by insect-pests and diseases in horticulture crops have special relevance in our country where much still needs to be done in the modernization of management measures against plant diseases. Long-term Sustainable horticultural crop production approach should be considered in association with other agricultural crop production programs to check poverty, improve nutrition, and sustain economic growth. The increasing usage of agrochemicals in the management of pests of vegetables and fruits is the cause of serious concern because of the adverse effects on human health that could result from the consumption of pesticide contaminated agro produce. To minimalize the necessity on noxious pesticides, ecological bio-agents like *Trichoderma* spp. are applied to succeed soil-borne pathogens like Fusarium, Rhizoctoria, Pythium, Phytophthora and plant parasitic nematodes in horticultural crops. Multiple benefits of *Trichoderma* spp. inoculation for improving holistic crop health have been recognised, with growth and yield improvement and the mitigation of biotic and abiotic stresses (Pandey et al., 2021). For fruitful diseases control, delivery system and establishment of Trichoderma to the site of act is very significant. *Trichoderma* strains produce wide spectrum of extra cellular hydrolytic enzymes *i.e.* cellulases, chitinases, proteases, glycoside, β-1,3-glucanase, β-1,6-glucanase, N-acetyl-β-D- hydrolases, etc., infecting mycelia of fungi (Harman et al., 2004). Endophytic *Trichoderma* changes the plant physiology due to alteration in plant gene expression and contributions improved uptake of the nutrients and resistance to diseases. *T. asperellum, T. harzianum,* etc. applied through soil and seeds, induced SAR against plant pathogens by stimulating metabolic changes (De Meyer et al., 1998). The great estshared claims tyles of Trichoderma are seed treatment, seed biopriming, seedling dip, soil application and wound dressing. Therefore, in this chapter, we have focused upon biological mechanism, methods for applications, plant growth promoter and mediated horticultural crop diseases management.

2. Biological Roles of *Trichoderma* spp.

It is well documented that the *Trichoderma*-based biocontrol devices mostly trust on mycoparasitism, production of antibiotic and/or hydrolytic enzymes, competition for nutrients, as well as induced plant resistance; numerous subordinate metabolites formed by *Trichoderma* species could directly inhibit the growth of several plant pathogens. These mechanisms may act straight or circuitously against the beleaguered plant pathogen. Mycoparasitism is a direct mechanism for biological control that works by parasitizing, detecting, growing, and colonizing pathogen.

The delivery of *Trichoderma* spp. results in important vegetative development on a extensive array of crop plants. Interaction between *Trichoderma* spp. and the plant activates enhanced immunity against plant diseases, thus improving plant health. Plant growth improvement is showed by the increase in productivity, nutrient uptake, biomass, resistance to stress, and improvement of plant health. *Trichoderma* spp. were described to persuade the amalgamation of regulatory proteins in plants particularly under certain disease stress, where these governing proteins detect microbe effecters and trigger the plant's defense systems. *Trichoderma* spp. are ironic and significant foundations of secondary metabolites (SMs) used for organic controller of plant diseases. The crucial biocontrol tactics that *Trichoderma* grows in direct battle through mycological pathogens are mycoparasitism, struggle and antibiosis.

2.1. Mycoparasitism

Mycoparasitism contains the thru attack of one fungal kind on alternative and is among the most substantial antagonistic mechanisms uttered by *Trichoderma* spp. About 75 *Trichoderma* species with mycoparasitic latent have been hitherto informed. There are several explorations which specify that many straining of *Trichoderma* bout and smash plant morbific fungi, e.g., *Rhizoctonia solani, Alternaria alternata, Sclerotinia sclerotiorum, Fusarium* spp., *Botrytis cinerea* and *Pythium* spp. (Harman et al., 2004).

2.2. Competition

The inadequate accessibility of and competition for nutrients chief to the ordinary management of fungal groups and phytopathogen development. Competition for micro- and macronutrients such as C, N and Fe plays aessential role during interactions of beneficial and harmful fungi and is coupled with the biocontrol systems (Vinale et al., 2008). It has been well recognized that *Trichoderma* kind competition for nutrients, living niches or infection parts with pathogens in plant rhizosphere (Ahluwalia et al., 2015). *Trichoderma* shows a better capability to mobilize and absorb nutrients from the soil in dissimilarity to other rhizospheric microbes; therefore, the control management of some pathogens (e.g., *B. cinerea*) by using *Trichoderma* includes the synchronization of plentiful approaches, such as the rivalry for nutrients, which is measured amid the most important. The active utilization of nutrients be contingent upon the ability of *Trichoderma* spp. to get vigor copied from the metabolism of carbohydrates like cellulose, chitin, glucan and glucose, which are often present in the mycelial environment. Root exudates and the rhizosphere are mainly ironic in nutrients like carbohydrates, amino acids, organic acids, vitamins, Fe, etc., but the struggle for C between *Trichoderma* and pathogenic fungi like *Rhizoctonia solani, F. oxysporum*, etc. was measured to be most extraordinary (Alabouvette et al., 2009). As likened to other microbes in the soil, the competent mobilization of immobile nutrients and their use delivers supremacy to *Trichoderma*. For this drive, *Trichoderma* induces the reduction of soil pH via the biosynthesis and release of organic acids like gluconic, citric and fumaric. Besides, these organic acids moreaid the solubilization of micronutrients and mineral cations such as phosphates, Fe, Mn and Mg (Sarrocco et al., 2009).

2.3. Antibiosis

Antibiosis is the technique by which diffusible low-molecular heft amalgams interrelate and decrease the advance of other microbes. Mainly, antibiosis emphases on the creation of subordinate metabolites, which show an inhibitory or noxious product on a parasitic fungus. More than 180 subordinate metabolites representative discrete curricula of chemical products have been isolated after fungal species fitting to genus *Trichoderma* (Masi et al., 2018).

3. Methods of Applications

Distribution of bioagents with suitable and appropriate delivery approaches, at the cite of action is important thought to deliver proper active propagules and colony forming units (CFU)/ gram of the product. The following transference systems have been developed for the use of *Trichoderma* spp.

3.1. Seed Treatment

Trichoderma products having least $2x10^6$cfu/ g @ of 5-10 g/ kg seed at the time of propagating.

3.2. Nursery Soil Treatment

Apply *Trichoderma* augmented FYM (1kg *Trichoderma* in 100kg of FYM) in the nursery soils @ 100g per m^2 at the time of sowing or drenching *Trichoderma* @10g mixed with 100g of well decomposed FYM per litre of water at the disease beginning.

3.3. Soil Treatment

Apply talc-based formulation of *Trichoderma* (1%) @ 5Kg/ hectare in the soil after plowing of sunhemp grown for green manuring.

3.4. Treatment of Planting Materials

Convinced vegetable, floriculture crops, medicinal, aromatic and fruit plants etc. are elevated complete planting materials viz. tubers, corm, cuttings, roots, etc. These planting materials can be treated with *Trichoderma* formulations alone @ 1 or 1.5kg per 10 lit. of water, or along with 100g of well decomposed FYM/ liter of water by dipping them for 15 minutes, then dried in shade and used for establishing.

3.5. Seedlings Root DIP

The seedlings of vegetables can be treated by talc-based *Trichoderma* formulations @ 1 or 1.5kg per 10 lit. of water and dipping the roots of seedlings for 15 minutes, at the time of transplant.

3.6. Seed Biopriming

It is the topical and beneficial technique of seed treatment with bioagents for the managing of the diseases in crop plants (Pandey, 2017; Jaisani et al., 2016). In seed biopriming improve the preliminary step of plant development by augmented seed germination and provide protection before seedling rise.

3.7. Plant Treatment

Giving plants near collar region with *Trichoderma* @ 10g / litre of water; or sideways with sand @ 5 Kg /ha can help to achieve the stem or collar rot of crops.

3.8. Through Drip Irrigation Water

T. harzianum and *T. viride* completed decent to succeed white mold (*S. sclerotiorum*) of tomato, with three applications @ 1.0 L ha^{-1} of formulation having 1×10^9 CFU mL^{-1} starting from 30 days after transplanting and at 10-day intervals, applied through drip irrigation (Aguiar et al., 2014).

3.9. Through Hydroponic Systems

Trichoderma spp. have been created to perform well in the management of the diseases of crops grown in soil less medium ie. Peat saw dust, rockwood, sand, etc. *Trichoderma* spp. has efficiently managed wilt of tomato, damping-off of cucumber and other crops due to *Pythium* sp. and *Rhizoctonia* sp. (Paulitz, 1997).

4. *Trichoderma* Spp. For Disease Managing in Horticultural Crops

Trichoderma species also have positive effects on different plant growth parameters, which help in increasing the proliferation of secondary roots, leaf area, shoot length, dry weight, and crop yield (Hermosa et al., 2013; Mukherjee et al., 2013). *Trichoderma* spp. are also known as plant growth-promoting fungi (PGPF) as they can secrete phosphate-solubilizing enzymes, siderophores, and a varied choice of plant growth promoting organic molecules, *i.e.,*phytohormones (Doni et al., 2013). Several Trichoderma species, *viz., T. asperelloides, T. hamatum, T. harzianum,* and *T. viride*, showed high qualitative and quantitative chitinase activity and recorded notable antagonistic action against *M. incognita* infecting tomato. For example, the Trichoderma species *T. asperelloides* successfully abridged the root galling to up to 92.81% and final nematode population to up to 97.81% (Sayed et al., 2019). The two antagonistic fungi like *Trichoderma harzianum* and *T. viride* caused remarkable reductions in nematode-related parameters such as galls, egg masses, and fecundity and reproduction factors of *M. incognita* (Mukhtar, 2018). Bairwa et al. (2017) described that neem cake (5 t/ ha) in combination with *T. viride* (5kg/ ha) produced significantly maximum yield of potato and reduced the potato cyst nematodes multiplication ratio.Many *Trichoderma* spp. have also been used to guard fruits and vegetables of commercial worth through post-harvest storage. Mortuza and Ilag, (1999) engaged 10 isolates of *T. harzianum* and *T. viride* against *Lasiodiplodia theobromae* (fruit rot pathogen of banana). Similarly, Batta (2004) useful the invert-emulsion devising of *T. harzianum* Rifai in opposition to apple blue mold infection to prevent post-harvest decline of fruit. *Trichoderma* spp. are well-recognized fungal antagonists of crop/seed pathogens. Management of *Colletotrichum truncatum*, causing brown blotch of cowpea, has been done via the pre-treatment of seeds in *T. viride* microorganism suspension (Bankole *et al.,* 1996). Muthulakshmi et al. (2010) testified seed treatment with *Trichoderma viride* at 4g/kg of seed + soil application of *T. viride* at 5 kg/ha reduced the rot disease in fenugreek. Early blight of potato caused by *Alternaria alternata* is one of the major diseases causing considerable yield loss. Efficacy of bio-agents was considered and exposed the all four Trichoderma species significantly compact the mycelial growth of the pathogen over the control by dual culture method. Significantly extreme inhibition of mycelial growth was obtained with *T. viride* (57.08%) followed by *T. azospirillum* (50.00%), *T. harzianum* (43.26%) and *T. virens* (37.08%) of the pathogen after 48 hours of incubation. The *in-vitro* evaluation of five different bioagents against *Macrophomina phaseolina*, incitant of root and collar rot of okra, discovered that *Trichoderma viride* was the most effective with maximum growth inhibition (73.06%) followed by *T. harzianum* (68.89%). Out of the eight different treatments experienced in vivo through pot culture studies, seed dealing with *T. viride* mutual with soil delivery of *T. viride* enriched FYM recorded the minimum mortality percentage (26.55%) and highest vigour index (1643).

Table 1: Examples of *Trichoderma* spp. aganist plant pathogens

Sr. No.	Name of disease	Crop	Causal agent	Biocontrol strain	Reference
1	Root rot	Pepper plants (*Capsicum annuum*)	*Rhizoctoniasolani*	*T. harzianum*	Ahmed et al. (2003)
2	Root rot	Egg plant (*Solanum melongena L.*)	*Macrophomina phaseolina*	*T. harzianum, T. polysporum, T. viride.*	Ramezani (2008)
3	Damping off	Cucumber (*Cucumis sativus*)	*Pythium*sp.	*T. harzianum*	Paulitz et al. (1990)
4	Wilt	Tomato (*Solanum lycopersicum*)	*Fusarium oxysporum* f. sp. *lycopersici*	*T. asperellum*	El Komy et al. (2015)
5	Anthracnose and grey mold	Strawberry (*Fragaria ananassa*)	*Colletotrichum acutatum Botrytis cinerea*	*T. hamatum, T. atroviride, T. longibrachiatum*	Freeman et al. (2004)
6	Stem gall	Coriander (*Coriandrum sativum* L.)	*Protomyces macrosporus*	*T. viride*	Khan and Parveen, 2018 and Dabbas et al. 2010.
7	Root Rot	*Papaya (Carica papaya L.)*	*Fusarium solani*	*T. viride*and, *T. harzianum*	*Gupta* et al. *2020.*
8	Damping off	Brinjal (*Solanum melongena* L.)	*P. aphanidermatum*	*T. harzianum*	Nirmalkar et al. (2018)
9	Black-scurf	Potato (*Solanum tuberosum*)	*R. solani*	*T. virens, T. atroviride*	Hicks et al. (2014)

5. *Trichoderma* Spp. as Plant Growth Promoter Agent

Trichoderma species are commonly reported as plant growth promoters. *Trichoderma* spp. significantly suppress the growth of plant pathogenic microorganisms and adjust the rate of plant growth. Recent works have revealed that common plant disease such as root rot disease, damping off, wilt, fruit rot and other plant diseases can be controlled by *Trichoderma* spp. (Begum et al., 2010; El Komy et al., 2015; Yadav et al., 2021). The subordinate metabolites concealed by *Trichoderma* spp. have verified its role in suppressing the growth of pathogenic microorganisms and inspiring the plant growth (Contreras-Cornejo et al., 2015). Besides, the interaction between plant and *Trichoderma* spp. successfully regulates root manner, increase the length of lateral and primary root that result in the efficiency of nutrient uptake by the plant (Naseby, 2000). *Trichoderma* spp. Are multi functional plant symbionts, acting as enhancer of plant growth, and productivity; and as biocontrol agents for plant disease management (Harman, 2011). *Trichoderma* treated plants have shown enhanced growth and yield in numerous crops *viz.* rice, maize, etc. due to production of phytohormones (Chowdappa et al., 2013), cytokinin like molecules (Zeatin) and gibberellins related molecultes (GA3 or GA4) (Kashyap et al., 2017); higher photosynthetic rate (Doni et al., 2014), increased nutrient uptake, and acceptance to abiotic and biotic stress, etc. which helped in enhancing growth parameters and crop growth. The significant optimistic effect of Trichoderma isolates was found on the root weight of pepper seedlings. Besides root weight, some isolates positively affected the shoot weight as well. Root enhancement caused by *T. harzianum* treatment has been also reported for sweet pepper (Idowu et al., 2016) and cucumber plants with 24% of roots dry weight and 45% of root length growth (Yedidia et al., 2001).

6. Conclusion

Trichoderma spp. is a remedy for crucial sorrows of plants due to biotic and abiotic factors. Trichoderm good options for soil, seed and air borne diseases in organic cultivation of horticultural crops through its smart mechanism. Its well known as plant growth promoters through its opportunistic virulent plant symbionts in sustainable farming system. Its proved preventive activities against various methods of applications for crop health management. Its uniue identity for soil borne with mix nematic diseases of horticultural crops. *Trichoderma* spp. enriched with FYM and vermicompost are very beneficial for increased spore population and multiplication for management of diseases in horticultural crops.

References

Aguiar, R.A., Cunha, M.G., Murillo L. Jr. 2014. Management of white mold in processing tomatoes by *Trichoderma* spp. and chemical fungicides applied by drip irrigation. Biological control 74, 1-5.

Ahluwalia, V., Kumar, J., Rana, V.S., Sati, O.P., Walia, S. 2015. Comparative evaluation of two *Trichoderma harzianum* strains for major secondary metabolite production and antifungal activity. Nat. Prod. Res. 29, 914–920.

Ahmed, A.S., Ezziyyani, M., Sánchez, C.P., Candela, M.E. 2003. Effect of chitin on biological control activity of *Bacillus* spp. and *Trichoderma harzianum* against root rot disease in pepper (*Capsicum annuum*) plants. Eur. J. Plant Pathol. 109, 633-637.

Alabouvette, C., Olivain, C., Migheli, Q., Steinberg, C. 2009. Microbiological control of soil-borne phytopathogenic fungi with special emphasis on wilt-inducing *Fusarium oxysporum*. New Phytol. 184, 529–544.

Aravind, T., Brahmbhatt, A. 2018. Management of root and collar rot (*Macrophomina phaseolina* (Tassi) Goid.) of okra (*Abelmoschus esculentus* (L.) Moench) through bioagents, oil cakes and fungicides. Journal of Pharmacognosy and Phytochemistry. 7(4), 631-635.

Bairwa, A., Venkatasalam, E., Umamaheswari, R., Singh, B., Colleg, J. 2017. Effect of cultural practices on potato cyst nematode population dynamics and potato tuber yield. Indian J Hortic. 74(1), 91–96.

Bankole, S.A., Adebanjo, A. 1996. Biocontrol of brown blotch of cowpea caused by *Colletotrichum truncatum* with *Trichoderma viride*. Crop Prot. 15, 633–636.

Batta, Y.A. 2004. Effect of treatment with *Trichoderma harzianum* Rifai formulated in invert emulsion on postharvest decay of apple blue mold. Int. J. Food microbiol. 96, 281–288.

Begum, M.F., Rahman, M.A., Alam, M.F. 2010. Biological control of Alternaria fruit rot of chili by *Trichoderma* species under field conditions. Myco. 38, 113-117.

Chowdappa, P., Mohan Kumar, S.P., Jyothi Lakshmi, M., Upreti, K.K. 2013. Growth stimulation and induction of systemic resistance in tomato against early and late blight by *Bacillus subtilis* OTPB1 or *Trichoderma harzianum* OTPB3. Biological Control. 65(1), 109–117. ·

Contreras-Cornejo, H.A., Macías-Rodríguez, L., Vergara, A.G., López-Bucio,J. 2015. *Trichoderma* modulates stomatal aperture and leaf transpiration through an abscisic acid-dependent mechanism in Arabidopsis. J. Plant Growth Regul. 34, 425-432.

Dabbas, M.R., Singh, D.P., Prakash, H.G. 2010. Management of stem gall of coriander through IDPM practice. Int. J. Plant Prot. 2(2), 255-257.

De Meyer, G., Bigirimana, J., Elad, Y., Hofte, M. 1998. Induced systemic resistance in *Trichoderma harzianum* T39 biocontrol of *Botrytis cinerea*. Eur. J. Plant Pathol. 104, 279- 286.

Doni, F., Al-Shorgani, N., Tibin, E., Abuelhassan, N., Anizan, I., Che-Radziah, C. 2013. Microbial involvementin growth of paddy. Curr Res J Biol Sci.; 5(6), 285-290.

El Komy, M.H., Saleh, A.A., Eranthodi, A., Molan, Y.Y. 2015. Characterization of novel *Trichodermaasperellum* isolates to select effective biocontrol agents against tomato Fusarium wilt. Plant Pathol. J., 31, 50-60.

Freeman, S., Minz, D., Kolesnik, I., Barbul, O., Zveibil, A., Maymon, M., Nitzani, Y., Kirshner, B., Rav-David, D., Bilu A., Dag, A. 2004. Trichoderma biocontrol of *Colletotrichumacutatum* and *Botrytis cinerea* and survival in strawberry. Eur. J. Plant Pathol. 110, 361-370.

Gupta, P., Chitara, M., Singh, S. 2020. Antagonism of native and commercial *Trichoderma* spp. against *Fusarium solani* isolates causing root rot of papaya (*Carica papaya* L.). Int. J. Curr. Microbiol. App. Sci. 9(02), 1260-1269.

Hajek, A.E., Eilenberg, J. 2018. Natural enemies: an introduction to biological control (Second Ed), Cambridge University Press, New York.

Harman, G.E. 2011.*Trichoderma* not just for biocontrol anymore, Phytoparasitica. 39, 103–108.

Harman, G.E., Howell, C.R., Viterbo, A., Chet, I., Lorito, M. 2004. *Trichoderma* species—opportunistic, avirulent plant symbionts. Nat. Rev. Microbiol. 2, 43–56.

Hermosa, R., Rubio, M., Cardoza, R., Nicolas, C., Monte, E., Gutierrez, S. 2013. The contribution of *Trichoderma* to balancing the costs of plant growth and defense. Int. Microbiol. 16, 69–80.

Hicks, E., Bienkowski, D., Braithwaite, M., Mclean, K., Falloon, R. and Stewart, A. 2014. *Trichoderma* strains suppress Rhizoctonia diseases and promote growth of potato. Phytopathologia Mediterranea. 53(3), 502-514.

Idowu, O., Oni, A., Salami, A. 2016. The interactive effects of three Trichoderma species and damping-off causative pathogen *Pythium aphanidermatum* on emergence indices, infection incidence and growth performance of sweet pepper. Int J Recent Sci Res. 7(4), 339-347.

Jaisani, P., Prajapati, H. N., Yadav, D.L. and Pandey, R. N. 2016. Seed Biopriming and *Trichoderma* enriched FYM based soil application in management of chickpea (*Cicer arietinum* L.) wilt complex. Journal of Pure & Applied Microbiology. 10(3), 2453-2460.

Kashyap, P.L., Kumar, S., Srivastava, A. 2017. Nano diagnostics for plant pathogens. Environ Chem Lett. 15, 7-13.

Khan, R., Manzoor and Ghazala, P. 2018. Supplementing biocontrol agents with botanicals improved growth and yield of coriander (*Coriandrum sativum* L.) infected with *Protomyces macrosporus* Unger. Current Plant Biology. 15, 44-50.

Masi, M., Nocera, P., Reveglia, P., Cimmino, A., Evidente, A. 2018. Fungal metabolites antagonists towards plant pests and human pathogens: Structure-activity relationship studies. Molecules. 23, 834.

Mauthulkshmi, Palnisamy, S., Kandasamy, R., Kandasamy, S., Ramasamy. 2010. Eco-friendly approaches for the management of fenugreek root rot. Archives of Phytopathology and Plant Protection. 43, 1268-1272.

Meena, P.K., Yadav, D.L., Meena, C.B., Patidar, B.K., Pratap Singh. 2020. Efficacy of Bio-agents and Media against the Early Blight of Potato. Int.J.Curr.Microbiol.App.Sci. 9(7), 2208-2213.

Mortuza, M.G. and Ilag, L.L. 1999. Potential for biocontrol of *Lasiodiplodia theobromae* (Pat.) griff. & maubl in banana fruits by Trichoderma species. Biol. Control. 15, 235–240.

Mukherjee, P., Horwitz, B., Herrera-Estrella, A., Schmoll, M., Kenerley, C. 2013. *Trichoderma* research in the genome era. Annu Rev Phytopathol. 51, 105–129.

Mukhtar, T. 2018. Management of root-knot nematode, *Meloidogyne incognita*, in tomato with two *Trichoderma* species. Pak J Zool. 50(4), 1589–1592.

Naseby, D., Pascual, J., Lynch, J. 2000. Effect of biocontrol strains of *Trichoderma* on plant growth, *Pythium ultimum* populations, soil microbial communities and soil enzyme activities. J. Appl. Microbiol., 88, 161-169.

Nirmalkar, V.K., Tiwari, R.K.S., Singh, S. 2018. Efficacy of bio-agents against damping off in solanaceous crops under nursery conditions. Internat. J. Plant Protec. 11(1), 1-9.

Pandey, R. N. 2017. Seed bio-priming in the management of seed- and soil-borne diseases. Indian Phytopath, 70 (2), 164-168.

Pandey, R.N., Jaisani, P., Yadav, D.L. 2021. *Trichoderma* spp. in the management of stresses in plants and rural prosperity. Indian Phytopathology. 74, 453–467.

Paulitz, I.C. 1997. Biological control of root pathogens in soilless and hydroponic systems. Hort Science. 32, 193-196.

Priyanka K., Meena, P., Yadav, D.L., Meena, C.P., Patidar, B. K., Pratap Singh. 2020. Efficacy of Bio-agents and Media against the Early Blight of Potato. Int. J. Curr. Microbiol. App. Sci. 9(07), 2208-2213.

Ramezani, H. 2008. Biological control of root-rot of eggplant caused by *Macrophomina phaseolina.* Am. Eurasian J. Agric. Environ. Sci., 4, 218-220.

Sarrocco, S., Guidi, L., Fambrini, S., Degl'Innocenti, E., Vannacci, G. 2009. Competition for cellulose exploitation between *Rhizoctonia solani* and two Trichoderma isolates in the decomposition of wheat straw. J. Plant Pathol. 91, 331–338.

Sayed, M., Abdel-rahman, T., Ragab, A., Abdel latif, A. 2019. Biocontrol of root-knot nematode *Meloidogyne incognita* by chitinolytic *Trichoderma* spp. Egypt J. Agronematol. 18(1), 30–47.

Vinale, F., Sivasithamparam, K., Ghisalberti, E.L., Marra, R., Woo, S.L., Lorito, M. 2008. Trichoderma–plant–pathogen interactions. Soil Biol. Biochem. 40, 1–10.

Woo, S.L., Ruocco, M., Vinale, F., Nigro, M., Marra, R., Lombardi, N., Lorito, M. 2014. Trichoderma-based products and their widespread use in agriculture Open. Mycol. J., 8, 71-126.

Yadav, D.L., Jaisani, P., Kumar, V. 2021. Damping-off: A Dreadful Nursery Disease and Its Ecofriendly Management. In: Advances in Genetics and Plant Breeding. Akinik Publications, New Delhi; 12, 13-28.

Yedidia, I., Srivastva, A., Kapulnik, Y., Chet, I. 2001. Effect of *Trichoderma harzianum* on microelement concentrations and increased growth of cucumber plants. Plant Soil. 235(2), 235-42.

25

Recent Trends for Stem Gall Disease Management in Coriander

***D.L. Yadav*[1]*, Preeti Verma*[1]*, Pratap Singh*[1]*, Chaman K Jadon*[1] *and Sharda Choudhary*[2]**

[1]*Agriculture University, Kota, Rajasthan, India*
[2] *ICAR-National Research Centre on Seed Spices, Ajmer , Rajasthan India*

1. Introduction

Coriander is a vital spice crop with a grave role in delicious food test. It is a tropical crop and cultivated as a major in *Rabi* season in chill free extents through the blossoming phase. Especially, all portions of the plant are comestible, but the fresh shrubberies and the desiccated seeds are the parts conventionally used in culinary. Approximately of the diverse forms of coriander are greenery, blossom and capsules. Coriander leaves are mainly ironic in vitamin A, vitamin C and vitamin K, with moderate pleased of nutritional minerals. Though seeds usually have inferior happy of vitamins, they do widely provide significant quantities of nutritional fibre, calcium, selenium, iron, magnesium and manganese. Mostly, coriander plant productivity is around one and three-fourth tonnes per acre of seed. The root is crushed and used as a condiment in useable form. A crucial oil from the seed is used as a food flavouring agent for dietary. As a substance of statistic, this mixture is used in daily cookery in Indian pantries. The fruit of the coriander vegetable covers two seeds which, when desiccated, are the helpings used as the dried spice. Coriander seeds are obtainable as whole or in ground/ powder form for marketable. Madhya Pradesh formed the major capacity of coriander seeds in financial year 2021 crossways India. This amounts to more than 394 thousand metric tonnes spread across 293 thousand hectares. During that time period, the country's annual output of coriander seeds was around 822 thousand metric tonnes. Taking these parameters into consideration, it was discovered that the agro-ecological condition of this region, as well as the soil system atmosphere of Baran, Bundi, Jhalawar, and Kota, are most suited for coriander seed growing. Coriander suffers from a variety of fungal diseases, the most serious of which is stem gall disease that is caused by the fungus *Protomyces macrosporus* (Bilgrimi and Rizvi 1991).

The disease is frequently raised as tumor of coriander. The disease establishes in the form of tumor like swellings on all the above crushed plant parts counting seeds. The main aim of this chapter is to highlight symptoms, favourable condition of disease, behaviour against chemicals as well as bio agent for management and new resistant genotypes against stem gall disease of coriander.

2. Geographical Origin, Distribution and Losses

The disease has global distribution. It is prevalent in coriander growing regions. Stem gall disease was reported first time in India (Sydow and Butler, 1911). Later, Gupta (1954) reported this disease in northern part of Madhya Pradesh and contiguous regions of Rajasthan and Uttar Pradesh. Singh et al. (1984) conducted a survey of coriander fields in seven districts of Uttar Pradesh *viz*. Muzaffarnagar, Varanasi, Kanpur, Agra, Balia, Allahabad, and Jhansi and reported disease intensity between 7.5 to 24.4 % with extreme at Muzaffarnagar and least in Jhansi. Similarly, Paul (1992) inspected coriander fields in Himachal Pradesh and reported the range of disease intensity between 10.76 to 60 %, 0.32 to 7.3% and 0.2 to 2.35% in the district of Kangra, Mandi and Kullu, respectively. Bag (1998) reported the disease first time in Arunachal Pradesh. Later, Verma et al. (2014) noted its incidence from different sites of Jabalpur (Madhya Pradesh) and reported infection percentage between 5-9% with a mean of 3.6%.

Gupta et al. (1954) reported 23% disease strength of stem gall in coriander fields in Morar block (Gwalior, M.P.) resultant in 15% yield loss. Mundkur (1959) recorded that due to incidence of stem gall disease, fruits formation was almost nil. Gupta and Sinha (1973) reported up to 1.3% seed yield loss due to 3.8 to 21% disease intensity under field circumstances on eighteen cultivars of coriander. Singh et al. (1984) reported up to 33.8% losses in predictable yield due to 7.5 to 24.4% disease intensity in Uttar Pradesh. Lakra (1999) valued up to 50% seed yield loss due to 22 to 66& disease intensity in a field experimental showed in stem gall sick plot at Hisar (Haryana). Khan and Praveen (2016) described 6.04 to 27.87% yield loss due to 7 to 59% disease passion in Aligarh (Uttar Pradesh). Malhotra et al. (2016) also observed a mean loss of 15 % in seed yield with 23% infected plants. Mishra and Pandey (2017) reported 33-36% yield harm owing to this disease. Sandeep et al. (2019) reported 20.8% yield loss due to the disease in eastern part of Uttar Pradesh. 50% loss in yield happens owing to stem gall disease in Hadoti and contiguous zones in Rajasthan (Verma et al., 2019).

3. Identification and Isolation of *Protomyces macrosporus*

P. macrosporus was isolated from sick coriander plants that showed obvious galling indications. Surface sterilisation with 0.1 percent sodium hypochlorite solution was followed by washing with double distilled water. The damaged plant parts were chopped into small pieces and aseptically placed in petri dishes with potato dextrose agar (PDA) diet containing with yeast extract for 15 days at 25 °C. The pathogen was identified as *P. macrosporus* based on microscopic morphological characteristics (Pagvi and Mukhopadhyay, 1970). Preece and Hick (2001) described that *Protomyces macrosporus* was the associate of the Protomycetaceae and presently *Protomyces macrosporus* has been positioned in Kingdom: Fungi, Division: Ascomycota Class: Taphrinomycetes, Order: Taphrinales, Family: Protomycetaceae, Genus: Protomyces and Species: macrosporus (Krik et al., 2008). The hyphae are intercellular, densely septate, and wide, with irregular splitting. Dispersed cells in the hyphae inflate and form ellipsoidal or globose bodies, which eventually evolve into chlamydospores. As the chlamydospores grew, they were surrounded by a thick, hyaline, three-layered wall of 50 to 60m in circumference. Although the fungus's mycelium originates solely in tumours, the fungus's latent spores cause systemic infection.

3.1. Symptoms

The contagion of stem galls causes in initial phase of seed germination but the signs seem considerable advanced (Shrivastava, 1955). Pavgi and Mukhopadhyay (1972) reported that the infection develops systemic in the host stem earlier or during pre-flowering period, encouraging organogenic variations in flowers and fruits toward remain hypertrophied and hyper plasied. Chauhan et al. (1981) reported that *Protomyces macrosporus* cause variable degree of pollen sterility at different growth-stages. The plants immunized at the floral bud initiation stage remained entirely sterile. Redfern and Shirley (2002) reported that the disease usually seems in early mainspring when the significantly partial leaves first appear. On veins, petioles and mid-ribs, elongated blisters like gall appears as glowing yellow-white swellings. Depending on the stage of infection, diseased seeds become hypertrophied, lowering yield and quality of produce (Khan and Parveen, 2016). The top phase of the crops is to be more sensitive to the disease's harm. The illness appears as tumor-like bulges on the stem, leaf, peduncle, flowers, and seeds (Verma et al., 2019).

3.2. Disease Cycle and Epidemiology

The disease is seed-born. Key inoculant of the disease is allowable to the field through seed haggard from a diseased crop, fragments and/or untreated seed diverse through fruit galls (Leharwan and Gupta, 2019). The hibernated thick-walled chlamydospores resist unfavourable dry environment and perenate summer, which may prolong for up to several years in absence of host. Chlamydospores germinate during the early winter in the host. The disease is favoured by prolong cold temperature (8-22°C and 65% relative humidity) of winter season. Mukhopadhyay and Pavgi (1971) discovered that pathogen primary inoculum was transferred to the field via seed extracted from diseased plants, crop rubble, and uncleaned seed that varied with seed galls. The infection lasted season after season in the chlamydospore phase, and the spore sprouts to ensure fresh contagion in the following period. Mukhopadhyay and Pavgi (1971) discovered that the pathogen's major inoculum was transmitted to the field via seed collected from diseased plants, crop rubble, and unclean seed, all of which differed in terms of seed galls. The pathogen lasted in the chlamydospore phase from season to season and later sprouted spores to ensure repeated infection in the following time. The chlamydospores from fruit galls are induced to develop cool weather condition mainly in January in Rajasthan and nearby states on the soil surface. During January-February, the seedlings infected through the endospores and subsequent blastospores. Tripathi (2003) observed that the spread of stem gall diseases was fewer in initial and late sowing (September 16 and December 1) compared to other sowing dates at Gwaliar (Madhya Pradesh). Verma et al. (2017) observed higher coriander seed yield when the crop was sown between 30th October to 15th November in Zone V of Rajasthan, however the stem gall incidence was less in late sown crop.

3.3. Control Measures

The produce victims by stem gall disease in coriander can be minimalized through various practices such as cultural performs, biological, chemical control and resistant cultivars.

3.3.1. Cultural Control

According to Butler (1918), disease severity can be reduced by removing and obliterating gall-bearing plant parts. Some cultural practise like use of healthy seeds, field sanitation, soil solarization, weed and diseased plants eradication from field are very used full in controlling the occurrence of stem gall in coriander. The disease can be controlled by regulating the sowing timing as per available recommended conditions. Early sowing is better for control of stem gall in coriander (Lakra, 2000, Tripathi, 2003).

3.3.2. Biological Control

Management concluded bio control mediators have been extensively used in agricultural crop disease management due to its well-known biological device mechanism. Success to management of plant diseases, plant growth, decomposition process and bioremediation. A wettable powder is the most frequent preparation, followed by granules. Generally, Trichoderma are useful to seed or propagation substance at the time of planting, and then the secondary application is during plant development, proposing an affordable and ecologically friendly disease management strategy. Dabbas et al. (2009) reported an enhancement in yield and reduction in stem gall intensity of coriander with seed and soil behaviour of *T. viride*. According to Vinale et al., Trichoderma spp. produce plant growth stimulating effects and secondary metabolites that may function as auxin-like chemicals, resulting in plant growth development (2008). The results revealed that using biocontrol agents/botanicals alone or in combination reduced stem gall intensity significantly (20.33–84.82 percent) as compared to the control. Amongst the verified biocontrol agents, *T. viride* was create greater in attractive the growth and yield as well as decrease in stem gall intensity (44.37%). Among numerous mixtures tested *T. viride* + *Melia azedarach* and *T. viride* + *E. globulus* caused healthier rise in plant growth and yield characters both in pathogen inoculated as well as uninoculated plants noted by Khan and Parveen, (2018). For the management of coriander stem gall, Dabbas et al. (2009) recommended seed treatment with T. viride 4 g/kg seed + soil treatment with T. viride 2 kg/ha. Kumar et al. (2014) found that seed treatment with bio-fungicides such as T. viride and Pseudomonas @ (0.4 percent) and foliar spray with Trichoderma spp. and Pseudomonas spp. @ (0.4 percent) after 40, 60, and 75 days of sowing reduced the percentage occurrence of disease by 24.50 and 32.65, respectively. Biological management effectively suppresses coriander stem gall. Trichoderma viride and Pseudomonas fluorescens seed, soil, and foliar spray treatments 40, 60, and 75 days after planting reduce disease by up to 32%. (Leharwan and Gupta, 2019).

3.3.3. Chemical Control

Nene et al. (1966) recommended seed treatment with thiram @ 0.25 kg per 100 kg of seed for control of stem gall disease. Another approach is see and soil treatment with Thiram @ 8kg/acre in the soil. Gall growth can be reduced or controlled by using ceresan on seed and omnamycin as soil application (Pavgi and Mukhopadhyay, 1970). Eight different antifungal and their combinations were recommended by Paul (1992). Lakra (1997) stated that seed treatment with Thiram @ 2g/kg seed followed by three sprays of calixin @ 0.1% at diverse growth phases of crop, was best in controlling the intensity of stem gall disease. Lakra (2000) preserved seeds with captan, carbendazim, thiram and vitavax (carboxin) @ 2 and 4 g a.i. /kg seed before sowing. Carbendazim (0.2%), carboxin (0.1%), copper oxychloride (0.3%), mancozeb (0.25%), streptomycin (500ppm) and tridemorph (0.1%) were applied to foliage after 30, 45, 60 and 75 days of sowing. Control of the disease by seed treatments with thiram, vitavax, Bavistin and captan at 4 g/kg seed was 53.5, 50.0, 35.7 and 32.1 %, respectively, and 66.6, 62.5 and 58.3 % disease control was gained with streptomycin, carboxin and tridemorph, respectively, at 30 days after sowing. Shukla et al. (2006) found the combination of seed treatment with carbendazim (1g/kg) and foliar spray with Bayleton (0.04%) effective that reduced up to 57.66 % disease severity followed by seed treatment with carbendazim + foliar spray through karathane (0.04%) and foliar spray with Bayleton (0.4%). Kumar et al. (2014) testified least disease incidence of 44.16 % by seed treatment with hexaconazole (0.2%) and foliar spray of hexaconazole (0.2%) after 40, 60 and 75 days of sowing followed by disease incidence of 48.33 by seed treatment with propiconazole (0.2%) and foliar spray of propiconazole (0.2%) after 40, 60 and 75 days after sowing. Verma et al. (2017) tested four fungicides against stem gall disease and observed that foliar spray of 0.2% hexaconazole is active in managing of the disease.

3.3.4. Resistant Cultivars

To avoid yield and quality loss in any crop occurred due to diseases, the greatest preventive mode is to develop resistant varieties to the pathogens causing diseases. Use of resistant varieties against stem gall for cultivation is the calmest, possible, and eco-friendly approach to battle the disease and attain developed yields. Mishra and Pandey (2015) evaluated 27 genotypes against to stem gall disease and found K. Selection as resistant. Kumar et al. (2016) evaluated 102 germplasms against stem gall disease under accepted field condition and detected that out of 102 germplasm of coriander, 35 germplasm lines displayed moderately susceptible reaction to stem gall disease whereas, 67 germplasm lines were observed susceptible to stem gall disease of coriander. Khan and Parveen (2016) evaluated 27 varieties of coriander against stem gall disease and observed that out of 27 varieties evaluated, none of the variety was found free from stem gall infection. Four varieties viz., UD-125, UD-317, UD-749 and RCr-480 were graded as resistant against the disease. Malhotra et al. (2016) found ACr-1 variety resistant against stem gall disease. Ajmer Coriander-1 (ACr-1) is a new variety developed by ICAR-NRCSS, Ajmer. Some studies by Choudhry et al. (2019, 2020, 2021) showed that ACR01 is promising tolerance variety for stem gall disease. Vijaykumar et al. (2021) evaluated 28 genotypes for resistance against stem gall disease and observed that nobody of the genotypes was completely free from disease, while one genotype *viz.*, ICS-4 found resistant, eleven genotypes found moderately resistant and sixteen genotypes found moderately susceptible against the disease. Lately, Ajmer Coriander-2 (ACr-2) is a new variety developed by ICAR-NRCSS, Ajmer. To assess the genetic countenance of disease resistance-relevant genes in prevalently grown coriander cultivars in India such as ACr-1, Rajendra Swathi, ACr-2, CO-2 and CS-6 were used for LRR, GDSL, ANK, USP and PDR gene appearance using Real Time PCR along with 18S housekeeping gene as internal control for the standardization. Result exposed the different expression decoration of genes among the cultivars verified. Highest expression was exposed in cultivar AgCr-1 followed by Pant Haritma, Hisar Sugandh and ACr-1, and smallest expression in ACr-2, Hisar Anand, Rajendra Swathi, CS-6 and CO-2. Domain analysis revealed the genes' conserved domain relevance. This is the first study to look at the expression of stem gall resistance genes in coriander. The discovered genes play a dormant role in coriander and will be used more extensively in crop improvement programmes (Choudhary et al., 2021).

4. Conclusion

Coriander (*Coriandrum sativum* L.) is a well-known spice and aromatic crop that is grown all over the world. Stem gall infection is emerging as one of the key constraints for its leaf and fruit quality utilised for consumption, as well as negatively impacting productivity. The connecting fungus perennates in the form of chlamydospores in soil and in seed quantifiable for which soil solarization and seed treatment with effective fungicides or with bio-agents is passable. The flowering period seems to be the most dangerous for contagion; hence, antifungal behaviours essential tic earlier flowering. Significantly yield loss under South Eastern part of Rajasthan was observed under favourable environment conditions. *Trichoderma viride* and *Pseudomonas fluorescens* are good for management of this disease for organic cultivation of coriander. Triazole fungicide viz., hexaconazole and propiconazole effectively managed as seed treatment as well as foliar sprays of stem gall in coriander. Ajmer Coriander-1 & 2 (ACr-1& 2) are resistant varieties developed by ICAR-NRCSS, Ajmer.

References

Bag, T.K. 1998. Stem gall of coriander- a new record from Arunachal Pradesh. Indian J. Hill Farming, 11:1 -2, 119.

Bilgrimi, K.S., Rizvi, M.A. 1991. Fungi of India-Lists and references. Today and Tomorrow's Printers and Publishers, New Delhi p: 798.

Butler, E.J. 1918. Fungi and Diseases in Plants. Thacker Spink and Co., Calcutta. P. 547

Chauhan, S.V.S., Srivastava, J.N., Kinoshita, T. 1981. Histological and histochemical changes in the anthers of some diseased vegetable crops. Journal of the Faculty of Agriculture, Hokkaido University, 60, 152-157

Choudhary, S., Naika M.B.N., Meena R.D. 2020.Identification and expression analysis for candidate genes associated with stem gall disease in coriander (*Coriandrum sativum* L.) cultivars. Molecular Biology reports. 47, 5403–5409. https://doi.org/10.1007/s11033-020-05630-0.

Choudhary, S., Naika M.B.N., Meena R.D. 2021. Development and characterization of genic SSR FDM for stem gall disease resistance in coriander (*Coriandrum sativum* L.) and its cross species transferability. Molecular Biology Reports. https://doi.org/10.1007/s11033-021-06396-9.

Choudhary, S., Naika, M.B.N., Meena, R.D. 2021. Development and characterization of genic SSR-FDM for stem gall disease resistance in coriander (*Coriandrum sativum* L.) and its cross species transferability. 48(5), 3963-3970.

Choudhary, S., Naika, M.B.N., Sharma, R., Meena, R.D., Singh, R., Lal, G. 2019. Transcriptome profiling of coriander: a dual purpose crop unravels stem gall resistance genes. J. Genet. 98, 1–11.

Dabbas, M.R., Singh, D.P., Prakash, H.G. 2009. Management of stem gall of coriander through IDM practice. Intern. J. Pl. Protec. 2, 255-257.

Gupta, J.S. 1954. Disease appraisal of stem gall of coriander. Indian Phytopath. 7, 53-60.

Gupta, R.N., Sinha, S. 1973. Varietal field trials in the control of stem gall disease of coriander. *Indian Phytopath.* 26:337-340.

Khan, M.R., Parveen, G. 2016. Screening of some recently developed coriander varieties against stem gall disease caused by *Protomyces macrosporus*. J. Pl. Path. Microbio. 7(8), 1-3

Krik, P.M., Cannon, P.F., Minter, D.W., Stalpers, J.A. 2008. Ainsworth & Bisby's Dictionary of the Fungi, 10th Edition. CABI Europe-UK: 771.

Kumar, G., Yadav, S.K., Patel, J.S., Sarkar, A., Awasthi, L.P. 2014. Management of stem gall diseases in coriander using Pseudomonas and Trichoderma (bioagents) and fungicides. J. Pure Appl. Microbio. 8, 1-4.

Kumar, S., Chaudhary, V.P., Singh, V., Kumar, K. 2019. Biochemical changes in coriander plant parts infected by stem gall disease caused by *Protomyces macrosporus*. International Journal of Chemical Studies. 7(1), 2196- 2199.

Lakra, B.S. 1997. Stem gall disease of coriander-an introspection and strategies for control. J. Mycol. P1. Pathol. 28, pp. 98.

Lakra, B.S. 1999. Assessment of losses due to stem gall of coriander. Pl. Dis. Res. 14, 85-87.

Lakra, B.S. 2000. Management of stem gall of coriander (*Coriandrum sativum* L.) incited by *Protomyces macrosporus*. *Indian J. Agric. Sci.* 70, 338-340.

Leharwan, M., Gupta, M. 2019. Stem Gall of Coriander: A Review. Agricultural Reviews. 40(2), 121-128.

Malhotra, S.K., Kakani, R.K., Sharma, Y.K., Singh, D.K. 2016. Ajmer coriander- 1(NRCSS, ACR-1) resistant to stem gal disease- an innovative farming technology. Indian J. Arecanut, Spices Medic. Pl. 18, 1-7.

Mishra, R.S., Pandey, V.P. 2015. Evaluation of varieties of coriander (*Coriandrum sativum* L.) for resistance to stem gall disease and seed yield. Current Advances in Agriculturul Sciences. 7(2), 151-153.

Mishra, R.S., Pandey, V.P. 2017. Evaluation of coriander (*Coriandrum sativum* L.) genotypes for resistance to stem gall disease and seed yield. J. Spices Aromatic Crops. 26(1), 33-36.

Mukhopadhyay, A.N., Pavgi, M.S. 1971. Environment in relation infection of coriander by *Protomyces macrosporus* Unger. Ann. Phytopath. Soc. Japan. 37, 215-219.

Mundkur, B.B. 1959. *Fungi and Plant Diseases.* MacMillan and Co., Ltd., London. pp, 107-109.

Nene, Y.L., Siddiqui, I.A., Kharbanda, P.D. 1966. Control of stem gall of coriander by fungicides. Mycopathologia Mycologia Applicata. 29,142-144.

Pagvi, M.S., Mukhopadhyay, A.N. 1970. Cytology of chlamydospore germination in *protomyces macrosporus*. Cytologia. 35, 359-367

Paul, Y.S. 1992. Studies on seed-borne mycoflora of coriander with special reference to stem gall in Himachal Pradesh. Pl. Dis. Res. 7:19-23.

Pavgi, M.S., Mukhopadhyay, A.N. 1970. Cytology of chlamydospore germination in *Protomyces macrosporus* Unger. Cytologia. 35, 359-367.

Preece, T.F., Hick, A.J. 2001. An introduction to the Protomycetales: *Burenia inundata* on *Apium nodiflorum* and *Protomyces macrosporus* on *Anthriscus sylvestris*. Mycologist. 15, 118-125.

Redfern, M., Shirley, P. 2002. British plant galls: identification of galls on plants and fungi. Shrewsbury: field studies council, United Kingdom, 222p.

Shrivastava, H.C. 1955. Studies on tumour disease of *Coriandrum sativum* caused by *Protomyces macrosporus* Unger. Proceeding of 42nd Indian Science Congress part. III, 218p.

Shukla, S., Saxena, R.P., Awasthi, L.P., Singh, S. 2006. Comparative efficacy of some commercial fungicides and onion extract for the control of stem gall disease of coriander (*Coriandrum sativum* L.). In 58th Annual Meeting and National Symposium on Emerging Plant Diseases their Diagnosis and Management held at North Bengal University, Siliguri-734013, Jan. 31 – Feb. 2, 2006.

Singh, S.P., Gupta, J.S., Sharma, A.K. 1984. Disease appraisal and crop loss estimates in coriander attacked by *Protomyces macrosporus* Unger in Uttar Pradesh. *Geobios* 11:276-278.

Sydow, H., Butler, E.J. 1911. Fungi Indiae orientalis. Ann. Mycol. 9, 372-421.

Tripathi, A.K. 2003. Effect of sowing dates and weather parameters on stem gall disease of coriander. Indian Phytopath. 56, 191-193.

Tripathi, A.K., Bartaria, A.M., Pandya, R.K., Tripathi, M.L. 2002. Effect of cultural practices on stem gall disease of coriander. Ann. Agric. Res. 23,171-173.

Verma P., Mohan L.Y., Meena., C.B. Yadav, D.L. 2019. Stem Gall Disease of Coriander: Causes and Integrated Management. Int. J. Curr. Microbiol. App. Sci. 8(11), 2573- 2580.

Verma, P., Meena, D.S., Jadon, C., Singh, D.K. 2017. Effect of sowing time and management practices on incidence of stem gall and seed yield of coriander (*Coriandrum sativum* L.). International Journal of Seed Spices. 7(1),79-81.

Verma, R.K., Soni, K.K., Chourasiya, S., Singh, S. 2014. Stem gall of *Boerhaavia diffusa* caused by *Protomyces macrosporus*: A new host record. Indian J. Trop. Biod. 22, 91-94.

Vijaykumar, S., Raj, K., Tehlan, S.K. 2021. Screening of coriander genotypes against stem gall disease caused by *Protomycesmacrosporus.* International Journal of Current Microbiology and Applied Sciences. 10(03), 113-117.

Vinale F., Sivasithamparam, K., Ghilasberti, E.L., Marra, R., Barbeeti, M.J. 2008. *Trichoderma* secondary metabolites in the interactions with plants. Physiol. Mol. Plant. Path. 72, 80–86.

26

Chilli Diseases and Their Management

G.L. Kumawat[1], Jitendra Singh[1], A.K. Yadav[2], R.P. Ghasolia[1] R.K. Bagri[3], P.K.S. Gurjar[4] and D.L. Yadav[5]

[1]*Sri Karan Narendra Agriculture University, Jobner, Rajasthan, India*
[2]*College of Agriculture, SKNAU, Kishangarh Bas, Alwar, Rajasthan, India*
[3]*Rajasthan Agricultural Research Institute, SKNAU, Jaipur, Rajasthan, India*
[4]*Rajmata Vijayaraje Scindia Krishi Vishwa Vidyalaya, Gwalior, Madhya Pradesh, India*
[5]*Agricultural Research Station, Agriculture University, Kota, Rajasthan, India*

1. Introduction

Chilli (*Capsicum annuum* L.) is a major vegetable and spice crop due to its high consumption, nutritional and cash value to farmers and consumers. It is cultivated almost in whole world as consume in the form of green and red ripe fruits. China is the world leader in chilli production followed by India, Mexico, Morocco, Pakistan, Thailand, and Turkey. India is the second largest exporter in the world of dry chilli and its powder. Fruits of chilli are used as fresh salad, cooked vegetable, pickles, souces hot spices in powder form (Parey et al., 2013) and having many medicinal values (Nadkarni, 1927). Fresh green fruits of peppers and red peppers are the major source of vitamin A, C and antioxidant compounds, besides these also found minerals such as calcium, phosphorus, potassium, magnesium, iron, manganese, sodium, zinc and copper. There are about 15 different types of chilli characterized based colour, shape and pungency. The commonly grown species of chilli include *Capsicum annum* and *Capsicum frutescence.* It does not require complex farm practices for successful farming. Chilli crop is suffered by a number of biotic and abiotic diseases which leads to great loss in terms of quality as well as quantity (Majit et al., 2018). The major constraints to chilli production are diseases caused by fungi, bacteria, and viruses. The major diseases of the crop can be listed as.

Table 1: Major diseases of the chilli and causal organism

S.N.	Disease name	Causal organism	Stage of the crop
1	Damping off	*Pythium aphanidermatum*	Seedling
2	Cercospora leaf spot	*Cercospora capsici*	Vegetative
4	Fusarium wilt	*Fusarium solani*	Vegetative
5	Verticillium wilt	*Verticillium sp.*	Vegetative
6	Root rot	*Sclerotium rolfsii*	Vegetative
7	Powdery mildew	*Leveillula taurica*	Vegetative
8	Leaf blight	*Phytophthora nicotianae*	Vegetative
9	Choanephora blight	*Choanephora cucurbitarum*	Vegetative
10	Downy mildew	*Peronospora tabacina*	Vegetative
11	Fruit rot	*Alternaria solani*	Vegetative
12	Bacterial soft rot	*Erwinia carotovora subsp. carotovora*	Vegetative
13	Leaf curl	*Tobacco leaf curl virus*	Vegetative
14	Tobacco mosaic	*Tobacco mosaic virus*	Vegetative
15	Chilli mosaic	*Chilli mosaic virus*	Vegetative
16	Potato virus 'X' disease	*Potato virus X*	Vegetative
17	Potato virus 'Y' disease	*Potato virus Y*	Vegetative
18	Cucumber mosaic	*Cucumber mosaic virus*	Vegetative
19	Alfalfa mosaic virus/disease	*Alfalfa mosaic virus*	Vegetative
20	Pepper mottle	*Pepper mottle virus*	Vegetative
21	Pepper veinal mottle	*Pepper veinal mottle virus*	Vegetative
22	Tobacco streak	*Tobacco streak virus*	Vegetative
23	Tobacco etch	*Tobaco etch virus*	Vegetative
24	Broad bean wilt virus disease	*Broadbean wilt virus*	Vegetative
25	Fruit rot and die-back	*Collectotrichum capsici*	Reproductive
26	Bacterial leaf spot	*Xanthomonas vesicatoria*	Reproductive
27	Chilli mosaic	*Chilli mosaic virus*	Reproductive

Source: http://ikisan.com/tn-chilli-disease-management.html

2. Damping Off

Damping off is a serious problem in chilli nurseries. The disease may appear at any stage from seed germination to seedling maturity and up to 90 per cent nursery seedling death due to this disease (Sowmini, 1961).

2.1. Symptoms

Damping off disease develops in two phase namely, Pre-emergence damping off and Post-emergence damping off (Majit et al., 2018).

2.1.1. Pre-Emergence Damping Off

In this phase the seedlings unable to germinate i e seedlings killed before emergence and seedbed will be thinly populated. It is due to rotting of seed below soil surface or due to infection in the hypocotyls of the germinating seeds. Germinating seeds when shoots remain below soil surface (Horst, 2013).

2.1.2. Post Emergence Damping Off

Post-emergence damping-off symptoms occur when seedlings decay, wilt, and die after emergence (Horst, 2013). Lesions are developed on soft and succulent stems of the young seedlings nearby soil surface which leads to seedling topple down. In the advanced stage of infection the stem shows shriveling and brown discoloration near the soil which causes rotting under damp condition. If damping off caused by *Rhizoctonia*, black coloured lesion developed on seedling stem nearby soil surface and the seedlings get toppled down and the root portion may be look like a wire which is a typical symptom of Rhizoctonia damping off. Disease appears at nursery and fields in patches.

2.2. Causal Organism

The damping off disease is mainly caused by two different types of fungus namely *Pythium* and *Rhizoctonia*. Different species of *Pythium* namely *P. aphanidermatum*, *P. ultimum*, *P. debaryanum* etc. associated in the damping off disease but *P. aphanidermatum* is most common. Pythium is associated in the damping off disease under damp soil condition whereas *Rhizoctonia solani* is associated in damping off disease in dry soil condition.

2.3. Favourable Conditions

Both the pathogen soil borne in nature and survive with infected plant debris or alone in dormant stage. Excessive irrigation, poorly drained soil, heavy rainfall, high soil moisture and dense plant population with temperature around 25-30°C., high atmospheric moisture have a crucial role in infection of the disease (Martin and Loper, 1999). *P. aphanidermatum* prefers temperature above 20°C, whereas *P. debaryanum* and *P. ultimum* prefer cool temperature. (below 20°C). The disease cause by *R. solani* is favored by soil temperatures between 16 to 23°C. Dubey et al. (2020) reported that high cardinal temperature favored high disease incidence incited by *P. graminicola*.The disease also tends to be more severe on dry light soils.

2.4. Management

Nursery beds should be 15cm risen from the ground surface. Seedling planting in the well-drained field prevent damping off (Koike et al., 2000). The seed bed should be treated with formaline dust (15 parts formaline and 85% charcoal ash) @ 30 g/m^2 before seed sowing. Use low seed rate in the nursery beds. Seed dressing with Argosan or ceresan or captan or Thiram @ 2-3 g/kg. Ayub et al. (1998) revealed that metalaxyl was effectively controlled the tomato damping off. Soil drenching with 1% Bordeaux mixture or COC@ 0.3% or metalaxyl@0.2%. Biocontrol with *Trichoderma* viride and *Trichoderma harzianum*. Gomathi et al. (2011) observed that *Trichoderma* spp. having antibiotic ability against damping off. Soil drenching with 0.25% Copper oxychloride for young seedling protection from the soil borne pathogens.

3. Anthracnose, Ripe Fruit Rot and Dieback

3.1. Symptoms

The disease symptoms appear in two phases (i) Anthracnose and ripe fruit rot (ii) Die-back

3.1.1. Anthracnose and Ripe Fruit Rot

This is the most destructive phase of this disease. Ripen fruits are more susceptible than immature or green ones. Small, yellowish to pinkish, circular, sunken spots are seen on fruits having acervuli (fungal fruiting body). Pathogen causes infection on green as well as red stage, primarily on ripe fruits (Agrios, 2005). Severe affected fruits fall down and from such fruits discoloured seeds covered with fungal mycelial mat are produced. In advance infection fruits become shrivel and dry. Such infected fruits become whitish grey colored and reduce the pungency.

3.1.2. Die Back

Initially, small, brownish black, circular to irregular shaped, erratic spots appear on leaves. In advance stage of the disease defoliation of the leaves may be occurs. Subsequently the disease spreads up to the growing tips of the tender twigs which lead to tip backward necrosis of the branch and transformed the twig into a straw coloured tissue. Necrotic tissues appear grayish white with development of black dot like acervuli in the center (Saxena et al., 2016). Flower shedding takes place by infection on pedicel and tips of branches. The infection may also occur on succulent part of the stem and fruit stalk.

3.2. Causal Organism

Colletotrichum capsici (Syd.) Butler and Bishy.

3.3. Predisposing Factors

The fungus is seed borne as well as plant debris borne. The disease favoured by temperature of about 27°C and more than 97% relative humidity. Humid weather with rainfall at frequent intervals intensifies the disease development. Roberts et al. (2001) observed that around 27°C temperature and 80% atmospheric humidity leads the disease development.

3.4. Management

Collection and destruction of infected plant debris and always use healthy seeds. Removal and destruction of Solanaceous weed hosts. Spraying the captan @1.5% or mancozeb @0.25% thrice, first just before flowering, second at fruit formation stage and third spray 15 days after second spray. Spraying of Wettable sulphur @ 0.2%, Copper oxychloride @ 0.25% and Zineb 0.15% reduced the disease incidence. Use of fungicides both mancozeb and carbendazim found effective to control the anthracnose disease (Waller, 1992).

4. Cercospora Leaf Spot

Cercospora leaf spot (CLS) disease, also known as frogeye leaf spot disease, was first reported by Saccardo in 1876.

4.1. Symptoms

Symptoms appear on as small, water-soaked, circular, brownish spots on the leaves, tender stems, petioles and fruit stalk. Spots on stems and petioles are oblong rather than circular. Gradually the leaf spot enlarges and turn into grayish to black in colour with a light grey centre surrounded by a dark brown or reddish-purple border. Sometimes spots are surrounded by a diffuse yellow necrotic zone. Two or three spots may coalesce as they expand and develop blighted areas. Finally, the centre of the leaf spot falls down producing shot hole or frog eye symptom. Severely infected leaves drop off prematurely.

4.2. Causal Organism

Cercospora capsica Heald and Wolf.

4.3. Favorable Conditions

The fungus survives in or on seed and in old affected leaves in the soil. An average temperature of about 22.5-23.5°C, Relative humidity of about 77-85% and intermittent rains for a prolonged period favours the disease development. Cerkauskas (2004) stated that the disease is severe in wet weather and becomes in presence of high relative humidity.

4.4. Management

Use disease free seed. Seed treatment with captan or bavistin @ 2-4g/kg, Agrosan GN or Emisan @) 2g/kg of seed helps in eliminating the seed borne inoculum. Two sprayings of Mancozeb 0.25% or Chlorothalonil 0.1% or Benlate @ 0.1% twice at 10-15 days interval with gives good results. Seed treatment with carbendazim protect the chilli crop against Cercospora leaf spot (Islam et al., 2015) and Meah, (2006) revealed that spraying of carbendazim @ 0.02% was effective against the Cercospora leaf spot of chilli. Spraying of Tilt 25% EC @ 0.05% + *Eucalyptus* spp. @ 10% + *Pseudomonas fluorescens* @ 5g/litre reported to manage the disease efficiently.

5. Fusarium Wilt

5.1. Symptoms

Disease Symptoms of disease appear later stage of the crop and are first looks on the older (lower) leaves, later on disease spread towards younger leaves and at last plant may die. In many cases, only one branch or side of the plant shows symptoms. Systemic infection results in discoloration of the vascular system particularly in the rootsand lower stem. Plugging of the vessels of the plant system leads to severe water stress then wilting may occur (Khan et al., 2018). The roots turn into cortical decaying tissue. Restrict the flow of water in plant system due to interfere of pathogen with plant water conducting vessels resulting wilting and yellowing of the foliage parts.

5.2. Causal Organism

Fusarium oxysporum f.sp. *capsici*. Fusarium wilt is a soil-borne pathogen that attacks potato, tomato, eggplant and pepper plants.

5.3. Favorable Condition

Pathogen survives in soil as in chlamydospore form. In rare cases the pathogen may be seed borne. Hot and dry climatic conditions with temperature more than 25°C favour the disease. Excessive soil moisture and dry atmospheric conditions increase the disease development (Khan et al., 2018).

5.4. Management

Use wilt resistant varieties. Crop rotation adopt with non-host crop suppresses the disease in next season crop by reducing the pathogen inoculum. Drench 1% Bordeaux mixture in nursery seed bed as well as in main field. Treatment the seeds with *Trichoderma viride* culture @ 0.2% or Carbendazim @2g/kg seed is effective. Treat the seedlings before planting. Root dip treatment of the seedlings by dipping the roots in 0.2% solution of mancozeb reduces the disease effectively. Use of Metalaxyl + mancozeb 2.5 g/litre of water or Mancozeb @2.5 g/litre of water or Fosetyl aluminum @ 1.0 ml/litre of water to manage the disease incidence. Seed treat with carbendazim or captan or thiram before seed sowing, besides, seedling root dip in captan or benlate or carbendazim solution before transplanting found effective to manage the chilli wilt (Khan et al., 2018).

6. Bacterial Leaf Spot

6.1. Symptoms

The symptoms appear as brown, circular to irregular, water-soaked lesions on the leaves later on the numbers of leaves increases and they coalesce each other causing necrotic areas giving the plants a blighted appearance, become yellow, dry and fall off or remain on the plant. Diseased stems and petioles have elliptical, raised lesions. Flower infection results in severe blossom drop. On young fruit, lesions are small, circular, green spots reaching a diameter of 2-3mm. They become brown and have a raised, coarse, wart-like surface. Ripe fruit are rarely infected by the pathogen.

6.2. Causal Organism

Xanthomonas campestris pv. *vesicatoria*.

6.3. Predisposing Factors

The bacterium is seed borne in nature and may present either externally or internally. The bacterial propagules are also capable of surviving in the infected crop debris. Splashing rain drops are the chief means of dissemination. Temperature of about 22 to 34°C with high humidity is found to be congenial for maximum infection.

6.4. Management

Seeds should be collected from disease free fruits only. Seed treatment with 0.1% mercuric chloride solution or Agrosan GN or Cereson 2.5 g/kg seed for 2 to 5 minutes is beneficial. Spraying of 1% Bordeaux mixture or copper oxychloride @ 0.25% at the beginning the disease development. Streptomycin spraying twice or thrice gives good control of the disease.

7. Bacterial Wilt

7.1. Symptoms

Drooping of the leaves and branches gradually become bronzy and die and sudden wilting and death of the plant are typical symptoms of the disease. The vascular system of the affected stems turns brown. Affected plant when cut and put in a glass of clear water bacterial ooze is seen. Degrade of the xylem vessels and leads to death of adjacent tissues of the plant (Hayward, 2000).

7.2. Causal Agent

This disease caused by *Ralstonia solanacearum*

7.3. Predisposing Factors

The bacterium can survive in soil and are source of primary infection. The bacterium can survive in diseased crop debris. The disease is favored by temperature ranging from 15-35°C. High soil moisture with high temperature and pH below 6.2 is favorable for the bacterium. Bacterial wilt of tomato and pepper are increased by high soil temperature and moisture (Wang and Lin, 2005).

7.4. Management

Eradicate weed hosts. Remove wilted plants, root debris and volunteer hosts, and burn them to reduce spread of the disease from plant to plant. Treat the seeds with *Trichoderma* culture protects the seedlings in the initial stage. Bio-control agents like *Paenibacillus macerans*, *Bacillus subtilis* and *B. pumilus* found have been found effective against *R. solanacearum* and *Xanthomonas vesicatoria* in tomato (Liu et al., 2013; Wachowska et al., 2013). Bacterial wilt tolerant varieties should be used wherever the disease is very severe. Stable Bleaching Powder @ 15 kg/ha may be applied in furrows 10-15 days before planting. For effective disease control of bacterial wilt, 10 g of *Trichoderma viride* or *Pseudomonas* sp. per liter of water should be used as spraying. Pairashi (2007) observed in by their experimental trials that Frog eye leaf spot of tobacco was effectively controlled by *P. fluorescens* culture.

8. Mosaic

8.1. Symptoms

Initially appear alternate green and yellow patches in a mosaic pattern on leaves and thickening and mottling of the leaves is also common symptoms of the disease. Plant growth is severely reduced if the infection occurs in the early plant growth stage. Gradually the leaves become elongated and narrow which become filamentous. Yellowing, chlorotic ring spots are developed on leaves and fruits. The infected plants produce fewer flowers, deformed fruits of low quality.

8.2. Causal Agent

Cucumber mosaic virus (Cucumo virus).

The pathogen is transmitted by aphids *Aphis gossypii* and *Myzus persicae*. Cilia et al. (2012) expressed that insect *Bemisia tabaci*, *M. persicae* and Aphid can transmit the virus. Presence of infected wild host plants around the crop field enhances the disease severity. Temperature ranging from 26-30°C accompanied by enough vector population increases the disease infestation.

8.3. Management

The infected plants should be uprooted and burnt or buried to avoid further infection. Avoid monoculture of chilli crop. Barriers crops taller than the chilli crop should be used to avoid the vector transmitted virus infestation. Soaking seeds in a solution containing 150 g Trisodium orthro phosphate per liter of water for 30 minutes inhibits seed - borne inoculum. Protect the seedling from viral infection by covering the beds with nylon net or straw. Raise 2-3 rows of maize or sorghum as border crop to restrict the spread of aphid vectors. Apply Carbofuran 3G @ 4-5kg/acre in the mainfield to control sucking complex and insect vectors selectively. Foliar application of insecticides like Imidacloprid, Malathion, Cypermethrin or Dimethoate decreased the transmission percentage of the disease. Seed treatment with Imidacloprid and foliar application Imidacloprid, dimethoate and Acetamiprid found effective to control the insect vectors (Bhagat et al., 2018).

9. Leaf Curl

9.1. Symptoms

The disease characterized by rolling, curling, puckering of the leaves and interveinal blistering, vein thickening, shortening of the internodes and petioles. Older leaves become brittle and leathery, proliferation of number of lateral branches and whole plant remains stunted also the symptoms of the disease. Flower buds fall down before attaining full size and produce sterile deformed flowers. Fruits are of deformed shape and low quality.

9.2. Causal Agent

Chilli leaf curl virus (ChLCV)

9.3. Transmission

The chilli leaf curl virus is transmitted by the whitefly *Bemisia tabaci* in a persistent manner (Thakur et al., 2018). Besides chilli, the virus also infects tomato, brinjal, okra, tobacco, beans, cotton, cucurbits, citrus and different weeds. Dry weather with high relative humidity increases the disease due to active presence of the vector. During wet weather the spread of the disease become low.

9.4. Management

Seedling protect from white fly infestation by covering of nursery bed with nylon Plant 2-3 rows of border cropping with maize or sorghum or bajra 15-20 days before transplanting to reduce the spread of the disease. Rouging and destruction of affected to restrict the further infection. Avoid Monoculture and virus susceptible hybrids of chilli mainly in endemic areas. Seedling root-dip treatment with Imidacloprid 17.8% SL @ 0.5 ml/l for 15 min. before transplanting. The vector can also be controlled by the application of Dimethoate or Malathion at fortnightly interval. Seed treatments with Imidacloprid 70WS @ 5g/kg along with two foliar sprays with Imidacloprid 17.8 SL @ 0.24 ml/lit. water at 45 and 60 days after transplanting (Mishra and Chauvey, 2018). Foliar application of Azadirachtin 1000 ppm @ 20 ml per 10 litres of water along; with pesticide is highly effective.

10. Conclusion

The chilli (*Capsicum annum*) is a most important vegetable and spice crop grown widely throughout the world. The crop is vulnerable to enormous attack by a number of fungi, bacterial and virus that incited the disease Damping off, Fruit rot and die back, Powdery mildew, Root rot, Choanophora blight, Downey mildew, Verticillium wilt, Bacterial soft rot, Leaf curl and Mosaic, that's why crop yield is restricted in terms of quality as well as quantity. Lacking of awareness the disease diagnosis and their management in grower's community is strike to economic losses. Here approaches are recommended for management the chilli diseases with integrated disease management system, though the enhance productivity and reduce the losses come out by these diseases.

References

Agrios, G.N. 2005. Plant Pathology. St.Louis, MO: Academic Press.

Ayub, M., Khan, A., Amin, M. 1998. Efficacy of some of the fungicides for managing pre-emergence damping off in tomato and their effect on seedlings vigour. Sarhad Journal of Agriculture.14, 475-481.

Bhagat, S., Singh, S., Gupta,S., Sharma, D., Choskit, D. 2018. Management of mosaic disease of chilli (*Capsicum annuum*) through host resistance and chemicals. Pl. Dis. Res. 33(2), 213-216

Cerkauskas, R. 2004. Pepper Diseases: Cercospora Leaf Spot Published by AVRDC–The World Vegetable Center, Shanhua, Taiwan. pp. 741.

Cilia, B.M., Fish, T., MacCoss, M.J., Gray, S. 2012. Homoptera vector biomarkers for circulative efficient transmission plant viruses are conserved in multiple aphid species and the whitefly *Bemisia tabaci*. Journal of Integrative Agriculture. 11(2), 249-262.

Majeed, M., Mir, G.H., Hassan, M., Mohuiddin, F.A., Paswal, S., Farooq, S. 2018. Damping Off in Chilli and Its Biological Management-A Review. Int. J. Curr. Microbiol. App. Sci 7(4), 2175-2185.

Dubey, M.K., Zehra, A., Aamir, M., Yadav, M., Samal, S., Upadhyay, R.S. 2020. Isolation, identification, carbon utilization profile and control of *Pythium graminicola*, the causal agent of chilli damping-off. J. Phytopathol. 168, 88–102.

Gomathi, S., Ambikapathy, V., Pannerselvam, A. 2011. Potential strain of *Trichoderma* spp. to control damping-off of disease in chilli. International Journal of Applied Biology and Pharmaceutical Technology. 2, 64-69.

Hayward, A.C. 2000. *Ralstonia solanacearum*. In Encyclopedia of Microbiology ed. Lederberg, J. pp. 32–42. New York: Academic Press

Horst R.K. 2013. Damping-off. Westcott's plant disease handbook. Springer Netherlands, Dordrecht, p 177.

Islam, M.S., Fatema, K.K., Alam, M.B., Meah, M.B. 2015. Diagnosis and prescription for Cercospora leaf spot of chilli. J. Bangladesh Agril. Univ. 13(2), 191–196.

Khan, K.A, Nabi, S., Bhat, N.A., Bhat, F.A. 2018. Chilli Wilt Disease: A Serious problem in Chilli cultivation in India. Indian Farmer. 5(09). 988-991

Koike, S.T., Gaskell, M., Fouche, C., Smith, R., Mitchell, J. 2000. Plant disease management for organic crops. ANR Publication 7252, University of California, Davis, CA.

Liu, J., Sui, Y., Wisniewski, M., Droby, S., Liu, Y. 2013. Review: utilization of antagonistic yeasts to manage postharvest fungal diseases of fruit. Int J Food Microbiol. 167, 153–160.

Martin, F.N., Loper, J.E. 1999. Soilborne Plant Diseases Caused by *Pythium* spp.: Ecology, Epidemiology, and Prospects for Biological Control. Crit. Rev. Plant Sci., 18, 111–181.

Meah, M. B., Wick, .LR. 2006. Diagnostic plant pathology. Dept. of plant soil and insect sciences, University of Massachusetts. Amherst MA, USA. pp. 5-71.

Mishra, R.S., Chauvey, A.N. 2018. Chilli Leaf Curl Virus and its Management. Acta Scientific Agriculture 2.3, 24-28.

Nadkarni, K.M. 1927. The Indian MateriaMedica. Bombay. p xviii -1142.

Pairashi, M. 2007. Studies on frog eye leaf spot of bidi tobacco caused by *Cercospo ranicotianae* Ell. & Eve. M. Sc. (Agri.) Thesis, Univ. Agric. Sci., Dharwad, Karnataka (India).

Parey, M.A., Razdan, V.K., Sofi, T.A. 2013. Comparative study of different fungi associated with fruit rot of chilli and screening of chilli germplasm against *Colletotrichum capsici.* International Journal of Agriculture and Crop Sciences. 5, 723-730.

Roberts, P.D., Pernezny,K., Kucharek,T.A. 2001. Anthracnose caused by *Colletotrichum* sp. On Pepper. University of Florida /Institute of Food and Agricultural Sciences, Available online at:http://edis.ifas.ufl.edu/PP104

Saxena, A., Raghuwansh, R. V., Gupta, V.K., Singh, H.B. 2016. Chilli anthracnose: Epidemiology and Management. Frontiers in Microbiology. 7, 15-27.

Sowmini, R. 1961. Studies on phycomycetes in agricultural soils with special reference to *Pythiaceae*. M.Sc. (Agri.) Thesis, University of Madras, pp. 160.

Thakur, H., Jindal, S.K., Sharma, A., Dhaliwal, M.S. 2018. Chilli leaf curl virus disease: a serious threat for chilli cultivation. Journal of Plant Diseases and Protection. 125, 239-249

Wachowska, U., Kucharska, K., Jedryczka, M. and Łobik, N. 2013. Microorganisms as biological control agents against fusarium pathogens in winter wheat. Pol J Environ Stud. 22, 591-597.

Wang, J.F., Lin, C.H. 2005. Integrated Management of Tomato Bacterial Wilt, pp. 1–16. Taiwan: AVRDC-The World Vegetable Center.

27

Diseases of Brinjal Crop and Their Management

Jitendra Singh[1], G.L. Kumawat[1], A.K. Yadav[2] and R.K.Bagri[3]

[1]Sri Karan Narendra Agriculture University, Jobner, Rajasthan, India
[2]College of Agriculture, SKNAU, Kishangarh Bas, Alwar, Rajasthan, India
[3]Rajasthan Agricultural Research Institute, SKNAU, Jaipur, Rajasthan, India

1. Introduction

Eggplant or brinjal (*Solanum melongena* L. family. Solanaceae) is the major vegetable crop of wet and hot climatic zone of south Asia (Srinivasan and Huang, 2008). After China, India is the second largest brinjal producer country in the world (Majumdar and Das, 2020). It is nutritious vegetable and consumed in throughout the world. Brinjal is considered as Indian native crop and it has been cultivated for last 4000 year. India is known to be the Origin Centre of brinjal cultivation, from where it distributed to throughout the world (Choudhury and Malda, 1968). In India, brijnal crop grow almost throughout the year. Brinjal fruit contains moisture 92.7g, carbohydrates 4.0g, proteins 1.4g, minerals 0.3g, vitamins 124 I.U. and vitamin 12.0mg per 100 g of edible part (Kalra et al., 1988). It is used as vegetable purpose as well as pickle (Singh et al., 1963). It is a best source of liver suffering patients and have also important role to play for remedy of diabetes. Brinjal cultivation is suffered by a number of biotic and abiotic factors (Majumdar, and Das, 2020). Pathogenic diseases are one of the limited factors in brinjal cultivation and its production is greatly constrained by various diseases like damping off (*Pythium aphanidermatum, Phytophthora aphanidermatum*), phomopsis blight or fruit rot (*Phomopsis vaxans*), *l*eaf spot (*Alternaria melongenae* and *Cercospora melongenae*), collor rot (*Sclerotium rolfsii*), bacterial wilt (*Ralstonia solanacearum*), Mosaic (*Potato virus Y*), little leaf (*Mycoplama*) (Bhupendra et al., 2014; Edgerton and Moreland, 1921; Vishunavat, 1992; Pagoch et al., 2015; Singh et al., 2014). The various diseases attack on the plant during all growth stages (Ahmed and Hossain, 1985). At different growth stages the crop suffers from various diseases during all the growing season (Ahmed and Hossain, 1985).

Table 1: The common diseases of brinjal crop

S.N.	Name of the disease	Causal organism
1	Damping off	*Pythium aphanidermatum, Phytophthora palmivora*
2	Phomopsis blight	*Phomopsis vaxans*
3	Sclerotinia blight	*Sclerotinia sclerotiorum*
4	Cercospora leaf spot	*Cercospora melongenae*, C. *solani*
5	Alternaria leaf spot	*Alternaria melongenae, A. solani*
6	Verticillium wilt	*Verticillium dahlae*
7	Bacterial wilt	*Pseudomonas solanacearum*
8	Little leaf	*Phytoplasma*
9	Mosaic	*Potato virus Y*

Source: Kalita, M.K. (2019)

2. Damping Off

Damping off brinjal is serious problem in nursery, young seedlings plants are more susceptible, maximum damaged occurred 5-6 weeks after sowing. Favourable conditions for disease development are high soil moisture and moderate temperature along with higher relative humidity during rainy season. Many pathogenic fungi weakening and destroying the seed and seedlings of various vegetable, field and ornamental crops (Filer and Peterson, 1975; Kraft et al., 2000).

2.1. Symptoms

Damping-off symptoms can be observed from seeding until the fourth to sixth week post-sowing (Horst, 2013). The disease symptoms divided in two phases based on the time and its appearance.

2.1.1. Pre-Emergence Symptoms

Seed decays prior to emergence and can be occurs (i) before germination and (ii) They occur when seeds rotten prior to emergence. Germinating seeds when shoots remain below soil surface (Horst, 2013). In the first occurrence, seeds become soft, decayed, and break down to germinate and in second case, characteristic water soaked lesions which to reddish brown to black occurrence on germinating seed stems below the soil (Landis, 2013). Quick girdling young tender stems leads to wilt and die prior to emergence.

2.1.2. Post-Emergence Symptoms

Post-emergence damping-off symptoms occur when seedlings decay, wilt, and die after emergence (Horst, 2013). In most cases, all symptoms result in the collapse and death of at least some seedlings in any given seedling population. These symptoms can be also accompanied by leaf spotting and a complete root rot may occur. Overall, the water soaked sunken lesions symptoms on seedling stem nearby soil surface or slightly below the ground line i.e. on root portion which cause the seedling fall over (Filer and Peterson, 1975).

2.2. Causal Organism

This dieses ceased by *Pythium aphanidermatum, Phytophthora* spp. Both fungi lying in soil and water as in oospores form, in favourable conditions oospores liberate zoospores, which disseminate through wind current and can spread quickly from one nursery seed tray (or garden row), to another.

2.3. Management

Seed to be used for sowing should be healthy and certified. Seed treatment with chemicals is the major practices in management the damping off of seedlings (Howell, 2007). Treat the seeds before sowing with Thiram @ 2g/kg of seed. Raising of nursery should not be continuous in the same plot. Nursery soil should be treated with Thiram @ 5g/ m^2 and seedling nursery should also be drenched with the same fungicide solution 2g/litre of water at an interval of 15 days. The beds which are to be used as nursery should be cover with 250 gauge polythene sheet for 30 days for soil solarisation and soil treatment with bio-control agent *Trichoderma viride* culture @ 2Kg/ha found effective against damping off disease.

3. Phomopsis Blight

The phomopsis blight or fruit rot of brinjal is cosmopolitan in tropical and subtropical regions caused considerable loss to the crop, it start from the leaves but the devastating damage to the crop by fruit rot (Singh, 1992). The pathogen is distributed in all brinjal growing areas of the world (Bhanushree et al., 2021)

3.1. Symptoms

The pathogen attack to the plant generally on all growth stages i.e. symptoms producing in nursery as damping off and on young plant as collar rot. Circular to irregular greyish brown spots having light in the centre symptoms appear on the leaves. A number of lesions appear on the leaves and may cover entire leaf area (Manda et al., 2020). Leaves become yellowish in colour and may fall off later a number of black pycnidia are seed on older spots. On the stem, dark brown round to oval with greyish centre lesions appear and develops the pycnidia on the lesion. The fungus cause constriction on the base of stem which leads to development of canker and topple down the plant. Small pale sunken spots appear on the fruits, later on increase in size and cover entire surface of the fruit (Manda et al., 2020). The enlarged spots become watery and lead to soft rot of the characteristic disease where develop a number of dot like pycnidia on the surface. The fruit infection through calyx end leading dry fruit rot and fruit become black and mummified.

3.2. Pathogen

This disease caused by pathogen *Phomopsis vexans (P.S: Diaporthe vexans).* Primary infection: Pathogen is seed borne in nature and also survive in the soil with disease infected plant debris. Secondary infection by conidia through rain splashes, irrigation water, insect and agricultural implements. 26^0 C with 55% atmospheric relative humidity favours the disease development (Chowdhury and Hasija, 1979). Islam and Sitansu (1990) stated that maximum fruit rot develops at 30^0 C and more than 40^0 C is adverse condition for disease development.

3.3. Management

Disease infected plant debris should be removed and destroyed. Deep summer ploughing and crop rotation practices in the field are effective to reduce pathogenic inoculum. Singh, (1987) declared that reduction of inoculum through burying of infected crop debris by deep ploughing Seed to be used for sowing should be healthy (*i.e.* disease free) and certified. Seed treatment with hot water @ 50^0 C found effective against the pathogen. Seed treatment with fungicide Thiophanate methyle @ 1 g/kg of seed. Foliar spray with Thiophanate methyl or Carbendazim 0.1% at an interval of 20 days. Seed treatment with *Trichoderma harzanium* culture and *Pseudomonas fluorescens* culture is effective to control the disease (Srinivas et al., 2005).

4. Bacterial Wilt

Wilt caused by bacteria is the major disease of brinjal and tomato crops in tropical and subtropical regions (Ravishankar et al., 2021).

4.1. Symptoms

Yellowing of leaf surface, drooping the leaves and stunting plant growth and later on collapsing the complete plant are the considerable disease symptoms. Browning of vascular system of the affected root plant are seen when longitudinally split out the root and oozing of bacterial cells from the affected parts. Initially, wilting shows in plants at noon hours will recover at nights, but later die.

4.2. Pathogen

Bacterial wilt disease caused by bacteria called *Pseudomonas solanacearum.*

4.3. Mode of Spread and Survival

The bacterium having wide host range and can infects chilli, tomato, potato, radish, fennel, ginger etc., It is non spore former and can alive and viable more than 15 months in the laboratory conditions and in the field pathogen can alive upto 10 months with infected plant debris. The root knot nematode, *Meloidogyne javanica* injured the roots of plant which leads to increase disease incidence.

4.4. Management

Always use resistance variety. Follow the crop rotation with cruciferous crops like cabbage, cauliflower, which help in lowering the disease incidence. Field should be sanitized, disease infected plant debris are to be collected and destroy them by burning. Seed soaking in brine solution found increase nursery seedling vigour and become tolerance to *P. solanacearum* (*Ralstonia solanacearum*) (Nakaune et al., 2012). Foliar spray with fungicide Copper oxychloride @ 0.3%water solution. Management of root knot nematode (*Meloidogyne javanica)* will reduce the disease spreading, as the disease is more prevalent by infect the nematode.

5. Cercospora Leaf Spot

5.1. Symptoms

Angular to irregular, chlorotic lesions are characteristic symptoms. Later on become greyish- brown with exuberant sporulation at the centre of the spot. The spots on the upper surface of leaves are brown to steel-grey and on lower surface are light brown (Anon., 2013). In severe condition of the disease the chlorotic spots cover a large area of leaves and fall off the infected leaves prematurely. Resulting the adverse effect of flower and fruit setting on the plant. They are brown to steel-gray on the upper surface and light brown on the lower surface (Anon., 2011).

5.2. Pathogen

This disease caused by *Cercospora melongenae, C. solani.* The fungus produces stromata which are globular. Conidiophores in mass are medium dark and slightly olivaceous brown in colour and paler towards the tip. Conidia are sub hyaline to pale olivaceous.

5.3. Mode of Spread and Survival

The disease is spread by air borne conidia. The pathogen perpetuate in soil with infected plant debris. Primary infection through air borne conidia and secondary infection through conidia produced on primary infected plants. Ullasa and Sohi (1987) found that the atmospheric high humidity, hot days and night are cool favours the disease development.

5.4. Management

Destruction of weed plants from the brinjal fields to reduce relative humidity (Srivastava and Nelson, 2021). Use of resistant variety Pant Samrat against the leaf spot disease. Diseases can be managed by growing resistant varieties. Spraying 1 per cent Bordeaux mixture or 2 g Copper oxy chloride or 2.5 g Zineb per litre of water effectively controls leaf spots.

6. Alternaria Leaf Spot

6.1. Symptoms

Irregular brown coloured spots appear on the leaves; they are mostly 4-8 mm in diameter, later increase in size and may coalesce each other to cover large leaf surface area, concentric rings become in the spots. The characteristic symptoms of the disease are commonly caused by two species of *Alternaria* fungi. Cracking and shot hole become in the leaf spot. Severely affected leaves may become necrotic and drop off the leaves prematurely. The pathogen *Alternaria melongenae* also infects to fruits which cause large deep-seated spots and fruits become yellow and prematurely drop off from the plant. Bochalya et al. (2012) advocated that the pathogen initially infects on leaves and later on spread on fruits, where rotting may occur and decrease the quality to consume.

6.2. Pathogen

This disease is caused by *Alternaria melongenae, A. solani.*

6.3. Mode of Spread and Survival

The disease is spread by air borne conidia. The pathogen perpetuate in soil with infected plant debris. Primary infection through air borne conidia and secondary infection through conidia produced on primary infected plants.

6.4. Management

Spraying 1 per cent Bordeaux mixture or 2 g Copper oxy chloride or 2.5 g Zineb per litre of water effectively controls leaf spots. Kanna Reddy (2021) stated that the treatment of seedlings of brinjal with 2.5% *Trichoderma viride* culture + 2.5% neem oil was effective to manage the disease.

7. Little Leaf

Little leaf of brinjal (LB) is one of the major disease which cause tremendous loss to the cultivars (Mitra, 1993) and this disease was first reported in India from Coimbatore (Thomas and Krishnaswami, 1939). LLB incited by one phytoplasma which is lack of wall (wall-less) moliicutes (Varma et al., 1969)

7.1. Symptoms

The petioles of the leaves are very much short that shows leaves to be sticking with shoot or stem. The leaves become narrow, smooth, soft, small and yellowish in colour and newly emerged leaves remain much shorter. Shoot of the plants stunted due to shortened internode. Axillary buds are enlarged and their petioles and leaves remain short, so that plants appearing bushy. Proliferation of branches with size of leaves reduced and shortened internodes shows little leaf (Raj et al., 2008). Commonly there is not flower formation, if flowers formed they remain green and fruits not formed, if fruiting occur they become hard, tough and cannot mature. Young fruits become get mummified and cling to the plant for longer period.

7.2. Pathogen

The disease is caused by phytoplasma having ovoid to spherical body which is concentrated in the phloem sieve tubes.

7.3. Spread

The pathogen perennates in weed hosts. Jassid–*Hishimona sphycitis*. The disease is also transmitted by *Emposca devastans*, however, it is a less efficient vector. Mycoplasma also transmitted by leaf hopper (*Hisimonus phycitis)* (Singh et al., 2014; EFSA Panel on Plant Health, 2017). Brinjal little leaf disease is transmitted by insects (leaf hopper) such as *Hisimonus phycitis* which is common on brinjal crop. Pusa purple long and selection T varieties are highly susceptible. Collateral hosts: *Datura fastuosa*, D. stramonium, *Vinca rosea*, *Argemone mexicana*, chilli, tomato and tobacco. Among the two vectors, *A. biguttula* was dominantly prevalent in brinjal throughout the growth stages

7.4. Management

Tolerant variety: Pusa Purple Round, Pusa purple cluster and Arka Sheel. Destruction of affected plants. Eradication of solanaceous weed hosts. Spray methyl demeton, 2 ml/l or Malathion 3 ml/l or soil application of phorate granules. Seed dip in tetracyclines (10-50ppm).

8. Mosaic

8.1. Symptoms

Initially symptoms appear on leaves as mosaic mottling and stunted plant growth. Later on infected leaves become deformed, leathery and remain small later on develop blisters. In advance or severe case of the disease plant produce very few flowering and fruiting.

8.2. Pathogen

This disease caused by *Potato virus Y* (PVY) and *Tobacco mosaic virus* (TMV). The Potato virus Y is sap transmitted easily. And the pathogen transmitted through aphids (*Aphis gossypii* and *Myzus persicae*) in the field and survive on the wild host plant *Solanum xanthocarpum and S. nigrum*). Virus which is non-sap transmissible in nature also produce mosaic symptoms (Raychaudhyri, 1947). Different strains of cucumber mosaic virus infect to brinjal (Dale, 1956; Seth et al., 1967; Rana and Vovlas, 1971) and Tobacco ring spot virus also infect to brinjal plants (Gibbs and Harrison, 1969). TMV is spread by sap, soil debris, farm implements and hands and foot of the labour. Virus have wide host range and can be perpetuate on legumes, tomato, tobacco, pepper and cucumbers.

8.3. Management

Weed should be destroyed from the field. Avoid raising crops of legumes, tomato, tobacco, pepper and cucumbers near brinjal nursery or field. Clean the hand with soap and water before operational activity in the nursery. Install the yellow sticky sheets three weeks after transplanting (Pandey and Chaturvedi, 2019).Foliar spray with insecticide Diamethoate @ 2ml/little of water to control the aphids.

9. Conclusion

The brinjal a major vegetable cultivated widely throughout the world. The crop is vulnerable to enormous attack by many pathogens fungi, bacterial, virus and mycoplasma that incited the diseases, that's why production of the crop is restricted in terms of quality as well as quantity. Lacking of awareness the disease diagnosis and their management in grower's community is strike to economic losses. Here approaches are recommended for management the brinjal diseases with integrated disease management system, though the enhance productivity and reduce the losses come out by these diseases.

References

Ahmed, H.U., Hossain, M.M. 1985.Crop disease survey and establishment of a herbarium at BARI, Plant Pathology Division, BARI, Joydebpur, Gazipur.p: 107.

Anonymous. 2011. Small fruit and vegetable IPM advisory – Integrated pest management, Utah State University Cooperative Extension Web portal. http:// utahpests.usu.edu/ipm/htm/ advisories/small-fruitand-vegetable-advisory/articleID=13472 – leaf spot (accessed 28 February 2012).

Bhanushree, N., Partha, S., Tomar, B.S., Munshi, A.D. 2021. Phomopsis blight in eggplant and strategies to manage through resistance breeding. The J. of Hort Sci. and Biotech

Bhupendra, K.S., Singh S., Yadav, S.M. 2014. Some important plant pathogenic diseases of brinjal (*Solanum melongena* L.) and their management. Plant Pathology. 13(13), 208-213.

Bochalya, S., Shekhawat, S., Singh, R., Chohan, P.K. 2012. Studies on age of fruit, age of inoculum and inoculum load on development of Alternaria fruit rot of brinjal caused by *Alternaria alternata*. Annals of Agri Bio Research.18, 388-390.

Chowdhury, S.R., Hasija, S.K. 1979. Pathological studies on phomopsis vexans causing soft rot of brinjal fruits. Indian Phytopathology. 32, 495-496.

Choudhury, B., Malada, T.S. 1968. Brinjal: A vegetable of masses. Indian Horticulture 12: 21-22.

Dale, W.T. 1956. Virus diseases of solanaceous crops. Trop. Agric. 33:35-50.

Edgerton, C.W., Moreland, C.C. 1921. Eggplant blight Louisiana Agricultural Experiment Station Bulletin. 178, pp 1-44.

Jeger, M., Bragard, C., Caffier, D., Candresse, T., Chatzivassiliou, E., Dehnen, S. K., Gilioli, G., Gregoire, J.C., Anton. J., Miret, J., Navarro, M.N., Niere, B., Parnell, S., Potting, R., Rafoss, T., Rossi, V., Urek, G., Bruggen, A.V., van der Werf, W., West, J., Winter, S., Gardi, C., Aukhojee, M., Bergeretti, F., MacLeod, A. 2017. Pest categorisation of Hishimonusphycitis. EFSA Panel on Plant Health (PLH), EFSA Journal.15, 5037.

Filer, T.H.J., Peterson, G.W. 1975. Damping-off. In: Peterson G.W, Smith R.S (Eds) For. Nurs. Dis. United States. USDA Forest Service. Agriculture Handbook No. 470, Washington DC, pp 6–8.

Horst, R.K. 2013. Damping-off. Westcott's plant disease handbook. Springer Netherlands, Dordrecht, p 177.

Howell, C.R. 2007. Effect of seed quality and combination fungicide-*Trichoderma* spp. seed treatments on pre- and post-emergence damping-off in cotton. Phytopathology. 97, 66–71.

Islam, S.J., Sitansu, P. 1990. Variabilities among isolates of *Phomopsis vexans*. Environment and Ecology 8(1B):315-9.

Kalita, M.S. 2019. Diseases of horticultural and field crop and their management-I, Kalyani publishers, New Delhi pp 102.

Kalra, C.L., Berry, S.K., Sehgal, R.C. 1988. A resume on brinjal (*Solanum melongena* L.), a most common vegetable. Indian Fedration Packer. 2, 46-59.

Kanna, R., Tiwari, S., Rohini, T. 2021. Eco-friendly Management of Leaf Spot of Brinjal (*Solanum melongena* L.) caused by Alternaria spp. International Journal of Current Microbiology and Applied Science. 10(05), 132- 142.

Kraft, J.M., Haware, M.P., Halila, H. 2000. Soilborne diseases and theircontrol. In: Knight R (ed) Link. Res. Mark. Oppor. Pulses 21st Century. Kluwer Academic Publishers, Dordrecht, pp 457–466.

Landis, T.D. 2013. Forest nursery pests: damping-off. For Nurs Notes. 2, 25–32

Majumdar, S., Das, B.K. 2020. Studies on little leaf of brinjal and morpho taxonomy of the leafhopper species associated from Bengal. Journal of Entomology and Zoology Studies. 8(3), 514-521.

Manda, R.R., Addanki, V.A., Srivastava. 2020. Phomopsis blight of *solanum melongena* – brinjal/eggplant. Plant Cell Biotechnology and Molecular Biology. 21(55&56), 7-12.

Mitra, D.K. 1993. Little leaf, a serious disease of eggplant (*Solanum melongena*). In: Management of Plant Diseases Caused by Fastidious Prokaryotes. Raychaudhuri S.P and Teakle D.S. (Eds), Associated Publishing Co, New Delhi, India, 73-78.

Nakaune, M., Tsukazawa, K., Uga, H., Asamizu, E., ImanishiS, M. C., Ezura, H. 2012. Low sodium chloride priming increases seedling vigor and stress tolerance to *Ralstonia solanacearum* in tomato. Plant Biotech. 29, 9–18.

Pagoch, K., Srivastava, J.N., Singh, A.K. 2015. Damping-off diseases of seedlings in Solanaceous vegetables: current status and disease management. In: Awasthi, L.P. (Ed.), Recent Advances in the Diagnosis and Management of Plant Diseases. Springer, India. pp. 35–46.

Pandey, R., Chaturvedi, A.K. 2019. Effective management strategy of little leaf disease in brinjal (*Solanum melongena*L.). Journal of Experimental Biology and Agricultural Sciences. 7(3), 295-300.

Raj, S.K., Snehi, S.K., Kumar, S., Khan, M.S. 2008. First finding of '*Candidatus* Phytoplasmatrifolii' (16SrVI group) associated with little leaf disease of *Datura inoxia* in India. New Disease Reports. 18, 44.

Rana, G.L., Volvas, C. 1971. Le virosidellepianteortensi in Puglia. V.N. mosaicodella Melanzana. Phypath. Medit. 9,187-191

Ravishankar, M., Chen, J.R., Paola, S.C., Lawrence, K., Srinivasan, R. 2021. Evaluation of different bacterial wilt resistant egg plant rootstocks for grafting tomato. Plants *(Basel)* 10(1), 75.

Raychaudhuri, S.P. 1947. A mosaic disease of brinjal (*Solanum melongena* L.).Current Sci. 16:149- 150.

Seth, M.L., Chaudhuri, S.P., Nath, R. 1967. A new mosaic disease of brinjal (*Solanum melongena* L.). Phytopath. 59, 385-389.

Singh, B.K., Singh, S., Singh, B.K., Yadav, S.M. 2014. Some Important Plant Pathogenic Disease of Brinjal (*Solanum melongena* L.) and their Management. Plant Pathology Journal. 13, 208-213.

Singh, B.K., Singh, S., Singh, B.K., Yadav, S.M. 2014. Some important plant pathogenic diseases of brinjal (*Solanum melongena*L.) and their management. Plant Pathol. J. 13, 208-213.

Singh, R.S. 1987. Plant pathogens. Second Edition. New Delhi, India: Oxford and IBH Publishing Company.

Singh, R.S. 1999. Diseases of vegetable crops (third edition) Oxford & IBH Pub. Co. Pvt. Ltd., New Delhi and Calcutta, 4061.

Singh. S., Krishnakumar, S., Katyal, S.L. 1963. Fruit culture in India. Indian Council of Agriculture Research, New Delhi. pp: 412.

Srinivas, C., Niranjana, S.R. Shetty, H.S. 2005. Effect of bioagents and fungicides against Phomoposis vexans and on seed quality of brinjal .Crop Improvement. 32(1),95- 101.

Srinivasan, R., Huang, C.C. 2008. Effect of simulated borer infested shoot pruning on yield parameters of eggplants. Journal of Asia Pacific Entomology. 12, 41-43.

Srivastava, S., Nelson, S. 2012. Cercospora leaf spot of eggplant. Plant Disease (UH- CTAHR). PD-82.

Thomas, K.M., Krishnaswami, C.S. 1939 Little-leaf- a transmissible disease of brinjal. Proceedings of the Indian Academy of Science. 10,201-212.

Ullasa, B.A., Sohi, H.S. 1987. Studies on Cercospora leaf spot disease of brinjal in Karnataka. Vegetable Science. 14(1), 92- 97.

Varma, A., Chenulu, V.V., Raychaudhuri, S.P., Prakash N., Rao, P.S. 1969. Mycoplasma like bodies in tissues infected with sandal spike and brinjal little leaf. Indian Phytopathology. 22,289-291.

Vishunavat, K. 1992. Phomopsis blight and fruit rot of eggplant. In: Plant Diseases of International Importance Vol II. Diseases of Vegetables and Oil Seed Crops.

28

Nursery Disease Management of Horticultural Crops Under Protected Condition

D.L. Yadav[1], Rakesh Kumar Yadav[2], G.L. Kumawat[3], Praveen Kumar Singh Gujar[4], Rajesh Sharma[1] and Arvind Kumar Verma[5]

[1]Agricultural Research Station, Agriculture University, Kota, Rajasthan, India
[2]College of Agriculture, Ummedganj, Agriculture University, Kota Rajasthan India
[3]Sri Karan Narendra Agriculture University, Jobner, Rajasthan, India
[4]Rajmata Vijayaraje Scindia Krishi Vishwa Vidyalaya, Gwalior, Madhya Pradesh, India
[5]ICAR-National Research Centre on Seed Spices, Ajmer, Rajasthan

1. Introduction

India's horticulture production has exposed multiple rises in the historical two decades and the country is ranking second in the world next to China. The horticultural crops viz., fruits, vegetables, flowers, spices, plantation crops and medicinal and aromatic plants have become a vital motorist for economy development in numerous states in India. Horticulture contributes 30.4 per cent to GDP of agriculture, which calls for the need of research and technical backstopping in this sector. Horticulture is a division which has the potential to create multiple springs of income, thereby boosting the economic growth of a country. Horticulture can be undertaken as market driven cultivation of vegetables, fruits and flowers, as ecotourism, as therapeutic medicinal plant harvesting and as a part of multiple farming to accompaniment the main source of income.

The production of horticultural crops through nursery raising is advantageous forhealthier germination and establishment and also safeguards time and labour saving. Nursery is a small area where the plants are spread and continued in the initial years. Generally, horticultural crops are raised in nurseries and then transplanted in the field. The establishment of nursery needstechnical know-how of propagation methods and possessions such as land, mother plants, plant propagation structures, growing media, ampules, containers etc. The planting material for orchards are higher from seeds or vegetative propagules. The result of nursery depends on eminence and dependability of mother plants. It should be designated on the basis of its genetic traits and other factors like availability and adaptation to the particular environment. Keeping in mind, the increasing tendency in horticulture production, the purpose of this chapter was to study the nursery disease management in horticultural crops under protected cultivation which, is important for raising disease free plants.

2. Advantages of Nursery Management Under Protected Cultivation

- It providesbettergermination and growth conditions.
- Hybrid seeds of vegetables crop are very costly. Seeds sown in open conditions have higher chances of mortality due to unfavourable environmental conditions. Therefore, the farmers generally opt for higher seed rate which subsequently increases the charge of production. Further, nursery raising under protected conditions exhibits higher survival. As a result, the seed rate and cost of production is relatively low.
- Nursery raising under protected conditions provides protection against pathogenic infection to a great extent.
- Nursery grown under protected conditions is quite fast growing as compare to open field and fetches higher price in the market. Thus, it is economically more profitable.
- There is land and labour saving as main field will be occupied by the crop for lesser duration. Hence intensive crop rotation as well as other operations of crop cultivation can be followed.
- More time is available for the preparation of main field because nursery is grown separately.

3. Protagonist of Nursery in Horticultural Development

- It is critical for healthy and vigorous plant growth to produce pure nursery material or racially superior planting material. Both the rootstock and the scion plant must be genetically pure. The planting material ought to be sufficient in quantity and quality, as well as easily available for widespread multiplication.
- Globalization has increased the possibility of exporting high-quality plantlets to other countries. Special processes are used to produce high-quality planting materials for export. Similarly, when imported nursery planting material from outside, extreme caution is required.

- Skilled workers are in high demand for potting, potting medium, grafting, budding, and supplemental nursery activities. Nursery jobs include technical, semi-skilled, skilled, and unskilled labour. In the shifting national landscape, nursery can be a very profitable business.
- Like India, there are so many countries in this world, which expression drought every year. Growing and propagating drought tolerant varieties of fruit crops provide guaranteed income to the farmers.

4. Kinds of Nurseries Rendering to Kind of Plants Grownup

4.1. Fruit Crop Nurseries

Fruit crops are chiefly propagated by vegetative means and need particular approaches for propagations as well as repairment. Pomegranate, mango, guava, sapota, citrus etc. are propagated through vegetative methods. Fruit nurseries are indispensable for production of grafted and blossomed plants else the mother plants of cuttings and rootstocks.

4.2. Vegetable Nurseries

Vegetables like; tomato, capsicum, chilli, onion, cabbage, cauliflower etc. are raised under nursery. The protected nursery raising allows off-season raising of vegetable seedlings for advance crop production.

4.3. Ornamental Plant Nurseries

Gerbera, roses, carnation, gladiolus, lilies, and other floricultural and aesthetic crops that are propagated vegetatively are available. There is also a significant range of attractive plants that are propagated through seeds, such as asters, marigolds, and salvia.

5. HI-tech Nursery Production

Nearby, there has been an unanticipated increase in demand for proven profitable plants. Tissue cultured banana, strawberry, gerbera, and carnation, for example. Normal or usual nursery methods are unlikely to meet this obligation. To meet the increasing demand, special techniques and approaches are required. This type of need may be met by high-tech nurseries. These nurseries raise plants in greenhouses, which are clad with glass or a plastic sheet to protect young plants from harsh weather while providing access to light and ventilation. Temperature, ventilation, light, watering, and fertilisation may all be controlled automatically in modern greenhouses. Some have fold-back roofs to enable for "hardening-off" of plants without the requirement for human transfer to outdoor beds.

5.1. Site Selection for Nursery

Land is the rudimentary and fundamental physical resource for plant nursery. The area available must be measured before planning the nursery and the products.

- To avoid problematic and unmanaged soils, soil sampling at the chosen site should be done.
- The chosen location should have well-drained, absorbent soil with a medium texture. The pH of the soil should be between 6.5 and 7.5. Heavy, black cotton soil, sandy, poorly drained soils, and soils with a pH greater than 8 are severely evaded. The soils must be devoid of salts and other potentially hazardous components.
- The irrigation water at the chosen location should also be analysed for pH and the presence of salts.
- Low lying land should not be chosen to avoid flooding.
- The chosen location should be close to a train or bus terminal.
- Prior to planting nursery plants, windbreaks and shelter belts should be installed.

5.2. Media for Nursery

There are several media and media combinations that are available for use in plant propagation particularly for rooting and growth of vessel plants.

5.3. Characteristics of Good Nursery Media

- The growing media should have the following characteristics for better results.
- The growing medium must be hard enough to hold the cutting in place during roots.
- When it is dry or moist, the volume of nursery media must be generally constant.
- It must hold enough moisture to reduce the frequency of irrigation.
- The media should be permeable in order for excess water to be drained.
- It should be free of weed seeds, diseases, termites, nematodes, and other pests.
- The media should be sterilizable or acceptable for sterilisation without causing any adverse effects.

5.4. Some Important Nursery Media

5.4.1. Soil

Soil is very common, easily available and cheaper media used in nursery. It is made up of both organic and inorganic matter. Organic matter includes the leftovers of living and dead plant, animal, and microbial parts, whereas inorganic matter includes mineral nutrients. The soil solution, which contains water, dissolved minerals, as well as O2 and CO2, is the liquid component of the soil. The gaseous component of the soil is critical for maintaining proper air-water balance. The texture of the soil is determined by the quantities of silt, sand, and clay. Soils are classed as sandy, loamy sand, sandy loam, silt loam, clay loam, or clayey based on their texture. The arrangement of their particles in the soil mass is referred to as the soil structure. The nursery soil should be of good consistency and structure.

5.4.2. Sand

Weathering on parent rocks produces sand, it is commonly used as a plant propagation medium. Sand particle sizes typically range from 0.05 to 2.0 mm. The sand used in plastering is excellent for rooting cuttings. Before using sand as a media, it should be heated or fumigated. Sand, in general, doesn't really contain any mineral nutrients.

5.4.3. Peat

Peat media is made out of marsh swamp leftovers. Vegetative peat moss is available, however it must be finely ground before being utilised in various combinations or as media. It includes organic nitrogen and is beneficial for cutting roots or seed germination.

5.4.4. Sphagnum Moss

Sphagnum is a dehydrated remnant of acid hog plants that grows in three genera in humid forest regions. It is generally sterile, lightweight, and has a very high water capacity. In general, it absorbs and holds 20 times its weight in water. Sphagnum moss has an acidic pH of roughly 3.5 in nature. It also contains a fungistatic ingredient that helps to prevent damping off. Before being employed for propagation, moss is soaked in fungicide solution and saturated with nutrient solution. It, like pomegranate and figs, is used for air layering.

5.4.5. Cocopeat

Cocopeat is widely used to prepare soilless nursery media. Cocopeat prepared from coconut husk. It has good water holding capacity and provides aeration for better growth of seedlings.

5.4.6. Vermiculite

Vermiculite is a micaceous mineral that expands dramatically when heated. It is composed of hydrated magnesium, aluminium, iron, and silicate. It is very light in weight when enlarged. It has a neutral reaction, high buffering characteristics, and is insoluble in water. It is offered in four grades, with Horticulture Grade No. 2 recommended for rooting and Horticultural Grade No. 4 recommended for seed germination.

5.4.7. Perlite

It is grey white material having volcanic origin. It is neutral and has no cushioning reaction and contains no mineral nutrients. Perlite mainly provides aeration to the roots of developing seedlings.

5.4.8. Leaf mould

It is formed of fallen leaves from many tree species found locally, such as oak, silver oak, maple, neem, and ficus. It is finished by stacking many layers of leaves and topping them with a thin coating of dirt and cow dung slurry. To hasten the decomposition process, a live culture of decaying critters is added. After around 12 to 18 months of breakdown, the medium is suitable for usage.

5.4.9. Saw Dust

It is a byproduct of saw mills. The volume and quality are determined by the type of wood used. It is commonly used as part of nursery material for rooting cuttings. In nurseries, a combination of few media is consistently preferred and used. As nursery media, FYM, sand, and garden soil (3:1:1) are commonly used. Nowadays, soilless medium consisting of cocopeat, perlite, and vermiculite (3:1:1) are popular. The availability, affordability, and simplicity of handling of media must all be considered.

6. Common Nursery Diseases

6.1. Damping Off (*Pythium aphanidermatum*)

Out of different diseases, damping-off disease is ubiquitous in nature and is a major constraint for horticultural crops in nurseries and during early establishment phase of seedlings in field resulting in considerable loss of crops.

6.2. Fusarium Wilt

The incidence of fungal wilt is common in the vegetable crops mentioned below:

(i) Tomato -*Fusarium oxysporum* f. sp. *lycopersici*
(ii) Chilly-*Fusarium oxysporum* f. sp. *vesicatoria*
(iii) Okra- *Fusarium oxysporum* f. sp. *vasinfectum*
(iv) Water melon- *Fusarium oxysporum* f. sp. *niveum*

6.3. Leaf Spots

Leaf spot is caused by *Cercospora* sp., *Cercospora* spp., *Alternaria* sp. and *Botrytis* sp. It is mainly found in following crops:

(i) Chillies: *Cercospora capsisci*
(ii) Tomato: *Septoria lycopersici*
(iii) Okra: *Cercospora malayensis*
(iv) Brinjal: *Cercospora solani* and *Alternaria Solani*
(vi) Onion leaf blight and Purple blotch: *Botrytis* sp
(vii) Water melon: *Alternaria cucumerina, Colletotrichum orbiculare*

6.4 Viral Diseases

6.4.1. Mosaic

It is transferred by aphids. Theaffected plantsappear stunted with lightand dark green mottling on the leaves that gradually turn distorted, puckered andsmaller than normal leaves. It is common in tomato, chilly, brinjal, cucurbits etc.

6.4.2. Leaf Curl

Leaf curl is generally transmitted by whitefly. Reduced leaf size with margins and curl upward (giving leaves a cup-like appearance), interveinal chlorosis, shortened internodes giving the plant a stunted appearance are the major symptoms associated with this disease. This disease is common in tomato and chilly.

6.4.3. Root Knot Nematode

The root knot nematode viz. *Meloidogyne incognita* drastically affects the plant growth in nursery stage. The affected plants exhibit stunted growth, marginal necrosis followed by leaf yellowing, necrotic lesions on the root surface. The roots of infected plants develop characteristic knots or galls on the roots. Severely affected plant show wilting.

7. Significance Of Diseases

Damping-off of seedlings occurs worldwide in valleys and forest soils, in tropical and temperate climates and in every greenhouse. Itmoves seeds, seedlings and roots of plants. Affected seedlings in seed beds are completely destroyed or die soon after transplanting (Agrios, 2004). This disease is economically very important due to its wide host range (Sharma and Sain, 2005). Under favourable conditions it causes severe seedling losses from 25 to 75 % in cauliflower (Sofi et al., 2007), 80-90 % in ginger (Kannan and Nair, 1965), 90 % in chili (Bohra et al., 2006), 100 % in tobacco and tomato (Pandey and Dubey, 1997) and 100 % in bidi tobacco grown by hydroponic methods (Yadav and Joshi, 2012). Several species of *Pythium* cause damping-off. Species of convinced other soil borne fungi, such as *Rhizoctonia, Phytophthora, Fusarium, Sclerotium, Phoma, Sclerotinia* etc. often cause damping-off (Singh, 2001; Rahman *et al.*, 2003; Agrios, 2004). There are too many fungi that can cause seed and seedling rot but species of *Pythium* are generally known as damping-off fungi (Singh, 2007).

8. Favourable Conditions for Disease Infestation

Plant disease agents are numerous and include practically every kind of life. Plant diseases are generally caused by a variety of microorganisms such as fungi, bacteria, viruses, mycoplasmas, and others. However, the prevalence of physiological variables such as temperature variations, a lack or excess of soil moisture and aeration, a lack or excess of plant nutrients, soil acidity or alkalinity, and so on favour disease development. Furthermore, air pollution, naturally produced toxins, and incorrect pesticide application may exacerbate disease infection in plants.

9. Diseases Management in Nursery

Plant diseases in various crops cause an annual loss of around 20 to 30 percent of global agricultural production. Plant infections are a major bottleneck in horticulture production, affecting irrigated crops, monoculture cultivation techniques, widely produced rainfed crops, and plant nurseries. The following are the most prevalent methods for treating disease infection in nurseries:

9.1. Soil Solarization

Solarization is a non-chemical way for controlling soil-borne diseases, insects, nematodes and weeds in soil prior to planting crops. Soil solarization detentions the radiant energy of sunlight under clear polyethylene film. When soil moisture is passable, the heat produced pasteurizes the soil, killing plant pathogens and insect and weed pests, but leaving intact populations of many beneficial soil micro-organisms that promote plant growth. In desert environments, soil temperatures in the top six inches of soil reach 140°F. The effect of soil solarization is greatest at the surface and decreases with depth. However, under optimal conditions, heating in the top 12 inches of soil is effective against a large range of pathogens and pests (Stapleton et al., 2008). A study by Rahman et al., 2003 found that soil solarization with white polythene sheet for 30 days during peak summer was most effective and reduced damping-off of tomato, brinjal and chilli. Sofi et al. (2007) reported that soil solarization in nursery beds, significantly reduced damping-off in cauliflower. Bohra *et al.* (1996) observed root rot incidence of jojoba seedlings (caused by *F. solani*) was 3 % in solarized plots compared to 13 % in non-solarized plots. Due to the influence of hot summers on soil temperature (temperatures reached above 40 °C in topsoil layer) of altered irrigated plots, integrated agricultural residue of Indian mustard with one irrigation and without plasticulture induced a 70-80 percent reduction in population of *F. oxysporum fsp. cumini* (Mawar and Lodha, 2002). Soil solarization has been confirmed to control diseases caused by many fungal pathogens such as *Rhizoctonia solani, Fusarium* spp., *Pythium* spp., *Phytophthora* spp., *Verticillium* spp., *Bipolaris sorokiniana, Plasmodiophora brassicae, Sclerotium rolfsii* etc. in many crops (Katan et al., 1983; Raoof and Rao, 1997; Puri, 2016).

9.2. Organic Amendments

Jha *et al.* (2007) tested neem, castor, mustard, groundnut cakes and saw dust against *Sclerotinia sclerotiorum* and obtained the minimum mycelial growth under water extract of saw dust amended plate followed by castor cake. Similarly, extract of neem cake was found to be inhibitory to the mycelial development of *Fusarium oxysporum* f. sp. *cubense* followed by groundnut cake. The lesser growth of the pathogen was recorded under farm yard manure (Saravanan *et al.*, 2004). Muthukumar (2009) evaluated seven oil cakes *in vitro viz.*, neem, coconut, sesame, pungum (*Pongamia pinnata*), illupai/ mahua (*Madhuca indica*), cotton and groundnut against *F. oxysporum* f. sp. *polyanthi*-1causing tuber rot of tuberose and found that water extract of all the amendments were inhibitory to the growth of *F. oxysporum* f. sp. *polyanthi*-1. Neem and illupai cakes recorded the lowest mycelial growth. Pandya *et al.* (2009) revealed that maximum growth inhibition of *F. solani* was recorded under FYM (53.03 %) followed by castor (48.37%), mustard cake (42.28%), poultry manure (28.29%), press mud (20.78%) and neem cake (9.05%). Narayanaswamy *et al.* (1998) found that the minimum mortality of seedlings (40/m^2) were in plots where neem cake @ 1 kg/m^2 was applied followed by pongamia cake @ 1 kg/m^2, cow urine @ 3 L/m^2 (50/ m^2) and tank silt @ 5 kg/m^2 (54/ m^2). Mathur *et al.* (2006) conducted studies for the management of wilt disease caused by *F. oxysporum*on fenugreek through soil amendment with three oil cakes *i.e.* mustard, sesamum and cotton with vermicompost and FYM in inoculated condition. They found vermicompost, mustard and sesamum oil cakes were significantly superior in reducing the disease incidence with lowest percent disease severity index (PDI) in vermicompost. Yadav and Joshi (2012) revealed that application of jatropha and castor cake significantly inhibited the pathogen *P. aphanidermatum* recording negligible mycelial growth.

9.3. Biological Control

The employment of a living cell to control or conquer another living entity is defined as biological control. Parasites, predators, fungus, nematodes, and viruses are examples of natural enemies. The majority of biological control objectives are focused on insects and mites. Plant disease exploitive composts, on the other hand, are being used to efficiently reduce problems with specific root rot pathogens. In addition, new products for biological management of root and foliar infections are being introduced.

Seed dressing with fungal antagonist *Trichoderma viride* significantly succeeded pre and post emergence damping-off of chilli (Ghosh, 2002). Bohra *et al.* (2006) reported that three native isolates of *Bacillus subtilis,* and *Streptomyces* sp. as seed treatment effectively managed damping-off of tomato, onion, cabbage and cauliflower. Seed and soil treatment with *T. viride* and *Pseudomonas fluorescens* reduced both pre and post emergence damping-off in brinjal (Ramesh and Korikanthimath, 2004). Application of talc-based formulation of *T. viride* and *P. fluorescens* in nursery beds before sowing significantly decreases damping-off of tomato and chilli (Manoranjitham *et al.,* 2001). Ghosh (2002) studied the biocontrol efficacy of nine fungal antagonists for management of damping-off of chilli caused by *P. aphanidermatum* and found that antagonist *Trichoderma viride*-2 inhibited the maximum growth of the pathogen followed by *T. harzianum*-3 and *Gliocladium virens (T. virens)*. Kumar and Hooda (2007) evaluated seven isolates of *Trichoderma* spp. for their antagonism against *P. aphanidermatum* through dual culture technique *in vitro* and listed *T. viride*-I as the most effective and *T. viride* III isolate, *T. harzianum* and *T. virens* as the least effective. Ramesh (2004) evaluated *in vitro* antagonist against *P. aphanidermatum* the causal agent of damping-off of brinjal. The percent inhibition was high with *T. viride* isolates Tv 2 (78 %) and Tv 3 (68 %) and in *Pseudomonas fluorescens* isolates Pf 1 (56 %) and Pf 7 (45 %). Sharma *et al.* (2003) testified more than 50 % inhibition of growth of *P. aphanidermatumin vitro* by *T. harzianum* isolates. Manoranjitham *et al.* (2001) suggested that damping off of tomato caused by *P. aphanidermatum* significantly reduced pre and post-emergence damping-off by soil application of *T. viride*and, *P. fluorescens* in nursery beds before sowing either individually or in combination. Yadav and Joshi (2012) reported that *Pseudomonas fluorescens* and *Bacillus subtilis* were more inhibitory than *Trichoderma* spp. Seed biopriming with *T. viride*along with soil application of FYM enriched *T. viride* @ 100 g/m^2 caused in better seedling emergence (97.0%) and highest vigour index (3496.0), followed by seed biopriming with *T. viride* alone with seedling emergence (96.0%) and high vigour index (3235.2) (Jaisani *et al.,* 2016).

9.4. Botanicals

Bhat and Shukla (2001) discovered that a leaves extract of Oxyspora paniculata modified in pots resulted in the least amount of post-emergence damping-off of tomato. Lantana alba and Chenopodium ambrosioides essential oils were evaluated against different phytopathogenic fungi and found to have antifungal efficacy against *Colletotrichum* sp., *Fusarium moniliformae*, *Helminthosporium oryzae, Penicillium italicum*, and *Rhizoctonia solani* (Kishore and Dubey, 2002). Kishore and Dwivedi (1991) found that at a concentration of 2000 ppm, the essential oil of Tagetes erecta totally inhibited the growth of *P. aphanidermatum* and controlled damping-off in tomato seedlings by up to 50% in pot trials. Bhatt et al. (1994) discovered that cold water leaf isolates of *Polyalthia longifolia* at 5%, *Caesalpinia pauciflora*, and *Manilkara kauki* at 10% inhibited *P. aphanidermatum*, the causal agent of brinjal damping-off. Ram and Thakore (2005) investigated the fungitoxic effects of five commercial neem formulations (Amrutguard, Achook, Neemazal, Nimbicidine, and Neem oil) against two ginger storage pathogenic fungi (*Fusarium solani* and *P. aphanidermatum*) in vitro and in vivo. A 30-minute pre-storage treatment with Achook and Amrutguard at 0.1 percent reduced the incidence of storage rot and enhanced rhizome recovery. According to Bansal and Gupta (2000), *Azadirachta indica* extract totally prevented mycelial development and spore germination of *F. oxysporum Schlecht*, which causes fenugreek wilt, whereas *Ocimum bacillicum* and Lantana camara were likewise poisonous to *F. oxysporum*. Sharma and Gupta (2003) investigated the antifungal activity of Allium sativum, Ocimum sanctum, and seed kernels of *Lantana camara, Azadirachta indica*, and *Melia azedarch* against *Rhizoctonia solani*, which causes root rot and web blight in French beans. Soaking tomato seeds in chenopodium and lippia oil derived from *Chenopodium ambrosioides* and *Lippia alba*, respectively, was found to minimise damping-off disease and promote germination (Kishore and Dubey, 2002).

9.5. Resistant Sources

Plant resistance is a significant and ecologically safe strategy for the management of plant diseases. However, it depends on availability of resistant cultivars of crops. Plant disease resistance provide plantprotection againstpathogens in two habits: by pre-formed structures and chemicals, and by infection-induced responses of the immune system. Compared to a susceptible plant, disease resistant plant exhibits reduced pathogen growth on or in the plant. The wild families of the cultivated crop species are often hardy and act as source of disease resistance genes for varietal improvement.

9.6. Chemical Control

Chemical control of plant diseases is classified under three categories: seed treatments, soil treatments and protective sprays.

9.6.1. Seed Treatments

Chemical seed treatments may be effective in controlling infections in, on, and around sown seed. When bacteria or fungi infect the embryos, cotyledons, or endosperms beneath the seed coat, seed therapy is therapeutic. When seed treatment destroys fungus spores that pollute seed surfaces, it is referred to as eradicative. Seed treatment is beneficial because it prevents soil-borne fungus from penetrating seedling stems. There are two methods of seed treatment: physical and chemical. Hot water treatment, solar heat treatment, and other physical treatments are examples. Fungicides and bactericides are examples of chemical treatments. The seed-dip approach entails creating fungicide postponement in water and then immersing the seed in it for a set period of time.

9.6.2. Soil Treatments

Soil-borne infections multiply rapidly as soils are cropped repeatedly, eventually reaching such levels that contaminated soils are unsuited for agricultural cultivation.

Chemical soil treatments eradicate plant diseases and allow for continuous agricultural use by rapidly reclaiming infected soils. To remove weeds, insects, and plant pathogens, pre-planting chemical treatment of field soils for the management of nematode-induced illnesses, fumigation of seedbed and greenhouses soils with methyl bromide, and other practises are commonly used.

Soil applications of soil treatment chemicals for fungal control are often confined to furrow treatments. Many nursery seedlings are afflicted by seedling blights, stem rots, and root rots caused by sclerotia-producing fungi.

Due to the high cost of fungicides and the tiny proportion of treatments licenced for particular crops, notably those for decorative use, chemical management of damping-off as foliar application is limited to a few active components (Garzon et al., 2011). Among the most commonly used fungicides are etridiazole and metalaxyl, which are active against *Phytopthora* and *Pythium* spp. ; benomyl and thiophanate methyl, which are active against *Fusarium* and *Rhizoctonia* spp. ; mancozeb, which is active against *Fusarium* and *Pythium* spp. ; and captan, which is active against common damping- Many formerly available fungicides are rapidly becoming unavailable on the market, significantly limiting access to chemical treatments in many countries, particularly in the European Union (Lamichhane et al., 2016). On the other hand, resistance to routinely used fungicides developed by various pathogen strains has called into question the long-term viability of chemical control (Taylor et al., 2002; Reeleder et al., 2007; Weiland et al., 2014). All of these studies clearly demonstrate the necessity of non-chemical approaches for damping-off management, particularly for post-emergence aces.

10. Integrated Disease Management

Wajid *et al.* (1995) reported that solarization of the soil amended with neem cake @ 400 g/m2 + a foliar spray of Ridomil MZ @ 0.2% at 30 DAS effectively controls damping-off. The damping-off incidence in brinjal was reduced when seeds treated with *T. viride* or *T. harzianum* (4g/kg) and soil amended with FYM and neem cake (20t/h) before sowing (Karthikeyan *et al.*, 2001). Application of metalaxyl 0.1% as rhizome treatment and collective application of neem cake (250kg/h) with *P. fluorescens* reduced incidence of rhizome rot of ginger (Jayasekhar, 2004). Kumar and Hooda (2007) revealed that when antagonist *T. viride*-1 were in incorporation with metalaxyl (1g/kg), their effectiveness to control damping-off of tomato increased extremely compared to individual application. Patel *et al.* (2003) recommended that sunnhemp green manuring + soil solarization + metalaxyl MZ as effective and economical for the management of damping-off and nematodes management of nursery.

11. Defences to be Occupied During the Nursery Groundwork

- A nursery bed needs to be equippedprudentlyfor production of even and healthy seedlings.
- Besides nutrition, enough moisture and aeration must be maintained in nursery to ensure seedling growth.
- The nursery media must be readyusing fertile soil (rich in organic matter content), having adequate drainage and ventilation.
- Excess irrigation in sunken or flatbed may main to water-logging which, aggravates the incidence of damping-off disease. The frequency of irrigation of depends on the soil type. Sandy soil needs frequent watering while, soils having higher water retention bulk does not need frequent irrigation.
- Soil-borne infections caused by nematodes, insect-pests and pathogens should be achievedby treating the soil in different possible ways.

12. Conclusion

Nursery diseases are economically important due to extensive host array. The diseases are soil as well as seed borne difficult to control. Recently systemic fungicide is found to be effective against the diseases and widely used, there exists a possibility for development of resistance. Therefore, integrated management approach with soil solarization, organic amendments and bio-agents will give satisfactory control of nursery diseases.

References

Agrios, G.N. 2004. Plant diseases caused by fungi. In: Plant Pathology IV[th]Ed. Academic Press, USA.; pp 248-249.

Bansal, R.K., Gupta, R.K. 2000. Evaluation of plant extracts against *Fusarium oxysporum*, wilt pathogen of fenugreek. Indian Phytopath; 53(1): 107-108.

Bhatt, N., Sivaprakasam, M.K., Jeyarajan, R. 1994. Antifungal activity of some plant extracts. Indian J. Forest. 17(1), 10-14.

Bohra, B., Vyas, B.N., Mistry, K.B. 2006. Biocontrol agents and neem formulations for management of dumping off of brinjal and Chilli. Indian Phytopath. 59(2), 223-226.

Garzón, C.D., Molineros, J.E., Yánez, J.M., Flores, F.J., Jiménez-GascoMdM, Moorman, G.W. 2011. Sublethal doses of mefenoxam enhance *Pythium* damping-off of geranium. Plant Disease 95, 1233–1238.

Ghosh, S.K. 2002. Biocontrol of damping-off disease of chili caused by *Pythium aphanidermatum*. J. Mycopathol. Res. 40(2), 117-119.

Israel, S., Lodha, S. 2004. Factors influencing population dynamics of *Fusarium oxysporum*f. sp. *cumini* in the presence and absence of cumin crop in arid soils. Phytopathol. Mediter, 43, 3-13.

Jaisani, P., Prajapati, H.N., Yadav, D.L., Pandey, R.N. 2016. Seed biopriming and *Trichoderma* enriched FYM based soil application in management of chickpea wilt complex. J. Pure Appl. Microbiol. 10(3, 2453-2460.

Jha, A.K., Upadhyay, J.P., Lal, H.C. 2007. Evaluation of organic amendments against white mold of *Phaseolus vulgaris*. J. Mycol. Plant Pathol. 37(1), 141-142.

Kannan, K., Nair, P.V. 1965. *Zingiber officinale* (ginger) in Kerala. Madras Agric. J, 52, 168-176.

Katan, J. 1983. Soil solarization. In: Innovative approaches to plant diseases control. Chet, I. (ed.). John Wiley & Sons New York.; 77-105.

Kishore, N., Dubey, N.K. 2002. Fungitoxic potency of some essential oils in management of damping-off disease in soil infested with *Pythium aphanidermatum* and *P. debaryanum*. Indian J. Forest. 25(4), 463-468.

Kumar, M.R., Hooda, I. 2007. Evaluation of antagonistic properties of *Trichoderma* species against *Pythium aphanidermatum* causing damping-off of tomato. J. Mycol. Plant Pathol. 37(2), 240-243.

Lamichhane, J.R., Dachbrodt-Saaydeh, S., Kudsk, P., Messéan, A. 2016. Toward a reduced reliance on conventional pesticides in European agriculture. Plant Dis. 100,10–24.

Manoranjitham, S.K., Prakasam, V., Rajappan, K. 2001. Biocontrol of damping-off of tomato caused by *Pythium aphanidermatum*. Indian Phytopath. 54(1), 59-61.

Mathur, K., Bansal, R.K., Gurjar, R.B.S. 2006. Organic management of *Fusarium* wilt of fenugreek (*Trigonellafoenum-graecum* L.) a seed spice. J. Mycol. Plant Pathol. 36(1), 94-95.

Mawar, R., Lodha, S. 2002. Brassica amendments and summer irrigation for the control of *Macrophominaphaseolina* and *Fusarium oxysporum*f.sp. *cumini* in hot arid regions. Phytopathol. Mediterr. 41, 45-54.

Muthukumar, A. 2009. Evaluation of organic amendments on tuber rot of tuberose caused by *F. oxysporum* f. sp. *polyanthi*. *J. Mycol.* Plant Pathol. 39(2), 286-289.

Narayanaswamy, H., Syamrao, J., Kumar, M.D., Karigowda, C., Ravindra, H., Rudraradhya, M., Jagirdar, S. 1998. Management of damping-off disease in FCV tobacco nursery through organic amendments. Tob. Res. 24(2), 106-108.

Puri, S. 2016. Eco-friendly management strategies for soil borne plant pathogens. Int. J. Adv. Res. Biol. Sci. 3(1), 69-75.

Ram, J., Thakore, B.B.L. 2005. Fungitoxic effect of neem formulation against storage rot of ginger. J. Mycol. Plant Pathol. 35(1), 159-160.

Ramesh, R. 2004. Management of damping-off in brinjal using biocontrol agents. J. Mycol. Plant Pathol, 34 (2), 666-669.

Raoof, M.A., NageshwarRao, T.G. 1997. Effect of soil solarization on castor wilt. Indian J. Plant Prot. 25, 154-159.

Reeleder, R.D., Miller, J., Capell, B., Schooley, J. 2007. Mefenoxam sensitivity and the impact of fumigation on Pythium species and *Phytophthora cactorum* in ginseng soils. Can J Plant Pathol. 29, 427–436.

Saravanan, T., Muthusamy, M., Loganathan, M., Prabakar, K. 2004. Antifungal effects of organic amendments against *Fusarium* wilt pathogen in banana. Madras Agric. J. 91(7-12), 526-529.

Sharma, M., Gupta, S.K. 2003. Ecofriendly methods for the management of root rot and web blight (*Rhizoctonia solani*) of French bean. J. Mycol. Plant Pathol. 33(3): 345-361.

Sharma, P., Sain, S.K. 2005. Use of biotic agents and abiotic compounds against damping-off of cauliflower caused by *Pythium aphanidermatum*. Indian Phytopath. 58 (4), 395- 401.

Sofi, T.A., Tewari, A.K., Razdan, V.K. 2007. Soil solarization reduces damping-off caused by soil-borne fungal pathogens and improves vigour of cauliflower seedling. J. Mycol. Plant Pathol. 37(1),68-71.

Taylor, R.J., Salas, B., Secor, G.A. 2002. Sensitivity of North American isolates of *Phytophthora erythroseptica* and *Pythium ultimum* to mefenoxam (metalaxyl). Plant Dis. 86, 797–802.

Wajid, S.M.A., Shenoi, M.M., Sreenivasa S.S. 1995. Seed bed solarization as a component of integrated disease management in FCV tobacco nurseries of Karnataka. Tobacco Res, 21(12), 58-65.

Weiland, J.E., Santamaria, L., Grünwald, N.J. 2014. Sensitivity of *Pythium irregulare, P. sylvaticum*, and *P. ultimum* from forest nurseries to mefenoxam and fosetyl-Al, and control of Pythium damping-off. Plant Dis. 98, 937–942.

Yadav, D.L., Joshi, K.R. 2012. Bait technique for isolation of *Pythium aphanidermatum* (Edson) Fitzpatrick from soil. Indian Phytopath. 65(3), 2-6.

Yadav, D.L., Joshi, K.R. 2012. Efficacy of agrochemicals against *Pythium aphanidermatum* cause damping-off in bidi tobacco. J. Plant Dis. Sci. 7(1), 77-80.

Yadav, D.L., Joshi, K.R. 2012. Evaluation of various botanicals against *Pythiumaphanidermatum* incitant of damping-off tobacco. Green Farming. 3(2), 233-234.

Yadav, D.L., Joshi, K.R. 2011. Fungal and bacterial antagonists for the management of damping-off disease of bidi tobacco caused by *Pythium aphanidermatum*. Green Farming 2(6), 698-701.

29

Prospects of Value Addition in Flower Crops

***Kiran Kumari*[1], *Alok Nath*[1] *and Arvind Kumar Verma*[2]**

[1]*College of Horticulture, S.D. Agricultural University, Dantiwada Gujarat, India*
[2]*ICAR- National Research Centre on Seed Spices, Ajmer, Rajasthan, India*

1. Introduction

As purveyors of ornamentation, inciters of attraction, symbols of inspiration and peace as well as repositories of beauty, flowers hold special roles worldwide. Flowers have always been associated with mankind since time immemorial because they were used in religious offerings and other social ceremonies. Today; floriculture has emerged as one of the most lucrative business profession and as a major diversification options in the agri-businessin many countries of the world due to higher income per unit area. Floriculture is also known as sunrise sector of agriculture because of its fast growing and increasing demand. In India, about 3,09,000 hectares across the country was used for floriculture and produced 23,88,000 metric tonnes (MT) of loose flowers. India is second largest flower producer after China. According to a report of APEDA,India's total export of floriculture was Rs. 507.31 crores in 2017-18. The major importing countries are United States, Netherlands, United Kingdom, Germany, and United Arab Emirates. Flowers are the utmost perishable horticultural farm produce which is major impediment in its marketing and post harvest handling. The beauty and fresh look of flowers and cut foliage lost due to microbial activities and biochemical changes and can be retained only for a few days even by using the best technique of post-harvest management (Datta, 1999). Therefore, value addition is another important alternative to the correct use of fresh ornamental plants either in fresh or processed form.

Value-added floriculture is a process that enhances the economic value and attractiveness of flowers and ornamentals products through genetic modification, processing and diversification (Verma et al., 2012). Due to unstable prices for raw commodities and changing consumer preferences there is need for product diversification. Product diversification includes strategies and tools that will be used to attract new groups of customers by producing new products for new or existing target markets. The main objectives of value addition are utilization of excess produce and elimination of waste*i.e.* reduction in post harvest losses and utilization of unsold flowers, enhancing and stabilizing farm income by adding profits and finally affording achannel for creativity. In addition to floristry and flower arrangements various types of value added products are formulated and marketed nowadays like essential oils, gulkand, rose water, flavours, fragrance, natural dye, pharmaceutical and nutraceutical derivativse and insecticidal compounds etc. Farmers can get more returns by converting an indistinctive raw material into a unique and appealing product.For instance, establishing small oil extraction units in rose growing regions and export of rose oil rather than the rose flowers or petals (raw material) can help in generating considerable revenue for producers in national and international market (Kalam, 2015).

2. Need for Value Addition in Flowers

Flowers are the utmost perishable horticultural farm produce which is major impediment in its marketing and post harvest handling. The beauty and fresh look of flowers and cut foliage lost due to microbial activities and biochemical changes and can be retained only for a few days even by using the best technique of post-harvest management (Datta, 1999). Due to unstable prices for raw commodities and changing consumer preferences there is need for product diversification. Product diversification includes strategies and tools that will be used to attract new groups of customers by producing new products for new or existing target markets. The main objectives of value addition are utilization of excess produce and elimination of waste *i.e.* reduction in post harvest losses and utilization of unsold flowers to enhancing and stabilizing farm income by adding profits and finally affording achannel for creativity. In addition to floristry and flower arrangements various types of value added products are formulated and marketed nowadays like essential oils, gulkand, rose water, flavours, fragrance, natural dye, pharmaceutical and nutraceutical compounds and insecticidal compounds *etc.* Farmers can get more returns by converting an indistinctive raw material into a unique and appealing product. For instance, establishing small oil extraction units in rose growing regions and export of rose oil rather than the rose flowers or petals (raw material) can help in generating considerable revenue for producers in national and international market.

3. Value Added Products from Flowers

Flowers can be converted into a variety of beautiful value added products. There are different modes of value addition like creative skills in floristry, flower arrangements, tinting, flower dehydration and dry flower products, preservation and processing of flowers for making various products *viz.* essential oils, floral tea, *sharbat*, jam, pigments and dyes, pharmaceutical and nutraceutical compounds *etc*. So, for easy understanding of value addition in flowers, it can be classified broadly in following types:

3.1. Fresh Flower Value Addition

Among all other forms of value addition in flowers, the art of floral arrangements hold the key position. This needs lots of creativity and fewer inputs as compared to other forms of value addition. There is around 20-50% monetary value addition through flower arrangements in different forms. The popularity and commercialization of floral craft is rapidly increasing in our country especially during wedding and religious ceremonies. Flower bouquets are available on various websites also. Fresh flowers can be classified into two i.e. cut flowers and loose flowers. Value addition in fresh flowers can be done bycut flower products like bouquets, baskets, bunch, buttonaire, corsage etc. and loose flower products like garland, floral strings (toran), pomanders, wreaths, floral jewelry etc (De, 2020).

3.1.1. Flower Arrangement

Flowers have been used by man since time immemorial. Archeological findings suggests that the earliest remains of floral design was found in grave thousands of years ago. There has been documented evidences of cut flowers been used in floral designs through civilization of the Ezyptians, Romans, Greek, Indians, Chinese and Japanese. As civilization progressed, society changed so did the art forms and the use of flowers. From being used as offerings to gods. To a mark of honor with laurel wreaths, to a form of adornment, flowers have evolved over time to what we today know as floral design. Each civilization developed its own style of floral designs. The floral arrangements are broadly classified into Eastern and Western floral arrangements that influence today's floral industry. The Eastern style known as Ikebana is a minimalistic, disciplined floral art form which incorporates and symbolizes heaven, man and earth. Ikebana is an offshoot of Buddhism, came into being some 1000 years ago. First conceived by Buddhist priests for arranging flowers in temples. Thousands of schools exist in Japan today, but the most popular are Sogestu, Ohara and Ikenoba. In contrast to Eastern flower arrangement, western style emphasizes colour and uses a variety of botanical material in mass gathering of multiple flowers. The Western style is characterized by symmetrical, asymmetrical, horizontal and vertical style of arrangements. Different types of western flower arrangements are Triangular, round, oval, L-shape, S-shape and crescent shaped flower arrangements etc.

3.1.2. Garlands and Strings (*Toran*)

Garlands are prepared by tying flowers together with the use of needle and thread. Generally, sweet smelling flowers like rose, marigold and jasmine are used for making garlands. Garland can be prepared by only one type flower or combination of different flowers and foliage. Loose flowers along with foliage are used to make string or *toran* which is used to decorate doors during festivals, religious ceremony and marriages.

3.1.3. Floral Ornaments

The floral jewelry concept has become very popular among brides especially in metropolitan cities on mehandi event of marriage, as the floral jewelry will be matching to dress and light in weight and convenient to use.Floral bengals, crown, *veni*and*gajra* are made from flowers like jasmine, tuberose, crossandra, barleria and *Tabernae montana.* Floral ornaments of orchids are very popular in Thailand.

3.1.4. Wreath

It is a band of flowers or foliage inter-twined into a ring, usually palced on graves as amemorial.

3.1.5. Rangoli

The original name of Rangoliis, Rangavali, which is Ranga means colour and avail meansline in Sanskrit. Rangoli is a colourful two dimensional display in the form of an attractive design created on the floor to welcome the visitors, worship God and to mark special festive occasions and marriages.Rangoli when done with flowers is called floral rangoli and not only adds life and grace to the display but also provides elegance and beauty to the surroundings.

3.1.6. Speaking Flowers

It is a type of value addition in cut flowers. The speaking flowers are those in which the flowers petals are embossed with images, photos or short messages without harming or shortening the life of the flower. This refreshing twist of the embossed on flower enhances the emotional responses in both the giver and the receiver of the flower and has garnished international attention from celebrities as well as media. Speaking flowers revolutionized the floral industry and is making its mark on many markets that traditionally have not had interest in flowers, like gift market, greeting card market and promotional industry.

3.1.7 Tinting

Flowers are appreciated for their beautiful colours but all the colours are not available in all flowers like blue colour in rose. So, to have a flower colour of one's choice, they have to be tinted. Tinting means colouring flowers artificiallyafter harvesting. Tinting is one of the essential value addition technologies to introduced desired shades of colours to flowers and is generally done in crops where colour pigments are not present or dull e.g. tuberose, spider lily, white roses, white or light colour carnation etc. Other flowers with the possibility for tinting are white gladiolus, different orchid species, loose flowers of jasmine, crossandra, chandni, etc (De, 2011). The process of tinting of flowers involves the theory of capillary action. Due to transpiration, the cut end of the stem absorbs more water to replenish the water that was lost. Water and dye, travels through the xylem and gives colour to flowers. The stem of flower is cut by sharp knife and about 3-4 cm of cut stem dipped in a jar containing solution of food colouring dye (De, 2020).

3.2. Dry Flower Value Addition

The basic principle of dehydrating flowers is to slowly remove water while preserving the original shape and texture as much as possible. Dry flowers are the important products of present day floriculture due to their long lasting quality, year round accessibility, easy handling, low transportation cost, eco-friendly. India is fifth largest exporter of dried flowers and dried flowers have been exported for the last 40 years USA and Asian countries. The dry flower export from India has a share of around 70 percent of total export; therefore there is a great scope of drying the flowers grown in local area and to convert them into everlasting value added products like dry flower arrangements, baskets, bouquets, table arrangements and press dried flower products including greeting cards, bookmarks, swaths, paper weights, wall hangings, table tops, table matsetc. Among different dry flower products out pot pourri is a major dry flower export item. The majority of these dry flower products have a shelf life of 6- 9 months therefore, consumer can have the benefit of these flower products for comparatively longer time and utilize maximum benefit of the money spent. For successful value-addition it is recommended that carefully identify goods that exploit local resources and fulfill the gap in the market. Adding value adds to the cost of production, but careful planning and test marketing can significantly increase the net return of a small-scale floriculture enterprise. Some of the dry flower exporters are Natural Products Export Coporation Ltd. (NPEC) Tuticorin, Singhri flowers Pvt. Ltd., Kolkata, Ramesh flowers Pvt. Ltd, Tuticorin. Value-addition ensures high premium to the grower, while providing more acceptable quality products for the domestic and export market, and promote the consumers to buy such products. By making use of value addition technology, wastage can be avoided during the glut period and farmers involved in the flower cultivation can earn better remuneration. Value-addition ensures high premium to the grower, while providing more good enough quality products for the domestic and export.

3.2.1. Potpourri

Potpourri is a mixture of dried, naturally fragrant material, used to provide a gentle natural scent in houses. It is usually placed in a decorative wooden bowl, or tied in small sachets made from sheer fabric. Pot pourri is used inside the home to give air a pleasant smell. The fragrance is achieved by mixing spices and scented fixatives to the dried petals and herbs. The basis of potpourri is the aromatic oil contained in plants. Therefore, the design and manufacture of potpourri is based on the principle that these aromatic oils are slowly released into the atmosphere, giving the environment a pleasant scent. The two types of potpourri available are wet and moist, both of which require a fixative to absorb the aromatic oil and are then gradually released. The materials selected to make this particular type of dried flower arrangement must be mildew resistant, non-toxic and free of harmful odors (De, 2020).

3.3. Processed Products

Value addition also offers an opportunity of making processed flowers products like essential oilsjam, jelly, beverages, rose water, gulkand, gulroban, floral dyes, floral tea. Besides this a number of products like essential oil, perfumes, insect repellants, poultry feed, insect repellent, floral dyes,cosmetics, pigments like xanthophylls, carotenoids, cosmetics like calendula creamluteins, anthocyanins, etc. which have neutraceutical properties and pharmaceutical compounds can be prepared from the flowers (Jain, 2016).

3.3.1. Essential Oils and Fragrance

Essential oils are concentrated hydrophobic liquids containing volatile aromatic compounds of plant origin. Essential oils are also known as volatile oils used for aromatherapy, medicines, perfumes, soaps, cosmetics and confectionary etc. These oils are extracted using solvent extraction method, distillation, maceration, enfleurage and expression (Rajamani, 2003). The rise in consumer demand for natural food products, coupled with the associated need for flavouring compounds from natural sources, have provided major boost to the essential oils and oleoresin market. The world production of essential oils is growing at more than 10 per cent annually. India contributes 16 per cent to the world production of essential oils and next only to Brazil and USA. Next to rose, jasmine is the most important perfumery raw material used since the beginning of human civilization. Egypt is the largest producer of jasmine oil (10-15 tonnes concrete/year) followed by Morocco, France, Algeria, Italy and India. Various toiletry and other products are made of jasmine using jasmine oils, essence and plant material etc., such as soaps, moisturizers, massage oils, hair spa products, face washes, face packs, candles, incense sticks and reed diffusers etc. The companies dealing and manufacturing these products are Cowshed (UK), Soulflower India Pvt. Ltd. (Mumbai), Khadi Natural Healthcare (India), Faith in Nature (UK), Beaumont Products, Inc. (USA), Bain de Terre (UK) and K Hall Designs (USA) etc. Oil of tuberose is used on heavier types of scents and used in non- alcoholic beverages, ice cream, candy, baked goods etc. Fragrant flowers are added along with other ingredients to beverages prepared from chocolate and served hot or cold. In Java, the flowers are eaten along with the juices of the vegetables. In India, essential oils areusually extracted from chamomile, clary sage jasmine, lavender and rose for export.

3.3.2. Aromatherapy

Aromatherapy treats anxiety and mild illnesses by rubbing a pleasantly scented natural oil on the skin or inhaling the scent. It is the use of aromatic essential oils for the benefit of the body-in emotional and physical health and beauty. Science has discovered that our sense of smell plays an important role in our overall health. Essential oils are inhaled and transported directly through the sensory system to the limbic system of the brain. In fact, the brain responds to scents that affect our emotions and chemical balance. Essential oils are also absorbed by the skin and are carried throughout the body through the circulatory system to reach all internal organs. For example, many essential oils have antiseptic properties, but some are stronger than other essential oils. In addition, different essential oils have different properties, but in many cases they have an uplifting effect.

3.3.3. Edible Products

Beyond their aesthetic value and beauty flowers are also exploited for making various edible products. Rose and Rhododendron used for making squash. Rose petals are used for making gulkand, jam and marmalade. Chrysanthemum, chamomile, jasmin, hibiscus are used in herbal tea. Jasmine tea is not tea made from jasmine. Brewed Green tea is infused with jasmine buds. The resulting flavor of jasmine tea is subtly sweet and highly fragrant. It is the most famous scented tea in China and most popular drink of Japan. It reduces cholesterol level. Jasmine species commonly used for Jasmine tea are *Jasminum officinale* and *Jasminum sambac*

3.3.4. Phytochemicals

Pharmaceutical and neutraceutical compounds: Due to globalization international competition is increasing, people are becoming more conscious, therefore use of natural ingredients is increasing in food and pharmaceutical industry. Therefore, the flowers pigments like carotenoid, xanthophyll, anthocyanin etc. are being used in food industry and compounds like lutein, vincristine, catharanthine, vitamin C extracted from rose hips are being used in pharmaceutical industry. Carotenoids are the major source of pigment for poultry feed. The pigment is used in feed for intensification of yellow colour of egg yolk and broiler skin and increase the lutein content which is good for health.

Phytochemicals are secondary metabolites that give foods flavor, aroma, and color, and also have the characteristics of pharmaceuticals and dietary supplements used in medicine and agriculture. Chrysanthemum contains pyrethrin alkaloids, the active ingredient that can be extracted in the form of oleoresin by squeeze the flowers of *Chrysanthemum cinerariifolium* (pyrethrum). They are biodegradable and non-persistent, so they are used as safe pesticides. Xanthophiles, carotenoids, flavonoids, anthocyanins, flavonols and flavanones are especially known as phytochemicals because they give foods flavor, aroma and colour and have the properties of pharmaceuticals and dietary supplements that can be used in medicine preparation. They are powerful antioxidants and are primarily involved in the reception and protection from UV rays. Marigolds are used to procure important phytochemicals such as lutein and zeaxanthin. They are effective against cancer, heart disease, cataracts, and age-related nuclear degeneration and now commercially available in the market. Lycorine and tazetin from dried *Narcissus tazetta* (narcissus) bulbs have antiviral and anti-leukemic properties. Dahlia tubers are rich in insulin and fructose, phytic acid and benzoic acid, and are anti-diabetic drugs. *Calendula officinalis* contains celandine, flavonoids, polysaccharides and pectic substances and is used to treat fever, ulcers, burns and wounds. Dendrobine, malaxine and phalaenopsine are another phytochemicals obtained from orchid species (De, 2011).

3.3.5. Pharmaceutical and Neutracutical Products

The term "neutraceutical" was coined in 1989 by Stephen De Felice, founder and chairman of the Foundation for innovation in Medicine in America. Nutraceuticals, which have also been called medical foods, designer foods, functional food and nutritional supplements can be defined as a food or part of food, that provide medical or health benefits, including the prevention and treatment of disease. Over the past few years, an increasing number of dietary supplements have become available. The term nutraceutical is used to describe these medically or nutritionally functional foods. Nutraseuticals, in contrast to pharmaceuticals, are substances, which usually have not patent protection.

4. Dyes From Flowers

The flowers are enriched with various pigments viz. chlorophyll, flavonoids, xanthophylls, carotenoids, betalains and anthocyanins that generally used in colouring food, beverage, textile and paper industries. Natural dyes, dye stuff and dyeing are as old as textiltiles themselves. Man has always been interested in colours and the art of dyeing has a long past history which can be isolated and used for varied applications including dying the fabric and paper. Many native flowers possess valuable pigments like Anthocyanins, Flavonoids, carotnoids and Xanthophyll that are used in food, beverage, textile, paper and confectionary industry. Now a days there is a trend towards the use of natural coloursthroughout the world. Owing to the non-toxic effects of natural dyes people prefer more of natural colours. These colours are used in food processing, meat canning, manufacturing of saices, soft drinks, confectionary, ice cream, dairy products, cosmetic accessories, ink, paints, lipsticks, soap, etc. Some pigments are also nutritional supplements in food e.g. carotenoides are used as vitamin suppliments as carotene is the precursor of vitamin A. The natural colour yielding plants are Marigold, saffron, Rosemary, Henna, Germen Chamomile, *Salvia officinalis*, *Bixa orellana*, *Indigofera tinctoria*. *Butea monosperma*, *carthamus tinctoria* (safflower) are used for colouring wool and silk.

5. Floral Waste Conversion to Value-Added Products

Solid waste disposal is major hitch in the world. Agricultural residues, temple waste, domestic waste are enriched with carbon content. Floral waste is one of the major concerns. Flowers have applications in many industries viz; perfumes, cosmetics, food, liquor and textile industries. Disposal of flowers in rivers, oceans, etc. leads to water pollution as well as affects the living organisms present in the waters. The execution of floral wastes by solid state fermentation for the conversion intodifferent value-added products viz; compost, biofuels, biogas, bioethanol, organic acids, pigments, dyes, polyhydroxybutyrate -co-hydroxyvalerate production, food, products; biosurfactants production, sugar syrup, incensesticks etc (Waghmode et al., 2016). Many self help group and NGOs in India working on flower waste management and has several success stories. One of this is to convert flower waste into the incense sticks. In this process flowers waste is collected from the temples and flowers are segregated by hand and plastics or paper are weeded out. The incense sticks made from these waste flowers. The incense sticks are natural and free from any chemical or harmful substance.

6. Value Addition Through Gene Editing

Use of molecular and genetic engineering technology for novelty important ornamental characters such as flower color, fragrance, and flower shape can be clarified through isolation and analysis of related genes. Modification technology for these characters by genetic transformation has also been developed. Different improvement strategies are being applied like hybridization, mutation, polyploidy, genetic engineering to breed the cultivars with new flower forms (Chavan et al., 2019).

7. Conclusion

Value addition is a business strategy for creating new market demands from the set of conventional customers. Exploration of value addition avenues can answer the two major issues *viz*. Un- employment and poverty of the country by provided that a steady source of income to the producers. Broadening the mindset towards use of the flowers for other industries and not only ornamental purpose can help us create a source of additional income. There is an urgent need for value addition in floricultural products through processing, packaging and supply chain management to increase the farm income and generate employment. Shifting focus towards creation of varied products in allied industries can help popularize the plant, creating market demand. To achieve this goal, concerted efforts on the part of planners and policy makers as well as producers will be needed. In addition, the extension service network of the agricultural university and agricultural research stations in the region could also be used for this transfer. value-added products from non- conventional floricultural crops like essential oil of rose, tube rose, jasmine, marigold etc. and plants extracts used in medicines and pharmaceutical industry are unique and likely to face less competition in the international market after the post WTO scenario and thus have the potential for export and import substitution. But the challenges encountered are synergy in the plan, technology for value added products, knowledge for standards to be followed, availability of varieties as per market preferences, and approach of various agencies at various levels in different functional areas i.e., research, finance, quality assurance and certification.

References

Chavan, V.B., S.P. Khedkar, B.R. Salvi, Y.R. Parulekar and Dalvi, N.V. 2019. Novel Innovations In: Floriculture Industry. Int. J. Curr. Microbiol. App. Sci. 8(06), 1855-1863.

Datta, S.K. 1999. 'Dehydration flowers and foliage and floral crafts'. In: Floriculture and Landscaping (Eds) Bose, T.K., Maiti, R.G., Dua, R.S. and Das, P. Kolkata, India: NayaProkash, pp. 696-703,

De, L.C. 2011. Value addition in flowers and orchids. New India Publishing Agency, New Delhi.

De, L.C. 2020. Value Addition in Flowers. VigyanVarta. 1 (3), 10-15.

Jain, R. 2016. Floral Value-Added Products for Employment Generation. https://www.biotecharticles. com/Agricu lture-Article/Floral-Value-AddedProducts-for-Employment -Generation 3673.html

Kalam, A.P.J. 2015. Mission India: A Vision for Indian Youth.Penguin Publishers, UK.

Verma, A.K., Gupta, A., Kumar, D., Dhiman, M.R. 2012. Post harvest technologies for commercial floriculture. New India Publishing Agency, New Delhi

Waghmode, M.S., Gunjal, A. B., Nawani, N.N., Patil, N.N. 2016. Management of floral waste by conversion to value-added products and their other applications. Waste Biomass Valor. DOI 10.1007/s12649-016-9763-2

30

Value of Underutilized Vegetables for Nutritional Safety

Brijesh K. Maurya[1], Neetu[2], Nirankar[1], Arvind K. Verma[3]

[1]*Department of Vegetable Science, C.S. Azad University of Agriculture and Technology, Kanpur, Uttar Pradesh, India*
[2]*Department of Vegetable Science, Banda University of Agriculture and Technology, Banda, Uttar Pradesh, India*
[3]*ICAR-National Research Centre on Seed Spices, Ajmer, Rajasthan*

1. Introduction

Underutilized vegetables are essential to the lives of the poor in rural and urban areas, as they provide poverty reduction and environmental sustainability for healthy societies. Traditionally, underutilized vegetables have been known to play an important role in the nutrition of rural communities, especially during droughts, famines and the dry season. Some of these undeveloped resources have been domesticated, but most are still wild and worthless. The potential of these underutilized vegetables has been overlooked as modern agriculture has progressed. These vegetables are also called by different names including minor crops, orphan crops, neglected species, Underused Vegetables, under developed vegetables, local traditional, alternative vegetables and also promising species. Underutilized species have a lot of genetic variety, and they're tied to a lot of indigenous knowledge. Some cultivated species are replaced by native /underutilized plants. More than 120 crores people (or 15% of the global population) are malnourished and hungry today (Bal, 2003). 925 million people do not have enough to eat. In developing countries, Under nutrition contributes to 53% of the 97 lakh deaths of children under five each year. Now a days, deficiency of vitamins and minerals in human diet is causing severe diseases especially in children, young and pregnant women. Approximately 3 million deaths occur in a year from non-communicable diseases due to insufficient fruit & vegetable intake. The most common approach in addressing such type problems in supplementation and so called "food based strategies", which include basic education, awareness nutrition and food fortification. Some of the underutilized vegetables are also endangered. As we know, some species in vegetables are highly popularized, like okra, brinjal and cucurbits, whilst others are semi-cultivated or still growing wild. Little attention is paid to vegetables that are traditional / underutilized as valuable foods and medicines. Many of these plants are rich in medicinal properties and nutritional value, especially proteins, vitamins and minerals. There are several promising annuals and perennials that are used as vegetables around the world and can play an important role in improving human nutrition.

2. Underutilized Vegetable

According to IPGRI (International Plant Genetic Resources Institute) "Underutilized vegetable are those species which is grown in local production systems, where they are highly adapted to a range of ecological niches". Underutilized vegetables are grown, sold and consumed locally. The condition of these vegetable crops varies from crop to crop and from region to region. Species having potential for contributing food security, nutritional and medicinal importance, income generation and environmental services are comes under underutilized vegetable (Bull, 2000). whereas indigenous vegetables refers to species or variety genuinely traditional in area where over a period of time it has evolved, although the species may not be traditional.

2.1. Other Name of Underutilized Vegetable

Minor vegetable	Neglected vegetable	Lesser known vegetable
Underexploited vegetable	Rare vegetable	Wild vegetable

3. Importances of Underutilized Vegetable

Underutilized vegetable crops have many advantages

1) These vegetables are a promising source of vitamins, minerals, and other nutrients such as carbohydrates, proteins, and fats.
2) These are naturally strong and very easy to grow.
3) Production takes place under adverse soil and climatic conditions.
4) These vegetables have own history of consumption by local people & they are well know about their medicinal and nutritional properties.
5) These vegetables are cheap and readily available with less cultural practices.

4. Scope of Underutilized Vegetable

4.1. Consumer Demand

Seeing health benefits, the Majority of population, Young and old are beginning to choosing such vegetables.

4.2. Marketability

In the early stages of development, APMC (Agriculture Produce Market Committee) and the government need to take serious developments in this regard and provide appropriate channels for selling products.

4.3. Buyer

In future, shift of more wholesale buyers towards producers of such vegetables show great prospect of growth.

4.4. Storability of Crop

Focus on newly introduced varieties of such plants and extend their shelf life. This was once the growth of the space market.

4.5. Logistics and Supply Chain

This is a major issue related to agricultural products, especially vegetable crops. Because of their supply chain managements need to be improved to meet growing demand.

4.6. Protective Farming

Cultivation of Underutilized vegetable crops are not require proper care and maintenance. But for avoiding uncontrolled natural phenomena protected structures like green houses, polyhouses, Naturally Ventilated poly house, glass houses, etc can help in increasing yield.

5. Phytochemicals In Underutilized Vegetables

Word "Phyto" is a Greek word which stands for plant. Phytochemicals, in addition to vitamins, minerals and fiber tannins, phenolic compounds, mucus, etc., are plant-produced compounds that protect plants from sunlight, wind, and other natural damage. These are also responsible for the taste, colour, aroma & texture in vegetables. Phytochemicals such as flavonoids, anthocyanins, carotenoids, and antioxidant vitamins (C & E) have powerful antioxidant capacity to remove free radicals that contribute to aging and disease through cell damage. They can reduce the incidence of various degenerative diseases. Many vegetables are currently given the status of functional foods because they not only meet basic nutritional needs, but also provide health benefits. Consuming green leafy vegetables for health benefits is a common habit among Indians. Numbers of products are regularly consumed in this group. The major and acceptable green leafy vegetable of Kerala, Tamil Nadu and Karnataka are the leaves of *Moringa oleifera,* commonly known as drum sticks are generally recommended in diseases of liver, spleen etc. They are also mentioned for the treatment of arthritis, rheumatism and anaemia as indigenous medicine. Wild leek (*Allium ampeloprasum*) is considered a low energy source of Zinc and Fiber. Wild bulb antioxidant activity was better characterized by reducing power. *A. ampeloprasum* increases vegetable diversity enhancing the quality of current diets. Likewise, Punarnava is an underutilized leafy vegetable and leaves are rich source of protein, carbohydrate, fat, carotene, Vitamin C, riboflavin, niacin and various minerals and it is popular for its medicinal properties.

6. Nutraceuticals of Underutilized Vegetables

Vegetables provide food diversity and beneficial nutrients and phytochemicals in diets. According to De Felice, nutraceutical can be defined as, "A food (or a part of food) that provides medical or health benefits, including the prevention or treatment of a disease". Assessment of a wide range of vegetable dietary supplements provides important information for the selection and promotion of local vegetables for better use. Compared to popular vegetables, underutilized vegetables such as green onions, moringa, spinach, malabar spinach, and chow chow are easier to grow and more nutritious. Eating this vegetable helps improve the nutrition of humans around the world.

Table 1: List of underutilized vegetables crops and nutraceuticals

Underutilized Vegetables	Nutraceuticals
Leek, Shallot, Chive	Allylic sulphides
Parsley, Spinach	Carotenoids
Bitter Gourd	Cucurbitacins
Green leafy Vegetabes	Dietary fibre
Parsely	Flavonoids
Cruciferous Vegetables	Glucosinates
Brussels sprout, Broccoli, Chine Cabbage	Indoles & Sulfuraphane
Spinach	Lutein

7. Underutilized Vegetable Grown in India

Brief introduction of some popular underutilized vegetables which grown in India are given below-

Table 2: List of some importance underutilized vegetables

S. No.	Name of the Crop	Botanical Name	Family	2n No.	Edible Part	Sowing time	Propagation by	Harvesting time	Yield
CUCURBITACEOUS VEGETABLES									
1	Ash Gourd	*Benincasa hispida*	Cucurbitaceae	24	Immature fruit	June-July	Seed	Nov- Dec	10-15 t/ha
2	Kachari	*Cucumis callosus*	Cucurbitaceae	-	Mature fruit	June-July	Seed	Sep- Oct	2.5-3 t/ha
3	Pointed Gourd	*Trichosanthes dioica*	Cucurbitaceae	22	Immature fruit	Aug-Nov	Stem & Root Cutting	March-April	8-10 t/ha
4	Chow-Chow	*Sechium edule*	Cucurbitaceae	28	Fruit	April-May	Whole Fruit	September	20-25 t/ha
5	Spine Gourd	*Momordica dioca*	Cucurbitaceae	28	Immature fruit	June	Seed	Sept -Oct	10-15t/ha
6	Gherkins	*Cucumis anguria*	Cucurbitaceae	24	Young fruit	June-July	Seed	Sep- Nov	8-10t/ha
7	Snap melon	*Cucumis melo var. momordica*	Cucurbitaceae	24	Immature and ripe fruit	June-July	Seed	August	20-25t/ha
8	Snake gourd	*Trichosanthes anguina*	Cucurbitaceae	22	Immature fruit	June-July	Seed	Aug-Sep	8-10 t/ha
9	Ivy gourd	*Coccinia grandis*	Cucurbitaceae	24	Immature fruits	June-July	Stem Cutting	Nov-March	20-28 t/ha
UNDERUTILIZED BEANS									
10	Dolichus Bean	*Lablab purpureus*	Leguminosae	22	Pod and Seed	May-June	Seed	Oct-Nov	4-5 t/ha
11	Sword Bean	*Canavalia gladiate*	Leguminosae	22	Tender pod	May-June	Seed	Oct-Nov	2.5 kg pod/ plant
12	Lima Bean	*Phaseolus lanatus*	Leguminosae	22	Pod and Seed	May-June	Seed	Oct-Nov	2.0 kg pod/ plant
13	Clustered Bean	*Cyamopsis tetragonolobus*	Leguminosae	14	Pod and Seed	May-June	Seed	Oct-Nov	1.2 t/ha
14	Velvet Bean	*Mucuna pruriens*	Leguminosae	22	Pod and Seed	May-June	Seed	Oct-Nov	1.2-1.5 t/ha
15	Broad Bean	*Vicia faba*	Leguminosae	12	Green pod and Seed	Nov-Dec.	Seed	Jan-Feb	2.0 t/ha

UNDERUTILIZED COLE CROPS									
16	Broccoli	*Brassica olerasia var. italic*	Cruciferae/ Brassicaceae	18	Flower bud	Sep-Nov	Seed	Dec-Feb	15-16 t/ha
17	Brussels Sprout	*Brassica olerasia var. gemmifera*	Cruciferae/ Brassicaceae	18	Head (Mini cabbage)	Sep-Nov	Seed	Jan-March	8-10 t/ha
18	Chinese Cabbage	*Brassica chinensis*	Cruciferae/ Brassicaceae	20	Leafy head	October	Seed	Dec-Feb	10-12 t/ha
19	Kale	*Brassica olerasia var. acephala*	Cruciferae/ Brassicaceae	18	Shoots/ Leaves	Sep-Nov	Seed	Dec-March	7-8 t/ha
UNDERUTILIZED BULB CROPS									
20	Leek	*Allium porrum*	Amaryllidaceae	32	Blanched stem and Leaves	June	Seed	Dec-Jan	9.25 t/ha
UNDERUTILIZED LEAFY VEGETABLE									
21	Basella	*Basella rubra/alba*	Basellaceae	24	Fleshy stem and Leaves	June-July	Seed & Cutting	Dec-Jan	15-20 t/ha
22	Amaranthus	*Amaranthus spp.*	Amaranthaceae	32	Stem and Leaves	Feb-March	Seed	March-April	10-14 t/ha
23	Lettuce	*Lactuca sativa*	Asteraceae	18	Leaves	Sep-Nov	Seed	Jan-Feb	8-12 t/ha
24	Kulfa	*Portulaca oleracea*	Portulacaceae	16	Stem and Leaves	May-June	Seed	August	10 t/ha
25	Bathua	*Chenopodium album*	Amaranthaceae	54	Stem and Leaves	Sep-Oct	Seed	Nov-Dec	15-20 t/ha
UNDERUTILIZED WATER VEGETABLE									
26	Water spinach	*Ipomoea aquatic*	Convolvulaceae	30	Fresh Leaves	June-July	Stem Cutting	-	40-90 t/ha
27	Water Chest Nut	*Trapa natus*	Trapaceae	96	Kernel	June-July	Root	Feb-March	4-14 t/ha
28	Lotus	*Nelumbo nucifera*	Nelumbonaceae	-	Lotus Root	-	Rhizome	-	-

UNDERUTILIZEDTREE VEGETABLES									
29	Moringa	*Moringa oleifera*	Moringaceae	28	Leaf, fruit & Flower	June- July	Seed/ Limb cutting	March-Aug	70-80 t/ha
30	Curry Leaf	*Murraya koenigi*	Rutaceae	18	Leaves	July-Aug	Seed	-	20-25 t/ha
31	Chekurmanis	*Sauropus* androgynus	Euphorbiaceae	-	Leaves	Sep-Oct	Seed	-	30-50 t/ha
32	Casawa	*Manihot esculenta*	Euphorbiaceae	36	Tuberous Root	Oct-Nov	Stem Cutting	Aug-Sep	12.8 t/ha
33	Aghati	*Sesbania grandiflora*	Leguminosae	24	Flower and Leaves	July	Seed	Nov-Dec	4-6 kg/ha
UDERUTILIZED TUBERS									
34	Grater yam	*Discorea alata*	Discoreaceae	40	Underground Stem Tuber	March-April	Tuber	January	30-35 t/ha
35	Lesser Yam	*Discorea esculenta*	Discoreaceae	40	Underground Stem Tuber	March-April	Tuber	Jan-Feb	20-25 t/ha
OTHERS									
36	Globe Artichoke	*Cynara scolymus*	Compositae	34	Flower Bud	March-May	Suckers/Off shoots	May-July	10-15 t/ha
37	Parsnip	*Pestinaca sativa*	Apiaceae	22	Fleshy axil	March-April	Seed	June-July	25-35 t/ha

8.1. Underutilized Cucurbitaceous Vegetables

8.1.1. Snap Melon

Snap melon (*Cucumis melo var. Momordica*) belongs to family cucurbitaceae and chromosomes number are 2n=2x=24. It is originated to Africa Locally called as "Phoot" which means 'to split'. The fruits are oval or elliptic, with smooth skin, orange or light yellow in color with weight about 900 g. After removing the coat, seeds are used in bakery products and in preparation of traditional drink 'thandai'. Snap melon is good source of vitamin C, iron, Calcium. Snap melon tolerant to biotic and abiotic stresses.

8.1.2. Chow-Chow

Sechium edule indigenous to Mexico and West Indies is widely grown in Nilgiri and Madurai. In India chow–chow was introduced by western missionaries. Mizoram is a leading state in chow- chow production. The plant of chow-Chow is a hairy climber, making a growth of about 15-20 cm. The fruit colour of Chow -Chow varies light to dark green. The chow-chow contains moisture (92.50), protein (0.7 g), fat (0.1 g), mineral (0.4 g), fiber (0.6 g), carbohydrates (5.7 g), energy (27 k cal), calcium (140 mg), phosphorus (30 mg), iron (0.6 mg), riboflavin (0.04 mg), vitamin C (4 mg), niacin (0.4 mg) per 100 g of edible portion. The fruits of chow-chow are rich source of carbohydrates, proteins, fat, minerals and vitamins, particularly vitamin A and Vitamin C. In Khasi language it is called **piskot**. Owing to carbohydrate and minerals chow- chow fruits can also be used preparation of pickles, candy and flour products. The woody stems furnish a fine fiber. Fruits, vines and tubers are also excellent fodder for stock. The crop grows well in regions with a comparatively low temperature and high humidity.

8.1.3. Kachari

Cucumis callosus (*Kachri*) belongs to family Cucurbitaceae. It is native of India. In Sanskrit, it is known as chitra, chitraphala, mrigakshi whereas in English it is known as cucumber pubescent. It is a very hardy plant, monoecious in nature and highly cross-pollinated. The skin colour varies completely white to dark green and its various shades. At immature stage, it is generally green and bitter in taste but turns to pale yellow and sweet at ripening. Mature fruits are consumed in various forms as cooked vegetables, chutney and pickles. Powdered type of kachari is used in various types of spices on commercial scale. The dehydrated kachari is diuretic, gastric-stimulant, cooling refrigerant, coughicide and wormicide. Due to its medicinal properties chow -chow is used for improve digestion, facial paralysis, debility, bile, obesity, piles, and cough. Some work has been initiated at CIAH, Bikaner and two varieties have been developed which is AKH-119 and AKH 200. Fruit are harvested at maturity stage. At immaturity stage fruits are bitter in taste and these gradually becomes sweet and sour towards maturity. The Yield varies from 100-125q/ha.

8.1.4. IVY Gourd

Coccinia grandis (L.) (Ivy gourd) is also called little gourd or kundru or baby watermelon and gentleman's toes. India is considered centre of origin of Ivy gourd. It is dioecious, perennial climbing vine commercially propagated through vine cuttings. Tendrils are simple and long which are borne in leaves axils. The whole plant such as leaves, fruits, seeds and roots can be used for various medicinal purposes. Leaves are used in treatment of asthma, fever, skin and cough. The juice of the roots and leaves is used in the treatment of diabetes. The roots and stems are succulent and capable of storing water throughout the dry season.It is also used internally in the treatment of gonorrhoea.

8.1.5. Spine Gourd

Spine gourd (*Momordica dioca*) is a dioecious, perennial climber. Which are comes under family cucurbitaceae. Indo-Burma is believed to be its origin place. Tender, small, immature spiny fruits are used for culinary purpose. Sweet gourd is widely cultivated in Maharashtra, Gujarat, West Bengal, Jharkhand, Chhatishgarh and Orissa. It is good source of beta carotene, protein, carbohydrate and vitamin C. 100 g of edible fruit of spine gourd contain 84.1% moisture 7.7% carbohydrate, 3.1% protein, 3.1% fat, 3.0% fiber and 1.1% minerals (Singh 2009).

8.1.6. Sweet Gourd

Sweet gourd (*Momordica cochinchinensis*) is known by another named "Kakrol" is a cucurbitaceous vegetable having chromosome number 2n=2x=28. Fruits are oblong to round (15-20 cm), turn yellow to orange when mature. Numbers of spines are presents on fruit surface. Fruits of spine gourd are rich source of β-carotene, lycopene and zeaxanthin used to cure diabetes and several skin diseases.

8.2. Underutilized Leguminous Vegetables

8.2.1. Winged Bean

Winged bean (*Psophocarpus tetragonolobus)* popularly eulogized as goa bean, manila bean, asparagus bean and vegetable of 20th century. It is perennial in nature but horticulturally grown as annual. Its green pod dry seeds, leaves, flower and tuberous root are widely consumed. Winged bean contain 30-40% protein and 50-20% oil in dry seeds. Oil is rich in unsaturated fatty acids. Dry seed are also rich in lysine amino acid, anti-oxidant and tocopherol which improve human utilization of vitamin A. Due to low carbohydrates its tuberous roots are substitute of potato and good for diabetic patients.

8.2.2. Cluster Bean

Cluster Bean or guar (*Cyamopsis tetragonoloba*) is one of the hardiest vegetables among legumes successfully cultivated in north-west specially rajasthan and southern parts of India where rainfall is less than 750 mm. The green tender pods of guar are eaten as a vegetable and whole plants are also used in green manuring. Seeds of guar contain 68-70% galactomannan polysaccharides which are known as guar gum. Furthermore guar gum is used in textiles, paper, varnish and cosmetic industries. Green pods are rich source of vitamin A and iron. Dry seeds of cluster bean contain about 33.3% protein. Roasted guar is used as feed of pigs and poultry.

8.2.3. Sword Bean

Canavalia gladiata is a perennial climber popularly grown in Asia and Africa. Its tender pod and immature green beans are used as vegetable. Pods of sword bean look like sword in shape therefore it got named as sword bean. Each pod contains 10-12 red colour seeds. It has exceptionally high nutritive value such as carbohydrate and protein 55% & 23-34% respectively.

8.2.4. Velvet Bean

Velvet Bean (*Mucuna pruriens*) is an annual climber shrub which plant length may be over 15 meters. It is widespread tropical and sub-tropical parts of the world. Raw velvet bean seed contain about 29% protein some species of velvet bean have also been used for human consumption; such fruit are rich source of minerals, protein and amino acids. The roasted seeds of velvet bean are used as substitute of coffee. In pharmacologically study, velvet bean have several activities like anti-diabetic, anti-microbial, anti-neuoplastic and aphrodisiac. In Brazil its seed has been used in treatment of parkison's disease initially. Comparison to other pulse crops like soybean, rice bean and lima bean, velvet bean has high digestible protein (23–35%).

8.2.5. Broad Bean

Vicia faba L. (Broad Bean, Bakla bean, Faba Bean or Horse Bean) is a only frost resistant bean grown in winter season. It is cultivated in very limited scale in India but an important food crop of South America. The tender pods are cooked as vegetable and dried seeds are used as pulse. Pod are rich source of protein (26.12g/100g),carbohydrate (58.29%), calcium (103mg), vitamin-C (1.4mg) other mineral like potassium, Magnesium, phosphorus and zinc etc. Plant have distinguishing features like square and erect growing stems which grow to about 100 cm (tall varieties) and to 30 cm (dwarf varieties).Pod borne in cluster (5-7 pods) in the axils of the leaves which are 15 cm long and 2 cm thick.

8.3. Underutilized Leafy Vegetables

8.3.1. Kulfa

Portulaca oleracea is a succulent annual belonging to the Portulacaceae family with a green or purple stem. It grows as a weed throughout India. Carotene, folic acid, Vitamin C, and vital fatty acids are all abundant in it. Fresh kulfa leaves contain around 300-400 mg omega-3 fatty acids, 12.2 mg alpha-tocopherol, 26.6

mg ascorbic acid, 1.9 mg beta-carotene and 14.8 mg glutathione per 100 grams. Fresh herb is also used to treat skin wounds, bug bites and snake bites. Seed is the most common method of propagation of kulfa.

8.3.2. Amaranthus

Amaranthus is one of the most common and popular leafy vegetable in India. It belongs to the family Amarantheace, sub family Amaranthoideae, and genus Amaranthus. Amaranthus is C4 plant. There are 400 species of amaranthus are distributed in temperate, tropical and subtropical part of World. Leafy amaranthus is believed to be native of India. Amaranthus may be divided into two group basis on their consumption *viz.* grain amaranthus and leafy amaranthus. In India major leafy amaranthus growing states are Uttar Pradesh, Bihar, Kerala, Karnataka, Maharashtra, West Bengal, Orissa and Chhattisgarh (Kumar and Gupta, 2018). Grain amaranthus is very popular in Gujarat and Maharashtra and popularly known as Rajgirah "King seed" (Jangir et al., 2018). It contains three times more calcium and vitamin B than spinach as well as twenty times more iron than lettuce. Vegetable amaranths have many medicinal properties therefore it is recommended to patients of constipations, fever, hemorrage, anemia and kidney complaints.

8.3.3. Basella

Basella rubra (Indian spinach) is a cool season vegetable crop. Its tender leaves and stems are used for vegetable and medicinal purpose. Basella is also called by its other similar names like Indian spinach, Malabar spinach, Vine spinach. In southern and eastern part of India it is grown in every household. Basella contain high levels of vitamins and minerals such as calcium, phosphorus, zinc and sulfur. According to ICMR 100 g dried leaves of Basella fulfilled approximately 12 to 18% Recommended Dietary allowance (RDA) of protein vitamin C and iron for men and women. Plant juice of basella is used as a dye in food colouring.

8.3.4. Bathua

Bathua *(Chenopodium album* L.) is a herb belonging to family Chenopodiaceae. Its leaves and succulent stem are economical parts. Bathua leaves are cooked in form of Sag. It contains 11300 IU vitamin A, 35 mg Vitamin C and 4-6 gm Fiber at the rate per 100 gm fresh leaves. Its leaf powder is useful for treatment of peptic ulcer, general disability and splenopathy. There are many popular dishes are prepared from bathua leaves like bathua roti, bathua paratha and stuffed breads. When its leaf decoction is mixed with alcohol it gets pain relief in arthritis. Some fermented products are also prepared from bathua like idali, dosa and bread etc.

8.4. Underutilized Water Vegetables

8.4.1. Water Spinach

Ipomoea aquatica is a common vegetable which is eaten by all social groups. All plant parts of water spinach are edible although shoot tips and younger leaves are much preferred. Various medicinal effects are reported with the use of water spinach in different elements like headache, general weakness, stress, sleeplessness. The species contains several vitamins including vitamin A, B complex, C, E & Vitamin-K. Leaves and tender stems are often used for animal feeding. Due to high rich in iron doctors are suggest to patients who are suffering from anemia.

8.4.2. Water Chestnut

Trapa natans L. (Water chestnut) is a small free water floating herb belonging to the family Trapaceae recognized for its pharmaceutical and nutritional values. The kernels of Water chestnut are rich sources of carbohydrate, protein, and minerals, which are used in human diet. It can also use as anti microbial, analgesic, anti-inflammatory, anti-diabetic, antioxidant, antiulcer activity and antifungal. Water chestnut contains huge amounts of antioxidants, such as flavonoids and total phenol contents. Stem juice of water chestnut is used in eye disorders. Water chestnut plant is an important in water purification as they reduce current velocity and allow deposition of sediment and nutrient removal.

8.5. Underutilized Tree Vegetables

8.5.1. Drumstick

Moringa olefera L. (Drumstick), also known as horse radish tree or West Indian Bean, is a tree vegetable native to India that belongs to the Moringaceae family. Moringa is grown in kitchen garden and fields for its leaves, flower buds, seed, and tender fruits. Its tender fruits are valued for its distinct and attractive flavor. They are rich source of protein, minerals and vitamins. Moringa seed are consumed like peanuts in Malaysia. The seeds contain a lubricating agent called as ben or behen oil, which is used for illumination, soap preparation, lubricating watches, computers, and other expensive machinery. Seeds contain 38-40% non-drying oil that is transparent and odorless and never becomes rancid. Moringa has manifold medicinal properties which have described ayurvedic, sidhha and unani system of medicine such as anti-microbial, anti-asthamatic, analgesic, antipyretic, diuretic and cholesterol lowering. It is also suggested as a supplement of dietary minerals. Leaves and pods of Moringa contain good amount of sodium, copper, zinc, phosphorus and calcium (Farooq et al., 2012).

8.5.2. Curry Leaf

Murraya koenigii (L) is perennial, monoecious tree which belongs to family Rutaceae. It is commonly grown as backyard tree in Tamil Nadu, Kerala and Karnataka. The major chemicals compounds which are responsible for aroma and flavor in curry leaves are pinene, sabinene, cadinene, caryophyllene and cadinol. Many notable pharmacological activities of this plant such as anti-diabetic, anti-ulcer and anti-diarrhea are reported in human beings. It contains (12600 IU/100g) Vitamin-A, (2.3 mg/100g) nicotinic acid and (4 mg/100 g) Vitamin-C. Fresh leaves of curry leaves consist of volatile oils help to reduce the calcium deficiency.

8.5.3. Chekurmanis

Chekurmanis (*Sauropus androgynous*) eulogized as vegetable of 21st century, multivitamin green or madurakeera. In 1953 it was introduced in India form Malaysia. In southern states of country like Kerala, Tamil Nadu, Andhra Pradesh and Telangana are mainly consumed its leaves either in form of powder incorporated with spice powder or in form of cooked vegetables. Its tender shoots and leaves are used for culinary purposes. The leaves of chekurmanis contain 7.4% protein, vitamin A 4750IU, Vitamin B 51 IU and vitamin C 110 mg /100g. Besides, they also contain other vitamins such as D, F and K.

8.5.4. Sesbania

Sesbania grandiflora commonly known as agathi or humming bird tree is a small tree in the genus *Sesbania* and family Leguminosae. It is believed to be originated in Southeast Asia specially India, Indonesia, Malaysia and Philippines region. Agathi grows well primarily in hot and humid tropical areas of the world. Flowers and tender pods of agathi is used as vegetable purpose and it have various medicinal properties such as anti-diabetic, anti-microbial, anti-hyperlipidaemic and anti-bacterial. Apart from their acrid, bitter, and astringent taste sesbania is a good source of nutrients, vitamins and minerals. Consuming 20 gram of sesbania offers 14.6 mg of Vitamin c, 20 µg of Vitamin B9, 0.17 mg of Iron, 0.2 mg of Selenium, 0.017 mg of Vitamin B1 and 0.016 mg of Vitamin B2. Moreover many Amino acids like 0.004 g of Tryptophan, 0.011 g of Threonine, 0.014 g of Isoleucine, 0.021 g of Leucine and 0.013 g of Lysine are also found in 20 gram of sesbania.

9. Future Nutrition Strategy

In the past, nutrition was considered to be an individual's responsibility, however, more recently governments schemes to developed fortified varieties to full fill the nutrition requirement of human. Biodiversity is recognized as essential to food security and nutrition, and through the development of new vegetable varieties and the development of nutritional supplements and organisms through improved nutritional options and positive health benefits from traditional breeding approaches. Preferred genes for active compounds and edible colorants can be incorporated into cultivated varieties.

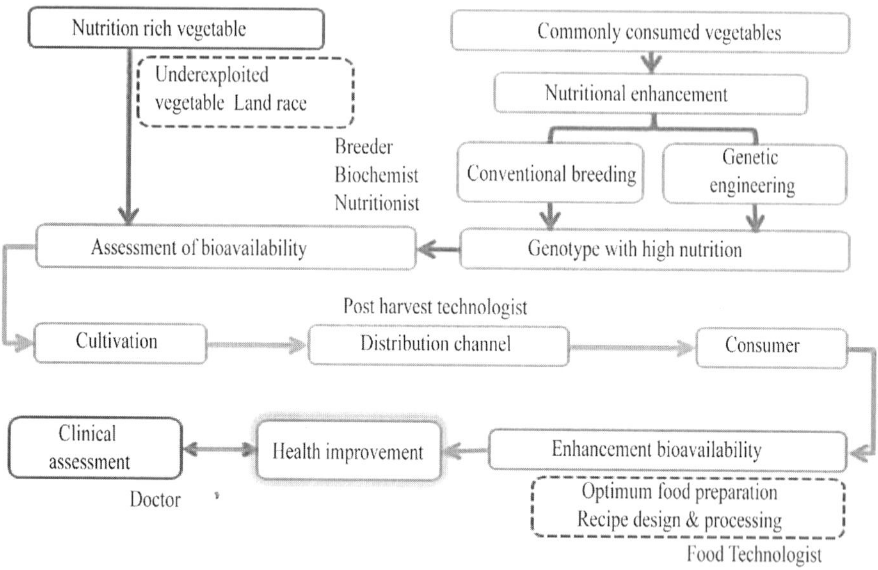

Fig. 1: A Flow Chart of Biofortification in Crops for Health Benefits

10. Conservation and Protection of Underutilized Vegetables

The National Bureau of Plant Genetic Resources (NBPGR) in New Delhi preserves active collections of numerous underutilized and under-exploited vegetables at its headquarters and sub stations across the country. In fact, the germplasm conservation of various plant species has been committed to the main centre of the All India Co-ordinated Research Project (AICRP).

10.1. Governments Policies for Development of Underutilized Vegetable Crops in India

The Indian Agricultural Research Institute (IARI) New Delhi started research on underutilized vegetable crops since 1960s. In 1982, All India Coordinated Research Project (AICRP), with its headquarters at the National Bureau of Plant Genetic Resources (NBPGR), New Delhi, enhanced this research to include the collecting, evaluation, utilization, and conservation of underutilized vegetable crops. Onwards, this work was initiated in many places of India. Underutilized crops were discussed as a way to improve India's food security and reduce poverty.

10.2. Schemes Launched by Government for Improvement of Underutilized Vegetable Crops

Beginning in 2014-15, a centrally funded program for the overall growth of the horticultural sector, including all horticultural crops and such underutilized vegetable crops, will be implemented during the XII project.

Objectives

1. Improved cultivar
2. Creation of water resources
3. Capacity building for farmers and technicians

10.2.1. Merchandize Export from India Scheme (MEIS)

The government has announced the Merchandize Export from India Scheme, which will take effective on 1.4.2015, as part of the revised Foreign Trade Policy 2015-2020. It not only replaces five similar incentive schemes that were available under the Foreign Trade Policy 2009-2014, but it also streamlines the incentives under the previous scheme, removes various restrictions, and greatly expands the scope of the previous scheme. Exporters/farmers are rewarded for exporting specific vegetables to specific markets, encouraging farmers to grow them because there is a large demand for them in the international market, and many countries rely on India to meet their vegetable needs.

10.2.2. Contribution of International Research Organization for Development of Underutilized Vegetable Crops

10.2.2.1. International Plant Genetic Resources Institute (IPGRI)

In recent years, IPGRI has spearheaded specialized national and international programmes aimed to increasing the conservation, promotion and use of these neglected species. IPGRI is leading institutions in the worldwide media to make people aware of underutilized crops and the development of systems to ensure that they receive their fair part of the benefits that result from their use.

10.2.2.2. International Centre for Underutilized Crops (ICUC)

The International Centre for Underutilized Crops (ICUC) was founded in 1948 to look at ways to increase the use of underutilized crops for food, medicine, and environmental improvement.

10.2.2.3. International Atomic Energy Agency (IAEA)

The FAO's IAEA is coordinating a project to use irradiation technologies to boost crop yields of underutilized species in low-income countries.

10.2.2.4. Global Facilitation Unit (GFU)

The GFU is a global multi-institutional initiative that supports and facilitates the work of various stakeholders in order to promote widespread usage of underutilized species.

11. Anti-Nutritional Compounds of Underutilized Vegetables

Vegetables contain a variety of anti-nutrient factors (ANFs) that affect metabolic processes and the bioavailability of nutrients in the human body (Deshmukh and Gaikwad, 2014; Devadas and Mallika, 1991).

Table 3: Antinutritional factor in underutilized vegetable crops

Underutilized Vegetable	Anti -nutritional compound
Celery	Bensofurans
Squashes	Cyanogens
Dioscorea bulbifera	3-Furadosine norditerpene
Lima Bean	Linamarin
Rhubarb	Oxalic acid
Leafy vegetables, Elephant foot Yam, Taro, Tannia	Oxalates
French Bean	Pisatin, Phaseottin Raw
Raw Spinach	Saponins
Brussels sprout, Broccoli	Thioglycosides
FabaBean	Vicines and Convicines
Coco Yam	Xanthosome
Legumes	Antitrypsin factors/Trypsin inhibitors

12. Hindrance In Development of Underutilized Vegetables

As for Food and Agriculture Organization report the World's Plant Genetic Resources for Food and Agriculture, 1998 find following hindrance:-

- Limited availability of genetic resources.
- Lack of technical information among farmers.
- Poor national policy.
- Lack of interest by scientists, agriculturist and extension professionals.
- Lack of producer interest to grow such type vegetables.
- Lack of seed distribution schemes.
- Lack of integration between conservation and used programs.

13. Conclusion

Therapeutic value of underutilized vegetables in our diet, along with their consumption and production are still limited by a widespread lack of understanding among society. Much needs to be done to encourage and promote this sector, including pilot projects, development projects, and involvement in entrepreneurial ventures. Research, development and conservation activity of these underutilized resources must needed to be launched for not only eliminate the hunger and malnutrition while also promoting human health. This vegetable should be prioritized for its genetic enhancement and molecular characterization. This helped identify the cause of resistance and related genes with nutritional and medicinal properties. Better communication with communities, farmers and villagers raises and improves awareness, importance and conservation of the diversity of these underutilized vegetables, which are slowly disappearing at an alarming rate due to the expansion of human development.

References

Ahirwar, J.R. 2015. Some Edible Plants of Bundelkhand Region of India. Research Journal of Recent Sciences.2277-2502.

Bal, J.S.2003. Genetic resources of under-utilized fruits in punjab subtropics. Acta Horticulture 623, 325-331.

Deshmukh, S.A., Gaikwad, D.K. 2014. A review of the taxonomy, ethnobotany, phytochemistry and pharmacology of *Basella alba* (Basellaceae). Journal of Applied Pharmaceutical Science.4(1), 153-165.

Devadas, V.S., Mallika, V.K. 1991. Review of research on vegetables and tuber crops - Amaranthus. Kerala Agricultural University, Thrissur.

Dubey, R.K., Singh, V., Singh, P.M. 2019. Underutilized Vegetables for Nutritional Security and Economic Prosperity.

Dubey, R.K., Singha, V., Devi, J., Gautam, K.K., Singh, P. M., Singh, J.2019. Climate Resilient Underutilized Vegetables for Nutritional and Economic Security. Innovative interventions for sustainable vegetable production under changing climate scenario(3rd September to 23rd September 2019) 93.

Farooq, F., Rai, M., Tiwari, A., Khan, A. A., Farooq, S. 2012. Medicinal properties of *Moringa olefera* : an overview of promising healer. Journal of medicinal plant Research. 6(27),4368-4374.

Jangir, R., Thanki, J. D., Tajane, D.and Kumar, S. 2018.Growth, yield and economic of grain amaranth (*Amaranthus hypochondriacus* L.) as affected by integrated nitrogen management. International Journal of Pure Appplide Bioscience7 (3), 329-334.

Kumar, Y., Prasad, R., Singh, P.R. 2018. Studies on collection and evaluation of Genetic variability available in amaranthus (*Amaranthus spp.*) under Chhatishgarh plain condition. Internal journal of current microbiology and applied science. 7(11), 540-3547.

Sharma, M., Chauhan, D., Banyal, A., Divya,S.K. 2020. Underutilized Vegetable crop: A lost Treasure. International Journal of Current Microbiology and Applied Sciences.9(4),2172-2178.

Pandey, A. K., Dubey, R.K., Singh, V. 2014. Aquatic vegetables-as source of Underutilized vegetables.

Pandey, A. K., Dubey, R. K., Singh, V., & Vida, E. (2014). Addressing the problem of micronutrient malnutrition in NEH region-Underutilized vegetables as a source of food. International Journal of Food and Nutritional Sciences.3(3), 77-83.

Rai, N., Asati, B.S., Patel, R.K., Patel, K.K., Yadav, D.S. 2005. Underutilized horticultural crops in north eastern region. ENVIS Bulletin: Himalayan Ecology 13.

Rohini, N., Shanmugasundaram, T. and Pandiarana, N. 2017. Review on Underutilized Vegetables for Human Health.Trends in Biosciences. 10(27), 5593-5603.

Salvi, J., & Katewa, S. S. (2016). A review: Underutilized wild edible plants as a potential source of alternative nutrition. International Journal of Botany Studies.1(4), 32-36.

Singh, D., Bahadur, V., Singh, D.B., Ghosh, G. 2009. Spine Gourd (*Momordica dioica*): An Underutilized Vegetable with High Nutritional and Medicinal Values. In: Proceeding of the International Symposium on the Socio-Economic Impact of Modern Vegetable Production Technology in Tropical Asia. 809.

Singh, S.C., Awasthi, P. 2020. Nutritional and livelihood security of small and marginal farmers in Bundelkhand region through processing of Underutilized fruit crops. Special issue of Indian Horticulture (May–June, 2020 Published by ICAR-IIVR,Varanasi).

31

Value Addition of Spice Crops for Economic Boost

Ritu Panwar, Ashok Kumar and Nirmal Kumar Meena

College of Horticulture and Forestry, Jhalawar Campus, Agriculture University Kota, Jhalrapatan, Rajasthan, India

1. Introduction

Spices play an important role in human life since time immemorial. Mainly in India, it is not only important as dietary or health product but few spices also have its religious significance. India is being not only largest producer and consumer but exporter of spices also and hence having potential to enhance the economic value of spices. According to Union Minister for Agriculture and Farmer's Welfare, Spices production in the country grew from 67.64 lakh tonnes in 2014-15 to 106.79 lakh tonnes around 2020-21 having annual increase of 7.90%, along with increase in area from 32.24 lakh hectares to 45.28 lakh hectares. With this increase in production and increasing demand of spices because of their health benefits, spices do have economic benefits and farmers can earn better through it.

Table 1: List of few common spices, its botanical name, family and parts used

S.N.	Common Name	Scientific Name	Economical Parts
1	Cardamom (Small)	*Elettaria cardamomum Maton*	Fruit, Seed
	Cardamom (Large)	*Amomum subulatum Roxb.*	Fruit, Seed
2	Pepper	*Piper nigrum L.*	Fruit
3	Chilli, Bird's Eye Capsicum, Chilli Paprika	*Capsicum frutescensL.* *Capsicum annuum L.*	Fruit
4	Ginger	*Zingiber officinale Rosc.*	Rhizome
5	Turmeric	*Curcuma longa L.*	Rhizome
6	Coriander	*Coriandrum sativum L.*	Leaf & Fruit
7	Cumin	*Cuminum cyminum L.*	Fruit
8	Fennel	*Foeniculum vulgare Mill.*	Fruit
9	Fenugreek	*Trigonella foenum-graecum L.*	Seed
10	Celery	*Apium graveolens L.*	Leaf, Fruit & Stem
11	Aniseed	*Pimpinella anisum L.*	Fruit

S.N.	Common Name	Scientific Name	Economical Parts
12	Ajowan	*Trachyspermum ammi L.*	Fruit
13	Caraway	*Carum carvi L.*	Fruit
14	Dill	*Anethum graveolens L.*	Fruit
15	Cinnamon	*Cinnamomum zeylanicum Breyn*	Bark
16	Cassia	*Cinnamomum cassia.Blume*	Bark
17	Garlic	*Allium sativum L.*	Bulb
18	Curry leaf	*Murraya koenigii(L) Sprengel*	Leaf
19	Kokam	*Garcinia indica Choisy*	Rind
20	Mint	*Mentha piperita L.*	Leaf
21	Mustard	*Brassica juncea L.Czern*	Seed
22	Parsley	*Petroselinum crispum Mill.*	Leaf
23	Pomegranate	*Punica granatum L.*	Seed
24	Saffron	*Crocus sativus L.*	Stigma
25	Vanilla	*Vanilla planifolia Andr.*	Pod
26	Tejpat	*Cinnamomum tamala (Buch Ham)*	Bark & Leaf
27	Pepper Long	*Piper longum L.*	Fruit
28	Star Anise	*Illicium verum Hook.*	Fruit
29	Sweet flag	*Acorus calamus L.*	Rhizome
30	Greater Galanga	*Alpinia galanga Willd.*	Rhizome
31	Horse Radish	*Armoracia rusticana Gaertn.*	Root
32	Caper	*Capparis spinosa L.*	Flower buds
33	Clove	*Syzygium aromaticum(L) Merr.& Perry*	Unopened Flower bud
34	Asafoetida	*Ferula asafoetida L*	Oleogum resin from rhizome and thickened root
35	Camboge	*Garcinia cambogia (Gaertn).Desr*	Rind
36	Hyssop	*Hyssopus officinalis L.*	Leaf
37	Juniper berry	*Juniperus communis L.*	Berry
38	Bay Leaf	*Laurus nobilis L.*	Leaf
39	Lovage	*Levisticum officinale Koth.*	Leaf & Stem
40	Marjoram	*Marjorana hortensis Moench.*	Leaf
41	Nutmeg	*Myristica fragrans Houtt.*	Seed
42	Mace	*Myristica fragrans Houtt.*	Aril
43	Basil	*Ocimum basilicum L.*	Leaf
44	Poppy seed	*Papaver somniferum L.*	Seed
45	Allspice	*Pimenta dioica (L) Merr.*	Fruit & Leaf
46	Rosemary	*Rosmarinus officinalis L.*	Leaf
47	Sage	*Salvia officinalis L.*	Leaf
48	Savory	*Satureja hortensis L.*	Leaf
49	Thyme	*Thymus vulgaris L.*	Leaf
50	Oregano	*Origanum vulgare L.*	Leaf
51	Tarragon	*Artemisia dracunculus L.*	Leaf
52	Tamarind	*Tamarindus indica L.*	Fruit

(Source: Spice Board of India)

Though spices as such are enough valuable but if value addition being done, it can be used more efficiently. At present many value added products of quite few spices are under daily use in almost every household of country. From red chilli powder to garlic paste, value added products of spices are of quite potential. Spices and herbs are not only being used as flavouring agents but as colouring and preservatives as well Spices are having medicinal properties and hence use of spices tremendously increased in recent few years which in turn holding great opportunities for industries and farmers.

2. Market of Spices

According to Spices Board India, export of spices continued to increase positively even during pandemic era (2020-21) and crossed around 3.5 billion US $ mark. Produce estimated from country in present year is around 1565000 tons valued Rs.27193.20 crores in contrast to the 1208400 tons valued Rs.22062.80 crores from the previous financial year. The total raise of 30% in volume, 23% in rupee terms and 17% in dollar terms is being noticed in comparison to previous financial year (Anonymous, Spice Board India, 2020-21).

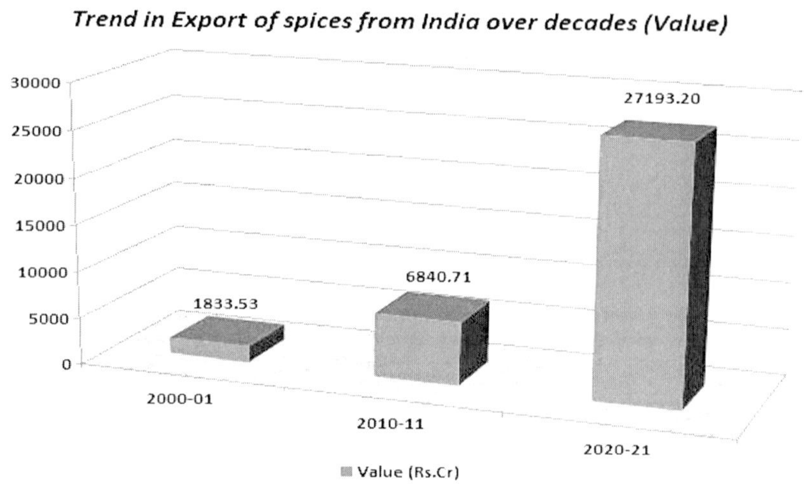

Source: Spice Board India, 2022

Table 2: Export of Spices from India

(Qty. Tonnes & Value Rs. Lakhs)										
	2016-17		2017-18		2018-19		2019-20		2020-21(EST)	
Item	Qty	Value	Qty	Value	Qty	Value	Qty	Value	Qty	Value
Pepper	17,600	1,14,312.60	16,840	82,078.48	13,540	56,868.00	17,000	57,370.94	16,300	54,445.50
Cardamom (S)	3,850	42,150.33	5,680	60,908.15	2,850	35,625.00	1,850	42,537.15	6,500	1,10,675.00
Cardamom (L)	780	8,265.45	760	5,646.60	860	6,106.00	1,310	7,090.17	1,325	9,126.25
Chilli	4,00,250	5,07,075.63	4,43,900	4,25,632.74	4,68,500	5,41,117.50	4,96,000	6,71,039.53	6,01,500	8,42,975.00
Ginger	24,950	25,704.85	22,605	21,607.49	18,150	19,602.00	60,410	52,905.00	1,25,700	75,665.00
Turmeric	1,16,500	1,24,190.65	1,07,300	1,03,567.63	1,33,600	1,41,616.00	1,37,650	1,28,690.53	1,83,000	1,67,660.00
Coriander	30,300	29,208.49	35,185	27,274.96	48,900	35,208.00	47,135	39,831.38	57,000	48,982.50
Cumin	1,19,000	1,96,320.14	1,43,670	2,41,798.78	1,80,300	2,88,480.00	2,14,190	3,32,806.00	2,99,000	4,25,310.00
Celery	6,250	6,246.11	6,480	5,950.30	6,100	6,649.00	6,230	6,903.85	7,650	9,983.50
Fennel	35,150	30,875.93	34,550	25,906.35	26,250	24,412.50	24,220	23,162.14	31,800	27,630.00
Fenugreek	34,680	18,276.49	29,280	12,688.57	27,150	13,846.50	26,570	15,690.38	38,300	24,642.00
Otherseeds(1)	18,100	15,455.86	22,175	16,045.55	29,740	18,736.20	37,580	22,080.72	48,800	30,008.00
Garlic	32,200	30,711.50	46,980	30,936.38	29,500	17,110.00	22,280	17,182.52	17,950	15,630.00
Nutmeg & Mace	5,070	23,641.65	5,500	22,094.31	3,300	15,015.00	2,900	13,280.00	3,875	19,000.00

Otherspices(2)	40,210	50,595.00	38,305	65,253.17	43,300	61,486.00	37,235	66,545.96	44,000	70,942.50
Currypowder/ Paste	28,500	59,910.43	30,150	61,619.55	33,850	74,470.00	38,370	81,278.66	38,450	89,145.00
Mint Products (3)	22,300	2,52,749.67	21,500	3,22,834.86	21,610	3,74,933.50	24,470	3,83,202.24	27,400	3,66,825.00
Spiceoils & Oleoresins	12,100	2,45,532.80	17,200	2,66,172.39	12,750	2,19,300.00	13,000	2,44,682.74	16,450	3,30,675.00
Total	9,47,790	1781223.59	10,28,060	17,98,016.24	11,00,250	19,50,581.20	12,08,400	2206279.91	15,65,000	27,19,320.25
Valueinmilli-onus$		2,655.29		2,789.35		2,805.50		3110.63		3,624.76

(Est):Estimate

(1)Include Bishops Weed (Ajwan Seed),Dill Seed, Poppy Seed, Aniseed, Mustard Etc.

(2)Include Asafoetida, Cinnamon, Cassia, Cambodge, Saffron, Spices (Nes) Etc.

(3)Include Menthol, Menthol Crystals And Mintoils.

Source: Dgci&S., Calcutta/ Shipping Bills/ Exporters'returns.

(*Source*: Spice Board India)

3. Health Benefits of Spices

Most of the spices bear antioxidant and anti-carcinogenic properties which not only enhances daily nutrient quality when consumed daily, but also aid in tackling health issues. Now a days, health being a prime concern and there is huge market demand of healthy consumables. Spices being potential consumable in respect of health benefits. Though there is only limited research being done on anti-carcinogenic property of few limited spices, but there are wide gene pools of spices that can be manipulated for human benefits. Few major health benefits of important spices are mentioned in Table 3.

Table 3: Medicinal properties of some important spices

Name	Medicinal aspects
Allspice	Stimulant,digestive,carminative,anodyne against rheumatism & neuralgia
Aniseed	Mild expectorant, stimulating, carminative, diuretic, diaphoretic, in asthma powders, in veterinary medicine.
Basil/Sweet basil	Stomachic, anthelmintic, diaphoretic, expectorant, antipyretic, carminative, stimulant, diuretic, demulcent, inskin diseases, asthma, opthalmia.
Bayleaves (laurel)	Stimulant in sprains, narcotic & in veterinary medicine
Bishop's weed (Ajowan)	Digestive, antispasmodic, stimulant, carminative, expectorant.
Caper	Diuretic, aspirant, expectorant, emmenagogue, tonic & in scurvy, rheumatism, gout, afflictions of liver and spleen.
Capsicum	Digestive, thermogenic, carminative, stimulant, cardiotonic, antipyretic, serdorific, rubefacient & sialagogue.
Caraway	Stomachic, carminative, anthemintic, lactagogue, adjuvant/ corrective for nauseating & griping effects of medicines.
Cardamom (small)	Stimulant, tonic, diuretic, carminative, digestive, expectorant, cardiotonic& used in several pharmaceutical preparations.
Cardamom(large)	Hypnotic, appetizer, astringent to bowels, tonic to heart and liver.
Cambodge	Astringent, digestive, thermogenic, constipating, used in haemorrhoids, diarrhea, & to control obesity.
Cassia	Astringent, stimulant, carminative, germicidal & for checking nausea & vomiting
Celery	Stimulant,tonic,diuretic,carminative,emmenagogue,anti-inflammatory
Cinnamon	Astringent, diuretic, carminative, aphrodisiac, deodorant, expectorant, febrifuge, stomachic.
Clove	Refrigerant, ophthalmic, digestive, carminative, stomachic, stimulant, antispasmodic, antibacterial, expectorant, rubefacient, Aphrodisiac, appetizer, emollient.
Coriander	Carminative,diuretic,tonic,stimulant,stomachic,refrigerant,aphrodisiac, analgesic,anti-inflammatory
Cumin	Digestive, carminative, astringent, anti-inflammatory, constipating, diuretic, revulsive, galactogogue, uterine & nerve stimulant.

Name	Medicinal aspects
Curry leaf	Astringent, anthelmintic, febrifuge, stomachic, appetizing, carminative, constipating, anti inflammatory, antiseptic, used in skin diseases, in diarrhea, ulcers.
Dill	Carminative, stomachic, antipyretic.
Fennel	Stimulant, carminative, stomachic, emmenagogue, refrigerant, cardiac stimulant, antiemetic, aphrodisiac, anthelmintic
Fenugreek	Carminative, tonic, aphrodisiac, emollient, antibacterial, used in vomiting, fever, anorexia, colonitis.
Garlic	Anticholestrol, antifungal, tonic, rubefacient, stimulant, thermogenic, aphrodisiac, used in cough, asthma, cardiopathy.
Ginger	Digestive, carminative, emollient, appetizer, stomachic, rubefacient, anodyne, expectorant, anthelmintic,stimulant.
Greatergalanga	Carminative, expectorant, digestive, vulnerary, febrifuge, stimulant, depurative, used in skin diseases, rheumatism, asthma, wounds, fever, haemorrhoids.
Horseradish	Thermogenic, appetizing, digestive, stomachic, laxative, anti inflammatory, anodyne, refreshing, antibacterial.
Hyssop Juniper	Stimulant, carminative, pectoral, used in nervous disorders, toothache, pulmonary & uterine troubles.
Berry	Carminative, stimulant, diuretic, useful in dropsy, leucorrhoea, urino genital disorders.
Kokam	Cooling, anthelmintic, cardiotonic, astringent, emollient, useful in piles, dysentery, heart complaints, bilious affections.
Mace	Cooling, febrifuge, expectorant.
Marjoram	Carminative, expectorant, tonic, astringent
Mint	Stimulant, stomachic, carminative, antiseptic, digestive, antispasmodic, contraceptive, used in vomiting, skin diseases, Amenorrhoea, dentalcaries.
Mustard	Thermogenic, anodyne, anti-inflammatory, carminative, digestive, anthelmintic, sudorphic, tonic, emetic, used in vomiting, abdominal colic, dyspepsia, flatulence, skin diseases.
Nutmeg	Astringent, sweet, thermogenic, aphrodisiac, antiinflammatory, anodyne, deodorant, digestive, expectorant, narcotic, anticonvulsant, antiseptic, constipating
Oregano	Stimulant, carminative, stomachic, diuretic, diaphoretic and emmenagogue
Parsley	Stimulant, diuretic, carminative, emmenagogue, antipyretic, anti-inflammatory, emetic, aphrodisiac, alexipharmic, refrigerant.
Pepper	Anthelmintic, carminative, alterant, antiperiodic, diuretic, digestive, emmenagogue, rubefacient, stimulant, stomachic, used infever, asthma, cough, dyspepsia, flatulence, arthritis.
Pepper long	Expectorant, thermogenic, diuretic, tonic, purgative, stomachic, digestive, emollient, antiseptic, used in bronchitis, fever, asthma.
Pomegranate	Astringent, cooling, tonic, aphrodisiac, laxative, diuretic, cardio tonic, used in pectoral diseases, dysentery, diarrhea, vomiting.

Name	Medicinal aspects
Poppy seeds	Expectorant, sudorfic, sedative, nervine tonic, constipating, aphrodisiac, used in internal haemorrhages, diarrhoea, dysentery.
Rosemary	Astringent, nervinetonic, stomachic, antibacterial, protistocidal, rubefacient, used in headaches and hardy menstruation.
Saffron	Stimulant, tonic, stomachic, aphrodisiac, anodyne, anti-spasmodic, emmenagogue, diuretic, laxative, used in bronchitis, fever, epilepsy, skin diseases, decolouration of skin.
Sage	Mild tonic, astringent, carminative, deodorant, insecticidal, antipyretic, used in gingivitis, dentifice, mouthwash, gargles.
Savory	Antispasmodic, antiatarrhal, astringent, carminative, laxative, diuretic, stomachic, sudorfic, vermifuge.
Star anise	Astringent, carminative, deodorant, expectorant, digestive.
Sweet flag	Thermogenic, constipating, emmenagogue, intellect promoting, emetic, carminative, stomachic, expectorant, sudorfic, antipyretic, resusticative, tranquilising, sedative, nervine tonic.
Tamarind	Refrigerant, digestive, carminative, laxative, antiscerbutic, febrifuge, ophthalmic useful in gastropathy, datura poisoning, alcoholic intoxication, scabies, constipation.
Tarragon	Aperient, stomachic, stimulant, febrifuge
Tejpat	Carminative, used in colic diarrhoea.
Thyme	Antispasmodic, carminative, emmenagogue, anthelmintic, spasmodic, laxative, stomachic, tonic, vermifuge
Turmeric	Thermogenic, emollient, anodyne, anti-inflammatory, vulnerary, depurative, antiseptic, appetizer, carminative, expectorant, stomachic, anthelmintic, stimulant, ophthalmic, tonic, used in skin diseases, dyspepsia, asthma, cough, bronchitis, inflammations, ulcers, worms, skin discolouration.
Vanilla	Aphrodisiac.

(*Source*: Spice Board India, 2012)

4. Regulatory Agencies Associated with Spices

Certain regulatory agencies regularly involved with spices, helps in maintaining proper quality, proper distribution and standard. Some such agencies are described below-

4.1. ASTA (American Spice Trade Association)

ASTA is responsible for spice trade and represent US spice industry. Its main aim is to ensure supply of pure and safe spices to the US consumer along with handling global policy matters regarding spice trades.

4.2. ESA (European Spice Association)

It is non-profit association which is associated with packaging, processing, quality assurance and food safety of spices and its related products.

4.3. BIS (Bureau of Indian Standards)

Bureau of Indian Standards is established under BIS Act 2016 for the consistent development of standardization, marking and quality certification of goods. It maintains quality and assures 3rd party guarantee.

4.4. FSSAI (Food Safety and Standards Authority of India)

It's an agency that maintains laws and regulations regarding food safety and standard quality of produce.

4.5. AGMARK

It is a third part quality certification agency which assures standard quality of a commodity.

5. Value Added Products of Spices

5.1. Paste

Paste is widely used product worldwide and is popularly used in almost every household not only in India but even in foreign countries. So far, garlic paste, ginger paste, mustard paste, etc is popular in market.

5.2. Powder

Several spices are commonly used in powder form in day to day life including red chili powder, turmeric powder, coriander powder, etc. Powdered form not only ensures the greater surface area of particle meaning more flavour release but it also is easy to handle. For onion powder methods like Hot-air, Vacuum and Freeze Drying Methods are being used. (Kang et al., 2007).This onion powder is being used in many food products, as flavouring agent or nutritional aspect. With increasing urbanization, demand of such flavouring agents and essence are gaining popularity. Tamarind pulp powder is also being prepared by drying the pulp and then turning this dried product into powder.

5.3. Oils

Oil content of spices posses medicinal value as well as aroma and hence many essential oils extracted from spices is being used in perfumery and cosmetics. Extraction method varies from commodity to commodity. Essential oil extracted from certain spices mainly cardamom, cinnamon, clove, coriander, etc., examined for its anti-microbial capabilities and cloves comes out to be having most inhibitory properties (Nanasombat et al., 2005).

Some essential oils and extracts also being used in preservation of meats. For extraction of essential oil from cumin, clove, etc, hydro distillation method is being used (Hernández-Ochoa et al., 2014). Oil extracted from onion bears Antibacterial, anti-dermatophytic and anti-toxigenic properties and hence it is being used in many cosmetics (Zohri et al., 1995).

Table 4: Characteristics of few essential oils

S.N.	Name	Oil per cent	Constituent
1.	*Acorus calamus*	1.5–3.0 %	Asarone (82 %)
2.	*Allium cepa*	6.01–0.015 %	d-n-Propyl disulphide, methyl n-propyl disulphide, vinyl disulphide
3.	*Alpinia galanga*	0.5–1.5 %	Methyl cinnamate (50 %), 1,8-cineol (20–25 %), α-pinene, d-pinene, sinigrin
4.	*Amomum subulatum*	2.5 %	α-Terpinyl acetate
5.	*Apium graveolens*	2–3 %	d-Limonene (19.5–68 %), B-selinene (8 %), n-butyl lidene phthalide (8 %)
6.	*Armoracia rusticana*	0.2–1.0 %	Sinigrin derived allyl isothiocyanate diallyl sulphide, phenyl propyl thiocyanate
7.	*Artemisia dracunculus*	0.5–0.7 %	Methyl chavicol (45–75 %), anethole (10 %) α- and β-pinenes, phellandrene, camphene
8.	*Carum carvi*	4.5–7.5 %	50–85 % d-carvone, α- and B-pinene and sabinene
9.	*Cassia leaf oil*	1.0–2 %	Eugenol (70–80 %), cinnamic aldehyde
10.	*Cinnamomum tamala*	2 %	Eugenol (80–85 %) and cinnamic aldehyde
11.	*Cinnamomum zeylinicum*	0.5–3.0 %	Cinnamaldehyde (65–75 %), eugenol, benzyl aldehyde
12.	*Crocus sativus*	0.6–1.4	Safranal-aldehyde, pinenes and cineol
13.	*Curcuma longa*	5–6 %	Turmerones (58 %) and 9 % tertiary alcohols
14.	*Eugenia caryophyllus*	15–17 %	Eugenol (81–83 %), eugenol acetate (7 %), B-caryophyllene (9 %), humulene
15.	*Illicium verum*	3–3.5 %	Anethole (85–90 %), α-pinene, phellandrene, p-cymene, limonene
16.	*Laurus nobilis*	1–3 % (bark)	1, 8 cineol (35 %), methyl eugenol (4 %), α-pinene (12 %) B-pinene (11 %), α-terpin (6 %), limonene (4 %), α-terpinyl acetate (6 %) sabinene (5 %) and eugenol (2 %)
17.	*Mentha arvensis (Japanese mint)*	1.5–4.0 %	Menthol (70–90 %), α-pinene, α-1 limonene, caryophyllene
18.	*Mentha citrata*	0.2–0.5 %	Carvone (50–70 %), dihydrocarvone, α-limonene, phellandrene, menthol, 1,8-dineol
19.	*Mentha piperit*	0.5–4.0 %	Menthol (30–55 %), menthone (15–35 %) methyl acetate (4–7 %), menthofuran, isomethane, limonene
20.	*Mentha spicata*	0.2–0.55 %	Linalool acetate (45–50 %) and linalool (7 %)
21.	*Murraya koenigii*	2–2.5 %	1-Sabinene (35 %), 1-α-dinene (27 %), diterpene (16 %), 1-terpenel, (7.7 %), 1-caryophyllene (6.7 %), cadinene (5.2 %)
22.	*Myristica fragrans*	6–16 %	Sabine (15–50 %, α-pinene (10–22 %) and B-pinene (7–18 %), limonene (3–4 %), 1,8-cineol (1.5–3.5 %), myristicin (1–13.5 %)

S.N.	Name	Oil per cent	Constituent
23.	*Ocimum bassilicum*	0.4–1.0 %	Linalool (40–65.3 %) and methyl chavicol (25 %)
24.	*Petroselinum crispum*	2.0–5.0 %	Apiol (36 %), myristicin (13 %), α-pinene
25.	*Pimpinella anisum*	1.5–6 %	Trans anethole (80–90 %), methyl chavicol (10–15 %) anethole (2 %) and anisaldehyde (12 %)
26.	*Rosmarinus officinalis*	1.5–2.5 %	α-Pinene (25 %), camphoo (15–25 %), boneol (15–20 %), 1,8-cineol, bornyl acetate
27.	*Salvia officinalis*	1.5–2.5 %	Thujone (25–45 %), linalyl acetate (10 %) α-pinene (2 %), 1,8-cineol, β-caryophyllene, limonene
28.	*Thymus vulgaris*	2–2.5 %	Thymol (45 %)

(*Souce*: Datta, 2007)

5.4. Colouring Material

Chilli is having diverse use as its powder, its dried flakes, oil extract, Capsacin i.e, compound responsible for pungency and capsanthin which is responsible for color along-with carotenoids. Some speicific varities of Chilli is being exported out mainly for Capsacin and Capsanthin.

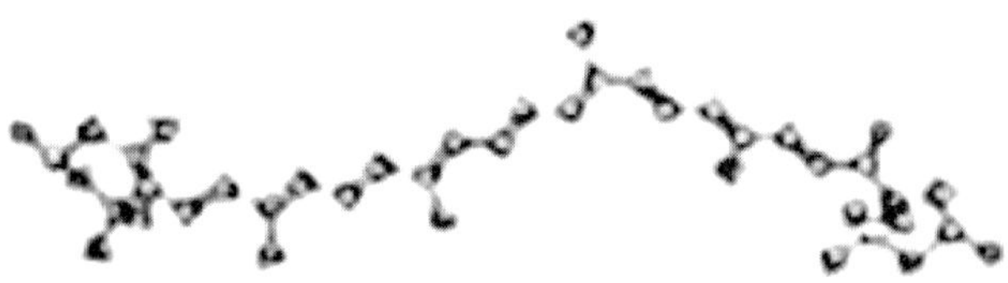

Fig. 1: Structure of capsanthin
(*Source* - National Center for Biotechnology Information)

5.5. Dried Products

Dry chilli flakes, oregano dried leaves, thyme leaves, basil leaves, etc is popularly used in restaurants now a days. It gives out specific aroma and hence its market demand is increasing.

5.6. Oleoresins

Oleoresins are naturally present flavouring agents in certain spices which mostly being extracted by solvent extraction method. Oleoresin can be used as flavouring compound, coloring compound or recently it is also being used in cosmetics and soaps.

Table 5: Study of Different solvents and extraction times affecting oleoresin yield (per cent) and effect on capsaicin and colouring matter of chilli oleoresin

Solvent	Oleoresin yield (per cent)	Extraction time (hours)	Oleoresin yield (per cent)	Capsaicin (percent)	Colouring matter (ASTA units)
Acetone	16.25	4	14.56	2.15	421
Ethyla cetate	14.21	5	16.02	1.75	364
Ethyl alcohol	16.18	6	16.06	2.06	310
C.D. at 5 %	0.560	–	0.560	0.12	11.8

(*Source*: Singh et al., 2001)

5.7. Beverages

From juice and RTS to wine can also be prepared by spices. Kokum RTS can be prepared keeping standard RTS specifications and same goes with squash where standard 25 percent juice content and 45 percent TSS is kept. Kokum wine can be prepared by fermenting kokum juice. Ginger extract can be fermented to produce ginger beer.

6. Conclusion

India being blessed with spices diversity. Spices having potential of being exported as whole but value addition results in fetching good price. Processing add certain extra income and economic value to particular product. Certain comparison between spices as whole and as its value added product - Coriander costs around Rs. 80/Kg, but when its being converted to powder, it's price hiked up to almost Rs. 200/Kg. Same goes with Chilli which is in market at price Rs. 200/Kg, its powder will be available at Rs. 300/Kg. This price rise trend is not only visible with powder products but also can be seen in dehydrated products, extracted oil and other by products obtained so far. Value addition helps in fetching good price to farmers and hence giving their economy a significant boost. On farm processing and packaging will cut the cost of transport to processing house but either way, it's a venture to earn many folds from the original raw product. In export market as well, demand of value added product is more and hence huge return can be expected from value added products of spices.

References

AGMARK. 2020. https://agmarknet.gov.in/

Anonymous. 2022. American Spice Trade Association, https://www.astaspice.org/

Anonymous. 2022. European Spice Association, https://www.esa-spices.org/

Anonymous. 2022. Spice Board India, Ministry of Commerce and Industry, Government of India, http://www.indianspices.com/

Bureau of Indian Standards, Ministry of Consumer Affairs, Food & Public Distribution, Government of India, https://www.bis.gov.in/

Datta, S. 2007 Value addition of spices with special reference to quality. In: Sharanghi AB, Acharya SK (Eds) Quality management in horticulture. Agrotech Publishing Academy (ATPA) Udaipur, pp, 1–360.

Datta, S., Guha, S., Sharangi, A.B. 2015. Value addition in spice crops. In Value Addition of Horticultural Crops: Recent Trends and Future Directions (pp. 59-82). Springer, New Delhi.

Food Safety and Standards Authority of India, Ministry of Health and Family Welfare, Government of India. https://www.fssai.gov.in/

Hernández-Ochoa, L., Aguirre-Prieto, Y.B., Nevárez-Moorillón, G.V., Gutierrez-Mendez, N., Salas-Muñoz, E. 2014. Use of essential oils and extracts from spices in meat protection. Journal of Food Science and Technology. 51(5), 957-963.

Kang, N.S., Kim, J.H., Kim, J.K. 2007. Modification of quality characteristics of onion powder by hot-air, vacuum and freeze drying methods. Korean Journal of Food Preservation. 14(1), 61-66.

Nanasombat, S., Lohasupthawee, P. 2005. Antibacterial activity of crude ethanolic extracts and essential oils of spices against salmonellae and other enterobacteria. Current Applied Science and Technology 5(3): 527-538.

National Center for Biotechnology Information. 2022. National Center for Biotechnology Information. PubChem Compound Summary for CID5281228, Capsanthin. https://pubchem.ncbi.nlm.nih.gov/compound/Capsanthin.Accessed Jan. 19, 2022.

Singh, R., Gill, B.S/, Hundal, J.S. 2001. Studies on extraction of chilli oleoresin. Spice India. 14(5), 13-15.

Tapsell, L.C., Hemphill, I., Cobiac, L., Patch, C.S., Sullivan, D.R., Fenech, M., Roodenrys, S., Keogh, J. B., Clifton, P. M., Williams, P. G., Fazio, V. A., Inge, K. E. 2006. Health benefits of herbs and spices: the past, the present, the future. The Medical journal of Australia. 185(S4), S1–S24.

Zohri, A.N., Abdel-Gawad, K., Saber, S. 1995. Antibacterial, antidermatophytic and antitoxigenic activities of onion (*Allium cepa* L.) oil. Microbiological Research. 150(2), 167-172.

32

Development of Jackfruit Based Value Added Products for Small and Medium Enterprises

Romila Xess[1] and Arunima Tripathy[2]

[1]Department of Horticulture, Fruit Science, Indira Gandhi Krishi Vishwavidyalaya, Raipur, Chhattisgarh, India
RMD, CARS, Indira Gandhi Krishi Vishwavidyalaya, Ambikapur Chhattisgarh, India

1. Introduction

Jackfruit is a fruit crop that is under appreciated and neglected. The jackfruit (*Artocarpus heterophyllus* Lam.) is a member of the Moraceae family. India is the world's greatest cultivator and producer of jackfruit, also known as the poor man's fruit because it grows more in rural than metropolitan areas. It is a pretty huge tree that bears the world's largest edible fruit. The jackfruit is a tropical fruit native to Southeast Asia. It is thought to have originated in India's rainforests, in the Western Ghats, and is now widely farmed throughout the region's lowlands. The leading jackfruit producers in the world include Bangladesh, India, Myanmar, Nepal, Thailand, Vietnam, China, the Philippines, Indonesia, Malaysia, and Sri Lanka. Jackfruit can be found in East Africa, Brazil, and Caribbean nations such as Jamaica. Bangladesh's national fruits are the jackfruit, mango, and banana.

When mature, it is eaten like a fruit, but when unripe, it is eaten as a vegetable. The tender and ripe fruits, as well as the seeds of jackfruit, are said to be high in nutrients and vitamins (Gopalan et al., 2000; Jagadeesh et al., 2007). It's also high in carbs, proteins, potassium, calcium, iron, vitamin A, B, and C, as well as dietary fibre (Jagadeesh et al., 2007; Rahim and Quddus, 2000) Jackfruit is commonly utilised as a complement to other staple foods during times of food scarcity in several places due to its high carbohydrate content. Because of the significant number of isoflavones, antioxidants, and phytonutrients found in its mature fruits, jackfruit possesses natural anti-carcinogenic properties (Shrikant et al., 2012). It is also said to aid in the treatment of ulcers and indigestion (APAARI,

2012). Jackfruit is also high in phytonutrients such as lignans, isoflavones, and saponins, all of which are beneficial to one's health. Anticancer, antihypertensive, antiulcer, and antiaging effects are all present in these phytonutrients. As a result, the phytonutrients present in jackfruit can inhibit the production of cancer cells in the body, decrease blood pressure, and slow the deterioration of cells that keep the skin looking young and beautiful. Antimicrobial properties in jackfruit seeds aid to avoid contamination with bacteria that cause foodborne illness (Theivasanthi and Alagar, 2011). Seed includes two lectins known as jacalin and artocarpin, according to Haq (2006). Jacalin could be used to assess the immunological state of patients infected with the human immuno deficiency virus. The jackfruit seeds are neglected despite their numerous therapeutic, nutritional, and physiological benefits. The seeds of the jackfruit are usually discarded or boiled and eaten as a snack or used in some local cuisines. People consume jackfruit raw, ripe, and cooked in a variety of delectable recipes. The seed of the jackfruit is also edible. Incorporating jackfruit seed into one's diet is a brilliant method to make food more nutritious (Jessica, 2015).

Value addition by processing and preservation is seen as the best option for decreasing post-harvest deprivation of this nutritive fruit and distributing it to the public during the off-season. Thousands of value-added goods may be manufactured from jackfruit, including flour for the manufacturing of acceptable products such as ready-to-eat snacks, which will assist to boost consumer acceptance of jackfruit. To combat jackfruit neglect, farmers can be educated, transformed into dealers of fresh fruit or processed products, and given trainings. It is necessary to promote scientific methods of post-harvest management, primary processing, and jackfruit preservation so that the raw materials are available to consumers and businesspeople throughout the year. The market potential of jackfruit can be better realised if the fruits are available to consumers year-round in ready-to-eat or ready-to-cook form. Through marketing and use at the home level, it can be a source of additional income for farm families and self-help organisations with value addition.

2. Value-Added Products from Jackfruit

2.1. Tender Jackfruit for Use as Vegetable

Green fruits that are tender and immature, with young seeds, are commonly used as a vegetable in curries and side dishes. At room temperature, Ready to Cook (RTC) products have a one-day shelf life. Browning is a concern with RTC; blanching prevents browning.

2.2. Ready-to-Serve Jackfruit Beverages

With a composition of 10% juice, 10% TSS, and 0.3 percent acidity, the ready-to-serve beverages can be made extremely quickly from fruits (Chopra and Chauhan, 2001). Singh et al. (2001) developed jackfruit pulp-based ready-to-drink beverages with a 10 percent pulp content, 12 percent TSS, and 0.3 percent acidity.

2.3. Jackfruit Nectar

Fruit nectar is a concentrated type of fruit pulp with a viscosity similar to honey. The pulp of the fruit is combined with sugars and citric acid to produce a 15-20°Brix product with a slight acid flavour. CFTRI (1977) standardized jackfruit pulp nectar. Jackfruit nectar was effectively standardized by combining it with other fruit pulp from two common types of jackfruit accessible in Kerala. The nectars improved in flavour and taste after being blended with other fruit pulps.

2.4. Jackfruit Wine

Fruit-based fermented beverages, particularly wine and vinegar, have a lot of potential in India. Wine and vinegar are two fermented items that can be made from jackfruit pulp. Jackfruit, in general, has a high concentration of easily fermentable sugars, making it an ideal substrate for the growth of wine yeasts. For commercial wine production, jackfruit can be a valuable resource. During the fermentation of jackfruit wine, pH, temperature, and inoculum concentration all have a role. Fermentation has also been observed to occur faster in jackfruit. Because of the progressive increase in phenolic content as fermentation occurs, jackfruit wine has high antioxidant activity.

2.5. Jackfruit Vinegar

Another fermented product that can be made from jackfruit is vinegar. Fermentation of jackfruit vinegar recovered from mature fruits yielded 7% alcohol and 6% acetic acid, according to the findings.

2.6. Dried Jackfruit Flakes

Dried jackfruit flakes are made by slicing jackfruit bulbs with a jackfruit bulb slicer and drying them in a combination dryer. An effective blancher-cum-drier can be used to dry mechanically cut jackfruits. KAU standardized dehydrated Jackfruit flakes with a one-year shelf life (1999). Vista Company in Sri Lanka has used dehydrated jackfruit flakes as a vegetable. Hardikars Food Processing in Pune, India, produces jackfruit flour that may be utilised as a raw ingredient for a variety of products. The flour made from dehydrated jackfruit flakes was shown to be acceptable for making chapattis, pazhampori, and bhaji by substituting 25% wheat flour, maida, or Bengal gram flour with jackfruit flour, respectively.

2.7. Preserved Jackfruit Bulbs

Fresh jackfruit bulbs are a popular consumer item that is enjoyed by people from all walks of life. Fresh jackfruit bulbs and seeds were treated with 1.5 percent KMS and 0.5 percent sodium benzoate and stored under vacuum (760 mm lbs pressure). The chemical contents of preserved bulbs showed little changes and were organoleptically stable for 15 days under refrigeration.

2.8. Dehydrated Jackfruit Bulbs

To increase the quality of the dried products, steeping jackfruit bulbs in 0.1 percent potassium metabisulphite solution for 30 minutes is the recommended method for producing dehydrated bulbs. When sulfured at a rate of 16 lbs sulfur/ton fruit/1000 cft space, good grade dehydrated products were obtained (drying ratio 3:1).

2.9. Canned Jackfruit Products

It is common knowledge that canning is commonly used to enhance the shelf life of fruits. Both raw and mature jackfruit bulbs might be effectively stored for later use in vegetable curries and for table usage. Raw Jackfruit bulbs are canned in a brine solution containing 0.5-0.75 percent citric acid in the Standardized method. When canned jackfruit is kept at room temperature (24-30°C), it retains its original colour, flavour, and aroma. However, after 19 weeks of storage at 37°C, the product showed signs of deterioration. Canned Jackfruit is a new industry in India, and the fruit has yet to be commercialised for canning. However, some major Western brands have canned jackfruit products on the market. The technology and know-how can be developed in India, and the product can be successfully marketed if the correct incentives are provided.

2.10. Dehydrated Jackfruit Slice

A technology for generating osmotically dehydrated jackfruit slices has been developed by the IIHR in Bangalore, India. Dehydrated fruit products based on a unique way to dehydration are known as Osmo-air dried fruits. Suitable fruits are chosen at their peak maturity (hard ripe stage), sliced, and then dipped in a sugar syrup containing citric acid, preservatives, and with or without maltodextrin. Following the immersion time, the slices are drained and dried in a cabinet dryer until the moisture level is around 15%. Dried slices are packaged in plastic punnets and can be stored for one year at room temperature. In terms of colour, flavour, and texture, the osmotically dehydrated product is comparable to fresh fruit. It is suitable for snacking. This product can be used in ready-to-eat dishes, ice creams, fruit salads, kheer, cakes, and bakery goods, among other things. One kg of Osmo-air dried slices requires approximately 11-12 kg of ripe fruits, and the product has a one-year shelf life in ambient circumstances.

2.11. Jackfruit Bar and Ice-Cream

Fruit bars that are ready to eat are well-known products that are professionally made and sold in our country. In India, jackfruit ice cream and jackfruit blended mango ice cream are becoming increasingly popular. Jackfruit bars stored in modified polypropylene packets (MPP) were found to have a higher percent of nutrient retention and a lower microbial count, as well as a distinct taste and flavour. Combining papaya pulp with jackfruit pulp improved appearance, colour, and texture, whereas blending with mango pulp improved flavour, taste, and overall acceptability.

2.12. Jackfruit Pickles

In 1977, the Central Food Technology Research Institute (CFTRI) conducted research in jackfruit preservation and found that tender jackfruit could be stored as pickles. Sweet oil pickle, spicy vinegar pickle, and plain vinegar pickle are all essential pickle preservations.

2.13. Jackfruit Chips

Raw bulbs are used to make jackfruit chips. The frying oil has an impact on the shelf life of Jackfruit chips. Antioxidants like butylated hydroxyl toluene and sorbic acid could improve the shelf life of jackfruit chips. Vacuum-fried Gokul chips from Kundapur, India, are immensely popular.

2.14. Jackfruit Papad

For making jackfruit papads, utilise jackfruit bulbs that are neither entirely mature nor completely raw. At room temperature (24-30°C), it was discovered that jackfruit papads wrapped in paper had a shelf life of 4-6 months. Mayura jackfruit papad from Kadamba Marketing Cooperative in Sirsi, Karnataka, and Ponsa Appolopapad from Karnataka are both well-known in India.

2.15. Jackfruit Sweets

Jackfruit bulbs can also be used to make a range of sweet dishes such as jackfruit halwa (varieties), pudding, jackfruit toffee, jackfruit barfi, elayappam, adda, muffin, and payasam, among others.

2.16. Jackfruit Jelly

In a blender, jackfruit pulp is combined and filtered via filter paper. The juice-containing filtrate is used to make jelly. Pectin and sugar are thoroughly combined. For cooking, the saucepan is put on a stove. The pectin and sugar combination is then mixed into it. The liquid is taken from the burner when TSS reached 66 to 67 percent, indicating the completion of the jelly production process.

2.17. Jackfruit Jam

Jam is a medium-moisture food made by cooking fruit pulp with sugar (sucrose), pectin, acid, and other ingredients (preservatives, colouring, and flavouring elements) to a thick consistency that holds the fruit tissues in place (Baker et al., 2005). The taste of jackfruit goods will be substantially improved by adding a synthetic flavouring ingredient such as ethyl or n-butyl ester of 4-hydroxybutyric acid at 100 and 120 ppm, respectively (ICUC, 2004). Other store fruit jams include a high amount of sugar, which raises the risk of diabetes. On the other hand, jackfruit jam is high in natural sugars and low in calories, making it an excellent food for weight loss.

2.18. Jackfruit Seed Flour

After drying out the anti-nutritional components, jackfruit seeds can be ground into flour. By combining the flour made from jackfruit seeds with wheat flour, chapattis can be made. Additionally, the resulting jackfruit flour can be employed as a thickening and binding agent in food systems. Because jackfruit seed flour is a high source of protein and has a low water and fat absorption capacity, it could be used in the preparation of deep-fried foods.

2.19. Roasted Nuts

The roasted jackfruit seeds are said to have a nutritional value and flavour similar to chestnuts and are appetizing. Fried seeds, on the other hand, have a short shelf life because they can only be stored for a few days at room temperature (24-30°C).

2.20. Value Added Products from Jackfruit Seed Flour

2.20.1. Chapaties

Chapatti's are unleavened flat flatbread prepared from whole wheat flour that are a staple dish in India. Chapatti is a popular lunch and dinner item in India. Adding jackfruit seed flour to the mix may boost the product's nutritional value (Afroza, 2013). Chapatti can be made using various proportions of wheat flour and jackfruit seed flour.

2.20.2. Biscuits

Biscuits are ready-to-eat, convenient, and affordable food products that incorporate important digestive and nutritional principles and contain less moisture than cakes and bread (Ayo et al., 2007). Biscuits have a long shelf life and are generally resistant to microbial degradation (Wang et al., 2002). They are nutritional treats made from unappealing dough that is converted into a tasty product by using heat in the oven (Olaoye et al. 2007). Biscuits can be made with various proportions of jackfruit seed flour and wheat flour.

2.20.3. Sattu

Sattu is a traditional dish that is enjoyed by people of all ages. Sattu is made with roasted flours made from solely cereals or legumes, or a combination of cereals and legumes, with the addition of flavouring ingredients (More and Housalmal, 2012).

2.20.4. Noodles

Noodles are a staple cuisine prepared from unleavened dough that is stretched, extruded, or rolled flat and cut into a variety of forms using extrusion technology before being cooked in boiling water, sometimes with the addition of cooking oil or salt. Noodles are widely consumed around the world, coming in second only to bread in terms of global consumption (Jayasena et al., 2008). To create a value-added noodle, jackfruit seed flour and wheat flour were blended in varying quantities.

3. Challenges in Marketing Jackfruit and Value-Added Products

3.1. Lack of 'Joint Action'

Each fruit matures at a different period, making it impossible for a farmer to bring it to market, especially because it is so large. They can, however, make money if they are well-organized and have direct market access.

3.2. Issues of 'Weight' of Whole Fruit

A market for both the ripe fruit and value added products can emerge and be sustainable if Jackfruit is made more accessible to consumers by providing convenient packaging rather than having to purchase a 7-10 kg fruit, and by implementing a well-thought-out and well-executed supply chain and go-to-market strategy.

3.3. Processing–Not Yet Widespread

Despite the fact that there is a demand for jackfruit value added products on the mainland, fruits are rotting under the trees due to transportation challenges, negligence, and a lack of value-added opportunities, such as awareness of the fruit's potential for income generation among both rural and urban communities.

3.4. Wastage

The country as a whole is projected to be wasting jackfruit worth Rs 2,000 crore. It's possible that the true loss is significantly higher.

3.5. Manufacturing

The unorganised sector accounts for a considerable portion of the value-added activity. The lack of interest demonstrated by the producers has resulted in a scarcity of processed items on the local market. However, only local consumption was documented for primary processing at the household level using traditional methods, and these traditional products are still uncommon on the market. Within the country, there are only a few commercial-scale processing plants.

3.6. Medium and Small-Scale Manufacturers and Exporters

Most MSMEs have only rudimentary marketing and distribution channels for their value-added products. Farmers do not earn their dues with the fruits fetching them limited returns in comparison to retail pricing in cities and across other states where the fruits or associated items are distributed because there are no regularised distribution channels and is heavily dependent on middlemen.

3.7. Supply Chain and Logistics-Challenges in the Current Scenario

India is the world's largest producer of many fruits and vegetables, but there is a huge gap between per capita demand and supply due to massive wastage during post-harvest storage and handling caused by improper bagging without crating, lack of temperature-controlled vehicles, unavailability of cold chain facilities in various parts of the country for preserving the produce, as well as significant processing of agricultural produce, all of which result in enormous losses to the farmers. To fulfil the country's demand, proper supply chain management in fruits and vegetables must be improved at all stages of the supply chain by adopting best worldwide practises in storage, packing, handling, shipping, value-added service, and so on.

3.8. Harvesting and Post Harvesting– Current Practices

Due to difficulties, supply, marketing, and distribution of Jackfruit is largely unorganised from the moment it is harvested until it reaches the market for final consumption.

Table 1: Nutritive value of jackfruit (100 g)

Constituent	Average value
Moisture (%)	76.2
Energy (cal)	88.0
Protein (g)	1.90
Fat (g)	0.10
Fibre (g)	1.10
Carbohydrates (g)	19.8
Potassium (mg)	107.0
Calcium (mg)	20.0
Phosphorus (mg)	41.0
Iron (mg)	0.56
β Carotene (mg)	175.0
Thiamine (mg)	0.03
Riboflavin (mg)	0.13
Niacin (mg)	0.40
Vitamin C (mg)	7.0

4. Conclusion

In India, both urban and rural areas are seeing an increase in demand for jackfruit value added products. From the immature stage to the fully ripened state, jack fruits can be used to make a variety of value-added goods. Each item has its own set of advantages in terms of flavour, preference, storage quality, and so on. As a result, building up jackfruit processing units in jackfruit-growing areas has a lot of potential.

References

Afroza, S. 2013. Physico-chemical and functional properties of jackfruit seed flour and quality enrichment of chapaties using composite flour M.Sc. thesis submitted to Department of Food Technology and Rural Industries Bangladesh Agricultural University Mymensingh.

APAARI 2012. Jackfruit Improvement in the Asia-Pacific Region – A Status Report. Asia- of Agricultural Research Institutions, Bangkok, Thailand, p. 182.

Ayo, J. A., Ayo, V.A., Nakam, I., Adewori, R. 2007. Physical, Invitro- digestibility and of Acha wheat biscuit supplemented with soyabean flour. Nigerian Food Journal 25:77-89.

Baker, R.A., Berry, N., Hui, Y.H., Barrett, D.M. 2005. Food Preserves and Jams. In: Processing Fruits, second edition. (Edited by Barrett, D.M., Somogyi, L., Ramaswamy, H.S.), CRC Press, Boca Raton, FL, USA. Pp.113-125.

CFTRI 1977. Home scale processing and preservation of food and vegetable. Central Food Technology Research Institute, Mysore. p. 288.

Devi, S. P., Talaulikar, S., Gupta, M.J., Thangam, M. Singh, N.P. 2014. A guide on jackfruit - cultivation and value addition. Technical Bulletin No. 41, ICAR (RC), Goa.

Gopalan, C., Rama Sastri, B.V., Balasubramanian, S.C. 2000. Nutritive value of Indian foods. National institute of nutrition, Hyderabad. pp. 51-71.

Haq, N. 2006. Jackfruit artocarpus heterophyllus, Southampton Centre for Underutilized crops, University of Southampton, Southampton. 17, 23.

International Centre for Underutilized Crops, 2004. UK .Technical Manual for Small- Scale Fruit Processors. [http://fao.org/fileadmin/templates/ inpho/documents/ AE618e.pdf] site visited on 20.04.2014.

Jagadeesh, S.L., Reddya, B.S., Swamya, G.S.K., Gorbala, K., Hegdea, L., Raghavanb, G.S.V. 2007. Chemical composition of jackfruit (*Artocarpus heterophyllus* Lam.) selections of Western Ghats of India. Food Chem. 102(1), 361-365.

Jayasena, V., Leung, P., Nasar□Abbas, S. M., Palta, J., erger, J. 2008. Development and quality evaluation of lupin fortified instant noodles.12th international lupin conference, Fremantle, Australia. 473-477.

Jessica, B. 2015. Nutrition in Boiled Jackfruit Seeds Cited from http://www.livestrong.com/article/546837-nutrition-in-boiled-jackfruit-seeds/ retrieve on 8/09/15. KAU (Kerala Agriculture University). 1999. Research Report-1996-1997. Directorate of Research, Kerala Agricultureal University, Thrissur. 89-91.

More, P., Housalmal, S. 2012. Effect of packaging on shelf-life of soy fortified Sattu. International Journal of Processing and Post-Harvest Technology, 3,227-230.

Olaoye, O.A., Onilude, A.A., and Oladoge, C.O. 2007. Breadfruit flour in biscuit making. African Journal of Food Science. 47, 20-23.

Rahim, M.A., Quaddus, M.A. 2000. Characterization and grafting performance of different accessions of jackfruit. MS Thesis (unpublished). Bangladesh Agricultural University, Mymensingh, Bangladesh.

Shrikant, B.S, Thakor, N.J., Haldankar, P.M., Kalse, S.B. 2012. Jackfruit and its many functional component as related to human health; A review. Comprehen. Rev. Fd. Sc. Safety. 11(6), 565-576.

Theivasanthi, T., Alagar M. 2011. An insight analysis of nano sized powder of jackfruit seed. Nano Biomedicine and Engineering. 3,163-8.

Wang, J., Rosell, C.M., Berber, C.B. 2002. Effect of the addition of different fibres on wheat dough performance and bread quality. Food Chemistery. 79, 221-226.

33

Metabolomics: A New Frontier in Phytochemical Research

Ravi Y.[1], Irene Vethamoni P.[2], Ashoka N.[3] N.K. Meena[1] Praveen Kumar Singh Gurjar[4]

[1]ICAR-National Reseah Centre on Seed Spices, Ajmer, Rajasthan, India
[2]Horticulture College and Research Institute, Tamil Nadu Agricultural University, Coimbatore, Tamil Nadu, India
[3]College of Horticulture, Sirsi, University of Horticultural Sciences, Bagalkot, Karnataka, India
[4]Rajmata Vijayaraje Scindia Krishi Vishwa Vidyalaya, Gwalior, Madhya Pradesh, India

1. Introduction

Plants offer a rich source of active compounds that have been exploited as scents, flavourings, pigments, pesticides, and medicinal compounds for thousands of years. Plants are thought to utilise these metabolites as a primary strategy for adapting to and surviving in a variety of ecological niches and environmental conditions, as well as to combat biotic intrusions in their environment (Weng, 2014). The increased interest in phytomedicine and medicinal supplements, as well as the increased regulatory obligations for their safety and reputable contents of active compounds, has prompted the development of methodologies for the quality control of multi - component concretes found in plant-derived nutraceuticals. Metabolite diversity in plants is thought to have evolved over time to help plants survive in harsh environments, and it is continually changing. Many of these metabolites are bioactive and poisonous to plant predators, giving them a competitive advantage. Plants create specific metabolites to overcome obstacles to their growth and survival, and these metabolites are also beneficial to humans. Plants that produce these particular bioactive metabolites have been utilised to cure a variety of diseases throughout human history, and they have served as the basis for the discovery and development of modern therapies (Butler, 2004).

The thorough examination of metabolites in a particular sample as a whole is the focus of metabolomics technology. It has been successfully applied to identify metabolites in plant materials despite intraspecific heterogeneity. Only recent advances in molecule separation and identification technology have made metabolic profiling or metabolomics viable. The enormous biosynthetic potential of medicinal and aromatic plants (MAPs) has hitherto to be fully revealed, and using cutting-edge technologies, biotechnology could be applied to create innovative active substances with superior or novel bioactivities. Metabolic engineering combined with phylogenetic analysis have both proved successful in quantifying the right amounts of metabolites in plants, resulting in a faster drug development process (Webster et al., 2016).

Plant-based drug development using traditional approaches is considerably more costly and require much time (DiMasi et al., 2003). The use of computational methodologies, on the other hand, aids in the acceleration of plant-based medicinal discovery (Harishchander, 2017). The involvement of numerous genes and proteins in metabolic progressions has recently been elucidated to the emergence of these omic sciences (Van Emon, 2015). A recent advancement in bioinformatics resources has resulted in a significant shift in contemporary MAPs research, including plant-based medication development. In the realm of plant metabolomics research, significant advances have been made, as shown in Table 1.

Table 1: Metabolites identified in different herbal plants

Plant Name	Secondary Metabolites	Parts Used	Method used for Metabolome profiling	References
Acorus calamus L.	β-Asarone, 9,12-octadecadienoicacid, α-asarone, *n*-hexadecanoicacid, shyobunone,	Rhizome	GC-MS	Kumar et al. (2010)
Carum carvi L.	Limonene, carvacrol, carvone, germacrene-D, *trans*-dihydrocarvone	Seeds	GC-MS	Iacobellis et al. (2005).
Citrus limon L.	Limonene, sabinene, γ-terpinene, β-pinene, α-pinene	Fruit	GC-MS	De Rodriguez et al. (1998).
Cuminum cyminum L.	1,8-cineole, limonene, α-terpineol, α-Pinene, linalool,	Seeds	GC-MS	Mohammadpour et al. (2012)
Foeniculum vulgare Mill.	Limonene, α-Pinene, fenchone, *p*-cymene, methyl chavicol	Root and schizocarps	GC-MS	Radulović and Blagojević, (2010)
Mentha piperata L.	Carvone, linalool, 3-octanol, *trans*-caryophyllene	Leaves	GC-MS	Sartoratto et al. (2004).
Mentha spicata L.	Carvone, 1,8-cineole, limonene, germacrene-D, β-caryophyllene	Leaf	GC-MS	Boukhebti et al. (2011)
Nigella sativa L.	Thymoquinone, limonene, *trans*-Anethole, *p*-cymene, longifolene	Seed	GC-MS	Gerigeetal. (2009)
Ocimum sanctum L.	Eugenol, cyclopentane, octadecane, caryophyllene, cyclopropylidene, benzene methanamine	Leaves	GC-MS	Devendran and Balasubramanian,(2011)
Origanum vulgare L.	Carvacrol, γ-terpinene, limonene, thymol, *p*-cymene	Leaves	GC-MS	Derwich et al. (2010)
*Piper longum*L.	Eugenol, D-limonene, β-Caryophyllene, 3-carene	Seed	GC-MS	Liuetal, (2007)
Tagetes minuta L.	*trans*-Tagetenone, *cis*-tagetenone, *trans*-pinocarvylacetate, carvone, dihydrotagetone,	Fruit	GC-MS	Vázquez et al. (2011)
Zingiber officinale Rosc.	α-Zingiberene, cyclohexane, α-farnesene, β-sesquiphellandrene, α-curcumen,	Rhizome	GC-MS	Setty et al. (2011)
Aegle marmelos L.	Umbelliferone, marmin, psoralene, skimmianine, imperatorin,	Root and stem bark	GC-MS	Gajbhiye et al. (2016)

Plant Name	Secondary Metabolites	Parts Used	Method used for Metabolome profiling	References
Andrographis paniculata Ness	Apigenin-7-O, β-D-glucuronide, andrographolide, deoxyandrographiside, 5,4′-dihydroxy-7,8-dimethoxyflavone, 14-deoxyandrographolide, 14-isoandrographolide,neoandrographolide,	Leaves	LC-MS	Song et al. (2013)
Artemisia annua L.	Artemisinin, artemisinic acid, artemisitene, arteannuin B, dihydroartemisinicacid	Leaves	LC-MS	Suberu et al. (2016)
Bacopa monnieri L.	Bacopaside III, bacopasaponin G, and bacopasides A	Whole plant	LC-MS	Hou et al. (2002)
Centella asiatica L.	Asiatic acid, madecassoside, asiaticoside	Leaves	LC-MS	Shen et al. (2009)
Commiphora wightii (Arnott) Bhand.	*E* and *Z* Guggulsterone	Resin	LC-MS	Haque et al. (2009)
Glycyrrhiza glabra L.	Liquiritigenin, liquiritin, isoononin, glycyrrhizic acid	Roots	LC-MS	Montoroaetal. (2011)
Gymnema sylvestre (Gur-Mar)	Gymnemagenin	Leaves	LC-MS	Kamble et al. (2013)
Hypericum perforatum	Hyperforin, polyprenylated derivatives	*In-vitro* tissue culture	LC-MS	Charchoglyan et al. (2007)
Impatiens balsamina L.	Kaempferol, Lawsone, quercetin and their glycosylate	Leaves	LC-MS	Chu (2016)
Panax ginseng L.	Ginsenosides, pseudoginsenoside, malonyl ginsenoside and ginsenoside	Rhizome	LC-MS	Xie et al. (2008)
Rauwolfia serpentine L.	Serpentine ajmalicine, ajmaline, and yohimbine	Roots	LC-MS	Kumar et al. (2016)
Rhodiol arosea L.	Flavonoids (gossypetin, herbacetin glycosides)	Arial part	LC-MS	Petsalo et al. (2006)
Withania somnifera (L.) Dunal	Withaferin A, withanolide D, withanolide sulfoxide withanoside IV or VI.	Roots	LC-MS	Trivedi et al. (2017)
Trigonella foenum-graecum L.	Apigenin-6, 8- C-dipentoside, tigogenin dihexosyl rhamnoside, Trigofoenoside B.	Seeds	UPLC-MS	Mohamad (2016)

Plant Name	Secondary Metabolites	Parts Used	Method used for Metabolome profiling	References
Dichocarpum kaiyangense	Berberine, palmatine, jatrorrhizine, magnoflorine, apigenin, dauricine, columbamine,	Roots	UPLC-Q-TOF-MS/MS	Pei et al. (2021)
Panax notoginseng Burk.	Ginsenoside, notoginsenoside, notoginsenoside-H, notoginsenoside	Flower buds, roots and rhizomes	UPLC–ESI-MS	Mo et al. (2008)
Justicia gendarussa Burm	Justidrusamide A, 6,8-di-C-α-L-arabinosylapigenin, 6-desmethylprazosin, and Justidrusamide C.	Leaves	UPLC-UHR-QTOF-MS	Indah et al. (2015)

2. Applications of Metabolomics

1. Generation of metabolomics databases
2. Isolation of phytochemicals and phytotherapy
3. Discovery of biochemical pathways, enzyme discovery and improve the efficiency of enzymes
4. Drug discovery and pharmacological studies
5. Drug toxicological studies
6. Biomarker discovery, identification and interpretation

3. Quantifying the Metabolome: Modern Analytical Technologies

Traditional approaches are expensive and/or time-consuming for natural product research, thus new tactics are needed to update them (Kinghorn et al., 2008). The most recent metabolomics knowledge are effective and precise for quickly identification of plant metabolites and producing a new understanding of their physiological and toxic consequences (Michel, 2013). Metabolomics is an interdisciplinary branch of study that integrates analytical chemistry, mass spectrometry, and nuclear magnetic resonance with advanced data analysis. It entails using sophisticated analytical tools to profile a biofluid's or tissues varied metabolic complement. Metabolomics provides a framework for comparing metabolites that reflect the dynamic mechanisms that keep cells in a state of homeostasis (Park, 2010). Recent improvements in analytical technology have paved the way for metabolite profiling to aid in the comprehension of complicated molecular developments and structure. The application of metabolomics has become popular in the last decade. Fourier transform infrared spectroscopy (FT-IR), MS connected to separation techniques, such as NMR, GC-MS, LC-MS, FT-MS, and UPLC-MS, and NMR have all been employed for metabolomics applications.

The identification of metabolites is an important step in metabolomics research. Metabolome analysis can be performed on a wide range of plant parts or based on economic importance, and it can be done on a variety of different expertise platforms. MS and NMR are two of the most recent metabolomics technologies, allowing for the quickest approach to metabolite identification and quantification.

3.1. Nuclear Magnetic Resonance-Based Metabolomics

Nuclear Magnetic Resonance (NMR) is the widely used spectroscopic analytical techniques for quantification and identification of a wide range of organic molecules in the micro molar array. NMR has been acquaint with the metabolomics sector, where it can able to provide unbiased information on metabolite profiles. Under certain settings, NMR-based metabolomics is a well-suited and effective method for metabolomic analysis since it can provide a "holistic perspective" of the metabolites (Wu et al., 2010). NMR is a simple, generally automated, and non-destructive technique that allows samples to be analysed further. It has been used for metabolite fingerprinting, profiling, and metabolic flux research.

3.1.1. Advantages

1. Provides a one-of-a-kind opportunity to improve the structural and functional characterization of metabolomes.
2. Is non-selective and aids in the simultaneous detection of low molecular weight substances in a single run
3. In biological samples, quantify the number of low-molecular-weight metabolites.
4. Aids in the identification of components in complicated mixes

3.1.2. Limitations

1. It has a low sensitivity, which makes it unsuitable for analysing huge quantities of low-abundance metabolites.
2. Quantities of putative biomarkers may be lower the recognition limit as compared to MS methods.

4. Mass Spectroscopy

Mass spectroscopy (MS) is becoming more popular in high-throughput metabolomics, and it is commonly used in concurrence with other techniques like chromatography-MS. MS has advanced significantly in recent decades and now has a prominent place in qualification in addition to separation techniques. The ability to monitor the amounts of hundreds of metabolites that are the end products of cellular regulation processes has been made possible by recent breakthroughs in MS-based metabolomics. Because of its high sensitivity and wide spectrum of chemicals, MS has become the preferred technology in many metabolomics studies. By infusing biological samples directly or after separation and purification, MS can be utilised to analyse them. Recent improvements in mass precision have expanded the proportion of metabolites that might have been investigated by MS and improved compound identification accuracy.

4.1. Advantages

1. Analyse the effects of time, stress, nutritional status, and environmental perturbation on hundreds of metabolites all at once, resulting in massive, complex data sets.
2. Has high sensitivity and resolution, allowing it to detect and quantify thousands of metabolites.
3. Helps with metabolic network reconstruction, biomarker discovery, and functional annotation.
4. MS-based technologies are becoming increasingly important in metabolomics research, and they are widely accessible to help researchers conduct effective investigations as follows.

4.2. Gas Chromatography and Mass Spectroscopy (GC-MS)

GC-MS has been utilised as a non-targeted analysis technique for hydrophilic metabolites in particular. In general, metabolomics based on GC-MS necessitates a high-throughput system able to handle a large sample size, as well as precise peak identification using standard time and mass spectra areas (Tsugawa et al., 2011). In metabolomics, GC-MS is usually applied because it produces accurate and repeatable data. Organic acids, most amino acids, polysaccharides, sugar alcohols, polyamines, and fatty acids can all be profiled at the same time using GC-MS, as well as precisely separate and characterize volatile metabolites.

4.2.1. Advantages

1. Separation of complex mixtures allows for the identification of organic components.
2. Quantitative identification of biomolecules
3. Organic contamination is measured at a trace level (low to mid-ppb for liquid matrices and low nano gram for solid matrices).

4.2.2. Limitations

1. Wafers, metal parts, oil, and other non-volatile matrices require special preparation (extraction, outgassing, etc.)
2. Plastic extracts are being evaluated.
3. The target chemicals must be either volatile or derivatize.
4. Difficult to deal with atmospheric gases (CO2, N2, O2, Ar, CO, H2O)

4.3. Liquid Chromatography and Mass Spectroscopy (LC-MS)

LC-MS is a technique that combines liquid chromatography with mass spectrometry to separate the components in a sample mixture using the detection power of mass spectrometry. A mobile phase running at high pressure pumps the sample through a stationary phase (LC column). The effluent is sent to the mass spectrometer after elution from the LC column. The mass determines the ion's mass-to-charge ratio; this information can be utilized to calculate the molecular mass and structural information about the sample.

4.3.1. Advantages

1. It is the most reliable analytical technique for detecting the exact molecular weight of a wide variety of samples since it has superior sensitivity and selectivity.
2. This can aid pharmaceutical manufacturers in streamlining their research and analytical procedures, as well as ensuring compliance with regulatory rules.
3. The use of both analytical techniques (MS and HPLC) together decreases experimental error and enhances precision.

4. It can separate and identify solutes in a complicated mixture at low concentrations (measured in parts per million, or PPM).
5. LC-MS is widely used in the pharmaceutical, biopharmaceutical, research, forensic, food, and environmental industries to ensure regulatory compliance.
6. In comparison to other chromatography techniques, the LC-MS provides great selectivity, resolution, exact mass, and specificity.

4.3.2. Limitations

1. It can only work with volatile buffers,
2. Leftover contaminants must be ionised before being examined.
3. When compared to other analytical instruments, the LC-MS has a high maintenance cost.

4.4. Ultra Performance Liquid Chromatography-Mass Spectroscopy (UPLC-MS)

UPLC-MS technology can be used in biomolecular research to quantify the action of signalling and metabolic processes in a diverse and comprehensive manner. It is based on porous, tiny particles (sub 2micron particles). The Van Deemter equation, which links the relationship between linear velocity and plate height, is the driving force behind this evolution. UPLC provides for a faster analysis without sacrificing resolution because the optimum linear velocity has a wider range. The integration of UPLC and MS identification enables the detection of a larger amount of polar metabolites, resulting in a higher total amount of analytes detected. Because the majority of primary metabolites are inherently polar, the UPLC-MS technique in metabolomics represents a suitable hyphenated micro separation platform. This technology has lately achieved significant recognition in the analytical world, with hyphenated UPLC-MS being used in metabolomic research for coupling to MS to provide a complementary tool that may be used in a variety of areas.

4.4.1. Advantages

1. More selective and sensitive with high-resolution performance and rapid resolving power
2. It minimises process cycle time and improves end-product quality by lowering operating costs and shortening run durations.
3. The use of a special column material with very fine particles improves sensitivity and enables for quick analysis.
4. It eases solvent usage, enhances sample throughput, and allows for real-time analysis in conjunction with industrial operations

4.4.2. Limitations

1. UPLC has higher back pressures than normal HPLC, which affects column life.
2. Particles less than 2μm are typically non-regenerable and so have a limited application.**7.**

5. Other Technologies

The metabolome is made up of a huge molecule with a variety of chemical compositions and abundances, requiring the use of analysis tools to fully cover it. Despite the reality that NMR, GC-MS, LC-MS, and UPLC-MS are the most often used techniques for large-scale research, metabolomics is not limited to these techniques. Other methods for studying metabolomics include thin-layer chromatography, HPLC with UV/visible absorbance, photodiode array or electrochemical detectors, FT-IR, MALDIFTICR-MS, and a variety of other enzymatic assays. Combining multiple techniques or detectors in an online or parallel analysis can greatly expand metabolite coverage and detection limits, as well as improve metabolite detection from a single biomolecule. FT-MS instruments are prominent among these analytical platforms because they deliver accurate and precise measurements with ppm and even sub-ppm errors, as well as high and ultra-high resolving power (Junot, 2010).

6. Statistics Analysis of Metabolomics Data

For investigation and efficient interpretation intricate analyte interactions, robust statistical analytic methods are beneficial. To comprehend high-throughput metabolomics data, two types of analysis are used: univariate and multivariate. Univariate techniques such as the t-test, fold-change analysis, Wilcoxon rank-sum test, analysis of variance (ANOVA), and others are used to establish the statistical significance of each peak. They are used to find the significantly changed metabolites in either unpaired or paired study. Univariate approaches, such as descriptive or permutation testing, are often used to obtain P-values.

Even if multi-molecular combinations would demarcate them on a systems level, univariate approaches fail to discern between subgroups when there are only modest changes at the single molecule level. Multivariate analysis aims to capture not only changes in single metabolites across groups, but also the relationships between distinct molecules. The most frequent multivariate analysis techniques utilised in the field of metabolomics include principal component analysis (PCA), cluster analysis, and partial least squares (PLS) regression. Multivariate analysis takes into consideration the combinatorial influence of multiple factors. There are two sorts of techniques: unsupervised and supervised procedures. PCA, which identifies a set of orthogonal projection directions that maximise the variance of the projected data, is one of the most common unsupervised techniques in LC–MS-based metabolomics studies. PCA has been utilised widely in a number of studies to elucidate metabolomics consequences in the investigation of phenotypic and genotypic links in pathological studies.

7. Challenges in Plant Metabolomics Studies

Two of the most major issues in plant metabolomics research are the immense complexity and diversity of the plant metabolome, and also the limited knowledge of plant metabolic pathways. Metabolome analysis needs a flexible appropriate statistical tool with a wide dynamic range because the plant contains a large number of main and secondary metabolites. The paucity of metabolite-specific libraries and known reference molecules for bioactive metabolite identification and/or structural characterization is a second challenge.

8. Conclusion

Plant-based products have traditionally been offered as a treatment for a wide range of ailments, and herbal medicines continue to be a key source of health care for a large portion of the world's population. In rural and tribal communities, traditional treatments focused on herbal drugs are still the primary source of health care. Even today, about 30% of modern medicines are derived from natural compounds from medicinal and therapeutic plants. Plant metabolomics gives a complete overview of the phytochemical contents of MAPs, and it is employed in a range of domains like medical science, synthetic biology, and so on. The most recent methodologies for broad-metabolite profiling have been developed as a result of developments in analytical chemistry, computation, and biotechnology.

References

Boukhebti, H., Chaker, A.N., Belhadj, H.S.F., Ramdhani, M., Laouer, H., Harzallah, D. 2011. Chemical composition and antibacterial activity of *Mentha pulegium* L. and *Mentha spicata* L. essential oils. Der. Pharm. Lett. 3 (4), 267-275.

Butler, M.S. 2004. The role of natural product chemistry in drug discovery. J. Nat. Prod. 67, 2141-2153.

Charchoglyan, A., Abrahamyan, A., Fujii, I., Boubakir, Z., Gulder, T.A.M., Kutchan, T.M., Vardapetyan, H., Bringmann, G., Ebizuka, Y., Beerhues, L. 2007. Differential accumulation of hyperforin and secohyperforin in Hypericum perforatum tissue cultures. Phytochemistry 68, 2670-2677.

Chu, L.S. 2016. Untargeted MS-based small metabolite identification from the plant leaves and stems of *Impatiens balsamina*. Plant Physiol. Biochem. 106, 16-22.

De Rodríguez, G.O., de Godoy, V.M., de Colmenares, N.G., Salas, L.C., de Ferrer, B.S. 1998. Composition of venezuelan lemon essential oil *Citrus limon* (L.) Burm. F. Rev. Fac. Agron. 15(4), 343-349.

Derwich, E., Benziane, Z., Manar, A., Boukir, A., Taouil, R. 2010. Phytochemical analysis and in vitro antibacterial activity of the essential oil of *Origanum vulgare* from Morocco. Am.Eur. J. Sci. Res. 5 (2), 120-129.

Devendran, G., Balasubramanian, U. 2011. Qualitative phytochemical screening and GC-MS analysis of *Ocimum sanctum* L. leaves. Asian J. Plant Sci. Res. 1 (4), 44-48.

DiMasi, J.A., Hansen, R.W., Grabowski, H.G. 2003.The price of innovation: New estimates of drug development costs. J. Health Econ. 22, 151-185.

Gajbhiye, N.A., Makasana, J., Dhanani, T., Saravanan, R. 2016. Development and validation of LC-ESI-MS/MS method for simultaneous determination of four coumarin derivatives and an alkaloid from root and stem bark of *Aegle marmelos* Correa. Acta Chromatogr. 28, 473-488.

Gerige, S.J., Gerige, M.K.Y., Rao, M. 2009. GC-MS analysis of *Xigella sativa* seeds and antimicrobial activity of its volatile oil. Braz. Arch. Biol. Technol. 52(5), 1189-1192.

Haque, I., Kumar, M., Mukhopadhyay, K. 2009. A rapid and simple UPLC-MS-MS based simultaneous determination of the medicinally important E- and Z-guggulsterone from oleogum-resins of naturally occurring *Commiphora wightii* plants. Chromatographia. 70, 1613.

Hou, C.C., Lin, S.J., Cheng, J.T., Hsu, F.L. 2002. Bacopaside III, bacopasaponin G, and bacopasides A, B, and C from*Bacopa monniera*. J. Nat. Prod. 65 (12), 1759-1763.

Iacobellis, N.S., Lo Cantore, P., Capasso, F., Senatore, F. 2005. Antibacterial activity of *Cuminum cyminum* L. and *Carum carvi* L. essential oils. J. Agric. Food Chem. 53(1), 57-61.

Indah, Y.N., Diah, I.P., Suwidji, W., Bambang E.W.P., Gunawan, I.M. 2015. Profiling of *Justicia gendarussa* Burm. Leaves Using UPLC-UHR-QTOF-MS. Sci. Pharm. 83, 489-500.

Junot, C., Madalinski, G., Tabet J.C., Ezan, E. 2010. Fourier transform mass spectrometry for metabolome analysis. Analyst. 135, 2203-2219.

Kamble, B., Gupta, A., Patil, D., Janrao, S., Khatal, L., Duraiswamy, B. 2013. Quantitative estimation of gymnemagenin in Gymnema sylvestre extract and its marketed formulations using the HPLC-ESI-MS/MS method. Phytochem. Anal. 24(2), 135-140.

Kinghorn, A., Norman, F., Djaja, S., Geoffrey, C., Steven, S., John, P., Mansukh, W., Monroe, W., Nicholas, O., David, K., Robert, K., William, R., Gregory, V., Craig, F., Russell, P., Robert, W. 2008. Novel Strategies for the Discovery of Plant-Derived Anticancer Agents. Pharm. Biol. 41, 53-67.

Kumar, S.S., Akram, A.S., Ahmed, T.S., Jaabir, M.S. 2010. Phytochemical analysis and antimicrobial activity of the ethanolic extract of *Acorus calamus* rhizome. Orient. J. Chem. 26(1), 223-227.

Kumar, S., Singha, A., Bajpaia, V., Srivastav, M., Singh, B.P., Kumar, B. 2016. Structural characterization of monoterpene indole alkaloids in ethanolic extracts of Rauwolfia species by liquid chromatography with quadrupole time-of-flight mass spectrometry. J. Pharm. Anal. 6 (6), 363-373.

Liu, L., Song, G., Hu, Y. 2007. GC-MS analysis of the essential oils of *Piper nigrum* L. and *Piper longum* L. Chromatographia. 66(9), 785-790.

Michel, T., Halabalaki, M., Skaltsounis, A.L. 2013. New concepts, experimental approaches, and dereplication strategies for the discovery of novel phytoestrogens from natural sources. Planta Med. 79(7), 514-532.

Mo, D., Mingming S., Xianfu G., Tie Z., Aihua, Z., Guoxiang, X., Yunping Q., Mingmei Z., Zhong, L., Wei, J. 2008. Metabolite profiling of *Panax notoginseng* using UPLC-ESI-MS. Phytochemistry 69,2237-2244.

Mohamed, A.F., Dalia, M.R., Matthias, K., Andreas G.H. 2016. Metabolite profiling in Trigonella seeds via UPLC-MS and GC-MS analyzed using multivariate data analyses. Anal. Bioanal. Chem. 408, 8065-8078.

Mohammadpour, H., Moghimipour, E., Rasooli, I., Fakoor, M.H., Astaneh, S.A., Moosaie, S.S., Jalili, Z. 2012. Chemical composition and antifungal activity of *Cuminum cyminum* L. essential oil from Alborz Mountain on growth of Aspergillus spp. Jundishapur J. Nat. Pharm. Prod. 7 (2), 50-55.

Montoroa, P., Maldinia, M., Russob, M., Postorinoc, S., Piacentea, S., Pizzaa, C. 2011. Metabolic profiling of roots of liquorice (*Glycyrrhiza glabra*) from different geographical areas by ESI/MS/MS and determination of major metabolites by LC-ESI/MS and LC-ESI/MS/MS. J. Pharm. Biomed. Anal. 54, 535-544.

Park, J.M. Kim T.Y. Lee, S.Y. 2010. Prediction of metabolic fluxes by incorporating genomic context and flux-converging pattern analyses. Proc. Natl. Acad. Sci. U. S. A. 107, 14931-14936.

Pei L., Jie S., Yue, L., Hui, Y., Meng Y., Chunnian, H., Peigen, X. 2021. Metabolite profiling based on UPLC-Q-TOF-MS/MS and the Biological Evaluation of Medicinal Plants of Chinese Dichocarpum (Ranunculaceae). Chem. and Biodi. 18(10), 1-17.

Petsalo, A., Jalonen, J., Tolonen, A. 2006. Identification of flavonoids of Rhodiola rosea by liquid chromatography-tandem mass spectrometry. J. Chromatogr. 1112, 224-231.

Radulović, N.S., Blagojević, P.D. 2010. A note on the volatile secondary metabolites of *Foeniculum vulgare* Mill. (Apiaceae). Phys. Chem. Technol. 8 (1), 25-37.

Sartoratto, A., Ana Lúcia, M., Delarmelina, M.C., Figueira, G.M., Duarte, M.C.T., Rehde, V.L.G. 2004. Composition and antimicrobial activity of es- sential oils from aromatic plants used in Brazil. Braz. J. Microbiol. 35(4), 275-280.

Setty, V.K., Santhosh, D., Rao, N.D. 2011. Preliminary phytochemical screening and antidiabetic activity of Zingiber officinale rhizomes. Int. J. Pharm. Life Sci. 2 (12), 1287-1292.

Shen, Y., Liu, A.L., Ye, M.D., Wang, L., Chen, J.H., Wang, X.R., Han, C. 2009. Analysis of biologically active constituents in Centella asiatica by microwave-assisted extraction combined with LC-MS. Chromatographia. 70, 431-438.

Song, Y.X., Liu, S.P., Jin, Z., Qin, J.F., Jiang, Z.Y. 2013. Quantitative analysis of *Andrographis paniculata* by rapid resolution liquid chromatography/ time-of-flight mass spectrometry. Molecules. 18, 12192-12207.

Suberu, J., Gromski, P.S., Nordon, A., Lapkin, A. 2016. Multivariate data analysis and metabolic profiling of artemisinin and related compounds in high yielding varieties of *Artemisia annua* field-grown in Madagascar. J. Pharm. Biomed. Anal. 117, 522-531.

Trivedi, M.K., Panda, P., Sethi, K.K., Jana, S. 2017. Metabolite profiling in *Withania somnifera* roots hydroalcoholic extract using LC/MS, GC/MS and NMR spectroscopy. Chem. Biodivers. 14,120-130.

Vázquez, A.M., Demmel, G.I., Criado, S.G., Aimar, M.L., Cantero, J.J., Rossi, L.I., Velasco, M.I. 2011. Phytochemistry of *Tagetes minuta* L. (Asteraceae) from Córdoba, Argentina: comparative study between essential oil and HS-SPME analyses. Bol. Latinoamer Caribe Plant Med. Aromat. 10 (4), 351-362.

Webster, G., Teh, A.Y.H., Ma, J.K.C. 2016. Synthetic gene design the rationale for codon optimization and implications for molecular pharming in plants. Biotechnol. Bioeng. 114(3), 492-502.

Weng, J.K. 2014. The evolutionary paths towards complexity: a metabolic perspective. New Phytol. 201, 1141-1149.

Wu, J.,An, Y., Yao, J., Wang Y., Tang, H. 2010. An optimised sample preparation method for NMR-based faecal metabolomic analysis. Analyst. 135, 1023-1030.

Xie, G.X., Plumb, R., Su, M.M., Xu, Z.H., Zhao, A.H., Qiu, M.F., Long, X.B., Liu, Z., Jia, W. 2008. Ultra-performance LC/TOF MS analysis of medicinal Panax herbs for metabolomic research. J. Sep. Sci. 31, 1015-1026.

34

Phytochemistry, Ethnobotany and Pharmacology of East Indian Arrowroot (*Curcuma angustifolia* Roxb.)

***Vijay Bahadur Singh Chauhan*[1]*, Samarendra Narayan Mallick*[1]*, Pinki Mohapatra*[1]*, Kalidas Pati*[1]*, R. Arutselvan*[1] *M. Nedunchezhiyan*[1] *and Arvind Kumar Verma*[2]**

[1]*Regional Centre, ICAR-Central Tuber Crops Research Institute Bhubaneswar, Odisha, India*
[2]*ICAR-National Research Centre on Seed Spices, Ajmer, Rajasthan, India*

1. Introduction

Therapeutic plants have different metabolites in them because of which they are generally utilized in the whole world by individuals to fix different illnesses. India has a rich history of involving many plants for therapeutic purposes. Therapeutic plants are assuming an exceptionally dynamic part in conventional medications for the treatment of different afflictions. India is a rich wellspring of restorative plants because of the varieties in soil, their heights, and other ecological and geological conditions. India's utilization of plants for medical services goes back nearly 5000 years, when Ayurveda systematized around 8000 natural cures by utilizing plants. Root and tuber crops possess a striking situation in the food security of the creating scene because of their high caloric worth and sugar content. Some of them are as of now under development; however different species are developing wild as a dismissed gathering of financial plants. A portion of the tuberous species is significant due to their restorative as well as modern application. East Indian arrowroot (*Curcuma angustifolia* Roxb) belongs to family Zingiberaceae, one of the wild consumable tuber plants utilized for palatable as well as therapeutic purposes. East Indian arrowroot is developed as restorative yield in many pieces of India under wet deciduous blended and sal timberland of Chhattisgarh, Madhya Pradesh, Odisha and Jharkhand where the rhizomes are utilized fundamentally for engendering (Misra and Dixit, 1983; Chauhan et al., 2020). It is customarily perceived as restorative plant and furthermore,contains starch in its rhizomes. *Curcuma angustifolia* is a quickly developing rhizomatous spice of Zingiberaceae family, generally known as East Indian arrowroot or white turmeric or Tikhur (Paikra et al., 2013). For the most part the ranchers and tribals living close to the backwoods or timberland occupants by and large gather the normally developed arrowroot rhizomes as a non wood woodland item (NTFP) while certain ranchers fill it industrially in their field for assortment of starches.

2. Taxonomic Position

Kingdom: Plantae

Class: Liliopsida

Subclass: Commelinidae

Order: Zingiberales

Suborder: Zingiberanae

Family: Zingiberaceae

Genus: *Curcuma*

Species: *angustifolia*

3. Vernacular Names

Sanskrit-Payaksheera ,Tavaksheera, Tavakshiri,Vamsalocana, Tavakshira

Odia-Palua

Bengali-Tikkur, Keturihalodhi

Hindi-Tekhur, Tikhur, Thikora, Theksura, Thavakheera, Thavsasheera.

English-East Indian arrowroot, *Curcuma*starch

Kannada-Kaadu arrowroot, Kovegida, Kove hittin gida, Thavakeela

Telugu-Gaddalu

Tamil- Kooa, Artimavu, Kookai, Araukizhangu, Kua,Kisangu

Malayalam-Koova,Kuva-kizhanna,

Konkani-Koovyajhaad.

Marathi-Tavakira,Thavakheera,Thavakil.

Gujarat-Tavkhir,Tikhur,

4. Distribution

C. angustifolia is an endemic to India, found in Northwestern, Central and South India. The rhizomatic herb is distributed throughout Eastern India, Southern and Central region of Indiabut most commonly reported from the Western Coastal Plains, Northeast region and hilly regions of India (Srivastava et al., 2006;Sharma, 2012; Chauhan et al., 2022). The plant grows wild in moist and cool conditions at altitudes of about 450 m.The plant grown as wild in the in hilly tracts of Odisha, Chhattisgarh, Madhya Pradesh, Bihar, West Bengal, Andhra Pradesh, Telengana, Maharashtra, Himachal Pradesh, Karnataka, Tamil Nadu, Kerala and Central Himalayas as well as cultivated in few amount for edible purposes. It is also cultivated in other countries such as Pakistan, China, Nepal, Burma, Laos, Sri Lanka, Africa, Cambodia and North Australia (Ravindran et al., 2007; Tiwari; Patel, 2013).

Fig.1: East Indian arrowroot crop in field and rhizomes

5. Botany (Morphology)

The East Indian arrowroot plant is a rhizomatous, slim spread herb can be developed upto 90-180 cm in stature with plump round and hollow rhizome and little rootstock. Stems as a rule supplanted by pseudo stems framed by leaf sheaths rather than genuine stems.The leaves are basic, green, disposed in two vertical rows, glabrous and lanceolate shaped with whole edges and inverse course of action. The leaves show fine equal venation with an unmistakable focal midrib. Leaves might develop to around 36-37 cm long and 8-10 cm in width. The dorsal side of the leaves is a more obscure and darker than the ventral surfaces. The leaves additionally smell and taste like turmeric (*Curcuma longa*) (Doble et al., 2011). The herb blossom is enduring, sexually unbiased, epigenous, zygomorphic, with humble and little spiked inflorescences. Three or four with yellow colored, channel formed roses are available inside bunches of pink terminal bracts, which are boat-molded and encase the whole perianth of the bloom. The blossoms of *C. angustifolia* have two fold anthers, a slim style, and a globular shame. Blossoms are generally sprouted toward the start of the storm time frame during July to August. Organic product is dry or fleshy, dehiscent or indehiscent in nature with ruddy earthy colored tone, little, arillate seeds. Rhizome of *C. angustifolia* can develop to really depend on 1.5 m fundamental part utilized generally for its nutritive and therapeutic properties (Elias et al., 2015).

6. Ayurvedic Properties

6.1. Guna Karma

East Indian arrowroot (*C. angust folia*) is madhura rasa, laghu and snigdhaguna and sitavirya properties which can lightens vata and pitta doshas. The plant is particularly valuable in kasa, shwasakushta, kamala, pandu, raktapitta, jwara, trishna, aruchi, daha, kshaya, vrana, and furthermore accommodating in charge of numerous different infections. Due to being snehana and laghu,the rhizome is utilized as pathyaahar against koshthagata vata, athisara, pravahika and grahni (Prakash, 2001; Tripathi, 1982; Ojha, 2004; Dwivedi, 2009; Vaishya, 2005).

6.2. Amlapitta

The rhizome powder called Palua or Tikhur has a most effective in the treatment of Amlapitta or acidic gastritis. Intake of starch of *C angustifolia* like *Maranta arundinacea* has potential to relive significantly from Shoola, Avipaka, ChhardiDaha, Tikta-amlodgara and the associated disease symptoms like Aruchi (aversion), Antrakujana (rumbling of the bowels), Vitbheda, Shiroruja, Angasada (flaccidity of limbs),Gaurava, Udaradhmana and Trit (Elias et al., 2015).It eases the pitta dosha and has astringent property, assists with capturing dying.

6.3 Adulterants

The starch called arrowroot collected from rhizome is adulterated by mixing with other starches prepared from flours of *Manihot esculentum, Ipomoea batata and Oryza sativa* (The wealth of India, 2001).

7. Phytochemistry

The oil ready from leaf contain 14-hydroxy-δ-cadinene (18.6%), curzerenone (33.2%), and γ-eudesmol acetic acid derivation (7.3%) (Jena et al., 2017). The rhizomes of *C. angustifolia* have been found to contain optional metabolites like flavonoids, terpenoids, alkaloids, tannins, saponins, phenols, curcumin, steroids, glycosides, and oils like Zingiberol present too.The rhizome for the most part contain sugar, starch,glucose, and gum. It addition, it also contains curcuminoids (curcumin, demethoxy curcumin and bisdemethoxy curcumin), curcumol, Curcumanolide A&B, zederone, fyranodiene, sesquiterpenoids,pyrocurzerenone, procurcumenol, (+) x-terpineol, (+) borheol (Kulkarni and Ansari, 2004; Itokawaet al., 2008; Elias et al., 2015). The oil collected from rhizome likewise contain β-pinene, α-terpineol, furanodienone, isofuranodienone,camphor, zingiberol, isoborneol,borneol, ar-curcumine, curzerenone,xanthorrhizol isomer, germacrone, 1,8-cineole, methyl eugenol, curdione, palmitic corrosive (Srivastava et al., 2006; Banerjee et al., 1980; Nguyen , 2001; Srivastava et al., 2006; Thongkhwan et al., 2017). The significant mixtures like α-amorphene, camphor, 2, 7-napthalenediol, trans-nerolidol, octadecanoic corrosive, butyl ester, humulen-6,7-epoxide have been additionally announced from rhizome from GC-MS analysis (Nayak et al., 2014). Root of arrow root contains intensifies D-cymarose, B-D-glucosy 1-Lthebetose, β-Elemenone, cinnamic corrosive and acidic corrosive. The pseudo stem of East Indian arrow root contains feronic corrosive, polyoxytregnanes, marstenacigenins A&B dresgenin, glycosides tenacissosidas A-E. The arrowroot seeds contain pregnanecissogenin, polyhydroxy, drebogenin, tenasogenin (Thongkhwan et al., 2017).

8. Physicochemical Properties

The physiochemical properties of East Indian arrowroot starch are presented in Table 1 where starch is the major constituent of East Indian arrowroot (97.43%) and it crude fat, protein and ash content present in low amount. The arrowroot starch is composed of 32.30% amylose and 67.70% amount of amylopectin. The amylose content in starches is important as presence of amylose determines the gelling ability of starches.East Indian arrowroot starch has high calorific value and this attribute can be utilized in gluten- free and infant food products to aid people with celiac disease.

Table 1: Physicochemical composition of East Indian arrowroot starch (g/100g) (Kumari et al., 2017)

Constituents	*C. angustifolia*
Moisture (db)	2.72 + 0.36
Crude fat	0.62 + 0.31
Crude protein	1.02 + 0.14
Ash	0.3 + 0.04
Starch (db)	97.43 + 0.52
Amylose (%)	32.30 + 0.44
Amylopectin (%)	67.70 + 0.38
Calorific value (cal/g)	3499 + 1.06
True density (g/cc)	1.59 + 0.03
Bulk density (g/cc)	0.51 + 0.27
Porosity (%)	68 + 1.

9. Food Values of East Indian Arrowroot

This arrowroot plant is beneficial to Indian culinary variety as a source of starch. The rhizomes of *C. angustifolia* are regularly ground into flour which can then be combined as one with milk or water to frame a nutritious feast (Ravindran, 2007). Tubers produce starch like arrowroot effectively edible and profoundly nutritious highly recommended for baby foods, frail youngsters and invalids. 10 gm rhizome powder is added with glass of milk and sugar treats and polished off as wellbeing drink to work on actual strength and further develop power (Ambasta, 1986). A beverage including East Indian arrowroot as a fixing is additionally utilized as a substitution of bosom milk or as a wholesome enhancement for children a brief time subsequent to weaning. Because the arrowroot starch is rich in energy, the starch powder can be taken by those are malnourished or starving (Upvash) during festivals.It is found as an essential fixing in cakes, natural product jelly, rolls and pudding (Doble et al., 2011). *C. angustifolia* have been accounted for to be a rich wellspring of proteins and carbs and are utilized in culinary arrangements as vegetables and food appetizers (Sanatombi and Sanatombi, 2017). Many varieties of sweet dishes are prepared from starch of rhizome such as barfi (Shankar et al., 2014), halwa (Banik et al., 2014) , Khoa-jalebi (Kumari et al., 2012) and sherbat (Singh and Palta, 2004) Because of its medicinal properties, The starchy flouris rich in medicinal properties for which the starch is used for preparation of weaning food called shotti (Sharma, 2012). The East Indian arrowroot starch is also used as a base for many food products like biscuits, cookies etc.

10. Therapeutic uses of East Indian Arrowroot

The rhizome of tikhur has great medicinal value (Jain, 1995;Sarkaret al., 2015). The It is well known for its use in ethnomedicine as the hizome is demulcent, nutritious, antipyretic and blood coagulant (Tushar et al., 2010).The arrowroot starch is utilized for kids due to effectively edible.The rhizome powder blended in with honey applied on the mucous layer of the oral cavity to fix stomatitis issue as well as the combination advances the recuperating of stomach ulcers.The rhizome is an astringent in nature as the utilization of rhizome results really great for looseness of the bowels, diarrhea and colitis. The diuretic action of starch because of cardiovascular weakness, it shows a compelling cure in Oedema issue. During diarrhea, dysuria, gonorrhea, the powder of rhizome starch go about as great eating regimen as conjee as the rhizome is energizer, carminative and stomachic (Kulkarni and Ansari, 2004). Utilization of arrowrrot is a lot of viable in typhoid fevers, ulceration of the guts and bladder. If there should arise an occurrence of troublesome and agonizing development it very well may be utilized as flimsy conjee arranged like grain water with milk and sugar added. It is nutritive and is utilized as a pleasant non disturbing eating routine in specific ongoing illnesses during strengthening structure fevers (Prakash, 2001). It is also reported that the boiled water extract of rhizome is used for treatment of diabetes (Sheikh et al., 2015) Root powder utilized with milk to treat consuming micturition or pee, trouble in micturition, fever, causticity. Root stock is valuable in infection, consuming sensation, dyspepsia, asthema, jaundice, leucoderma, paleness, kidney stones and illness of blood (Sharma, 2003). In Ayurveda, the plant is utilized to further develop strength and insusceptibility, further develops energy, sustaining, aphrodisiac,coolent, blood/skin/draining problems, jaundice, inordinate thirst, liver illnesses, asthma, TB, wt. misfortune, iron deficiency. The medicinal oil separated from various pieces of this species is utilized for antimycotic action, antifungal and antibacterial exercises, and cancer prevention agent movement (Doble et al., 2011; Shukla et al., 2011; Jena et al., 2017) .The rhizome of *C. angustifolia* Roxb. can be used to heal peptic ulcers, is used in treatments of dysentery, diarrhoea, and colitis and is often employed as a herbal tonic for patients suffering from tuberculosis. Its use for the treatment cough and bronchitis is very popular in the areas where it grows (Doble et al., 2011).

11. Pharmacological Activities

11.1. Antioxidant Activity of East Indian Arrowroot

In a DPPH extreme model, Nahak and Sahu (2011) investigated the antioxidant activity of five *Curcuma* species: *C. longa, C. zedoaria, C. angustifolia, C. aromatica, and C. amada,* based on their curcumin and phenol content where *C. longa* showed the most elevated cell reinforcement movement. To learn about antioxidant action, *C. angustifolia* showed a solid relationship with curcumin and phenol content other than the other four species aside from. Anyway *C. angustifolia* might be dynamic because of high fragrant oil content like eugenol, palmitic and camphor etc(Nahak and Sahu, 2011). The GC-MS investigation of oil removed from rhizome part and leaf section showed 32(rhizome) and 35(leaf) recognized constituents in oil. The cancer prevention agent property was more in leaf oil than rhizome oil as well as reference guidelines while the cell reinforcement limits was considered through various strategies like DPPH, ABTS and Reducing power capacity (RPA) (Jena et al., 2017).

11.2. Anti-Proliferative and Antioxidant Activity of East Indian Arrowroot

The movement and against proliferative action of aqueous extracts of maturing blossoms, leaves, full sprouted blossoms and combination of all pieces of *C. angustifolia* by utilizing MDCK, DPPH test, Vero and HeLa cell lines showed the outcome as potential enemy of proliferative and cancer prevention agent properties that validate the conviction of the native individuals to remember the plant for their dietary propensities which could be taken advantage of for use in planning of natural medications and enhancements (Assumi et al., 2017).

11.3. Anticancerous, Antimicrobialand Antioxidant, Activity of East Indian Arrowroot

The *in*-vitro antioxidant, antimicrobial and anticancerous activity of *C. angustifolia* methanolic extricates was explored from Odisha (Nayaket al., 2013). The methanolic extricate of rhizome was assessed in terms of add up to phenol substance, ascorbic corrosive add up to flavonoids substance, DPPH radical rummaging action for antioxidant movement whereas antimicrobial movement of plant extricate was decided by agar well dissemination strategy counting gram +ve and gram –tive microbes with six strains of microbes. The cytotoxicity activity of plant extract was studied by using methanolic extract of rhizome. on cervical cancer cell by MTT assay. The result showed the methanolic rhizome extract of *C. angustifolia* has highpotential asantimicrobial, antioxidant, and anticancerous properties for which the plant could be exploited in preparation of herbal drugs and could be rare on over exploitation.

11.4. Antimicrobialactivity of East Indian Arrowroot

Viji and Wilson (2017) confirmed the presence of nanoparticles by brine shrimp lethality assay while about the antimicrobial activity of methanolic, silver nano particle, chloroform rhizome extract and ethylacetate of *C.angustifolia* against *E.coli, B. subtilis, P. mirabilis, A. niger, M. indicus, K. pneumonia, S. aureus,* and *R. stolonifer* which confimed the cytotoxic potential of the plant extracts.

The methanolic extract from rhizomes of *C. angustifolia* resulted the antibacterial activity against *E.coli, S.arueus* and *S. flexneri* and *P. aeruginosa,* by well-plate method due to presence of varieties of secondary metabolites in the plant, such as flavonoid, tannin, glycoside, and phytosterol (Jadhao and Bhuktar, 2017). Dubey and Sao (2018) investigated stem extracts of *Z. officinale, C. angustifolia, C. longa* and *A. barbedensis* against skin pathogenic organisms like *E. faecalis, P. saruginosa, S. aureus,* and *C.albicans*. The ethyl acetic acid derivation from every one of the chose extricates has shown great action with *C. albicans* while all the plant extricates have not shown antibacterial movement with the tried microorganisms at 20µl against skin illness causing organisms.The extracts of *C. angustifolia* alongside *A. barbedensis* shown better antibacterial and antifungal activities at 50µl against skin illness causing microorganisms (Dubey and Sao, 2018).

11.5. Anti-Ulcerogenic Activity of East Indian Arrowroot

Rajashekhara et al. (2014) examined about the intense harmfulness potential and hostile to ulcerogenic movement from the rhizomes of the *C. angustifolia* and *M. arundinacea* on gastric ulcer in pale skinned person rodents. The intense harmfulness potential were examined for adaptogenic activity bydetermining and comparing the changes in rectal temperature, ulcer index and by histopathological parameters in the test drug group with that stress control groups of rats. The outcome showed that both the medications delivered to the rats showed no harmful movement or mortality rate even up to the most extreme portion level of 4400 mg/kg while the two test sedates fundamentally switched the pressure actuated gastric ulceration in contrast with pressure controlled rats. Both the test drugs proved the low chance of gastriculcer and anti-ulcer activity by showing significant decrease in the volume, increase in the pH level , reduced the freeacidity of gastric juice and decreased the peptic activity (Rajashekhara et al., 2014).

11.6. Hepatoprotective Activity of East Indian Arrowroot

The hepatoprotective activity of *Curcuma angustfolia* Roxb rhizome powder showed reversal effect on important parameters like SGOT, SGPT, ALP, serum total cholesterol and serum triglycerides.The intake of rhizome powder showed good protection and good cytoprotective effect (Elias et al., 2015).

12. Conclusion

The current review uncovered the *C. angustifoila* can possibly fix different infirmities uniquely swasa, kasa, kamala, raktapitta, kushta, jwara, pandu, trishna, daha, kshaya, vrana, aruchi and numerous different infections. Tikhur is a novel starch with restorative advantages and appropriate for use in various desserts as it confers no persistent flavor. Substance constituents like alkaloids, tannins, steroids, saponinsetc were accounted for in *C angustifolia.* The medium amylose content, lucidity and great dissolvability in water bestows it delicate and clear gel properties and the gelatinization of tikhur starch were higher than normal starches. Different tests demonstrated its enemy of dangerous, hepatoprotective and so on approves the justification for the native individuals to remember the plant for their dietary propensities to stay sound which could be taken advantage of in planning of natural medications for use in treating different illnesses. It very well may be inferred that arrowroot can be used for more extensive food and non-food applications and in light of these discoveries, FSSAI can set down quality principles like those for Tapioca Sago (IS: 899-1971) accordingly expanding the market worth of East Indian arrowroot starch.

References

Ambasta, S.P. 1986. The useful plants of India. (Publications of Information Directorate, CSIR, New Delhi), 51, 175-504

Assumi, S.R, Puro, K, Kasomva K, Duarah A, Jha A.K., Sen. 2017. Invitro anti-proliferative activity of Curcuma angustifolia and estimation of bioactive compounds for antioxidant activity. Journal of Food Science and Engineering. 7, 59-66.

Banerjee, A, Kual, V.K., Nigam, S. S. 1980. Riv. Ital. Essenze. Profumi, Piaute Off., Aromat., Syndets Saponi Cosmet. Aerosols. 62(2), 75-76.

Banik, A., Nema, S., Shankar, D. 2014. Wild edible tuber and root plants available in Bastar region of Chhattisgarh.Int. J. of Forestry and Crop Improv. 5(2), 85-89.

Chauhan, V. B. S., Pati, K., Bansode,V. V., Nedunchezhiyan, M. 2020. Status, diversity and potential of indigenous tropical tuber vegetables. Indian Horticulture 65(3), 78-82.

Chauhan, V.B. S., Mallick, S. N., Pati, K., Arutselvan, R., Nedunchezhiyan, M. 2022. Status and Importance of Underexploited Tuber Crops in Relation to Nutritional Security and Economic Prosperity. Compendium for Winter School on "Underexploited Vegetables: Unexplored Treasure Trove for Food, Nutritional and Economic Security". Published by ICAR-Indian Institute of Vegetable Research, Varanasi-221305, Uttar Pradesh. pp 246-264.

Doble, B., Dwivedi, S., Dubey, K., Joshi, H. 2011.Pharmacognostical and antimicrobial activity of leaf of *Curcuma angustifolia* Roxb.Int.J. of Drug Discov. and Herb. Res.1(2), 46-49.

Dubey, S., Sao, S. 2010. Antimicrobial activity of crude stem extracts of some medicinal plants against skin disease causing microbes from Chhattisgarh Region.Int. J. of Engg. Tech. Sci.and Res.5(1), 57-60.

Dwivedi, R.R. 2009. Sodala Nighantu.1sted. Varanasi: Krishnaas academy, 62.

Elias, L, Pradeep, Harini, A., Hegde, P.L. 2015. *Curcuma angustifolia* Roxb (Tavaksheeri): A Review. J. of Pharmacog. Phytochem.4(2),241-243.

Itokawa, H, Shi Q, Akiyama T, Morris-Natschke, S.L., Lee, K.H. 2008. Recent advances in the investigation of curcuminoids. Chin.Med. 11,3.

Jadhao, A.S., Bhuktar, A.S. 2017. Physicochemical and antibacterial activity of rhizomes of *Curcuma angustifolia* Roxb, (Zingiberaceae). Int. J.ofSci. andRes.6(8),198-200.

Jain, S.K. 1995. Ethnobotanical diversity in Zingibers of India.Ethnobotany. 7(1-2),83-88.

Jena, S, Ray, A, Banerjee, A, Sahoo,A, Nasim,N, Sahoo, S., Kar, B., Patnaik, J., Panda, P.C.,Nayak, S.2017. Chemical composition and antioxidant activity of essential oil from leaves and rhizomes of *Curcuma angustifolia* Roxb.Nat.Prod.Res,31(18),2188-2191.

Kulkarni, P.H., Ansari, S. 2004. Theayurvedic plants. Indian Medical Science Series No 132. Sri Satguru Publications,Delhi,126,261

Kumari, R., Shrivastava, S.L., Mishra, H.N. 2012.Optimization of Khoa and Tikhur mix preparation of Khoa-jalebi sweet. Open Access Scientific Reports. 1(10),1-4.

Kumari,R., Shrivastava, S. L., Mishra, H.N., Meghwal, M.2017. Physicochemical and functional properties of *Curcuma angustifolia* (Tikhur) - An underutilized starch, The PharmaInnov. Journal, 6(7), 114-119

Nahak, G.,Sahu R.K. 2011. Evaluation of antioxidant activity inethanolic extracts of five *Curcuma* species. Int.Res.J.ofPharm.2(12),243-248.

Nayak,S., Jena,A.K.,Mittal,D.K.,Joshi,D. 2014.GC-MS Analysis of phytoconstituents of some wild zingiberaceae plants methanolic rhizome extract, Res. in Plant Sci.2(1), 1-5.

Nguyen, T.B.T, Trinh, D.C., Do DR, Nguyen, X.D. 2001. Vietnam Hoa Hoc Va Cong Nghiep Hoa Chat, 5,11-14 .

Ojha, J.K. 2004. A Handbook of Ddravya guna. Chaukhamba Sanskrit Pratishtan Delhi.1st ed. 344.

Paikra, C.K., Patel, S., Naik, R., Ram, D.S.2013. Performance evaluation of tikhur (*Curcuma angustifolia*) starch extraction machine. Ind. J. of Sci. Res. and Tech.2(5), 34-38.

Panda, S.K., Patra, N., Sahoo, G., Bastia, A.K., Dutta, S.K. 2012. Antidiarrheal activities of medicinal plants of similipal Biosphere reserve, Odisha, India.Int. J. Med. Arom. Plants. 2(1), 123-134

Prakash, P.2001.Indian medicinal plants forgotten healers. Chaukhamba Sanskrit pratishtan. Delhi; 256.

Rajashekhara, N., Ashok, B.K., Sharma, P.P., Ravishankar, B. 2014. The evaluation of anti-ulcerogenic effect of rhizome starch of two source plants of Tugaksheeree (*Curcuma angustifolia* Roxb. and *Maranta arundinacea* Linn.) on pyloric ligated rats.Ayu. 35, 191-196

Rajashekhara, N., Ashok, B.K., Sharma, P.P., Ravishankar, B. 2014. Evaluation of acute toxicity and anti-ulcerogenic study of rhizome starch of two source plants of Tugaksheeree *Curcuma angustifolia* Roxb. and *Maranta arundinacea* Linn., AYU . 35(4),433-437.

Ravindran, P. N., Nahar, L, Sarker, S.D., Skornickova, J., Rehse,T., Sabu, M.,Ravindran, P.N. 2007.Turmeric: The Genus *Curcuma*. CRC Press Taylor and Francis Group London New York,4, 10-11

Sanatombi, R., Sanatombi,K. 2017..Nutritional value, phytochemical composition, and biological activities of edible *Curcuma* species: A review. Int. J. of Food Prop.20(S3),S2668-S2687.

Sarkar, P, Kumar, L.D.H., Dhumal, C., Panigrahi, S.S., Choudhary, R. 2015. Traditional and ayurvedic foods of Indian origin. J. of Ethnic Foods 2, 97-109.

Shankar, D., Shukla, N., Nag, J.,Sahu, M. 2014. Study on preparation procedure and standardization of recipe for tikhur Barfi. Int. J. of Process.AndPost HarvestTechn. 5(2),156-64.

Sharma, A. 2012.Traditional processing of Shotti (*Curcuma augustifolia* Roxb.) – arhizome based ethnic weaning food.Ind.J.ofTrad.Knowl.11(1),154-155.

Sharma,R. 2003.Medicinal plants of India. An Encyclopaedia. Daya Publishing House, Delhi,11002.

Sheikh, Y., Maibam, B.C., Biswas, D., Laisharm, S., Deb, L., Talukdar, N.C., Borah, J.C. 2015. Anti-diabetic potential of selected ethno-medicinal plants of north east India. J. of Ethnopharmacology, 171,37-41.

Shukla, A.C., Pandey, K.P., Mishra, R.K., Dikshit, A.,Shukla, N. 2011.Broad spectrum antimycotic plant as a potential source of therapeutic agent. Natural Product Journal 4, 42-50.

Singh, R., Palta, A. 2004. Foods and beverages consumed by abujhmarias –a primitive tribe of Bastar in Chhattisgarh. Tribal Health Bull. 10, 33-40.

Srivastava, A.K., Srivastava, S.K., Syamsundar, K.V. 2006. Volatile composition of *Curcuma angustifolia* Roxb. Rhizome from central and Southern India. Flav. and Fragnance J. 21,423-426.

Srivastava, S., Chitranshi, N., Srivastava, S., Dan, M., Rawat, A.K.S. and Pushpangadan, P. 2006. Pharmacognostic evaluation of *Curcuma aeruginosa* Roxb. Nat. Prod. Sci. 12,162-165.

The wealth of India.National institute of science communication council of scientific and industrial research.New Delhi, 2001; 2, 261.

Thongkhwan,P.,Chaibunga, T., Kwanboonjan, H.,Theanphong, O.2017.Essential oil constituents of the fresh root and rhizome of *Curcuma angustifolia* Roxb. From Thailand. Bull. Heal. Sci..Technol.15, 52-53.

Tiwari,S., Patel, S. 2013. A comparative study of tikhur traditional and partial mechanical processing and cost economics.Int. J. of Agric. Engg.6(1), 213-215.

Tripati, I.D. 1982. Raja Nighantu.1st ed. Varanasi: Krishnadas academy,171-172.

Tushar, Basak S, Sarma, G.C.,Rangan, L. 2010. Ethnomedical uses of Zingiberaceous plants of north east India. J. Ethnopharmacol. 132, 286-296.

Vaishya, S.S. 1995. Shaligrama Nighantu. Bombay: Khemraj Shrikrishnadas Prakashan, 120-121.

Viji, M.O., Wilson, N. 2017. *Curcuma angustifolia* rhizome extracts as reducing agent in synthesis of silvernano particles. Romanian J. Biophys. 27(1-2),L55-67.

35

Farm Mechanisation in Horticultural Crops

Sunil Kumar[1], Puneet Sharma[2] and Manu Tyagi[3]

[1]*Krishi Vigyan Kendra, Sangrur, Punjab Agricultural University, Punjab India*
[2]*Krishi Vigyan Kendra, Pathankot, Punjab Agricultural University, Punjab India*

1. Introduction

The status of mechanization in India is at a very low level for horticultural crops. The farmers usually go behind the use of traditional methods for different farm operations (Fig 1) and practices for cultivation of horticultural crops, their handling, storage, and processing. This results in less yields, high cost of cultivation and high crop losses from harvesting till processing. The use of different available hand tools, machinery and hi-tech technologies in the production and processing of the crops will lead to increased yields and decrease in losses to a great extent (Mandhar, 1999).

Tillage
Plough, Cultivator, spade, auger, fawda, khurpi etc.

↓

Sowing and Transplanting
Inclined plate planters, pneumatic planters, semiautomatic and fully automatic transplanters

↓

Irrigation
Drip irrigation, automated drip irrigation, sprinkler irrigation

↓

Weeding and Hoeing
Wheel hoes, power weeder, cultivator

↓

Spraying

Lever operated and power operated knapsack sprayers, mist blower, aero blast sprayers

↓

Harvesting

Diggers, robotic harvester, tree shakers, hydraulic ladder type harvesters

↓

Processing

Sorter, grader, packaging machine

Fig 1: Different Operations at Farm

2. Advantages of Farm Mechanisation

The major advantages of farm mechanisation are given below:

2.1. Higher Efficiency

Efficiency of the system increases as mechanisation of operations is observed to deliver more output as compared to manual operations. For example, in fruits computerised/mechanical grading is done on moving conveyer belts which removes the undesired or substandard quality of fruit from desired quality more efficiently as compared to manual grading which, is relatively more prone to error.

2.2. Large Scale Farm Management

Themechanisation ofoperations is economical at large farms or orchardsparticularly, in regions where labour cost is high. Although, entire mechanisation may not be possible for all operational needs, but partial mechanisation of farm operations is becoming the need of the horticultural crop production.

2.3. Prompt Execution of Farm Operations

Horticultural operations are mostly time bound operations which, require immediate attention of the farmer to act as soon as possible as it could significantly increase profits or decrease losses. The mechanisation of farm operations allows timely execution.

2.4. Uniformity in Operation

Operations like land levelling, irrigation and spraying require high rate of uniformity index for significant farm management. Manual execution of such operations on a commercial farm may likely produce irregularities. On the other hand, mechanised farm operations give uniform results.

2.5. Reduction in Drudgery

The implementation of mechanised farm operations in place of manual operationshas a positive impact onthe health of worker or labour. This in turn leads to efficient management of the working environment.

2.6. Improved Quality Produce

Farm Mechanisation contributes to improvement in quality of produce. For example, advanced packaging and grading equipment help in serving desirable quality horticultural produce to consumers.

2.7. Additional Income Generation Through Custom Hiring

Possibility of giving machinery, equipment, or systems on rental basis creates a viable option of additional income source.Machines such as tractors could be used on custom hiring basis or cooperative level. There are government financial schemes in India which supports farmers' groups to get benefit financially for purchase of subsidized agricultural tools, implements or machines for rentals.

2.8. Better Handling of Produce

Farm mechanisation enables better handling of crop produce such as fruits and vegetables aspackaging time is reduced significantly.

2.9. Reduced Labour

Labour cost has been increasing continuously with time which, makes it more economical to mechanise the operations.

3. Tools and Implements for Farm Mechanisation in Horticultural Crops

At present the quality of work depends on the skill of the worker as most of the horticultural operations are being done manually. This causes delaying of the operation that increases its cost due to low outputs. To improve precision and timeless in operations and reduction of damage to the highly perishable horticultural produce, latest efficient technologies are required.

In India farmers generally rely on hand tools and other farm equipment for completion of various field operations. Different types of tools and equipment used for different operations in cultivation of fruits and vegetables are discussed below:

3.1. Garden Tools

The common garden tools such as *fawada*(Spade), *gainti*(pick axes), *Kulhadi* (Machete), *Khurpi,* Knife, Pruning saw Sabbal (Crowbar)etc. are being used for raising nursery of horticultural crops andraising these fruit plants by the farmers (Ghosh et al., 1998). Other good quality tools being used for different operations are grafting knives, budding knives, pruning knife, weeding trowels, thinning scissor, weeding fork, bill hook, hand cultivator, spade, garden hoe, fruit gatherer, tree pruner, garden rake, digging fork, pruning saw, grass cutter, hedge cutter, hand wheel hoe, hand wheel hoe with seeder attachmentetc.

3.2. Farm Implements

3.2.1. Tractor Powered Post Hole Diggers

For planting fruit trees,manual digging of pits is accomplished using a spade (generally rectangular in shape), which is a time-consuming process(Bhardwaj and Srivastava, 1982)

To ease this process,tractor powered post hole diggers have been developed. These are available in market with different auger sizes. These can dig pits in much lesser time in comparison to manual digging. These can be used for digging pits for erection of different fruit plants as well as fencing poles. The pits of diameter range 40-100 cm and depth 30-100 cm can be easily excavated using this tractor operated machine(Poole, 1968).

3.2.2. Power Tiller Operated Rotavators

Weeding is an important operation in fruit crops. It should be carried out at specific intervals for proper growth of plants. Manual weeding operation consumes a lot of labour and also needs a lot of time. To conduct these operations power tiller operated rotavators can be used. Low height garden tractors are also available for such operations in orchards with high canopy. These tractors can easily perform weeding operation using matched disc harrows and rotavator.

3.2.3. Sprayers

In small fruit trees spraying is achieved using knapsack sprayer, foot operated sprayer, battery operated knap sack sprayer and stationary power sprayers. In case of tall trees, use of the earlier mentioned sprayers is not possible. To spray on such trees,tractor operated aero-blast sprayers have been developed. These sprayers are convenient and do not require large amount of water for spraying. In case of such sprayers, the chemical is sprayed over the crop foliage *via* high blast of air which aids in better penetration of the chemical in dense foliage and its application on both sides of leaves due to their fluttering in high wind velocity. Such sprayers save 30-50% chemicals. Electrostatic sprayers are also available that needs very less amount of water and are much more efficient in insectpest management as compared to other sprayers.

In vegetable crops spraying is accomplished using knapsack sprayer, foot sprayer, mist blower and power sprayer. Wide area coverage is being provided by tractor drawn sprayers.

3.2.4. Tractor Operated Saws and Pruners

Pruning is an important operation in grapes, peach, plum, pear, citrus, mango and other fruit plants. This operation is done to get quality and high yield. Tractor operated saws and pruners are available that can be utilised for this operation.

3.2.5. Tractor Operated Rotovators or Power Tillers for Seedbed Preparation

The available conventional tillage equipment which, are used in field crops can also be used in seedbed preparation for vegetablecrops.Seedbed can be easily prepared for vegetables using single or two passes of tractor operated rotovators or power tillers. It helps in saving time of different operations and saves fuel by reducing the number of operations for seed bed preparation.

3.2.6. Seed Drills, Planters and Transplanters

At present, the sowing of vegetable seeds and transplanting of vegetable seedlings is achieved manually. This involvesdrudgery and is also a time consuming operation. Vegetable seeds can be sown using tractor and power tiller operated seed drills and planters that helps in proper placement of seed and fertilizers in less time. Tractor operated semi-automaticand fully automatic potato planters are commercially available. Pneumatic planters are also available for precise placement of seeds at fixed equal intervals. These are available in 2,4 and 6 row units. The seed-to-seed distance is constant in these planters with uniform picking of seed number(as per requirement). Thisprevents the need for thinning after germination. Further, to aid vegetable seedling transplanting, vegetable transplanters are available. Amanual garlic planter has been developed at PAU, Ludhiana which works satisfactorily.

3.2.7. Weeders, Ridgers/Furrowers

Manual weeding and hoeing operations in vegetable crops is achieved using hand operated wheel hoe. These are 3-4 times more efficient than spade.In wide row crops weeding operation can be done using tractor operated sweep cultivators and power weeder. A tractor operated inter-row rotary weeder has been designed and developed by Punjab Agricultural University, Ludhiana for accomplishing weeding operation. In root crops, earthing up operation can be carried using ridgers/furrowers.

3.2.8. Harvesters and Diggers

Fruits are highly perishable commodity and must be harvested in a very narrow time range and should be handled carefully. Few fruit crops like jamun, bael do not mature uniformly and therefore needs selectivity while harvesting. This is a major hurdle in mechanical harvesting of fruit crops as more than one picking is required. At present most of the fruit harvesting is accomplished manually. The harvesting of apples and orange require skilled labour to reduce fruit damage. In case of fruit crops with large canopy like mango, jamun, aonla etc. fruits are plucked by persons climbing on the tree. In some of the fruit trees, branches are shaken manually or using shakers. Some fruits can be plucked using a bamboo stick on which ahook is fitted for plucking operation. In such methods fruits gets damaged when they fall. Harvesting of fruits from thorny plants such as ber, bael is quite difficult(Chadha, 1999). Owing to such problems, there is a wide need of harvesters that are suitable for harvesting of different types of fruits.

Tree shakers are available for mechanised harvesting however, their use is very limited. The basic principle involved in removing fruit by shaking is the acceleration of each fruit to such an extent that the inertia force developed in fruits due to acceleration becomes greater than the joining force of fruits and the tree. In developed nations tractor powered hydraulic ladder type fruit harvesters are used for fruit harvesting that helps in saving labour involved in harvesting operation,decreases harvesting cost and fruit damage. Harvesting of shrub fruit trees can also be completed using tractor operated harvesting machines (Peterson et al., 2003). These machines use toshake shrubs which, that leads to detaching and dropping of ripe fruits on a conveyor belt that is being fitted at machine bottom. These fruits are than cleaned further by using an air blast and collected in containers. The other different types of fruit harvesters are tree-shakers for shaking the whole tree and limb-shakers that use to shake the tree branches individually and this shaking causes ripe fruits to drop either on the ground or on platform. The efficient use of tree shakers demands the use of easy tooperate catching and collecting devices made using of canvas or nylon sheets.The other more sophisticated devices are inflatable cushions in which air is filled automatically when mattress spreads.Robotic harvesters are developed intechnologically superiornations that are used for harvesting of high value fruit crops (Peterson, 2005). These harvesters use to select the ripe fruits on the basis of colour, contrast, softness and other crop parameters. They are usually used at places where the labour costsinvolved aretoo high.

Potatoes are harvested using potato diggers. The equipments for uprooting are not available for root crops like tapioca, ginger, carrot, radish, sweet potato etc. A focussed research is needed to design and develop harvesting machinery for these crops.

3.2.9. Graders

Grading is an important operation in horticultural crops for fetching quality produce and high profits. It renders ease in handling of fruits and vegetables throughout the whole marketing chain. In India, fruits and vegetables are generally graded manually which, is labour intensive, time taking and moreover, lacks uniformity. On the other hand, mechanical grading is time saving and gives more accurate results. Grading machines may be categorised broadly into two categories on basisof diameter and weight. Machines, which grade by diameter, yields higher output over machines which, grade by weight. But, grading via weight exhibits higher accuracy over grading by diameter. Different types of size grading machines *viz.* screens, roller grader and diverging belt grader are available for use. Weight grading machine works on the principle of density and specific gravity (Londheet al., 2013).

4. Mechanization of Irrigation System in Horticultural Crops

Mechanization of horticultural crops is done by the installation of a drip irrigation system mainly to automate the operations involved. Automation of irrigation system is done mainly to control the amount of water supply to plants based on two approaches that are time-based approach and water flow measurement basis approach (Henten, 2006). The main advantage of automation is to give changing frequency of irrigation and fertilizer scheduling as there is variation in demand of water and fertilizer according to phenological stages of the crop. Also, the water supply can be varied remotely which saves time and energy. Due to automation system, operations like starting and stopping the irrigation system can be done instantly i.e.exactly when the farmer intends to start or stop. This is becausethe pumps start or stop as per the command of the operator(Benami and Ofen, 1995).

The timing approach is based on assumption that the flow of water is constant through the system. therefore, time of irrigation is increased by increasing water supply time of irrigation is increased by controlling Valves and similarly, the time of irrigation decreased for decreasing water supply. Low measurement involves volumetric measurements to observe the pipe pressure of the system to increase or decrease the discharge.

5. Maintenance and Repair of Farm Machinery

Machinery used in horticulture crops is often required for a specific short duration and remains unused for the remaining part of the year. Maintenance of such farm machinery is crucial. For example, intercultural crop operations are carried using small tractors. These perform tillage operation between inter-rows of the orchards which is executed during specific time of the year. It is observed that the machinery employed in the farm can be made to last for a longer duration in service by adopting following measures:

- After every operation,machinery should be cleaned and kept in clean and dry conditions after proper greasing.
- Periodic servicing of the various components of machinery and implements needs to be done according to manufacturer guidelines.
- It should be taken care that no one machinery should interfere or damage other machinery while operating in the farm. For example, the tillage operation by tractor should be avoided in the fields where subsurface drip irrigation is installed or should beconducted in a very careful manner so that damage to the lateral pipes of the irrigation system could be avoided.
- Farm-manager must understand how much mechanization is required to give best-optimized results in the long run.
- Stakeholders must calculate the break-even point to understand the cost benefit ratio of farm machinery. Thisaids in decision making process of farmers for investing in the mechanization of the farm.

6. Challenges in Mechanisation for Management of Horticultural Crops

Mechanisation offarm operations in horticultural crops partially or completely needs investment; as capital is required for purchase, installation and training.Also, if the system is advanced then hiring of skilled personnelmay becomenecessary.

Eventually repair and maintenance costs are also added to the operational cost. Perception of the farmer towards the mechanisation is also a big constraint in adoption of these technologies.

7. Conclusion

Mechanisation of farm operations in horticultural crops beginning from seedbed preparation till post-harvest handling is very important. It plays a major role in increasing production as well as productivity, improvement of quality, reduction in harvest and processing losses, decreasing drudgery with decrease in dependence on labour and all of these with a goal of maximisation of profits. Mechanisation of different operations and their automation will play an important role in producing quality end product for the user that will get reflected in terms of profits. It will improve timeliness of operations and reduce the consumption of resources as in automated drip irrigation system. However, a lot of research is needed in this field for indigenisation and cost reduction of various technologies available worldwide for their suitability to local crops.

References

Benami A., Ofen, A. 1995. Irrigation Engineering: Sprinkler, Trickle, surface Irrigation: Principles, design and agricultural practices. AGRIPRO -Agricultural projects (AGP) Israel.

Bhardwaj, K.C., Srivastava, N.S.L. 1982. Horticultural Tools used in India. Farmers Journal. 1(11), 35-45.

Chadha, K.L. 1999. Horticulture in the coming millennium: Issues and constraints. Agriculture Today. 11(4), 34-38.

Ghosh, S.P., Bhatnagar, P.S., Sukumaran, N.P. 1998. Fifty years of Horticultural Research, Division of Horticulture, ICAR, New Delhi.

Henten, E.J.V.2006. Greenhouse Mechanization: State of the Art and Future Perspective, Acta Hort. 710.

Londhe, D.H., Nalawade, S.M., Pawar, G.S., Atkari, V.T., Wandkar, S.V. 2013. Grader: A review of different methods of grading for fruits and vegetables. Agricultural Engineering International: CIGR Journal. 15(3), 217-230.

Mandhar, S.C. 1999. Thrust Areas for Mechanisation of Horticultural crops in India. Proc. Indo-Vzbek Workshop on Agricultural Research, held at CIAE, Bhopal during Nov. 15 - 16.

Peterson, D.L.2005. Harvest mechanization progress and prospects for fresh market quality deciduous tree fruits. Hort. Techn. 15(1), 72-75.

Peterson, D.L., Whiting, M.D., Wolford, S.D.2003. Fresh market quality tree fruit harvester Part I: Sweet cherry. Appl. Eng. Agri. 19(5),539-543.

Poole, W.D. 1968. The impact of mechanization of horticultural crop production in Louisiana on teaching, NACTA Journal, 82-84.

36

Deployment of Internet of Things (IoT) in Agriculture Enabling Technologies and Case Study on Startups in India

Aditya Sinha and Abhijeet Ghatak

Bihar Agricultural University, Sabour, Bhagalpur, Bihar, India

1. Introduction

The Indian economy is heavily dependent on agriculture (employing 53 % population). Around 84 % of the holdings are less than 2 ha while approximately 67% of the farms have less than 1 ha of operational farmland (Government of India, 2013; Pandey and Suganthi, 2015). A large fraction of the farming population is mostly reliant on monsoon rains for irrigation. The condition of the farmers is found to be deteriorating year after year with the revelation that 27% of farmers did not consider farming as a profitable enterprise and 40% were ready to migrate to other avenues for employment (NSSO, 2005; Pandey and Suganthi, 2015). With the rapid growth in internet penetration and affordable data availability in the country, it offers immense opportunities for start-ups to explore the information technology boom in the farming sector through the efficient deployment of smart-farming solutions. The share of urban and rural wireless subscribers in a total number of wireless subscribers was 55.40% and 44.60% respectively and a vast majority of the subscribers have access to the internet (TRAI, 2019). Recently, the agricultural sector is rejuvenated with the deployment of information and communication technologies (ICT) for the development of efficient value-chains, improving marketing opportunities, meeting the needs of organic farming etc. The sector is witnessing the participation of information technology professionals and budding entrepreneurs for harnessing the potential at their levels. The gap in integrating ICT with the agricultural sector is narrowing due to these efforts. The last three decades have experienced significant improvement in the use of ICT applications in farming particularly linked to food production and transportation coverage. The combination of IoT with other computing technologies (big data science, analytics, mobile application, cloud computing) could potentially have far-reaching benefits. An example is the application of Blockchain technology combined with IoT which was found to have immense utility in case of providing better value chain deployment at each stage of marketing. There is an urgent need to study the IoT applications in agriculture in the Indian context so that they could be disseminated to remote locations for meeting multiple objectives of increased productivity, enhanced farm income, sustainable production, and retaining youth in agriculture.

The purpose of this chapter is to discuss the applications of IoT in the agricultural sector with an emphasis on developing and emerging economies. The chapter is divided into five sections. Section 1 provides an introduction along with the need to conduct this review. Section 2 deals with the review of literature in the context of applications of IoT tools in different farming situations/contexts. Section 3 deals with the specific technologies in IoT currently available for use. The start-ups utilizing IoT in agriculture in the Indian context are discussed in Section 4. Section 5 provides concluding remarks and a roadmap of future studies on this aspect.

2. Utilization of IoT in the Agricultural Sector

This section provides brief information in the context of utilization of IoT in the agricultural sector mostly related to developing and emerging economies. In an emerging economy like India, efficient utilization of IoT for resource optimization on the farm, profit maximization, sharing of information and consultation could be performed through a combination of Wireless Sensor Network (WSN) for adaptive water control, cloud computing for sharing of information and numerical optimization through data science (Roy et al., 2017). Gill et al. (2017) proposed the Agriculture-as-a-Service (AaaS) platform utilizing cloud and big data technologies to gather and synthesize information using IoT sensors for transmitting required information to clients automatically in real-time using agricultural parameters like crop, weather, soil, pest, fertilizer and irrigation. The raspberry pi and cloud-based IoT system for obtaining real-time data on farmer's fields with the use of high precision monitoring tools and smart sensors could be effectively deployed for controlling the irrigation systems on the farm (Rao and Sridhar, 2018). Further, the irrigation system was combined with soil parameters for accessing the three major micro-nutrients, namely nitrogen (N), phosphorus (P), and potassium (K) by Raut et al. (2018). The application of big data and IoT coupled with cloud computing techniques are useful to assist farmers in providing weather-related information along with decision support tools in relation to agronomic, fertilizer and pest management (Chatuphale and Armstrong, 2017). The concepts related to the smart irrigation technique were further strengthened by Keswani et al. (2018) with the integration of IoT-enabled WSN nodes for timely data storage and acquisition with irrigation control systems. The use of radio frequency identification (RFID) has been used widely in the agricultural sector for the identification and tracking of animals. Currently, it has found increased use in the food chain and supply chain management for traceability. The sensors and tags have also been found of immense use in environmental monitoring, farm mechanization and growing high-value crops (Ruiz-Garciaand Lunadei, 2011). The application of RFID technology and sensors have also found utility in promoting irrigation management, control of temperature and humidity, and farm maintenance (Wasson et al., 2017). This manuscript emphasizes the different technologies using IoT in agriculture for meeting different goals. The technologies have the potential to increase reliability in the case of agricultural production leading to more sustainable and productive practices.

3. Tools in IoT Applicable for the Agricultural Sector

The current day agriculture is enhanced by several IoT tools like Geographical Positioning Systems (GPS), sensors for increased attention to details, cloud computing for better data analytics, estimation and storage (Luthra et al., 2018). The increased use of IoT in agriculture has paved a path for interconnected systems leading to better monitoring and control of production systems for increased efficiency and productivity. The section discusses key tools in IoT which is applicable in the agricultural context for a variety of situations:

3.1. Wireless Sensor Networks

The use of sensors finds increased applications in the modern agricultural practice like greenhouse technology for the year-round production of selected agricultural commodities. The controlled environmental conditions in the greenhouse along with environmental management are largely possible through the sensors. In recent years, the miniaturization of the sensor boards has found applicability in precision as a means to observe and control agricultural practices. It is used in a wide range of agricultural practices from dairy cattle management to field crop production (Baggio, 2005; Burrell et al., 2004; Mayer et al., 2004; Zhang et al., 2004).

In Australia, sensor networks are used to monitor and influence animals' positions in the landscape along with accessing the new level of information about the state of pastures. The use of solar-powered moisture nodes and low-resolution camera nodes is used initially for the process of pasture assessment. Along with it, herd behavior and the food intake by the livestock is also monitored for resource optimization (Wark and Corke, 2007).

3.2. Radio Frequency Identification (RFID)

RFID systems are able to track/ trace the items in real-time and provide important details on their current location and status. It is largely used in value chain systems (supply-chain management) in agriculture (Zhao et al., 2010). RFID is also finding increase use in promoting exports of products from the country where the RFID tags help in tracking the produce from their location. Also, the promotion of organic farming in the country and certification systems is increasingly using RFID for efficient monitoring and control. In countries like Australia, the RFID tag is also used in monitoring the cattle population with a unique RFID tag attached to the ears. It is useful in estimating the amount of food intake by the cattle in which the exact intake of the animal is tracked continuously and sent over a connected network (Wark and Corke, 2007).

3.3. Artificial Intelligence

The use of IoT in disease detection and control of pests widely uses artificial intelligence. Also, several mobile phone applications to detect the disease infestation through clicking a photograph of the affected leaf/ stem/ flower of the plant use such technology. The expert systems based on artificial intelligence are also tested and deployed at research and extension stations making use of the IoT technology. The use of artificial intelligence also becomes relevant at the food sorting stage in the production chain which makes use of artificial intelligence for securing the produce (Eerikäinen et al., 1993; Zhang et al., 2010; Luthra et al., 2018).

3.4. GPS-GIS-RS Technologies

The use of Geographical Positioning System and Geographical Information Systems along with Remote Sensing technologies along with IoT have led to the streamlining of extension activities. It is used for the collection and analysis of field data for better analysis and prediction. Along with it, it finds increasing use in precision agriculture, habitat analysis, and management of forests (Milla et al., 2005). The use of Unmanned Aerial Vehicles (UAVs) and drones have further increased the capability for data collection resulting in better ways of farm management.

3.5. Blockchain Technologies

The traceability systems in relation to the supply chain could be better managed with the use of blockchain technologies. The storage of data along the system could be performed on each and every node instead of a central entity. Thus, it promotes trust, transparency and traceability along the chain leading to an efficient transaction of data, goods, and financial transactions (Tian, 2017).

4. Case Studies of IoT-Based Agricultural Start-Ups in India

New technologies play a major role for the farmers to shift from input□intensive to knowledge intensive agriculture. The extension professionals are quite a few to meet the location-specific demands for the new generation farmers which require extension information ranging from weather forecasting to market penetration (Glendenning et al., 2010; Lele, and Goswami, 2017). The last decade has experienced a growing number of start-ups in agriculture based on IoT and smart applications in the Indian states of Karnataka, Tamil Nadu, Maharashtra and Uttar Pradesh. The initiatives of the government like the smart city project, digital India, incubation laboratories, and startup India have provided a suitable ground for entrepreneurs to benefit the most from this agricultural revolution. Few startups based on IoT working in a wide agricultural context like crop grading, value chain assessment, provision of crop advisories, irrigation management, yield estimation, farm mechanization etc. are discussed below.

4.1. Satsure

It is an Indo-European Union (EU) company that utilizes satellite imagery coupled with IoT enabled sensor technology and weather information to predict near-real-time early warning signals, monitoring options for crop suitability and vulnerability, prediction of production data and providing key intervention strategies for the farmers, government and financial partners (Satsure, 2019). The problems of farmers in relation to yield estimation are carefully tackled by the region-specific models for increased accuracy and reliability of the production estimates. The company exploits the potential of big data analytics, machine learning, cloud computing, and satellite and drone imagery to provide information pertaining to the specific location (Rajpal, 2018). Satsure offers a great solution to the farmers in relation to the yield estimation and the government and insurance companies to access the crop parameters accurately for claim disbursement.

4.2. Vassar Labs

The company uses Artificial Intelligence (AI) and IoT technologies to provides complete solutions to the farming community from agro-climatic zone planning to advisories related to crop-soil moisture stress, crop pests and diseases, data of sowing, acreage estimation, and soil health card analysis and planning (Kant, 2018). The state of Andhra Pradesh is making use of the services offered by Vassar labs in the agricultural domain. The advisories are disseminated to the farmers and agricultural field officers working with the farmers though text/ voice messages in vernacular languages as well as English (Vassar Labs, 2018).

4.3. Agricxlab Private Limited

Food wastages along the supply chain in developing economies like India poses serious problems in food availability and better price realization. About 18 % of the total amount of fruits and vegetables produced in the country are wasted from the post-harvest stage until they reach the final consumers (FASAR, 2014; Balaji and Arshinder, 2016). Agricx is making a difference in this sector with the facilitation of procurement, grading and certification of the produce from the farmers with the help of technology. It uses imaging and AI to quicken the process associated with assay and grading of agricultural produce (DIPP, 2019). The firm is currently procuring potatoes from farmers in various districts of Uttar Pradesh, India. It offers multi-chamber, modern cold storage with high volume capacity along with facilities of grading, inspection and weather control. It plans to introduce moisture and sugar assaying capability in the near future (AgricxLabPvt Ltd, 2019).

4.4. Cropin Technology

The firm provides software as a service (SaaS) solutions using technologies such as AI, IoT and drones in agriculture for digitizing the farm. The technologies are implanted to predict trends, achieve patterns and enable efficient farm management strategies for undertaking business decisions. The timely agricultural advisory based on satellite and weather input data, output predictability, the standard package of practices together with traceability and output predictability helps the farmers to maximize the per acre value on their farm (Cropin, 2019).

4.5. Oxen Farm Solutions

The startup uses Farming as a Service (FaaS) paradigm to connect farmers with advanced farm mechanization solutions. It is currently working in eight states across India offering efficient pay per use model in crop residue management, land preparation, planting, harvesting and post-harvest processing. It uses IoT to gather and process information from sensors in machinery providing useful information on the machinery performance. Also, satellite imagery and big data analytics are deployed for analyzing crop health dimensions.

4.6. Flybird Farm Innovations

The firm uses AI and IoT technologies to manage and conserve irrigation and fertilizer applications for small and marginal farmers of the country at affordable costs. It has developed an innovative irrigation controller called Siri to automate irrigation and fertigation systems leading to saving of water, conservation of electricity, reduction of weed infestation and increase in crop yield/ productivity (Maurya, 2016; PR Com, 2018).

4.7. Exabit Systems

The firm provides an end-to-end IoT platform in agriculture that is linked to precision farming and smart control options. It offers solutions to automate agribusiness processes with farm data such as purchase, processing, logistics and sales. Thus, it helps in resolving the problems associated with modern agricultural practices like land management, irrigation control etc.

5. Conclusion

The queries of the farmers concerning farm management and reaping maximum benefits from the land could be mostly solved by the emerging technologies in IoT. The case study of startups discussed in the chapter is providing a way for contemporary agriculture which is mostly mechanized, technology-driven, and less dependent on human labor. Precision farming and the use of geographical information systems in agriculture provide a great deal of opportunity in mapping each and every land resource of the country with respect to their production potential regarding various crops. The creation of an extensive knowledge base would result in proving better solutions to the farmers leading to efficient monitoring of crops, value chains, and market availability.

References

AgricxLabPvt Ltd. 2019. How do we certify. Retrieved from http://www.agricx.com/howdowecertify.php

Baggio, A. 2005. Wireless sensor networks in precision agriculture. In: ACM Workshop on Real-World Wireless Sensor Networks (REALWSN 2005), Stockholm, Sweden (Vol. 20).

Balaji, M., Arshinder, K. 2016. Modeling the causes of food wastage in Indian perishable food supply chain. Resources, Conservation and Recycling, 114, 153–167.

Bloomberg, 2019. Company Overview of eXabit Systems Pvt. Ltd. Retrieved from https://www.bloomberg.com/research/stocks/private/snapshot.asp?privcapid=422810469

Burrell, J., Brooke, T., Beckwith, R. 2004. Vineyard computing: Sensor networks in agricultural production. IEEE Pervasive computing, 3(1), 38–45.

Chatuphale, P., Armstrong, L. 2017. Indian Mobile Agricultural Services Using Big Data and Internet of Things (IoT). In: International Conference on Intelligent Systems Design and Applications. Springer, pp. 1028–1037.

Cropin, 2019. Smartfarm. Retrieved from http://www.cropin.com/smartfarm/

DIPP – Department of Industrial Policy and Promotion, 2019. Agriculture Grand Challenge Final Results. Retrieved from https://www.startupindia.gov.in/content/sih/en/agriculture-grand-challenge/agriculture-grand-challenge-results.html

Eerikäinen, T., Linko, P., Linko, S., Siimes, T., Zhu, Y. H. 1993. Fuzzy logic and neural network applications in food science and technology. Trends in Food Science & Technology, 4(8), 237–242.

FASAR – Food and Agribusiness Strategic Advisory and Research YES bank. 2014. Fruits & Vegetables Availability Maps of India, Retrieved from https://www.nicct.nl/wp-content/uploads/2018/12/Fruits-and-Vegetables-map-of-India.pdf

Gill, S.S., Chana, I., Buyya, R. 2017. IoT based agriculture as a cloud and big data service: The beginning of digital India. Journal of Organizational and End User Computing (JOEUC), 29(4), 1–23.

Glendenning, C.J., Babu, S., Asenso-Okyere, K. 2010. Review of agricultural extension in India: Are farmers' information needs being met? IFPRI Discussion Paper 01048, International Food Policy Research Institute, Washington, DC.

Government of India, 2013. Agricultural Statistics at a Glance. Directorate of Economics and Statistics, Ministry of Agriculture. Retrieved from http://agricoop.nic.in/agristatistics.htm

Kant, A. 2018. In an innovation nation. Retrieved from https://economictimes.indiatimes.com/blogs/et-commentary/in-an-innovation-nation/

Keswani, B. et al. 2018. Adapting weather conditions based IoT enabled smart irrigation technique in precision agriculture mechanisms. Neural Computing and Applications, 1–16.

Lele, U., Goswami, S. 2017. The fourth industrial revolution, agricultural and rural innovation, and implications for public policy and investments: a case of India. Agricultural Economics, 48(S1), 87–100.

Luthra, S., Mangla, S. K., Garg, D., Kumar, A. 2018. Internet of Things (IoT) in Agriculture Supply Chain Management: A Developing Country Perspective. In: Emerging Markets from a Multidisciplinary Perspective, Springer, pp. 209–220.

Luthra, S., Mangla, S. K., Garg, D., Kumar, A. 2018. Internet of Things (IoT) in Agriculture Supply Chain Management: A Developing Country Perspective. In: Emerging Markets from a Multidisciplinary Perspective, Springer, pp. 209–220.

Maurya, V. 2016. AgriTech Startup Flybird is Building Automation Devices for Farmers. Retrieved from http://www.iamwire.com/2016/06/agritech-startup-flybird-building-automation-devices-farmers/137369

Mayer, K., Ellis, K., Taylor, K. 2004. Cattle health monitoring using wireless sensor networks. In Proceedings of the Communication and Computer Networks Conference (CCN 2004) ACTA Press, pp. 8–10.

Milla, K.A., Lorenzo, A., Brown, C. 2005. GIS, GPS, and remote sensing technologies in extension services: Where to start, what to know. Journal of Extension, 43(3).

National Sample Survey Organisation (NSSO), 2005. Situation Assessment Survey of Farmers: Income, Expenditure and Productive Assets of Farmer Household. NSS 59th Round (January-December 2003), NSS Report No. 497 (59/33/5).

Oxen India, (2019). Farm Mechanization Services. Retrieved from http://www.oxenindia.com/

Pandey, V.L., Suganthi, D. 2015. Fueling agricultural growth in India: Some reflections. Land Use Policy, 42, 227–232.

Com PR (2018). Conser Water Technologies. Retrieved from https://www.pr.com/press-release/762047

Rajpal, S. 2018. This company will help farmers predict their yield and keep a track of their loans. Retrieved from https://www.edexlive.com/40-under-40/2018/may/15/cyber-hand-up-for-the-humble-farmer-2796.html

Rao, R.N., Sridhar, B. 2018. IoT based smart crop-field monitoring and automation irrigation system. In: 2nd International Conference on Inventive Systems and Control (ICISC), IEEE, pp. 478–483.

Raut, R., Varma, H., Mulla, C., Pawar, V.R. 2018. Soil Monitoring, Fertigation, and Irrigation System Using IoT for Agricultural Application. In: Intelligent Communication and Computational Technologies, Springer, pp. 67–73.

Roy, S. et al. 2017. IoT, big data science & analytics, cloud computing and mobile app based hybrid system for smart agriculture. In: Industrial Automation and Electromechanical Engineering Conference (IEMECON), 2017 8th Annual, IEEE, pp. 303–304.

Ruiz-Garcia, L., Lunadei, L. 2011. The role of RFID in agriculture: Applications, limitations and challenges. Computers and Electronics in Agriculture, 79(1), 42–50.

Satsure, 2019. About Us. Retrieved from https://www.satsure.co/

Tian, F. 2017. A supply chain traceability system for food safety based on HACCP, blockchain & Internet of things. In: Service Systems and Service Management (ICSSSM), 2017 International Conference on IEEE, pp. 1–6.

TRAI, 2019. Highlights of Telecom Subscription Data as on 31st October, 2018. Retrieved from https://www.trai.gov.in/sites/default/files/PRNo01Eng02012019.pdf

Vassar Labs, 2018. Case studies. Retrieved from http://www.vassarlabs.com/smart-agriculture-section.html

Wark, T., Corke, P. 2007. Transforming agriculture through pervasive wireless sensor networks. IEEE Pervasive Computing 6(2), 50–57.

Wasson, T., Choudhury, T., Sharma, S., Kumar, P. 2017. Integration of RFID and sensor in agriculture using IOT. In: Smart Technologies for Smart Nation (SmartTechCon), 2017 International Conference on IEEE, pp. 217–222.

Zhang, W., Kantor, G., Singh, S. 2004. Integrated wireless sensor/ actuator networks in an agricultural application. In: Proceedings of the 2nd international conference on embedded networked sensor systems, ACM, pp. 317–317.

Zhang, X., Zhang, J., Liu, F., Fu, Z., Mu, W. 2010. Strengths and limitations on the operating mechanisms of traceability system in agro food, China. Food Control 21(6), 825–829.

Zhao, J.C., Zhang, J. F., Feng, Y., Guo, J.X. 2010. The study and application of the IOT technology in agriculture. In: Computer Science and Information Technology (ICCSIT), 3rd IEEE International Conference on IEEE, 2, 462–465.

Colour Plates

Chapter 1: Taro (*Colocasia esculenta* (L.) Schoff.) for Nutritional Security and Health Benefits

Fig. 1: Taro plants in field and corms

Chapter 3: Chrysanthemum: A Versatile Crop of Commercial Significance

Floral Rangoli

Pots for Exhibition

Cut Stems for Bouquet

Garland Making

Flower Arrangement

Multiuser chrysanthemums

Classes of Chrysanthemum

Irregular IncurveReflex

Regular

Incurve

Decorative

Intermediate Incurve

Pompon/Button

Single and Semi Double

Anemone

Spoon

Quill

Spider

Brush or Thistle

Pinching

Disbudding

Staking with nets in field

Chapter 7: Cultivation of Underutilized Minor Fruit Plants in India

Other indigenous under utilized fruits of India.

A. Chironji, **B.** Khirni, C.Mahua, **D.** Rambutan, **E&F.** Mulberry, **G.** Custard apple, **G.** Wood apple

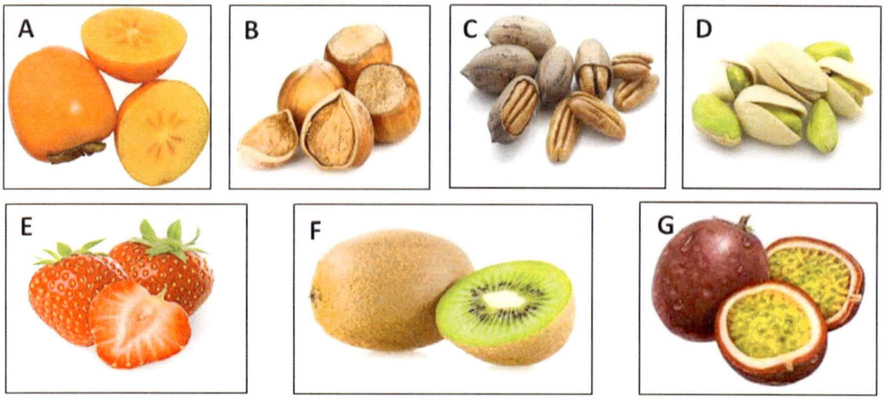

Potential exotic fruits of India.
A. Passion fruiti, **B.** Hazelnut, **C.** Pecanut, **D.** Pistachionut, **E.** Strawberry, **F.** Kiwi, **G.**

Indegineous underutilized fruits of India. (a) Bael, (b) Jamun, (c) Tamarind, (d) Karonda (e) Fig and F Phalsa

Chapter 12: Commercial Vegetable Nurseries Types, Establishment, Operations Management and Economics

Fig. 1: Upper panel - Conventional nursery. Raised nursery beds **(A)** sowing of seeds **(B)** and brinjal plants grown on raised bed **(C)**. Lower panel- Soilless media nursery. Black sheet to act as soil barrier, mulch and temperature regulator **(D)** sowing of seeds in protrays **(E)** and protrays covered with clear polysheet for faster germination of seeds **(F)**

Fig. 2: Hitech Nursery. Top panel **(A-C)**, Overview of nursery tables. Middle panel **(D-F)**, seedling of tomato **(D)**, cabbage **(E)** and brinjal **(F)** in protarys. Lower Panel, close p view of capsicum **(G)** and cabbage **(H)** seedling in protarys

Chapter 14: Plasticulture: A Hi-Tech Approach for Betterment of Horticulture

Fig. 1: Plants grow in plastic pots/bags/pro-tray

Fig. 2: Plastic lined farm ponds

Fig. 3: Drip irrigation installed on field

Fig. 4 : Sprinkler irrigation system installed on field

Fig. 5: Soil solarization on field

Fig. 6: Use of plastic mulch on field

Fig.7: Use of plastic material in greenhouse structure

Fig. 9: Use of plant protection nets on field

Fig. 11: Packaging of agricultural produce in different plastic containers

Chapter 17: Turfgrass Sod Production A Profitable Enterprise

Fig. 1: Machinery required for sod production business

Fig. 2: Pictorial representation of cultural operation for turfgrass sod production

Chapter 18: Multidimensional Benefits of Garden Cafeteria

Fig. 2: The diversity of a garden cafeteria established aside to the residential quarters. Images are showing plant diversity during winter (left) and summer to rainy (right) season. The area attracts pollinators, birds and other wild lives like squirrel, mongoose, chameleon etc.

Fig. 3: Variation in produce of obtained from a garden cafeteria. The vegetables and fruit (guava in blue tray) were harvested during Aril to September. Repeated harvesting was performed for the cucurbits, beans, okra and leafy vegetables.

Fig. 4: The glimpses of winter crops in a garden cafeteria before harvesting

Chapter 22: Role of Nanofertilizer for Sustainable Crop Production

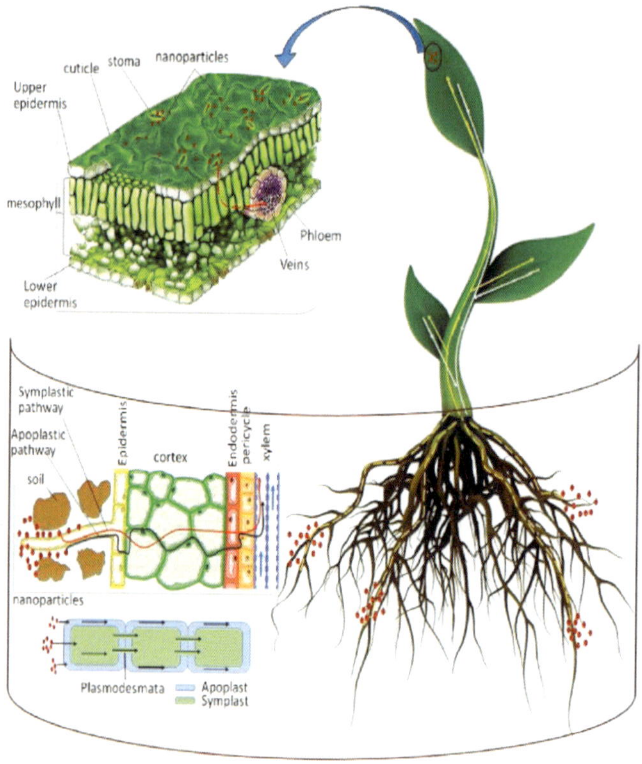

Fig. 2: Mechanism of action of nanofertilizer

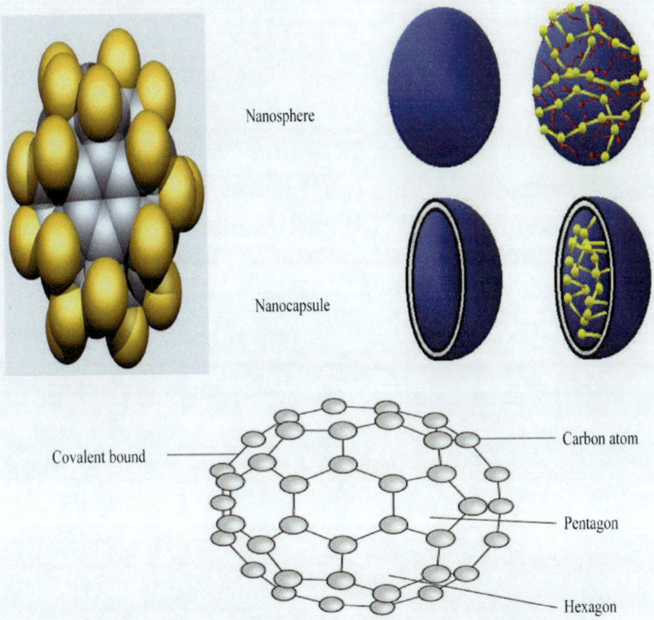

Fig. 4: Bucky Ball carries 60 carbon atoms (20 hexagons + 12 Pentagons)

Chapter 34: Phytochemistry, Ethnobotany and Pharmacology of East Indian Arrowroot (*Curcuma angustifolia* Roxb.)

Fig. 1: East Indian arrowroot crop in field and rhizomes

Index